HIGHER

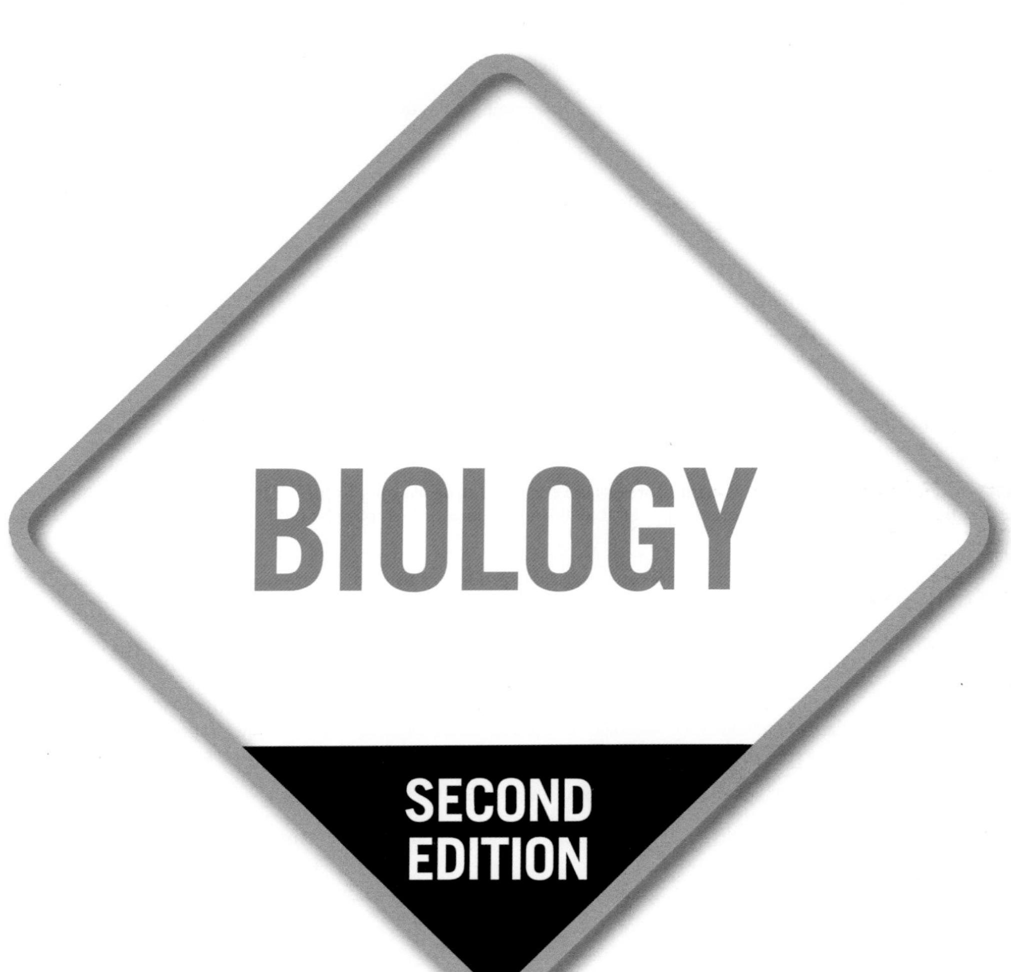

BIOLOGY

SECOND EDITION

James Torrance

with writing team: James Fullarton, Clare Marsh, James Simms, Caroline Stevenson

The publishers would like to thank the following for permission to reproduce copyright material:

Photo credits
p.1 (background) and Section 1 running head image © 2010 Steve Allen/Brand X Pictures/photolibrary.com; p.1 (inset left) © Scott Camazine/Alamy Stock Photo, (inset centre) © Dr Keith Wheeler/Science Photo Library, (inset right) © Ingram Publishing Limited; p.5 © Science Source/Science Photo Library; p.36 (top) © Eric Martz, (bottom) © Science Photo Library/Alamy Stock Photo; p.47 (left) © Dr Keith Wheeler/Science Photo Library, (right) © M.I. Walker/Science Photo Library; p.49 © Deco Images II/Alamy Stock Photo; p.50 (all) © Cindee Madison and Susan Landau, UC Berkeley; p.52 © James King-Holmes/Science Photo Library; p.58 James Torrance; p.66 © Mauro Fermariello/Science Photo Library; p.75 © Scott Camazine/Alamy Stock Photo; p.90 © Ingram Publishing Limited; p.103 (background) and Section 2 running head image © Loren Rodgers – Fotolia; p.103 (inset left) © Natural Visions/Alamy, (inset centre) © EcoView – Fotolia, (inset right) © Dr Jeremy Burgess/Science Photo Library; p.131 © Orlando Florin Rosu – Fotolia; p.143 © blickwinkel/Alamy Stock Photo; p.155 (top left) © outdoorsman – Fotolia, (top right) © EcoView – Fotolia, (bottom) © feathercollector – Fotolia; p.156 © Natural Visions/Alamy; p.157 (left) © FLPA/Linda Lewis, (right) © Steve Byland – Fotolia; p.159 (top) © FLPA/John Holmes, (bottom) © Mark Parisi, permission granted for use. www.offthemark.com; p.164 © Rafa Irusta/stock.adobe.com; p.169 (top) © Crown Copyright/Health & Safety Laboratory/Science Photo Library, (bottom) © Dr Jeremy Burgess/Science Photo Library; p.177 © Dr Jeremy Burgess/Science Photo Library; p.195 (background) and Section 3 running head image © pro6x7 – Fotolia; p.195 (inset left) © Anna – Fotolia, (inset centre) © corlaffra/stock.adobe.com, (inset right) © Avalon/Photoshot License/Alamy Stock Photo; p.213 James Torrance; p.214 © Nigel Cattlin/Alamy Stock Photo; p.219 (top) © Image used with permission from the Plant and Soil Sciences eLibrary at http:// passel.unl.edu, hosted by the Institute of Agriculture and Natural Resources at the University of Nebraska-Lincoln, (bottom) © Anna – Fotolia; p.220 © Margo Harrison – Fotolia; p.221 © Image used with permission from the Plant and Soil Sciences eLibrary at http:// passel.unl.edu, hosted by the Institute of Agriculture and Natural Resources at the University of Nebraska-Lincoln; p.223 © FLPA/Nigel Cattlin; p.233 © The Natural History Museum/Alamy Stock Photo; p.234 © FLPA/Nigel Cattlin; p235 © xalanx – Fotolia; p.241 © FLPA/Nigel Cattlin; p.244 (top) FLPA/Nigel Cattlin, (bottom) © Chris Ison/Alamy Stock Photo; p.245 (top left) © Science VU/J. R. Adams/Visuals Unlimited, Inc., (top right) © dragonraj – Fotolia, (bottom) © FLPA/Nigel Cattlin; p.253 (top) © levo – Fotolia, (bottom) © Eye Ubiquitous/REX/Shutterstock; p.255 (top) © Shaun Finch – Coyote-Photography.co.uk/Alamy Stock Photo, (bottom) © KtD – Fotolia; p.256 © Frank Monaco/REX/Shutterstock; p.262 (left) © FLPA/Ron Boardman/ Life Science Image, (right) © corlaffra/stock.adobe.com; p.265 © Oxford Scientific/Getty Images; p.270 © Ron Sanford/Corbis Documentary/Getty Images; p.277 (top left) © Juniors Bildarchiv GmbH/Alamy Stock Photo, (top right) © FLPA/imagebroker/ROM, (bottom) © FLPA/Jurgen & Christine Sohns; p.279 © Herbert Kehrer/Imagebroker/FLPA RF; p.289 © Avalon/Photoshot License/Alamy Stock Photo; p.296 (both) James Torrance; p.297 © dero2084/stock.adobe.com.

Acknowledgements
Every effort has been made to trace all copyright holders, but if any have been inadvertently overlooked the Publishers will be pleased to make the necessary arrangements at the first opportunity.

Whilst every effort has been made to check the instructions of the practical work in this book, it is still the duty and legal obligation of schools to carry out their own risk assessments.

Orders: please contact Bookpoint Ltd, 130 Park Drive, Milton Park, Abingdon, Oxon OX14 4SE. Telephone: (44) 01235 827827. Fax: (44) 01235 400454. Email education@bookpoint.co.uk Lines are open 9.00–5.00, Monday to Saturday, with a 24-hour message answering service.
Visit our website at www.hoddereducation.co.uk. Hodder Gibson can also be contacted directly at hoddergibson@hodder.co.uk

First published in 2012 © James Torrance, James Fullarton, Clare Marsh, James Simms, Caroline Stevenson
This second edition published in 2019 by
Hodder Gibson, an imprint of Hodder Education,
An Hachette UK Company
211 St Vincent Street
Glasgow G2 5QY

Impression number	5	4	3	2	1
Year	2023	2022	2021	2020	2019

ISBN: 978 1 5104 5767 6

Cover photo © michaelfitz - stock.adobe.com
Illustrations by James Torrance and Integra Software Services Pvt. Ltd., Pondicherry, India
Typeset in Minion Regular 11pt by Integra Software Services Pvt. Ltd., Pondicherry, India
Printed in Italy

A catalogue record for this title is available from the British Library

Contents

Preface

This book has been written to act as a valuable resource for students studying Higher Grade Biology. It provides a **core text** which adheres closely to the SQA syllabus for *Higher Biology (revised)* introduced in 2018. Each section of the book matches a section of the revised syllabus; each chapter corresponds to a content area. In addition to the core text, the book contains a variety of special features:

Suggested Learning Activities

Within each chapter there is an appropriate selection of suggested learning activities exactly as laid down in the SQA Course Support Notes. They take the form of *Case Studies*, *Related Topics*, *Research Topics*, *Related Activities* and *Investigations*. These non-essential activities are highlighted throughout in yellow for easy identification. They do not form part of the basic mandatory course content needed when preparing for the final exam but are intended to aid understanding and to support research tasks during course work.

Testing Your Knowledge

Key questions incorporated into the text of every chapter and designed to continuously assess *Knowledge and Understanding*. These are especially useful as homework and as instruments of diagnostic assessment to check that full understanding of course content has been achieved.

What You Should Know

Summaries of key facts and concepts as *'Cloze' Tests* accompanied by appropriate word banks. These feature at regular intervals throughout the book and provide an excellent source of material for consolidation and revision prior to the SQA examination.

Applying Your Knowledge and Skills

A variety of questions at the end of each chapter designed to give students practice in exam questions and foster the development of *Skills of Scientific Inquiry and Investigation* (for example, selection of relevant information, presentation of information, processing of information, planning experimental procedure, evaluating, drawing valid conclusions and making predictions and generalisations). These questions are especially useful as extensions to class work and as homework.

Updates and syllabus changes: important note to teachers and students from the publisher

This book covers all course arrangements for Revised Higher Biology but does not attempt to give advice on any 'added value assessments' or 'open assignments' that form part of a final grade in Higher Biology (2018 onwards).

Please remember that syllabus arrangements change from time to time. We make every effort to update our textbooks as soon as possible when this happens, but – especially if you are using an old copy of this book – it is always advisable to check whether there have been any alterations to the arrangements since this book was printed. You can check the latest arrangements at the SQA website (www.sqa.org.uk), and you can also check for any specific updates to this book at www.hoddereducation.co.uk/HigherScience.

We make every effort to ensure accuracy of content, but if you discover any mistakes please let us know as soon as possible – see contact details on imprint page.

1 DNA and the Genome

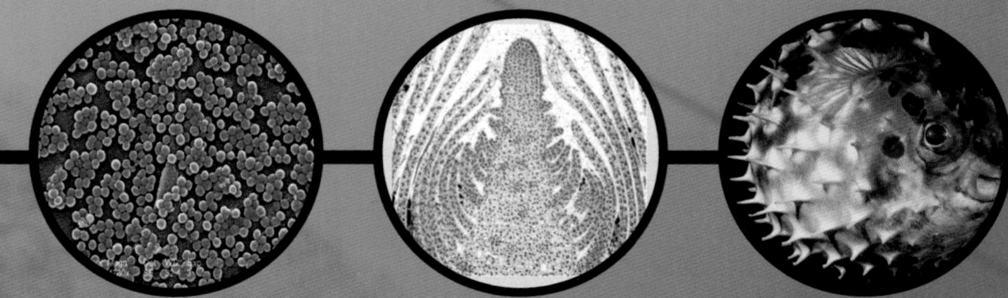

1 Structure of DNA

DNA (deoxyribonucleic acid) is a complex molecule present in all living cells. It stores genetic information in its sequence of bases which determines the organism's genotype and the structure of its proteins.

Structure of DNA

A molecule of DNA consists of two strands each composed of repeating units called **nucleotides**. Each DNA nucleotide consists of a molecule of **deoxyribose** sugar joined to a **phosphate** group and an organic **base**. Figure 1.1 shows the carbon skeleton of a molecule of deoxyribose. Figure 1.2 shows the four types of base present in DNA.

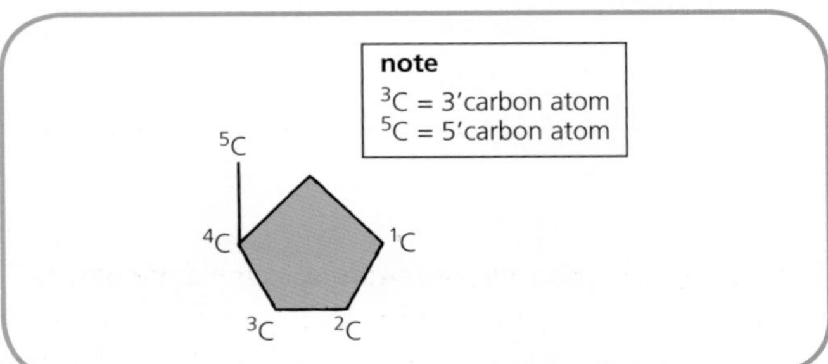

Figure 1.1 Deoxyribose

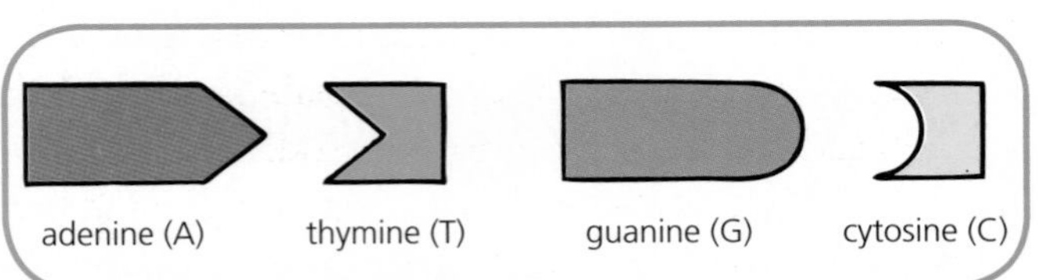

Figure 1.2 Four types of organic base

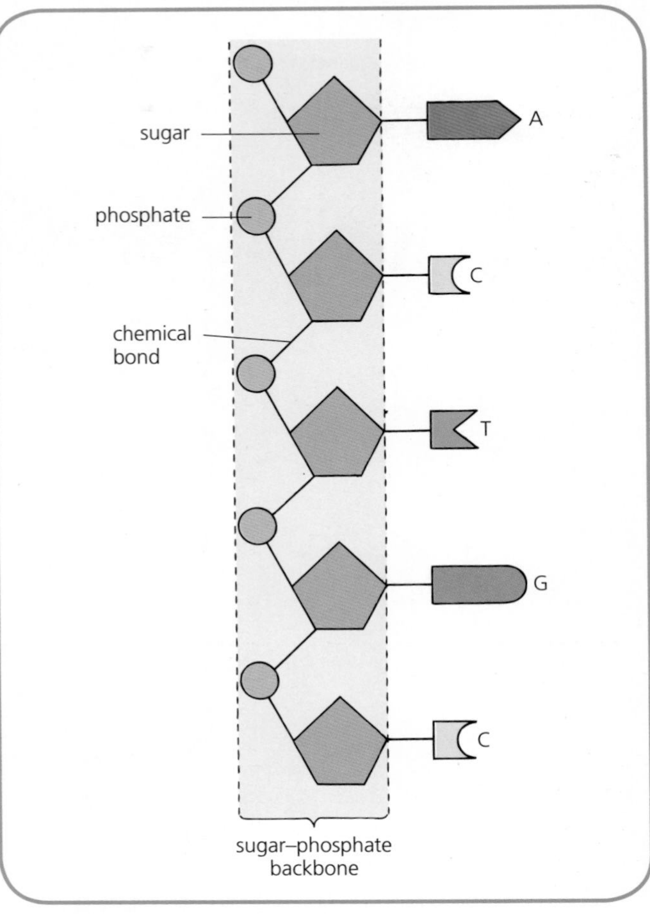

Figure 1.4 Sugar–phosphate backbone

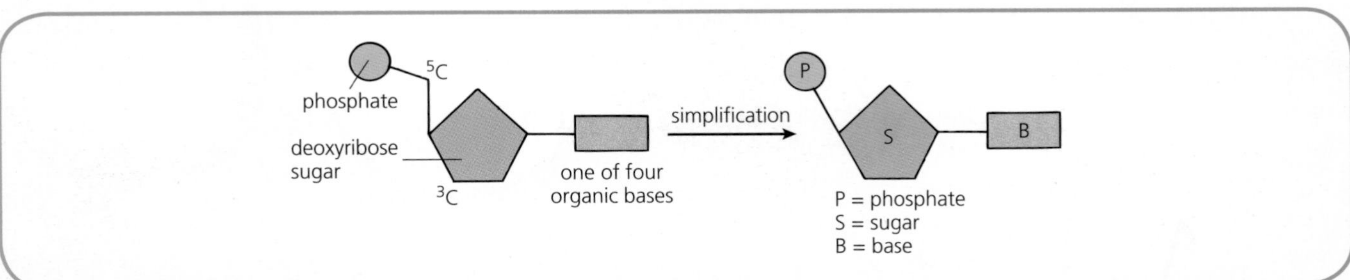

Figure 1.3 Structure of a DNA nucleotide

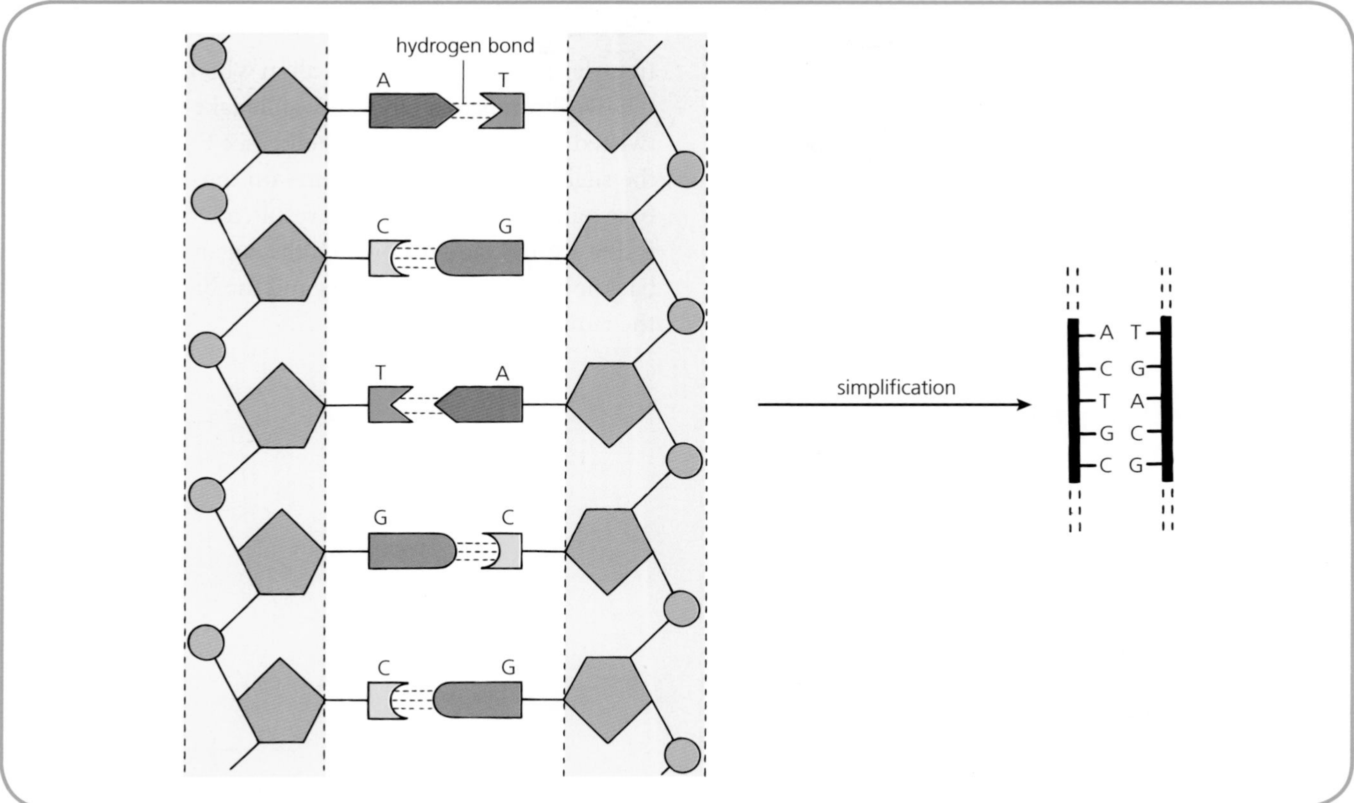

Figure 1.5 Base-pairing

Figure 1.3 shows how the deoxyribose molecule in a nucleotide has a base attached to its carbon 1 and a phosphate attached to its carbon 5. Since there are four types of base, there are four types of nucleotide.

Sugar–phosphate backbone

A **chemical bond** forms between the phosphate group of one nucleotide and the carbon 3 of the deoxyribose on another nucleotide (see Figure 1.4). By this means neighbouring nucleotides become joined together into a long permanent strand in which sugar molecules alternate with phosphate groups forming the DNA molecule's **sugar–phosphate backbone**.

Base-pairing

Two of these strands of nucleotides become joined together by **hydrogen bonds** forming between their bases (see Figure 1.5). However, the hydrogen bonds can be broken when it becomes necessary for the two strands to separate.

Each base can only join up with one other type of base: adenine (A) always bonds with thymine (T) and guanine (G) always bonds with cytosine (C). A–T and G–C are called **base pairs**.

Antiparallel strands

A DNA strand's **3′ end** on deoxyribose is distinct from its **5′ end** at a phosphate group. The chain is only able to 'grow' by adding nucleotides to its 3′ end. In Figure 1.6 the DNA strand on the left has its 3′ growing end at the bottom of the diagram and its 5′ end at the top. The reverse is true of its complementary strand on the right. This arrangement of the two strands with their sugar–phosphate backbones running in **opposite directions** is described as **antiparallel**. (For the sake of simplicity the letters in a diagram are normally all written the same way up.)

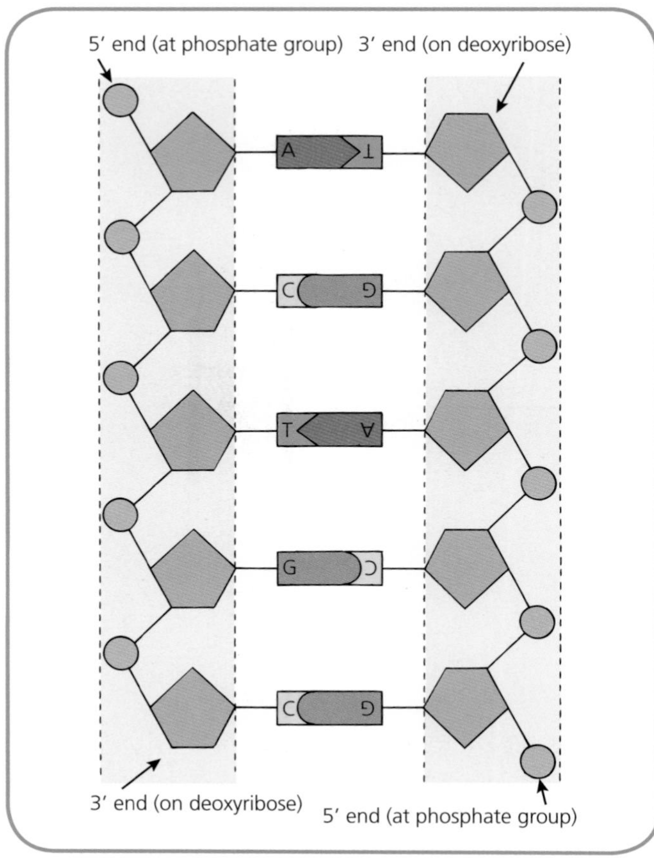

Figure 1.6 Antiparallel strands

Double helix

In order for the base pairs to align with each other, the two strands in a DNA molecule take the form of a twisted coil called a **double helix** (see Figure 1.7) with the sugar–phosphate backbones on the outside and the base pairs on the inside. As a result, a DNA molecule is like a spiral ladder in which the sugar–phosphate backbones form the uprights and the base pairs form the rungs.

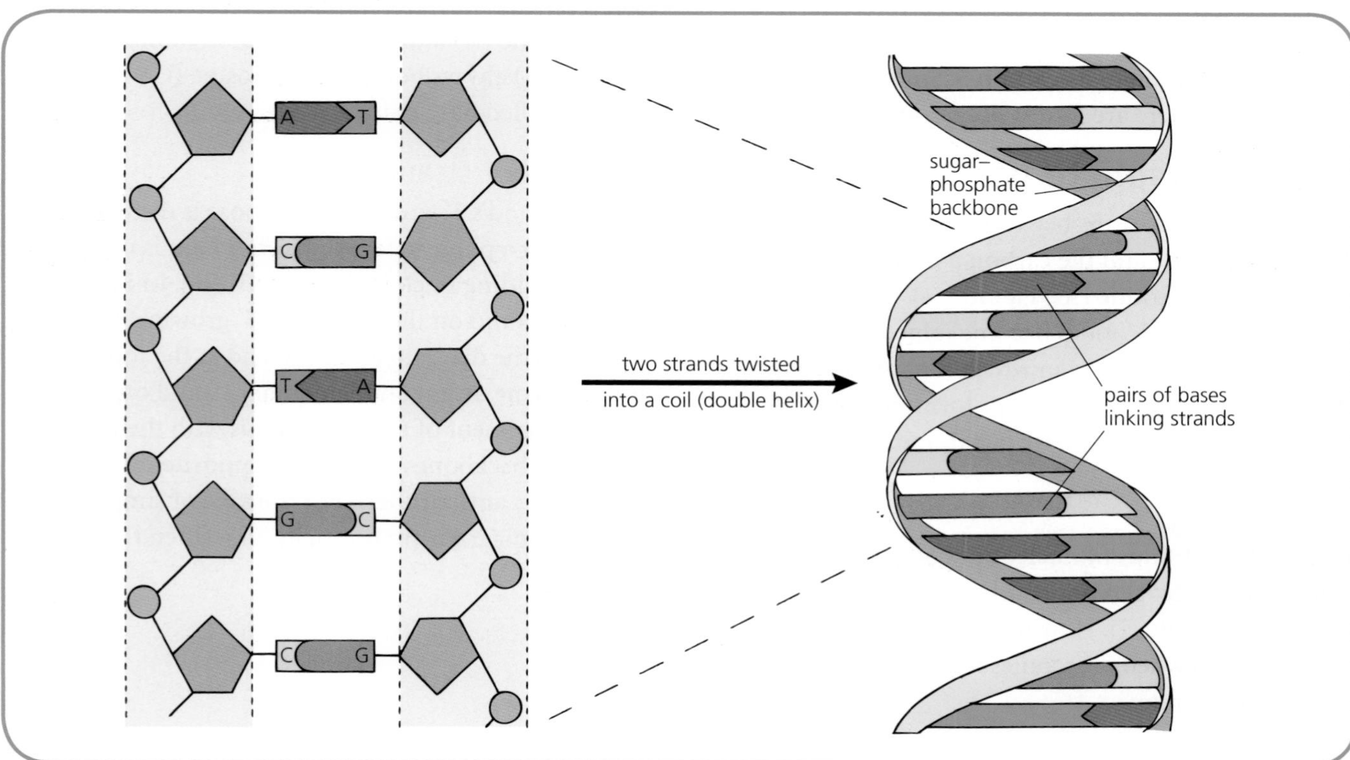

Figure 1.7 Double helix

Case Study Establishing the structure of DNA

During the first half of the twentieth century, the results from a series of experiments demonstrated conclusively that **DNA** was the genetic material present in cells. Next, scientists became keen to establish the three-dimensional structure of DNA.

Chemical analysis

In the late 1940s, Chargaff analysed the base composition of DNA extracted from a number of different species. He found that the quantities of the four bases were not all equal but that they always occurred in a **characteristic ratio** regardless of the source of the DNA. These findings, called Chargaff's rules, are summarised as follows:

The number of adenine bases = the number of thymine bases (i.e. A:T = 1:1).

The number of guanine bases = the number of cytosine bases (i.e. G:C = 1:1).

However, Chargaff's rules remained unexplained until the double helix was discovered.

X-ray crystallography

At around the time that Chargaff was carrying out chemical analysis of DNA, Wilkins and Franklin were employing **X-ray crystallography**. When X-rays are passed through a crystal of DNA, they become deflected (diffracted) into a **scatter pattern** which is determined by the arrangement of the atoms in the DNA molecule (see Figure 1.8).

When the scatter pattern of X-rays is recorded using a photographic plate, a **diffraction pattern** of spots is produced (see Figure 1.9). This reveals information

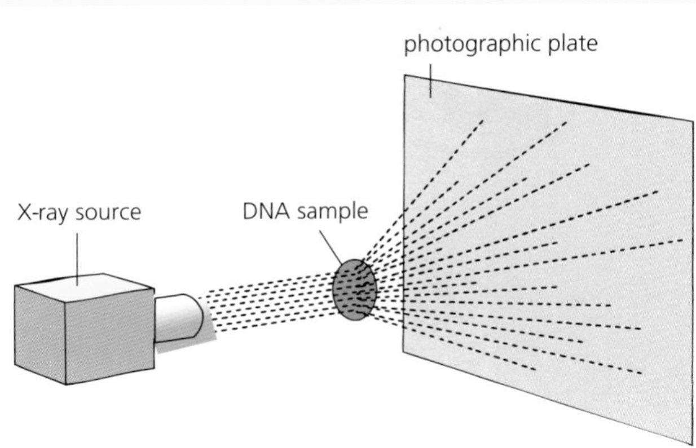

Figure 1.8 X-ray crystallography

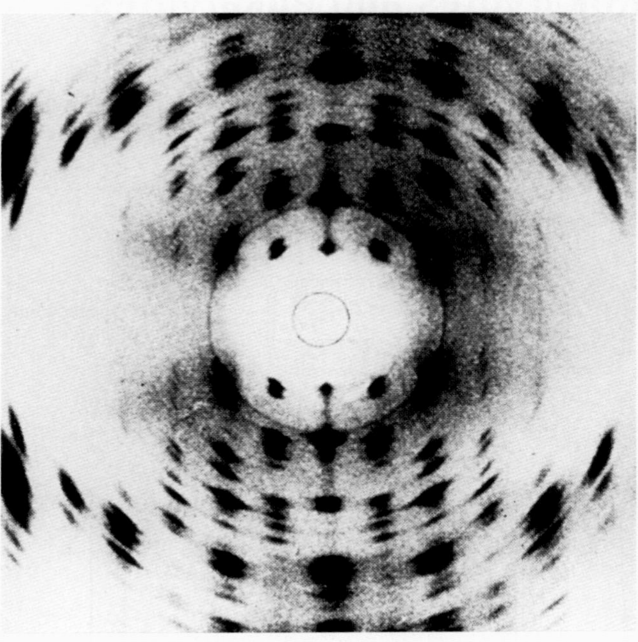

Figure 1.9 X-ray diffraction pattern of DNA

which can be used to build up a **three-dimensional picture** of the molecules in the crystal. Wilkins and Franklin found that the X-ray diffraction patterns of DNA from different species (such as bull, trout and bacteria) were identical.

Formation of an evidence-based conclusion

From the X-ray diffraction patterns of DNA (produced by Wilkins and Franklin), Watson and Crick figured out that the DNA must be a long, thin molecule of constant diameter coiled in the form of a **helix**. In addition, the density of the arrangement of the atoms indicated to them that the DNA must be composed of **two strands**.

From Chargaff's rules they deduced that base A must be paired with T and base G with C. They figured that this could only be possible if DNA consisted of two strands held together by specific **pairing of bases**. Taking into account further information about distances between atoms and angles of bonds, Watson and Crick set about building a wire model of DNA and in 1953 were first to establish the three-dimensional **double helix** structure of DNA.

Organisation of DNA in prokaryotes and eukaryotes

A molecule of DNA is double-stranded and can be **circular** or **linear** depending on the type of cell to which it belongs and on its location within the cell. Figure 1.10 compares a cell from a bacterium (a **prokaryote**) with a cell from a green plant (a type of **eukaryote**) with respect to the organisation of their DNA.

Some of a prokaryotic cell's DNA may be separate from the main circular chromosome in tiny rings called **plasmids**. Not all of a eukaryotic cell's DNA is located

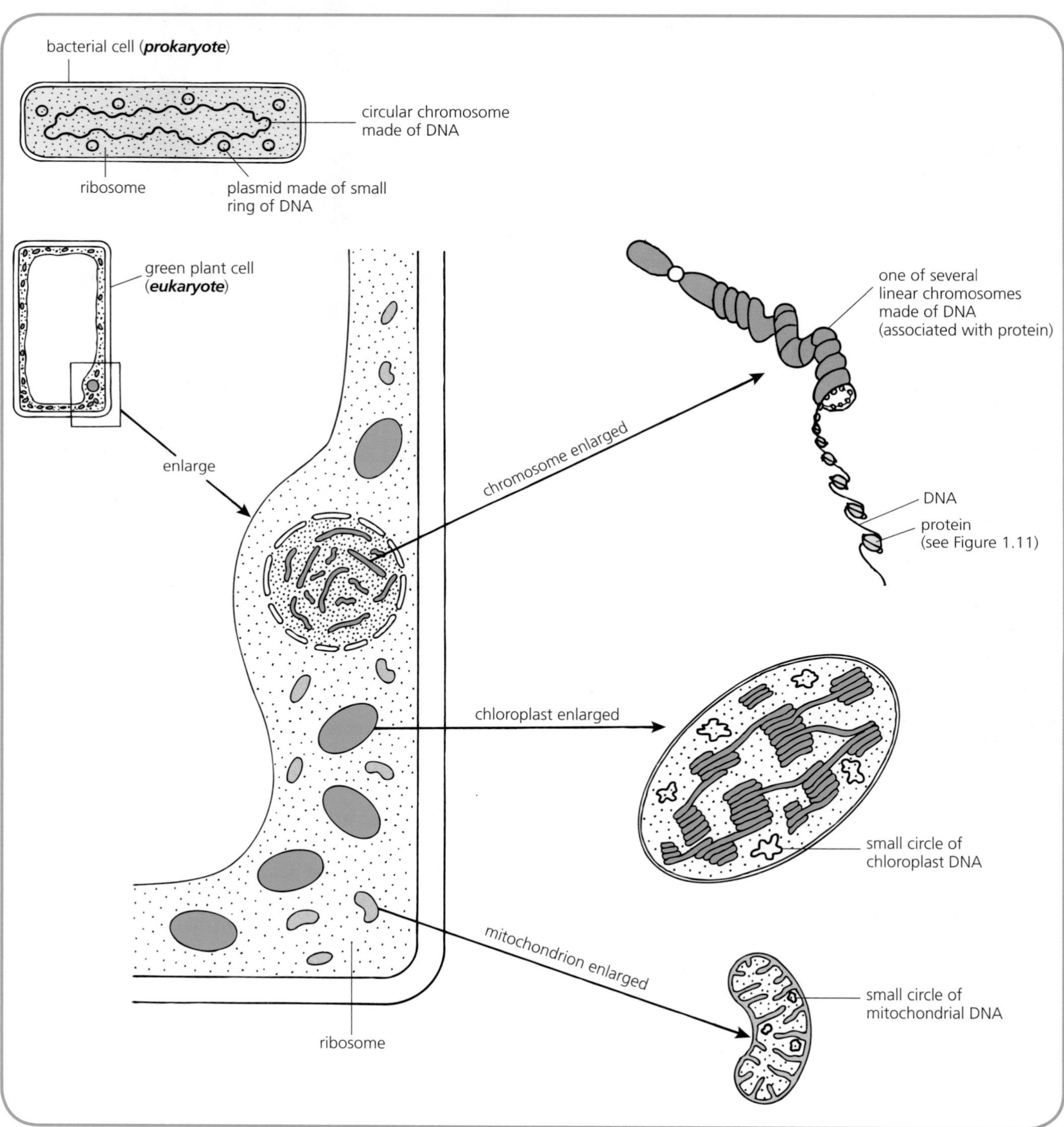

Figure 1.10 Prokaryotic cell and eukaryotic cell (not drawn to scale)

Characteristic	Prokaryotic cell	Eukaryotic cell
organism that has this type of cell	bacteria	fungi, green plants and animals
true nucleus bound by double membrane	absent	present
organisation of chromosomal DNA	composed of a ring of DNA associated with few or no proteins	composed of DNA in linear form associated with proteins
plasmids (each consisting of a small ring of DNA)	present in many types of bacterial cell	present in some yeasts; absent in plant and animal cells
chloroplasts (each containing several small circular chromosomes)	absent	present in green plant cells
mitochondria (each containing several small circular chromosomes)	absent	present
ribosomes	present	present

Table 1.1 Characteristics of prokaryotic and eukaryotic cells

in the nuclear chromosomes. Small **circles** of DNA are present in **chloroplasts** and **mitochondria**. The similarities and differences between the two types of cell are summarised in Table 1.1.

Arrangement of DNA in linear chromosomes

A strand of DNA is several thousand times longer than the length of the cell to which it belongs. Therefore it is essential that DNA molecules be organised in such a way that a chaotic tangle of strands in the nucleus is prevented. This is achieved by the molecules of DNA becoming **tightly coiled** and packaged around bundles of protein (called histones) like beads on a string (see Figure 1.11). They are able to unwind again when required to do so.

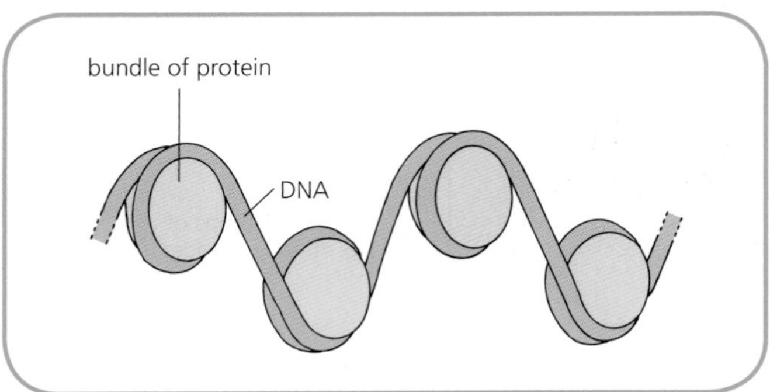

Figure 1.11 Structure of a linear chromosome

Related Activity

DNA extraction from plant tissue

DNA can be isolated from cells as shown in Figure 1.12 for pea seeds. When kiwi fruit is used instead of peas, most of the white strands that form as a precipitate in the upper layer of cold ethanol are made of pectin (a complex carbohydrate) rather than DNA. Since the DNA is obscured by pectin, this result is described as a **false positive**.

→

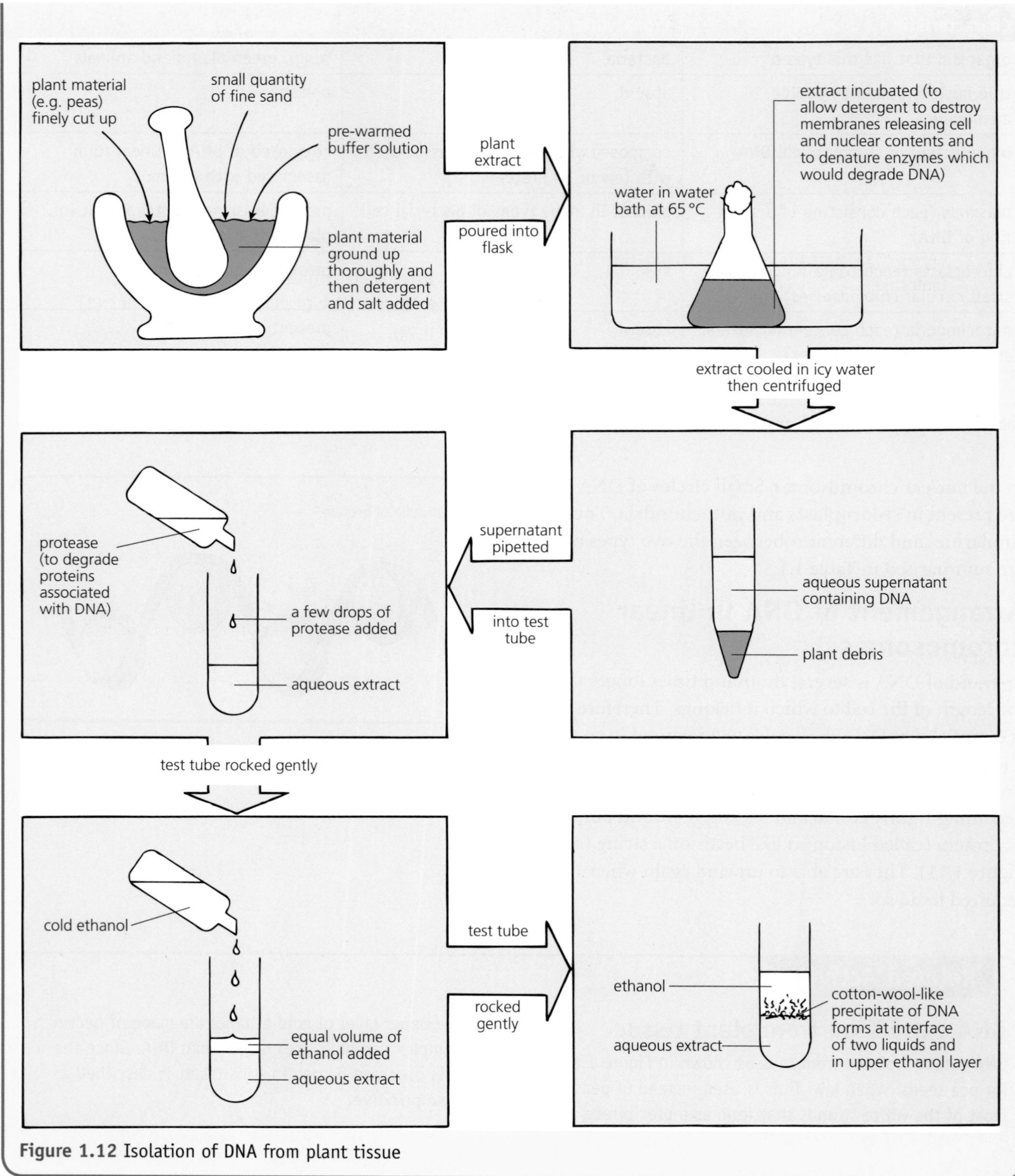

Figure 1.12 Isolation of DNA from plant tissue

Testing Your Knowledge

1 a) i) How many different types of base molecule are found in DNA?

ii) Name each type. (3)

b) Which type of bond forms between the bases of adjacent strands of a DNA molecule? (1)

c) Describe the base-pairing rule. (1)

2 a) Figure 1.13 shows part of one strand of a DNA molecule.

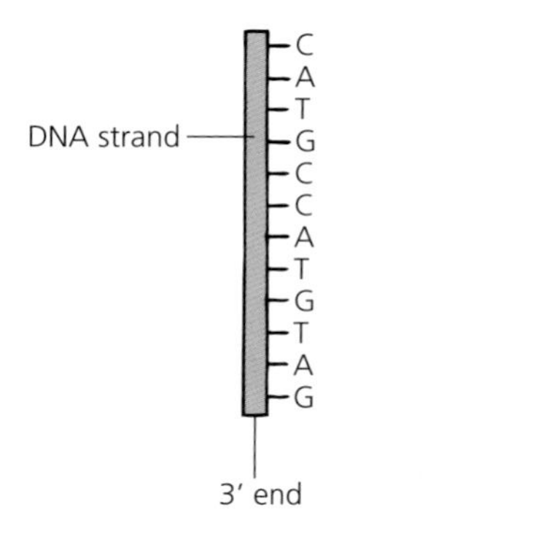

Figure 1.13

i) Redraw the strand and then draw the complementary strand alongside it.

ii) Label the 3′ end and the 5′ end on each strand. (2)

b) DNA consists of two strands whose backbones run in opposite directions. What term is used to describe this arrangement? (1)

3 a) What name is given to the twisted coil arrangement typical of a DNA molecule? (1)

b) If DNA is like a spiral ladder, which part of it corresponds to the ladder's

i) rungs?

ii) uprights? (2)

4 a) Identify a type of

i) prokaryotic cell

ii) eukaryotic cell that contains plasmids.

iii) What is a plasmid? (3)

b) State where:

i) a circular chromosome

ii) a linear chromosome

could be found in the same eukaryotic cell. (2)

Applying Your Knowledge and Skills

1 Table 1.2 shows a sample of Chargaff's data following the analysis of DNA extracted from several species.

a) Study the data and calculate the figures that should have been entered in boxes X and Y. (2)

b) i) State Chargaff's rules.
ii) Do the data in the table support these rules?
iii) Explain your answer. (3)

c) With respect to the number of the different bases in a DNA sample, which of the following is correct? (1)

A C = T B A = G

C A+G = C+T D A+T = G+C

2 a) Calculate the percentage of thymine molecules present in a DNA molecule containing 1000 bases of which 200 are guanine. (1)

b) State the number of cytosine bases present in a DNA molecule which contains 10 000 base molecules of which 18% are adenine. (1)

3 Figure 1.14 shows a cell's genetic material.

a) i) Name the parts enclosed in boxes 1, 2 and 3.
ii) Which of these boxed structures contains nucleic acid and consists of many different genes?
iii) Which of these structures is one of four basic units whose order determines the information held in a gene? (5)

b) The DNA helix of one of these chromosomes is found to be 5 cm long when fully uncoiled and 5 μm long when tightly coiled.
i) Express these data as a packing ratio of fully extended DNA : tightly coiled DNA.
ii) Suggest why scientists normally express the length of a chromosome in number of base pairs. (2)

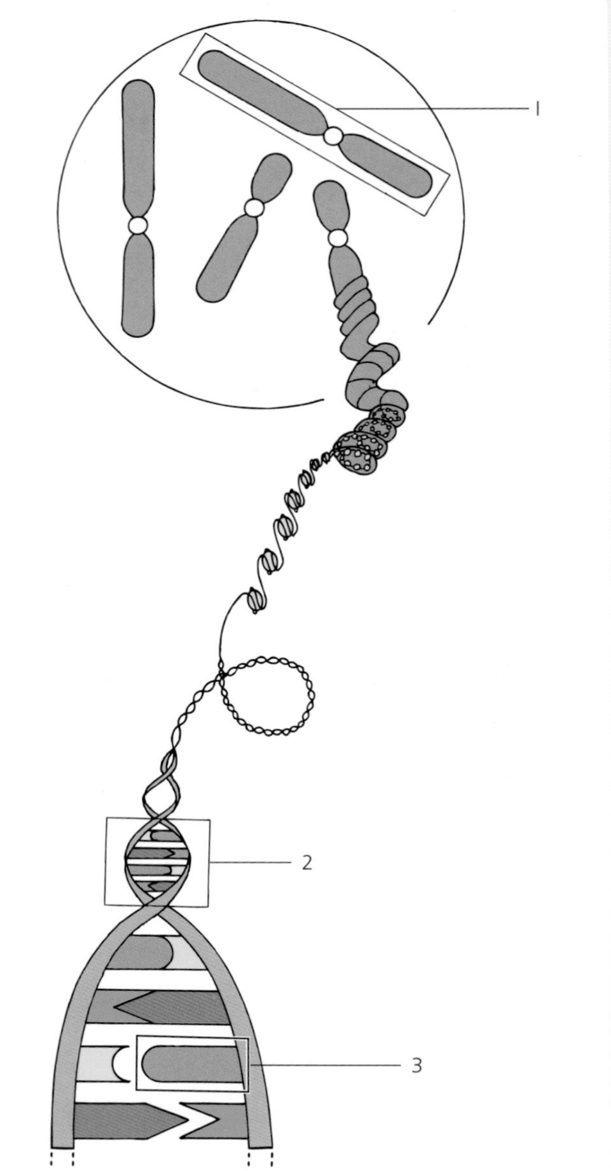

Figure 1.14

Species	%A	%C	%G	%T	A/T	G/C
chicken	28.0	21.6	**box X**	28.4	0.99	1.02
grasshopper	29.3	20.7	20.7	29.3	1.00	1.00
human	29.3	20.0	20.7	30.0	0.98	1.04
maize	26.8	23.2	22.8	27.2	0.99	**box Y**
wheat	27.3	22.8	22.7	27.2	1.00	1.00

Table 1.2

2 Replication of DNA

DNA is a unique molecule because it is able to direct its own **replication** and reproduce itself exactly. The replication process is shown in a simple way in Figure 2.1.

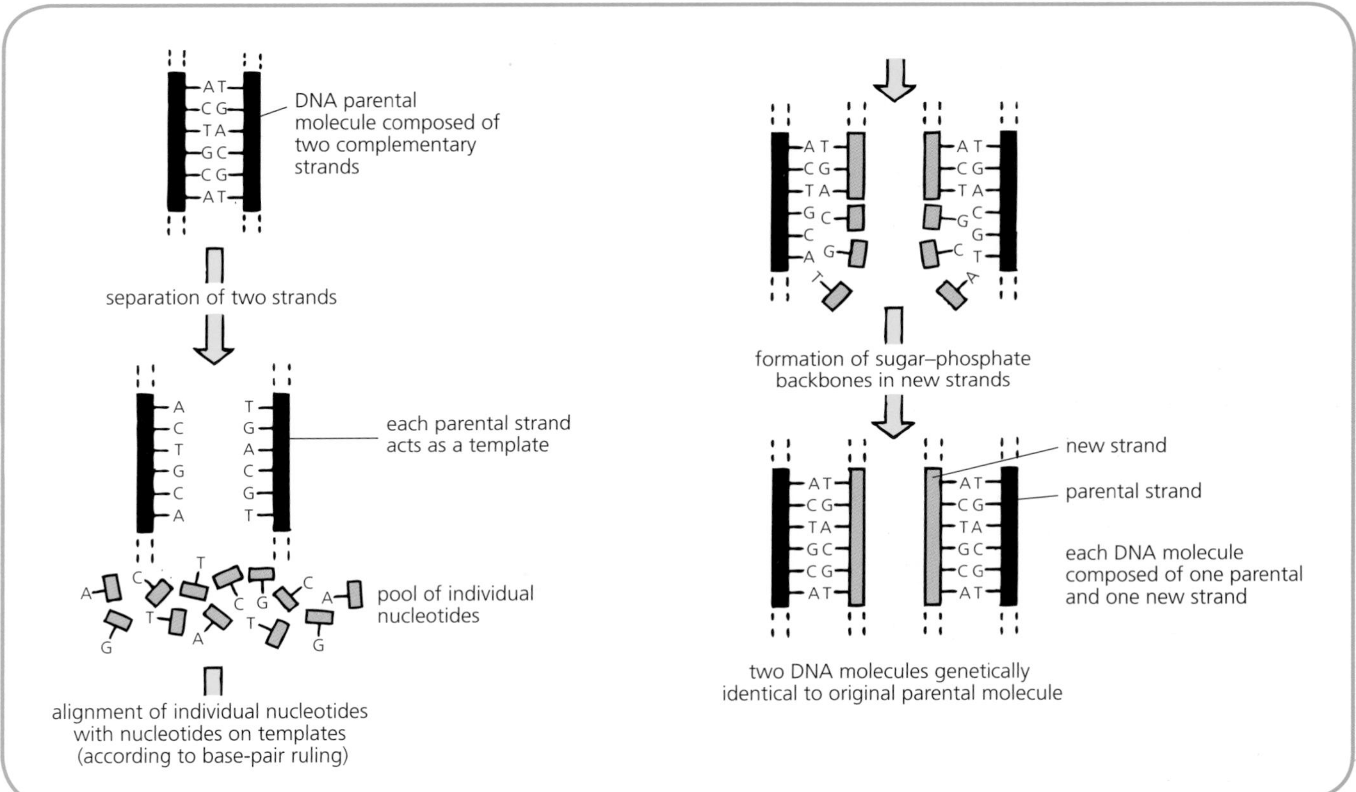

Figure 2.1 DNA replication

Case Study Establishing which theory of DNA replication is correct

Watson and Crick accompanied their model of DNA with a theory for the way in which it could replicate. They predicted that the two strands would unwind and each act as a template for the new complementary strand. This would produce two identical DNA molecules each containing one 'parental' strand and one newly synthesised strand. This so-called **semi-conservative** replication remained a theory until put to the test by Meselson and Stahl.

Hypotheses

Figure 2.2 shows three different hypotheses, each of which could explain DNA replication.

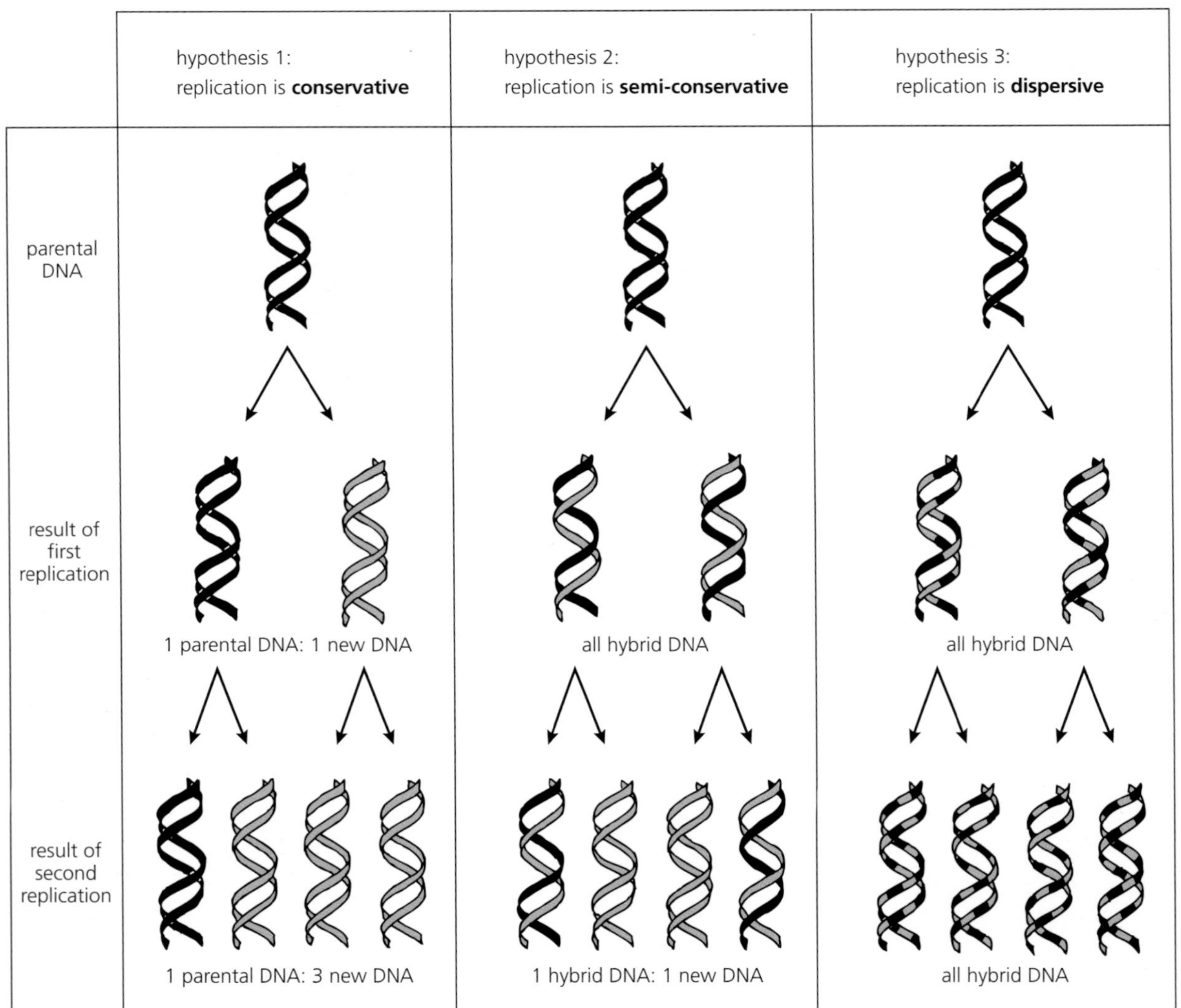

Figure 2.2 DNA replication hypotheses

→

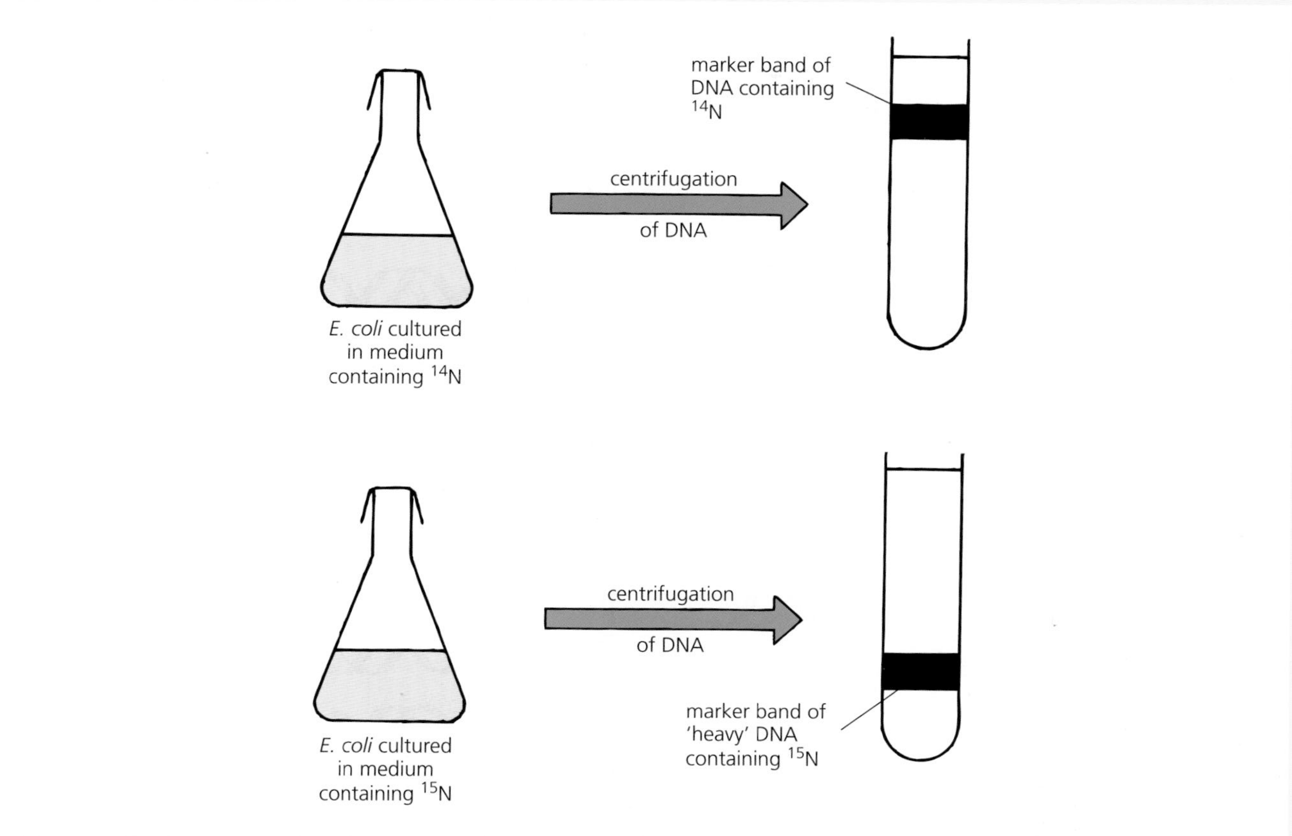

Figure 2.3 Labelling DNA with ^{14}N or ^{15}N

Testing the hypotheses

Background

When *E. coli* bacteria are cultured, they take up nitrogen from the surrounding medium and build it into DNA. They can be cultured for several generations in medium containing the common isotope of nitrogen (^{14}N) or the heavy isotope (^{15}N) as their only source of nitrogen. When DNA is extracted from each type of culture and centrifuged, the results shown in Figure 2.3 are obtained.

Putting the hypotheses to the test

Meselson and Stahl began with *E. coli* that had been grown for many generations in medium containing ^{15}N. They then cultured these bacteria in medium containing ^{14}N and sampled the culture after 20 minutes (the time needed by the bacteria to replicate DNA once). Figure 2.4 predicts the outcome of the experiment for each of the three hypotheses. Figure 2.5 shows the actual results of the experiment. From these results it is concluded that hypothesis 2 is supported and that the replication of DNA is **semi-conservative**.

→

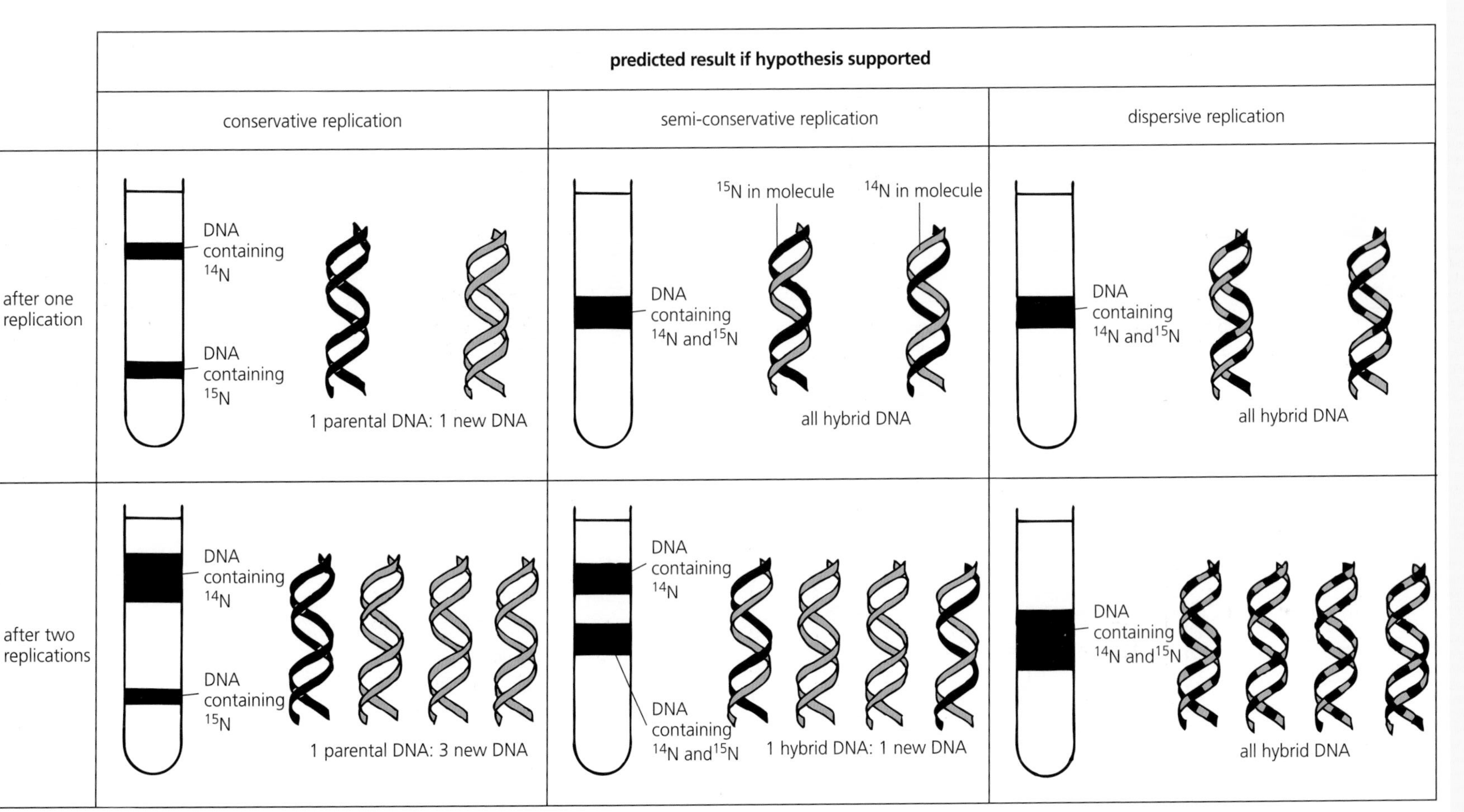

Figure 2.4 Predictions

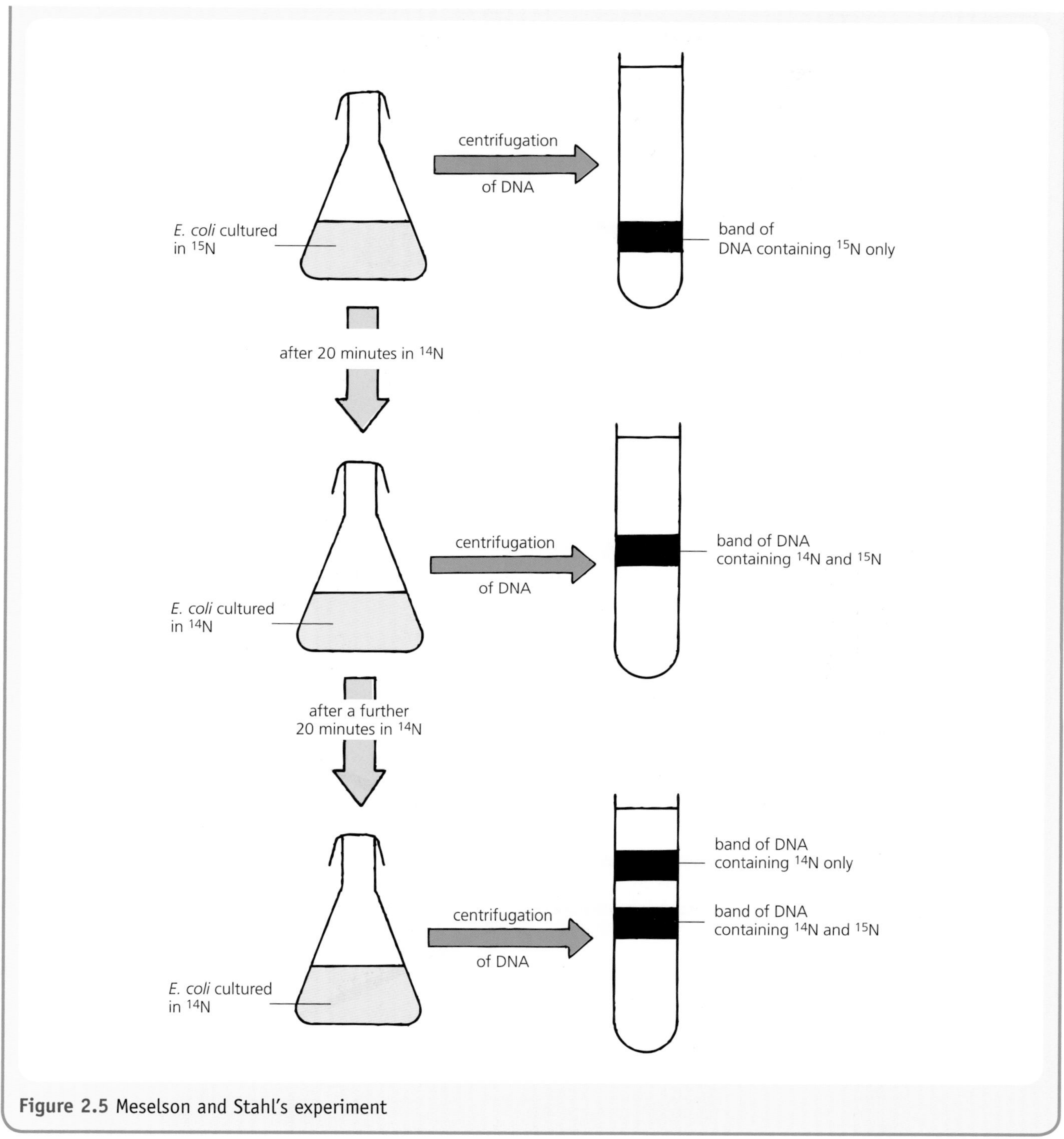

Figure 2.5 Meselson and Stahl's experiment

Enzyme control of DNA replication

DNA replication is a complex process involving many enzymes. It begins when a starting point on DNA is recognised. The DNA molecule unwinds and weak hydrogen bonds between base pairs break allowing the two strands to separate ('unzip'). These template strands become stabilised and expose their bases at a Y-shaped **replication fork** (see Figure 2.6).

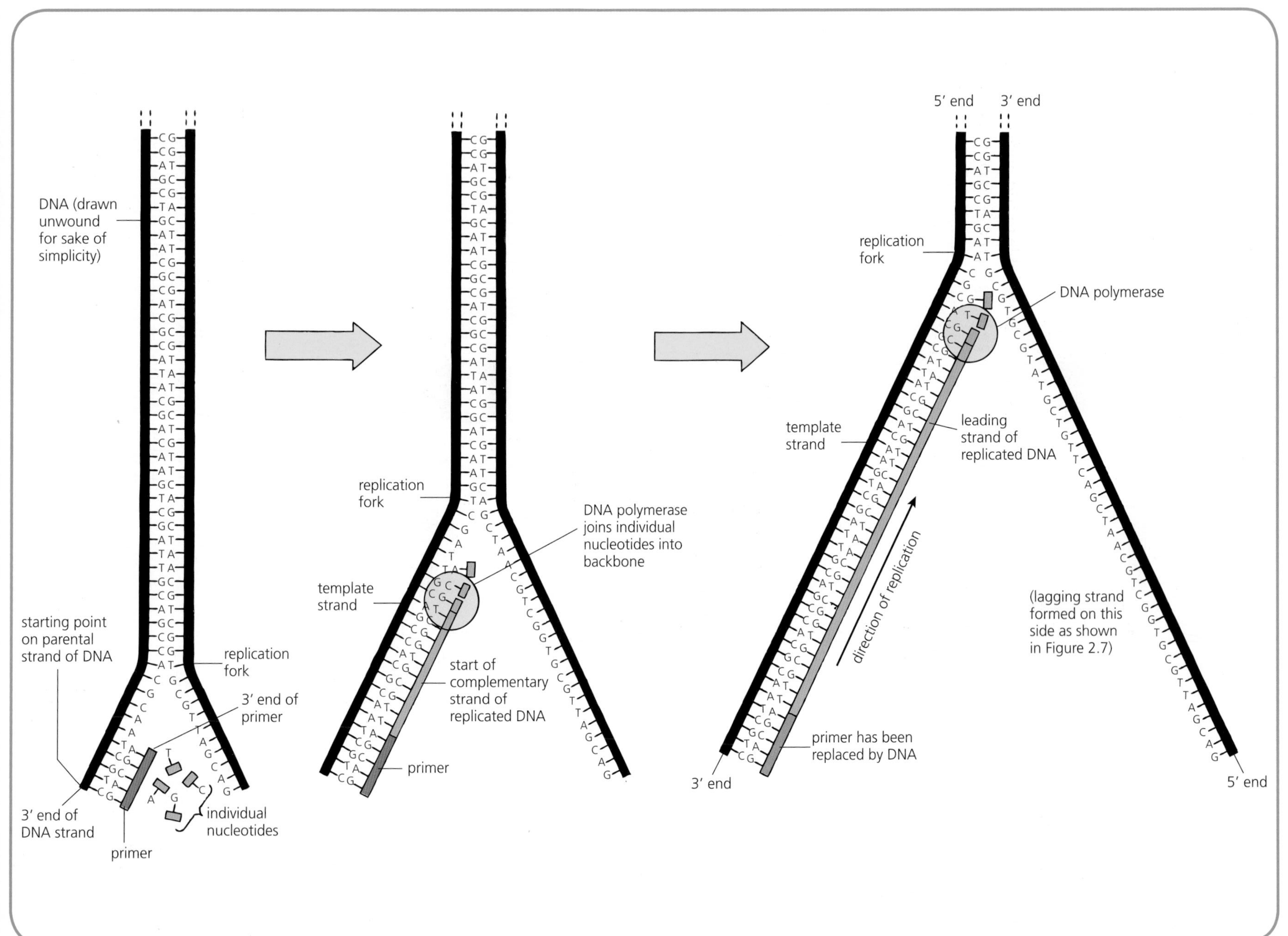

Figure 2.6 Formation of leading strand of replicated DNA

Formation of the leading DNA strand

The enzyme that controls the sugar–phosphate bonding of individual nucleotides into the new DNA strand is called **DNA polymerase**. This enzyme can only add nucleotides to a pre-existing chain. For it to begin to function, a **primer** must be present. This is a short sequence of nucleotides formed at the 3′ end of the parental DNA strand about to be replicated, as shown in Figure 2.6.

Once individual nucleotides have become aligned with their complementary partners on the template strand (by their bases following the base-pairing rules), they become bound to the 3′ end of the primer and formation of the complementary DNA strand begins. Formation of sugar–phosphate bonding between the primer and an individual nucleotide and between the individual nucleotides themselves is brought about by DNA polymerase. Replication of the parental DNA strand which has the 3′ end is **continuous** and forms the **leading** strand of the replicated DNA.

Formation of the lagging DNA strand

DNA polymerase is only able to add nucleotides to the free 3′ end of a growing strand. Therefore the DNA parental template strand which has the 5′ end has to be replicated in fragments each starting at the 3′ end of a primer as shown in Figure 2.7.

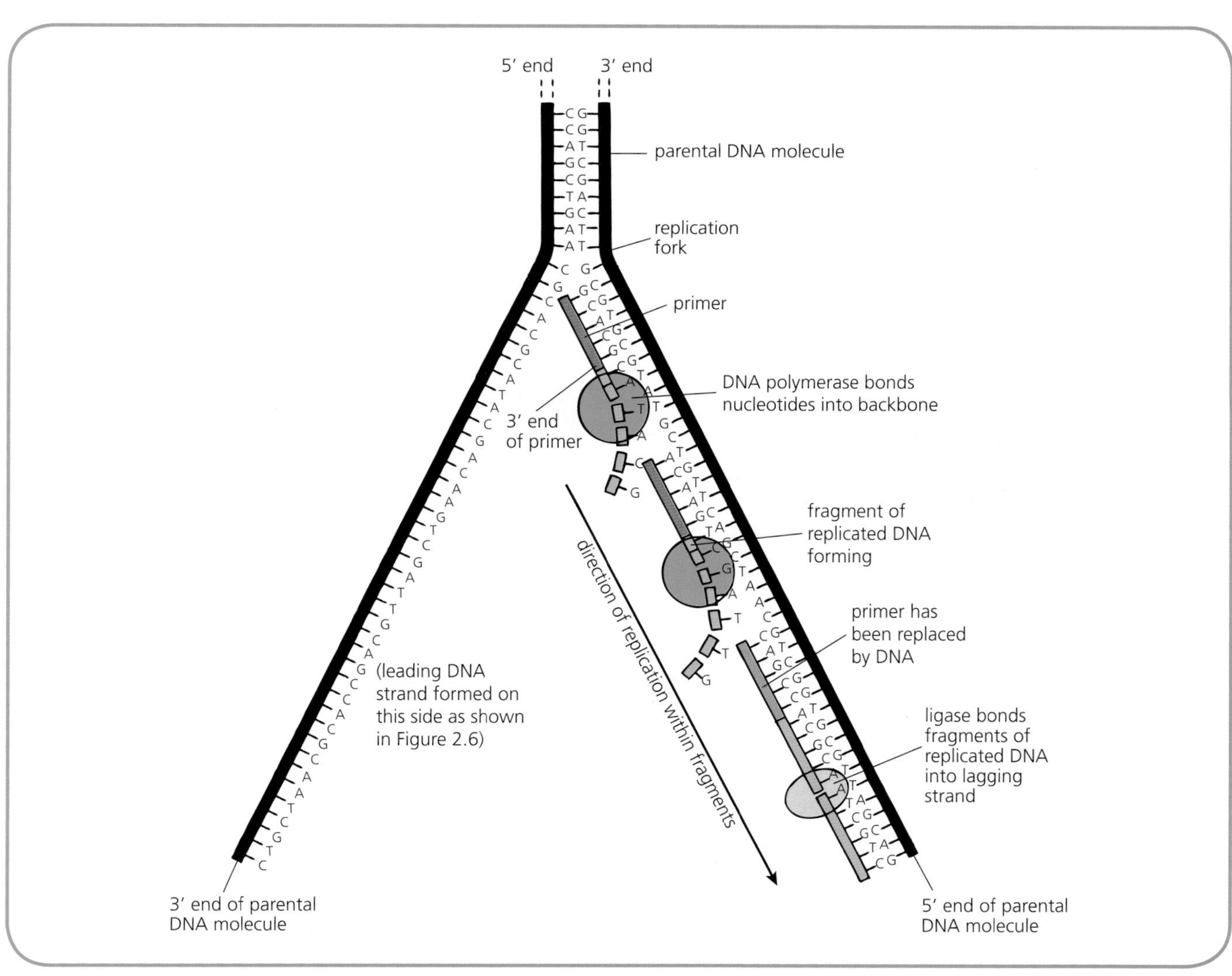

Figure 2.7 Formation of the lagging strand of replicated DNA

Each fragment must be primed as before to enable the DNA polymerase to bind individual nucleotides together. Once replication of a fragment is complete, its primer is replaced by DNA. Finally an enzyme called **ligase** joins the fragments together. The strand formed is called the **lagging** strand of replicated DNA and its formation is described as **discontinuous**.

Many replication forks

When a long chromosome (such as one from a mammalian cell) is being replicated, many replication forks operate simultaneously to ensure speedy copying of the lengthy DNA molecule (see Figure 2.8).

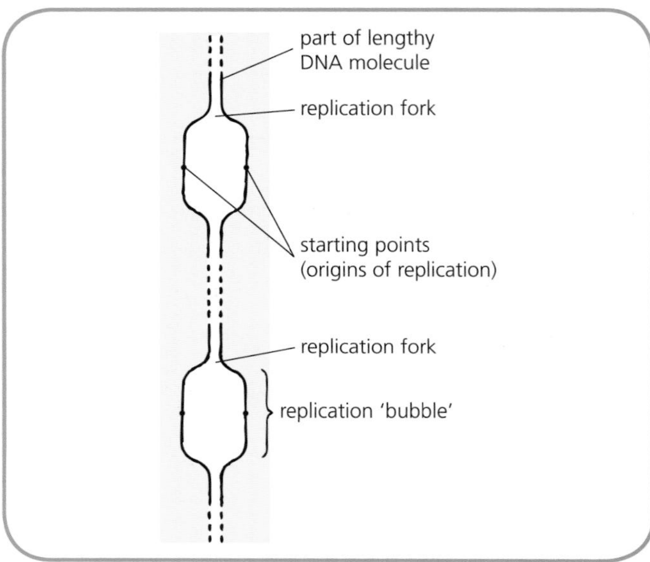

Figure 2.8 Replication forks

Requirements for DNA replication

For DNA replication to occur, the nucleus must contain:

- **DNA** (to act as a template)
- **primers**
- a supply of the four types of DNA **nucleotide**
- the appropriate **enzymes** (such as DNA polymerase and ligase)
- a supply of **ATP** (for energy – see page 131).

Importance of DNA

DNA is the molecule of inheritance and it encodes the hereditary information in a **chemical language**. This takes the form of a sequence of organic bases, unique to each species, which makes up its **genotype**. DNA replication ensures that an exact copy of a species' genetic information is passed on from cell to cell during growth and from generation to generation during reproduction. Therefore DNA is essential for the continuation of life.

Polymerase chain reaction

The **polymerase chain reaction** (PCR) is a technique (see Figure 2.9) that can be used to create many copies of a piece of DNA *in vitro* (in other words, outside the body of an organism). This **amplification** of DNA involves the use of **primers**. In this case, each primer is a short length of single-stranded DNA complementary to a specific target sequence at the 3′ end of the DNA strand to be replicated and amplified.

During a cycle of PCR, the DNA is heated to between 92 and 98 °C to break the hydrogen bonds between the base pairs and separate the two strands. The DNA is then cooled to between 50 and 65 °C to allow each primer to bind to its target sequence. Lastly the DNA is heated to between 70 and 80 °C to allow **heat-tolerant DNA polymerase** to replicate each strand of DNA by adding nucleotides to the primer at its 3′ end.

The first cycle of replication produces two identical molecules of DNA, the second cycle four identical molecules and so on giving an exponentially growing population of DNA molecules. By this means a tiny quantity of DNA can be greatly amplified and provide sufficient material for forensic purposes (see page 21).

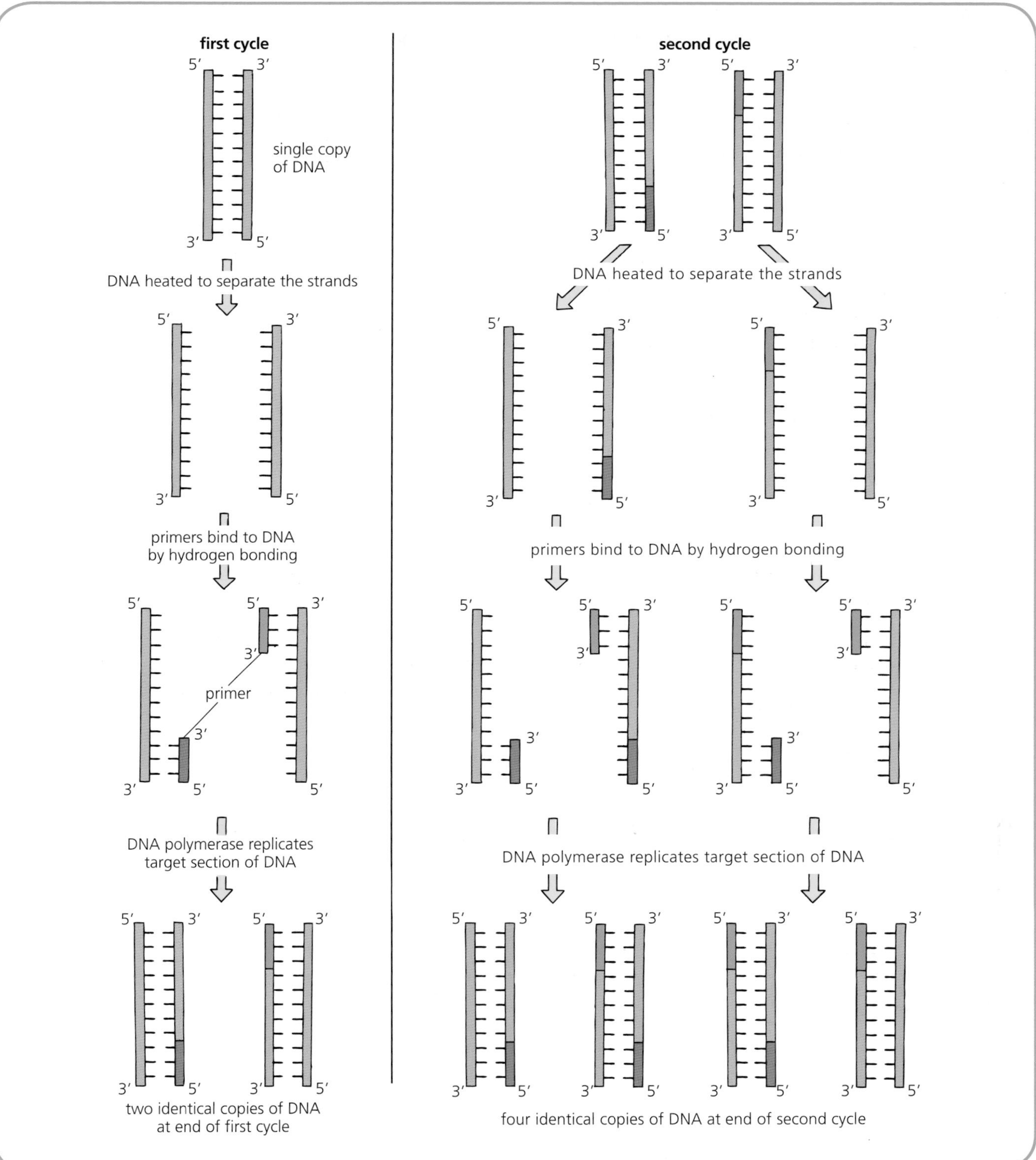

Figure 2.9 Polymerase chain reaction

Case Study Thermal cycling

PCR depends on a process called **thermal cycling**. A cycle consists of three steps each carried out at a different temperature. The earliest designs of this technique used three water baths and normal DNA polymerase. The latter was destroyed during the heating step in the cycle and had to be replaced for use in the next cycle.

The following two important innovations enabled PCR to become automated:

- the isolation of **heat-tolerant** DNA polymerase from a species of bacterium native to hot springs
- the invention of the **thermal cycler**, a computerised heating machine able to control the repetitive temperature changes needed for PCR.

Figure 2.10 shows a simplified version of the steps carried out during thermal cycling in order to amplify DNA.

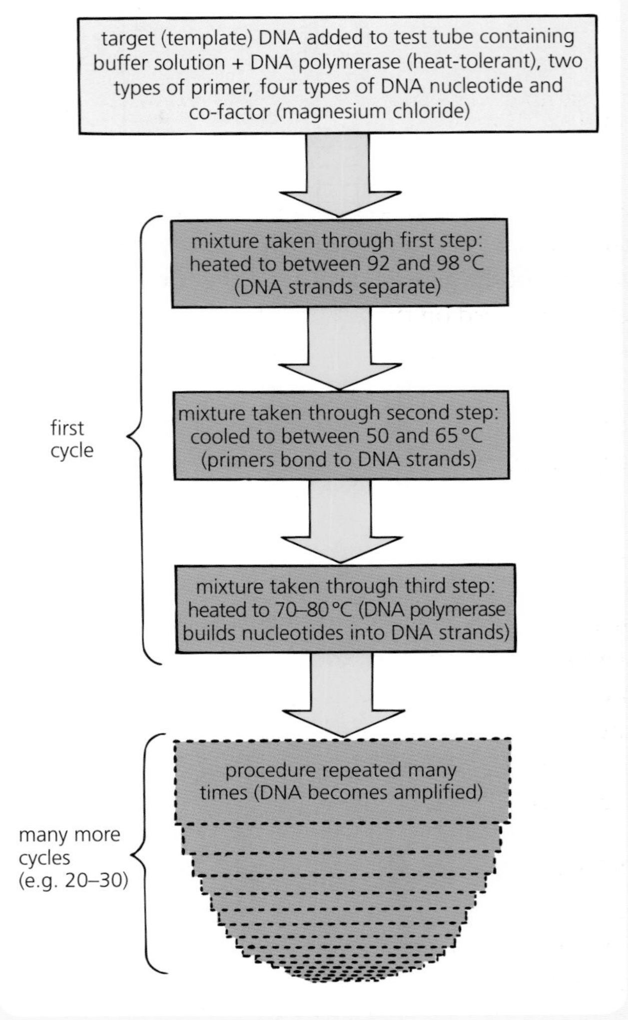

Figure 2.10 Amplification of DNA by thermal cycling

Related Topic

'Needles and haystacks'

Equally as impressive as the amplification of DNA by PCR is the **specificity** of the reaction. Each primer is a piece of single-stranded DNA synthesised as the exact complement of a short length of the DNA strand to which it is to become attached. This enables the primer to find 'the needle in the haystack'. In other words it is able to locate, among many different sites, the specific target DNA sequence that is to be amplified. The process then goes on to produce millions or even billions of copies of the DNA. Therefore this amplification of DNA by PCR is sometimes described as being like 'a haystack from the needle'.

Medical and forensic applications of amplified DNA

Medical

PCR can be used to amplify the genomic DNA from a cell sample taken from a patient. By this means sufficient DNA is generated to allow it to be **screened** for the presence or absence of a specific sequence known to be characteristic of a genetic disease or disorder. This enables medical experts to confirm a **diagnosis** of the genetic disorder if the condition is suspected (based on the patient's family history and/or the fact that the patient is showing early symptoms).

Forensic

DNA profile

The human genome possesses many short, **non-coding regions of DNA** composed of a number of **repetitive sequences.** These regions are found to be randomly distributed throughout the genome and to differ in length and number of repeats of the DNA sequences from person to person. Each region of repetitive sequences is **unique** to the individual who possesses it. Therefore these regions of genetic material can be used to construct a **DNA profile** for that person.

Crime scene

Forensic scientists make use of the PCR reaction to amplify DNA samples from a **crime scene.** DNA samples taken from the victim and the suspects are also amplified. Next the components of the samples are separated using gel electrophoresis (see page 22) and then compared. In the example shown in Figure 2.11, it is concluded that the DNA in sample 1 from the crime scene matches that of the victim and that the DNA in sample 2 from the crime scene matches that of suspect Q.

Paternity dispute

PCR followed by gel electrophoresis can also be employed to generate genetic profiles from DNA samples and **confirm genetic relationships** between individuals. Each person inherits 50% of their DNA from each parent, therefore every band in their DNA profile ('genetic fingerprint') must match one in that of their father or their mother. The fact that each person has 50% of their bands in common with each of their parents (see Figure 2.12) allows **paternity disputes** to be settled.

Evidence based on DNA amplified by PCR has also been used to identify missing people from human remains left at the site of a disaster and to secure the release of innocent people who have been wrongly imprisoned.

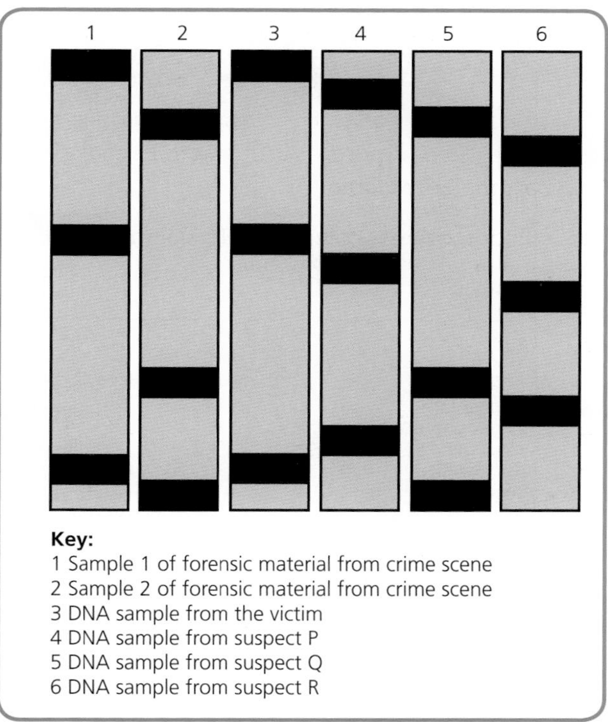

Figure 2.11 Forensic application

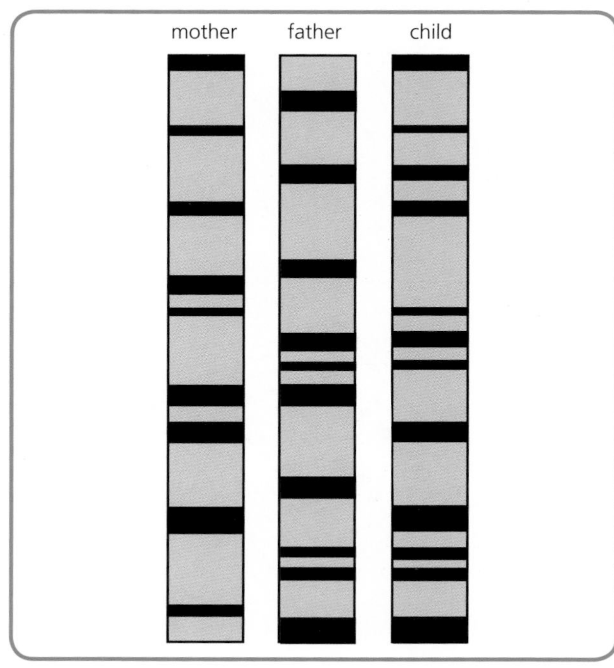

Figure 2.12 Genetic 'fingerprints'

Related Activity

Analysis of DNA using gel electrophoresis

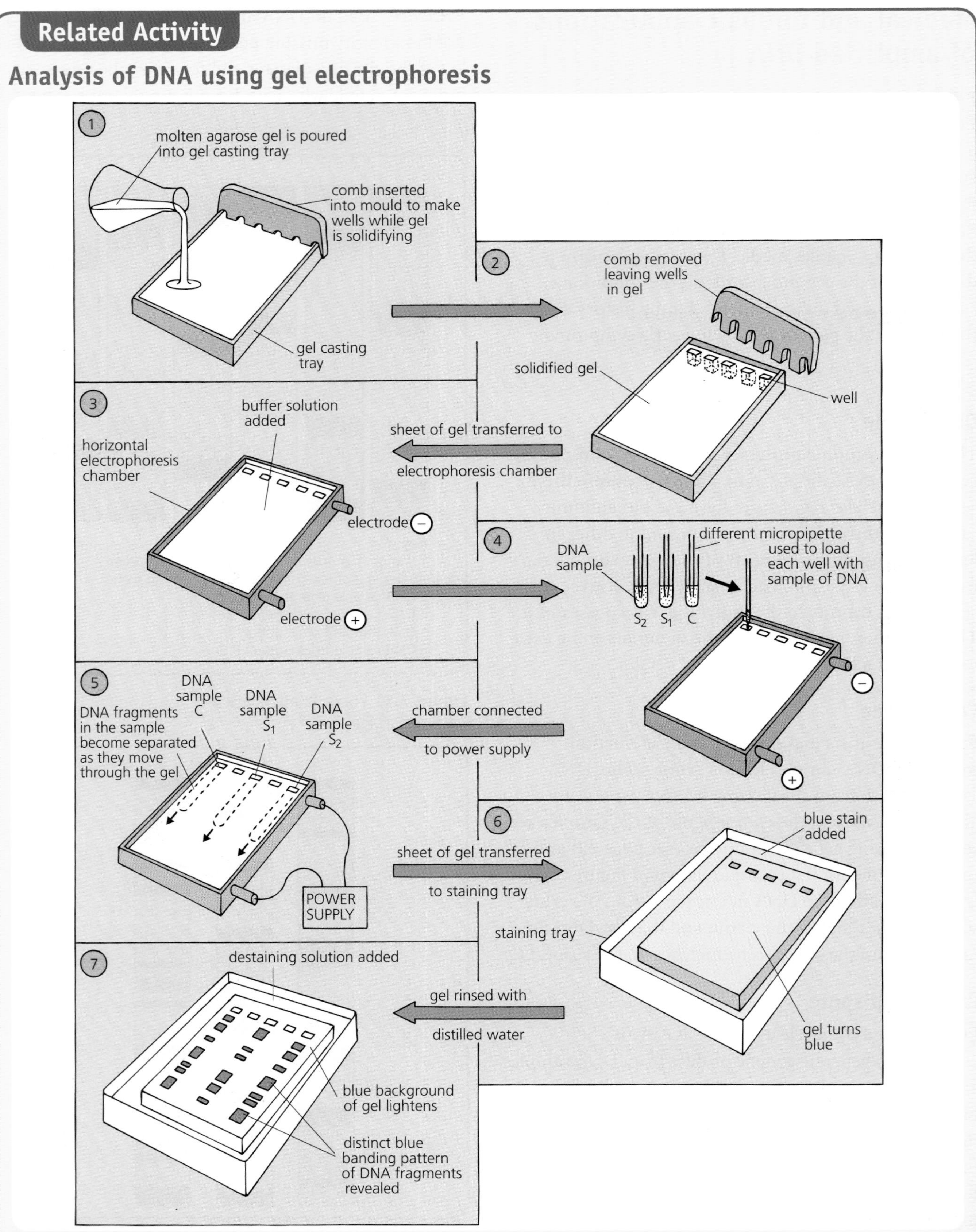

Figure 2.13 Separation of DNA by gel electrophoresis

→

Gel electrophoresis is a technique used to separate electrically charged molecules that vary in size by subjecting them to an electric current which forces them to move through a sheet of gel. When the molecules are negatively charged, they move towards the positively charged end of the gel. However, they do not all move at the same rate. Smaller molecules move at a faster rate and are therefore found to have **moved further** than larger molecules in a given period of time.

A **restriction enzyme** (also see page 185, chapter 13) is used to prepare each DNA sample by cutting it at specific sites into smaller pieces of varying lengths characteristic of that type of DNA. Figure 2.13 illustrates the separation of three samples of DNA: sample C (from the crime scene), sample S_1 (from suspect 1) and sample S_2 (from suspect 2) using gel electrophoresis. From the results it is concluded that the DNA from suspect 2 matches the DNA from the crime scene.

Testing Your Knowledge

1 Decide whether each of the following statements is true or false and then use T or F to indicate your choice. Where a statement is false, give the word that should have been used in place of the word in bold print. (5)

a) During DNA replication, each DNA parental strand acts as a **template**.

b) A guanine base can only pair up with a **thymine** base.

c) Complementary base pairs are held together by weak **antiparallel** bonds.

d) Each new DNA molecule formed by replication contains one **parental** and one new strand.

e) Many copies of a DNA sample can be synthesised using the **polymerase** chain reaction.

2 Figure 2.14 shows part of a DNA molecule undergoing replication. Match numbers 1–6 with the following statements. (6)

a) DNA polymerase promotes formation of a fragment of the lagging strand of replicating DNA.

b) Parental double helix unwinds.

c) DNA polymerase bonds nucleotide to primer.

d) Ligase joins fragments into the lagging strand of replicating DNA.

e) DNA polymerase promotes formation of the leading strand of replicating DNA.

f) DNA molecule becomes stabilised as two template strands.

3 a) Name FOUR substances that must be present in a nucleus for DNA replication to occur. (4)

b) Briefly explain why DNA replication is important. (2)

4 a) What can be produced *in vitro* by employing the polymerase chain reaction (PCR)? (1)

b) In PCR, what is a *primer*? (2)

c) Why is the DNA heated during the PCR process? (1)

d) What is the purpose of cooling the DNA sample? (1)

e) What characteristic of the DNA polymerase used in PCR prevents it from becoming denatured during the process? (1)

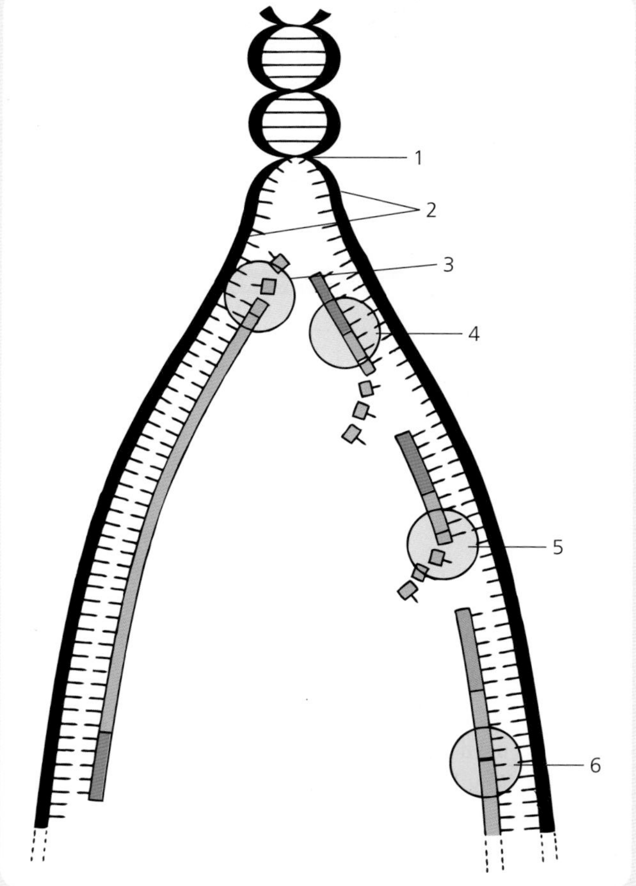

Figure 2.14

Applying Your Knowledge and Skills

1 Refer back to Figure 2.4 and then draw a labelled diagram to show both the DNA strands and the test tube contents that would result after three DNA replications involving semi-conservative replication only. (4)

2 Figure 2.15 shows a replication 'bubble' on a strand of DNA.

a) i) Redraw the diagram including the given labels and then mark '3′ end' and '5′ end' on the parental DNA strand for a second time.

ii) Draw in and label a starting point (origin of replication).

iii) Label one of the primer molecules.

iv) Label the leading DNA strand and a fragment of the lagging strand.

v) Use the letter P four times to indicate all the locations where DNA polymerase would be active. (6)

b) i) In this chromosome the replication fork moves at a rate of 2500 base pairs per minute. If this chromosome is 5×10^7 base pairs in length, how many minutes would one replication fork take to replicate the entire chromosome?

ii) In reality, replication of this chromosome's DNA only takes 3 minutes. Explain how this is achieved. (2)

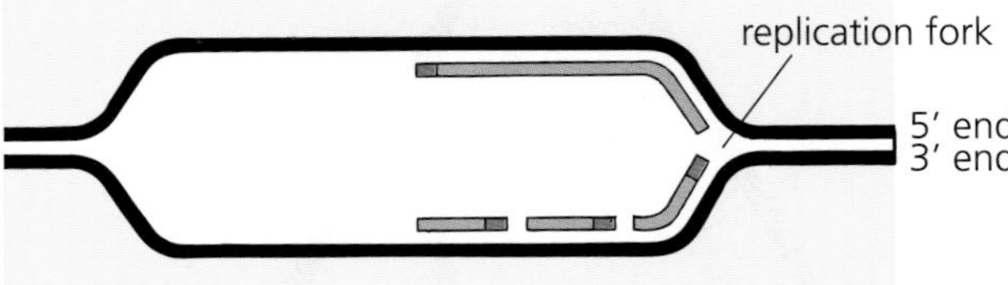

Figure 2.15

3 The graph in Figure 2.16 shows the expected number of copies of DNA that would be generated by the polymerase chain reaction (PCR) under ideal conditions.

a) i) What name is given to the type of graph paper used here to present the data?

ii) Why has this type of graph paper been used? (2)

b) How many cycles of PCR are required to produce 10 000 000 copies of the DNA? (1)

c) How many cycles are required to increase the number of copies of DNA already present at ten cycles by a factor of 10^3? (1)

d) i) How many copies of the DNA were present after 30 cycles?

ii) Now state your answer in words. (2)

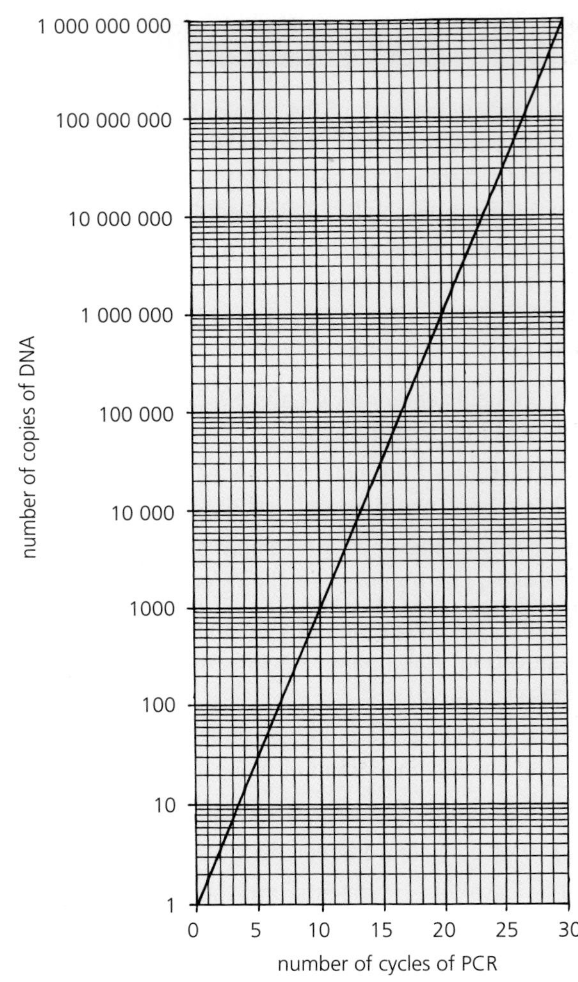

Figure 2.16

4 Describe the main processes that occur during the replication of a molecule of DNA. (9)

3 Gene expression

A cell's **genotype** (its genetic constitution) is determined by the sequence of the DNA bases in its genes (the **genetic code**). A cell's **phenotype** (its physical and chemical state) is determined by the proteins that are synthesised when the genes are expressed. Gene expression involves the processes of **transcription** and **translation** (discussed in this chapter). Only a fraction of the genes in a cell are expressed.

Structure of RNA

The second type of nucleic acid is called **RNA** (ribonucleic acid). Each nucleotide in an RNA molecule is composed of a molecule of **ribose** sugar, an organic base and a phosphate group (see Figure 3.1). In RNA, the base **uracil** (**U**) replaces thymine found in DNA and is complementary to adenine. Unlike DNA which consists of two strands, a molecule of RNA is a **single strand** as shown in Figure 3.2. The differences between RNA and DNA are summarised in Table 3.1.

Characteristic	RNA	DNA
number of nucleotide strands present in one molecule	one	two
complementary base partner of adenine	uracil	thymine
sugar present in a nucleotide	ribose	deoxyribose

Table 3.1 Differences between RNA and DNA

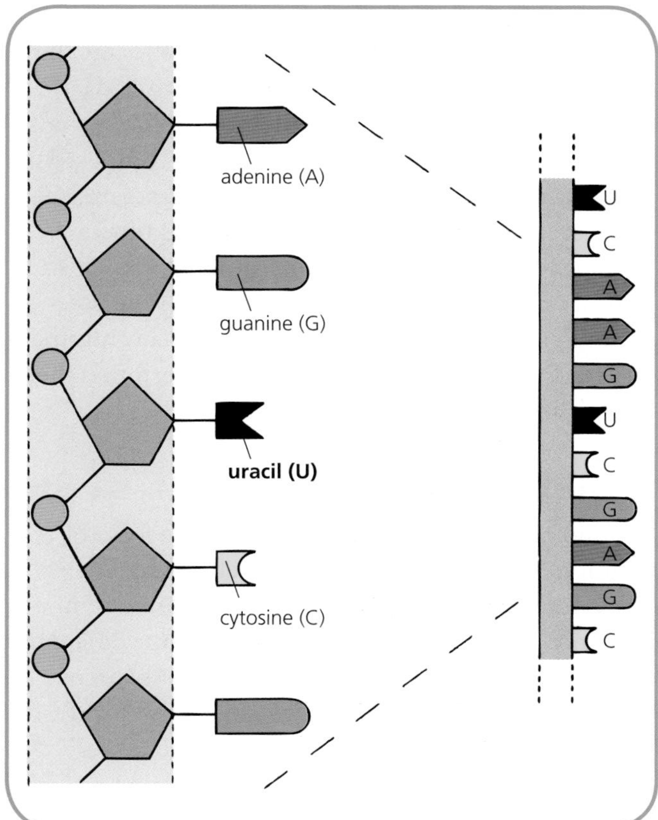

Figure 3.2 Structure of RNA

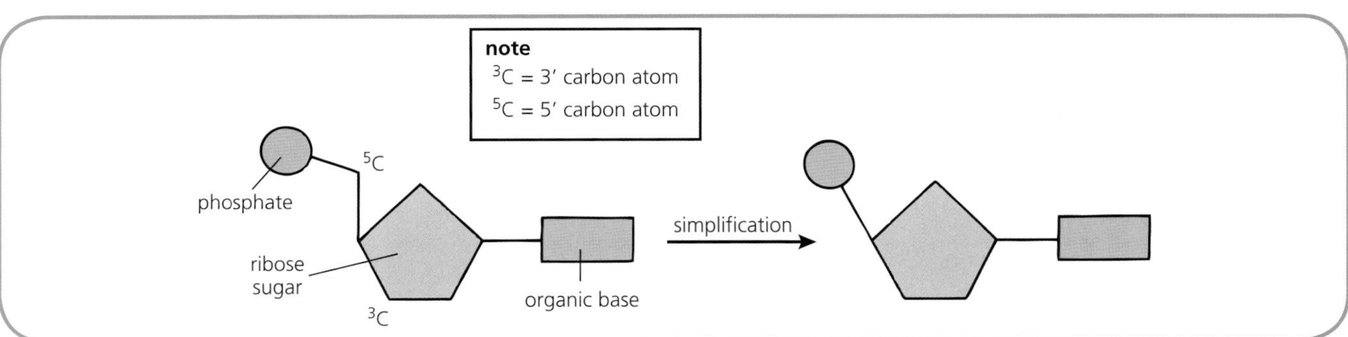

Figure 3.1 Structure of an RNA nucleotide

Control of inherited characteristics

The sequence of bases along the DNA strands contains the **genetic instructions** which control an organism's inherited characteristics. These characteristics are the result of many biochemical processes controlled by enzymes which are made of **protein**. Each protein is made of one or more **polypeptide** chains composed of subunits called **amino acids**. A protein's exact molecular structure, shape and ability to carry out its function all depend on the **sequence** of its amino acids. This critical order is determined by the order of the bases in the organism's DNA. By this means DNA controls the structure of enzymes and in doing so, determines the organism's inherited characteristics.

Genetic code

The information present in DNA takes the form of a molecular language called the **genetic code**. The sequence of bases along a DNA strand represents a sequence of 'codewords'. DNA possesses four different types of base. Proteins contain 20 different types of amino acid. If the bases are taken in groups of three then this gives 64 (4^3) different combinations (see Appendix 1). It is now known that each amino acid is coded for by one or more of these 64 **triplets** of bases. Thus a species' genetic information is encoded in its DNA with each strand bearing a series of base triplets arranged in a specific order for coding the particular proteins needed by that species.

Gene expression through protein synthesis

The genetic information for a particular polypeptide is carried on a section of DNA in the nucleus. However, the assembly of amino acids into a genetically determined sequence takes place in the cell's cytoplasm in tiny structures called **ribosomes**. Figure 3.3 gives an overview of gene expression through protein synthesis. A molecule of **mRNA** (messenger RNA) is formed (**transcribed**) from the appropriate section of the DNA strand and carries that information to ribosomes. There the mRNA meets **tRNA** (transfer RNA) and the genetic information is **translated** into protein.

Transcription

Transcription is the synthesis of mRNA from a section of DNA. A promoter is a region of DNA in a gene where transcription is initiated as shown in Figure 3.4 (where the DNA strand has been drawn uncoiled for the sake of simplicity).

RNA polymerase is the enzyme responsible for transcription. As it moves along the gene from the promoter, unwinding and opening up the DNA strand by breaking the hydrogen bonds between base pairs, it brings about the synthesis of an **mRNA** molecule. As a result of the base-pairing rule, the mRNA gets a nucleotide sequence complementary to one of the two DNA strands (the template strand) as shown in Figure 3.5.

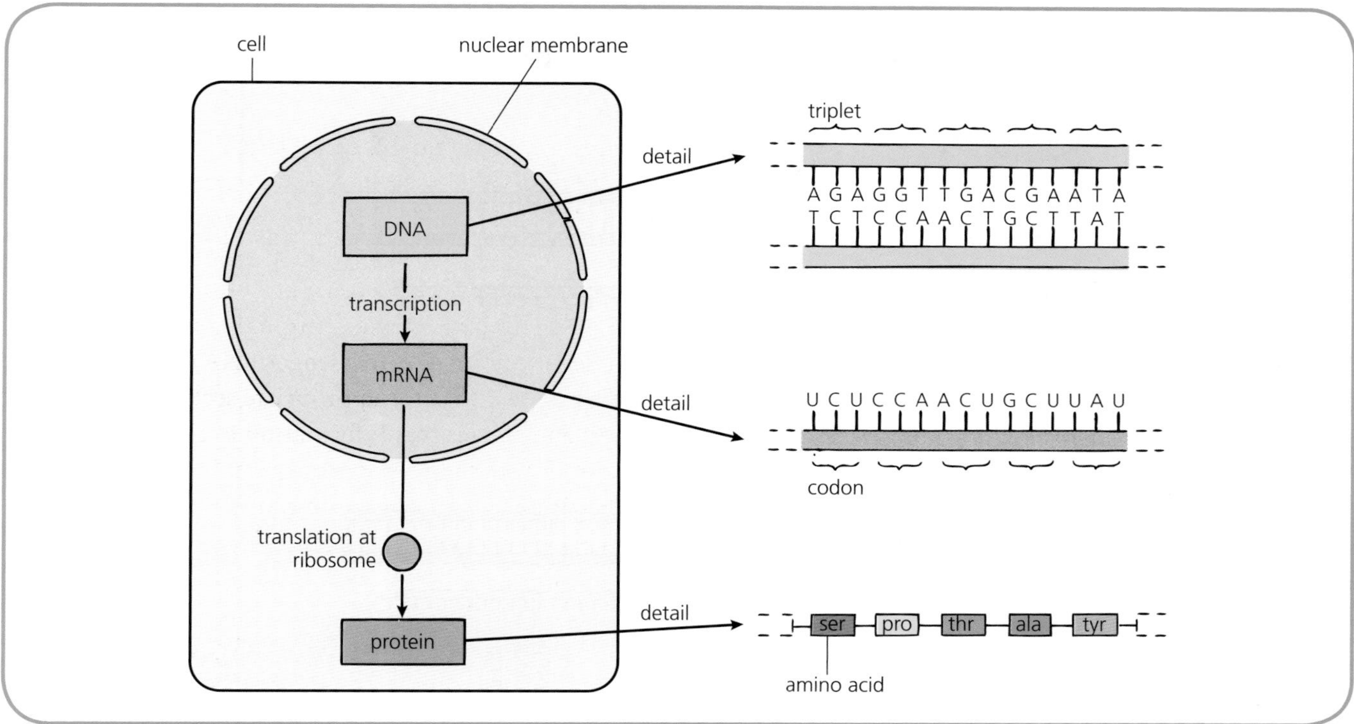

Figure 3.3 Overview of gene expression

RNA polymerase can only add nucleotides to the 3′ end of the growing mRNA molecule. The molecule elongates until a terminator sequence of nucleotides is reached on the DNA strand. The resultant mRNA strand which becomes separated from its DNA template is called a **primary transcript** of mRNA.

Modification of primary transcript

Normally the region of DNA transcribed to mRNA is about 8000 nucleotides long yet only about 1200 nucleotides are needed to code for an average-sized polypeptide chain. This is explained by the fact that in eukaryotes long stretches of DNA exist within a gene that do not play a part in the coding of the polypeptide. In addition these non-coding regions, called **introns**, are interspersed between the coding regions, called **exons**. Therefore the region in the primary transcript of mRNA responsible for coding the polypeptide is fragmented.

Splicing

Figure 3.6 shows how the introns are cut out and removed from the primary transcript of mRNA and the exons are **spliced** together to form mRNA with a continuous sequence of nucleotides. The order of exons remains unchanged during splicing. The modified mRNA called the **mature transcript** of mRNA passes out of the nucleus into the cytoplasm (see Figure 3.7) and moves on to the next stage of protein synthesis where it becomes translated into a sequence of amino acids.

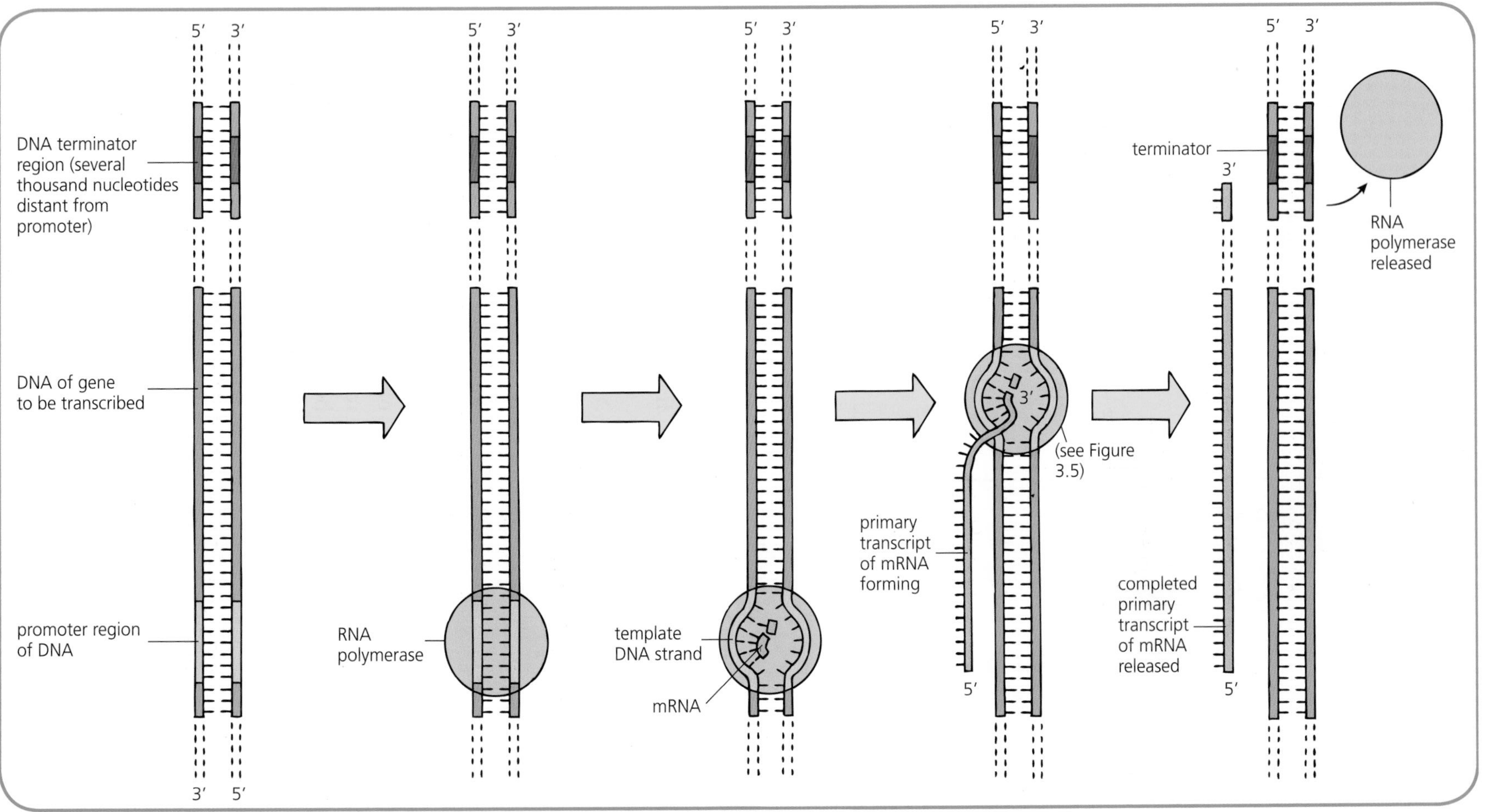

Figure 3.4 Transcription of mRNA

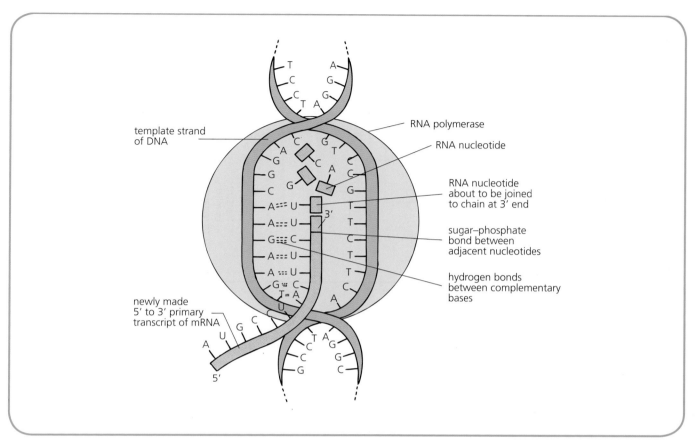

Figure 3.5 Detail of transcription

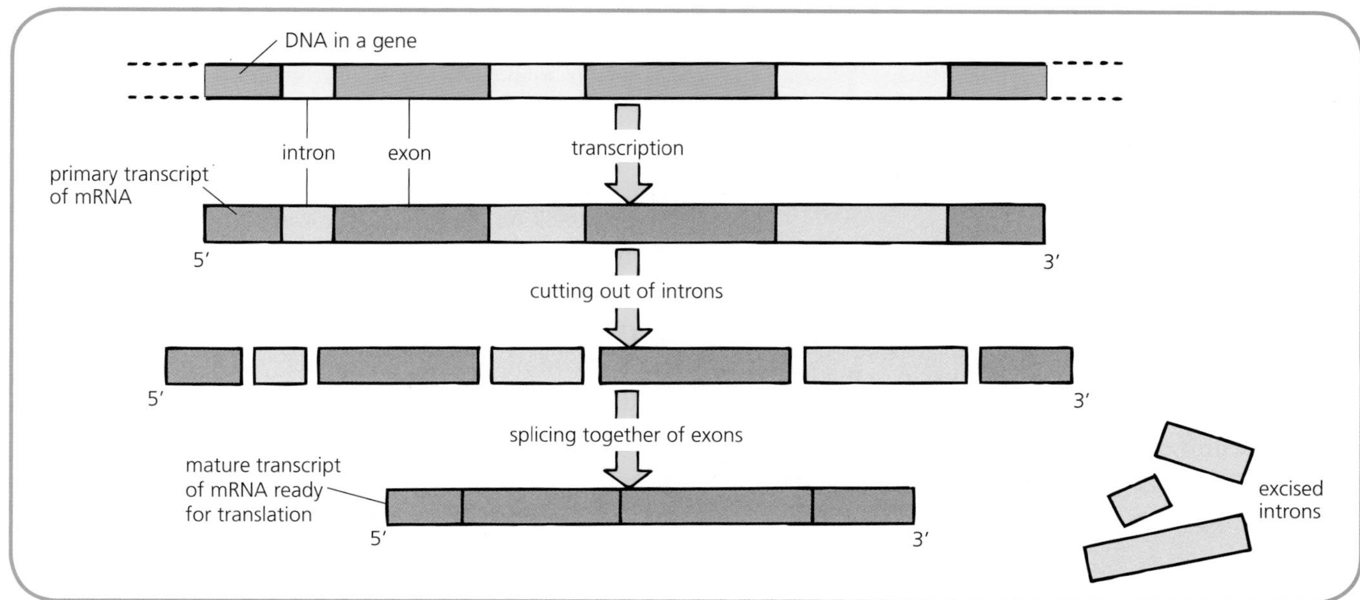

Figure 3.6 Modification of primary mRNA transcript

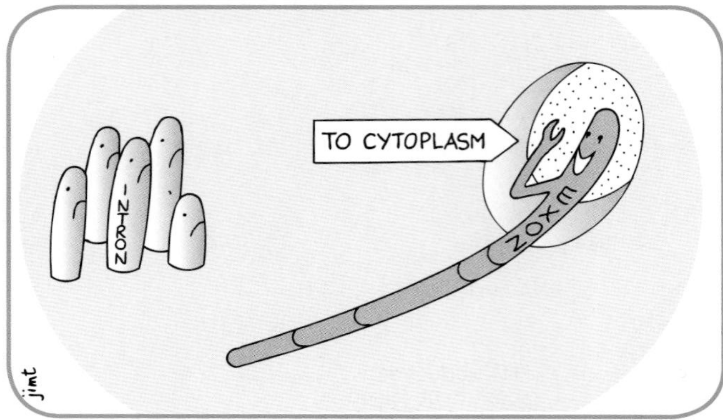

Figure 3.7 'Better luck next time, guys!'

Testing Your Knowledge 1

1 State THREE ways in which RNA and DNA differ in structure and chemical composition. (3)

2 **a)** In what way does the DNA of one species differ from that of another making each species unique? (1)

b) How many bases in the genetic code correspond to one amino acid? (1)

c) What name is given to the groups of bases that make up the genetic code? (1)

3 **a)** Draw a diagram of the mRNA strand that would be transcribed from section X of the DNA molecule shown in Figure 3.8. (2)

b) Name the enzyme that would direct this process. (1)

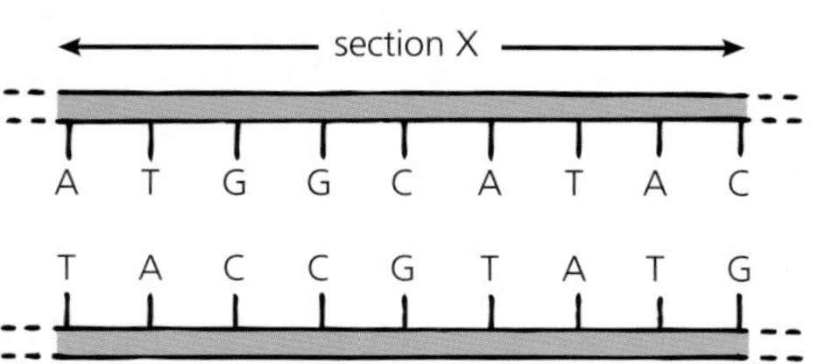

Figure 3.8

4 **a)** What is the difference between an *exon* and an *intron*?

b) Which of these must be removed from the primary transcript of mRNA?

c) By what process are they removed? (3)

Translation

Translation is the synthesis of **protein** as a polypeptide chain under the direction of **mRNA**. The genetic message carried by a molecule of mRNA is made up of a series of base triplets called **codons**. The codon is the **basic unit** of the genetic code. Each codon is complementary to a triplet of bases on the original template DNA strand.

Transfer RNA

A further type of RNA is found in the cell's cytoplasm. This is called **tRNA** (transfer RNA) and it is composed of a single strand of nucleotides. However, a molecule of tRNA has a three-dimensional structure because it is folded back on itself in such a way that hydrogen bonds form between many of its nucleotide bases as shown in Figure 3.9. Each molecule of tRNA has only one particular triplet of bases exposed. This triplet is called an **anticodon**. It is complementary to an mRNA codon and corresponds to a specific amino acid carried by that tRNA at its **attachment site**.

Table 3.2 shows the relationship between mRNA's codons, tRNA's anticodons and the amino acids coded. Many different types of tRNA are present in a cell, one or more for each type of amino acid. Each tRNA picks up its appropriate amino acid molecule from the cytoplasm's amino acid pool at its site of attachment. The amino acid is then carried by the

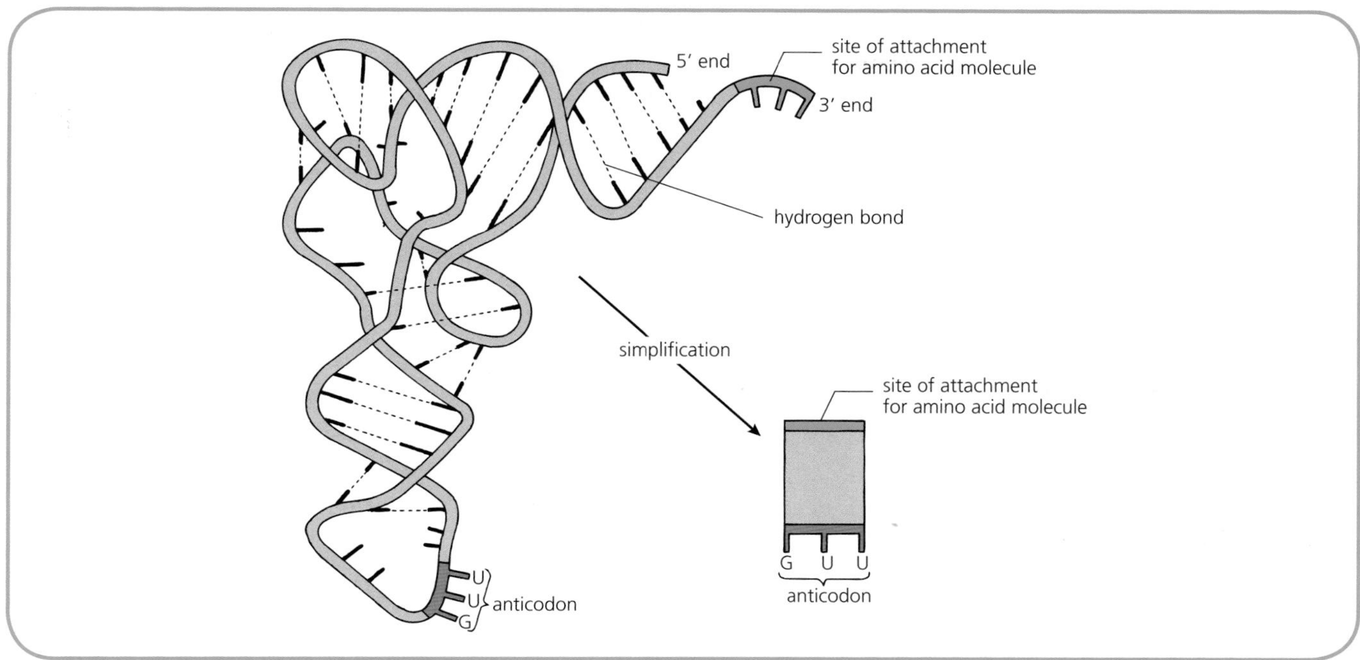

Figure 3.9 Structure of transfer RNA (tRNA)

Related Information

Codons and anticodons

Codon (mRNA)	Anti-codon (tRNA)	Amino acid	Codon (mRNA)	Anti-codon (tRNA)	Amino acid	Codon (mRNA)	Anti-codon (tRNA)	Amino acid	Codon (mRNA)	Anti-codon (tRNA)	Amino acid
UUU UUC UUA UUG	AAA AAG AAU AAC	} phe } leu	UCU UCC UCA UCG	AGA AGG AGU AGC	} ser	UAU UAC UAA UAG	AUA AUG AUU AUC	} tyr STOP STOP	UGU UGC UGA UGG	ACA ACG ACU ACC	} cys STOP trp
CUU CUC CUA CUG	GAA GAG GAU GAC	} leu	CCU CCC CCA CCG	GGA GGG GGU GGC	} pro	CAU CAC CAA CAG	GUA GUG GUU GUC	} his } gln	CGU CGC CGA CGG	GCA GCG GCU GCC	} arg
AUU AUC AUA AUG	UAA UAG UAU UAC	} ile met or START	ACU ACC ACA ACG	UGA UGG UGU UGC	} thr	AAU AAC AAA AAG	UUA UUG UUU UUC	} asn } lys	AGU AGC AGA AGG	UCA UCG UCU UCC	} ser } arg
GUU GUC GUA GUG	CAA CAG CAU CAC	} val	GCU GCC GCA GCG	CGA CGG CGU CGC	} ala	GAU GAC GAA GAG	CUA CUG CUU CUC	} asp } glu	GGU GGC GGA GGG	CCA CCG CCU CCC	} gly

Table 3.2 mRNA's codons, tRNA's anticodons and the amino acids coded (See Appendix 1 for full names of amino acids.)

tRNA to a ribosome and added to the growing end of a polypeptide chain. By this means, the genetic code is translated into a sequence of amino acids.

The mRNA codon AUG (complementary to tRNA anticodon UAC) is unusual in that it codes for methionine (met) *and* acts as the **start codon**. mRNA codons UAA, UAG and UGA do not code for amino acids but instead act as **stop codons**.

Ribosomes

Ribosomes are small, roughly spherical structures found in all cells. They contain ribosomal RNA (rRNA) and enzymes essential for protein synthesis. Many ribosomes are present in growing cells which need to produce large quantities of protein.

Binding sites

A ribosome's function is to bring tRNA molecules (bearing amino acids) into contact with mRNA. Ribosomes have one binding site for mRNA and three binding sites for tRNA as shown in Figure 3.10.

Of the tRNA binding sites:

- site P holds the tRNA carrying the growing polypeptide chain
- site A holds the tRNA carrying the next amino acid to be joined to the growing chain by a peptide bond
- site E discharges a tRNA from the ribosome once its amino acid has become part of the polypeptide chain.

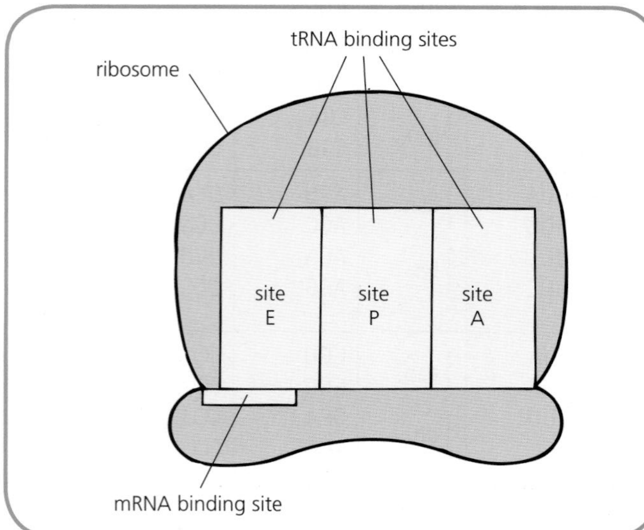

Figure 3.10 Binding sites on a ribosome

Start and stop codons in action

Before translation can begin, a ribosome must bind to the 5′ end of the mRNA template so that the mRNA's **start codon** (AUG) is in position at binding site P. Next a molecule of tRNA carrying its amino acid (methionine) becomes attached at site P by hydrogen bonds between its anticodon (UAC) and the start codon (see Figure 3.11).

The mRNA codon at site A recognises and then forms hydrogen bonds with the complementary anticodon on an appropriate tRNA molecule bearing its amino acid. When the first two amino acid molecules are adjacent to one another, they become joined by a **peptide bond**.

As the ribosome moves along one codon, the tRNA that was at site P is moved to site E and discharged from the ribosome to be reused. At the same time the tRNA that was at site A is moved to site P. The vacated site A becomes occupied by the next tRNA bearing its amino acid which becomes bonded to the growing peptide chain. The process is repeated many times allowing the mRNA to be translated into a complete **polypeptide chain**.

Eventually a **stop codon** (see Table 3.2) on the mRNA is reached. At this point, site A on the ribosome becomes occupied by a release factor which frees the polypeptide from the ribosome. The whole process needs energy from ATP (see Chapter 9).

One gene, many proteins

Alternative RNA splicing

Figure 3.6 on page 29 shows a primary transcript of mRNA being cut up and its exons being spliced together to form a molecule of mRNA ready for translation. This mature transcript of mRNA is not the only one that can be produced from that primary transcript. Depending on circumstances, **alternative segments of RNA** may be treated as the exons and introns. Therefore the same primary transcript has the potential to produce several mature mRNA transcripts, each with a different sequence of base triplets and each coding for a different polypeptide. In other words, one gene can code for several different proteins and a limited number of genes can give rise to a wide variety of proteins.

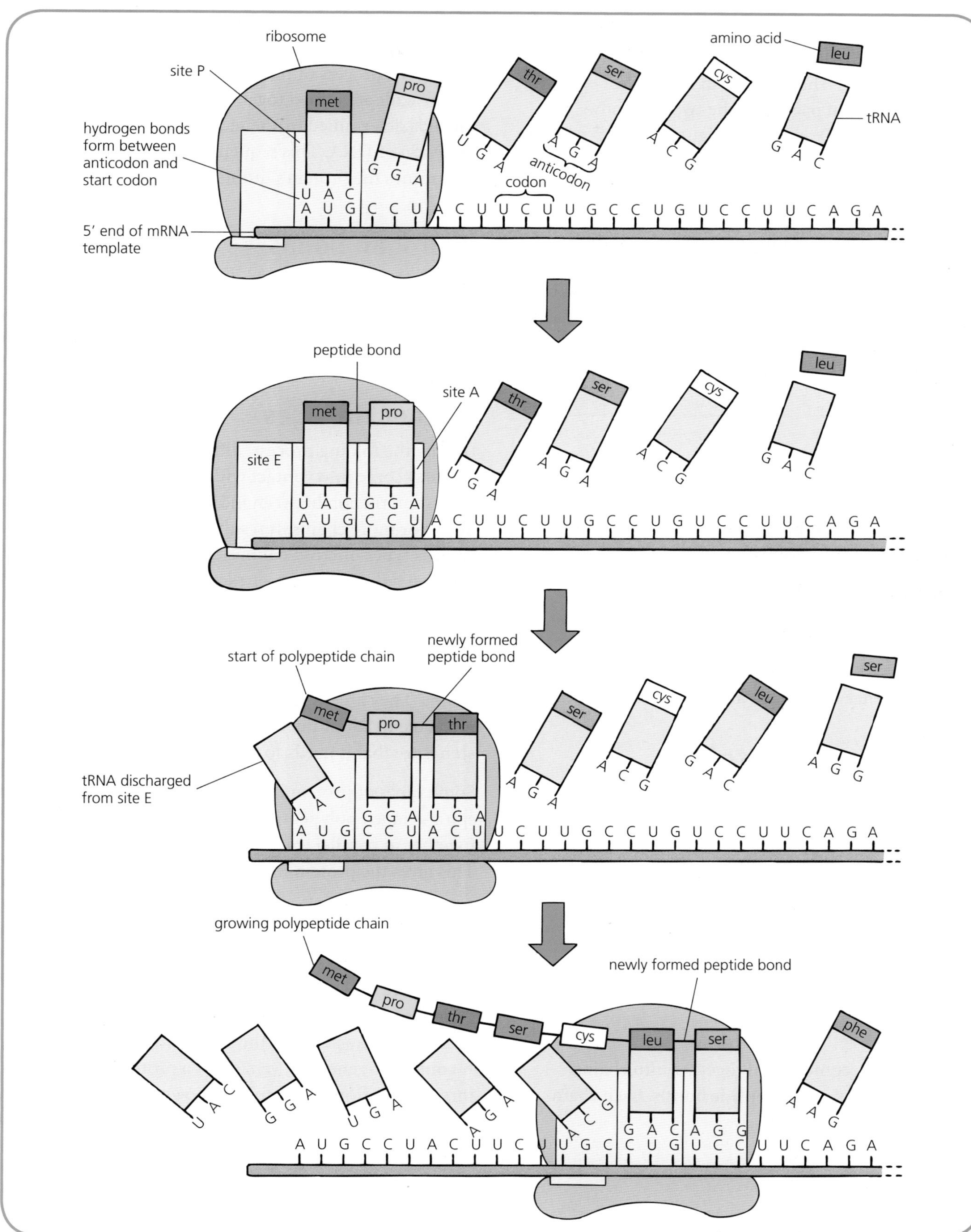

Figure 3.11 Translation of mRNA into polypeptide

One gene, two antibodies – an example of alternative splicing

An antibody is a Y-shaped protein molecule (see Figure 3.15 on page 36). The two antibody molecules (P and Q) shown in Figure 3.12 are coded for by the same gene yet they are different in structure. P possesses a membrane-anchoring unit coded for by an exon present in its mRNA. However, this membrane-anchoring unit is absent from Q because its mRNA lacks the necessary exon (discarded as an intron at the splicing stage). As a result, antibody P functions as a membrane-bound protein on the outer surface of a white blood cell whereas antibody Q operates freely in the bloodstream.

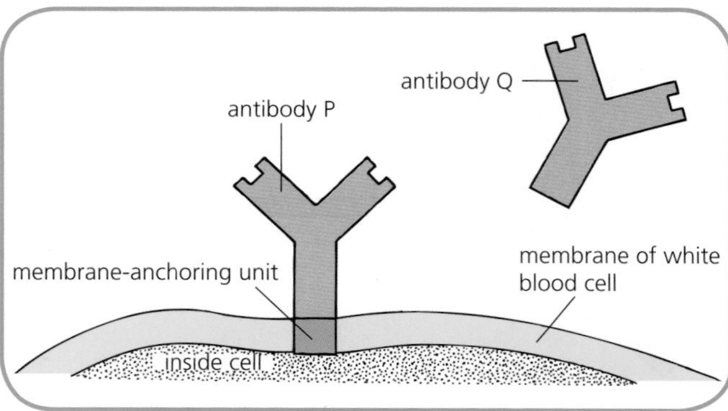

Figure 3.12 Products of alternative RNA splicing

Structure of proteins

All **proteins** contain the chemical elements carbon (C), hydrogen (H), oxygen (O) and nitrogen (N). Often they contain sulphur (S). Each protein is built up from a large number of subunits called **amino acids** of which there are 20 different types. The length of a protein molecule varies from many thousands of amino acids to just a few. Insulin, for example, contains only 51.

Polypeptides

Amino acids become linked together into chains by chemical links called **peptide bonds**. Each chain is called a **polypeptide** and it normally consists of hundreds of amino acid molecules linked together. During the process of protein synthesis (see page 26), amino acids are joined together in a **specific order** which is determined by the sequence of bases on a portion of DNA. This sequence of amino acids determines the protein's ultimate structure and function.

Hydrogen bonds

Chemical links known as **hydrogen bonds** form between certain amino acids in a polypeptide chain causing the chain to become coiled or folded as shown in Figure 3.13.

Further linkages

During the folding process, different regions of the chain(s) come into contact with one another. This allows interaction between individual amino acids in one or more chains. It results in the formation of various types of cross-connection including **bridges** between **sulphur** atoms, attraction between positive and negative charges and further hydrogen bonding.

These cross-connections occur between amino acids in the same polypeptide chain and those on adjacent chains. They are important because they cause the molecule to adopt the final **three-dimensional structure** that it needs to carry out its specific function.

Some types of protein molecule are formed by several spiral-shaped polypeptide molecules becoming linked together in parallel when bonds form between them. This gives the protein molecule a rope-like structure.

Other types of protein molecule consist of one or more polypeptide chains folded together into a roughly spherical shape like a tangled ball of string (see Figure 3.13). The exact form that the folding takes depends on the types of further linkage that form between amino acids on the same and adjacent chains. A computer-generated version of a protein molecule's three-dimensional structure is shown in Figure 3.14.

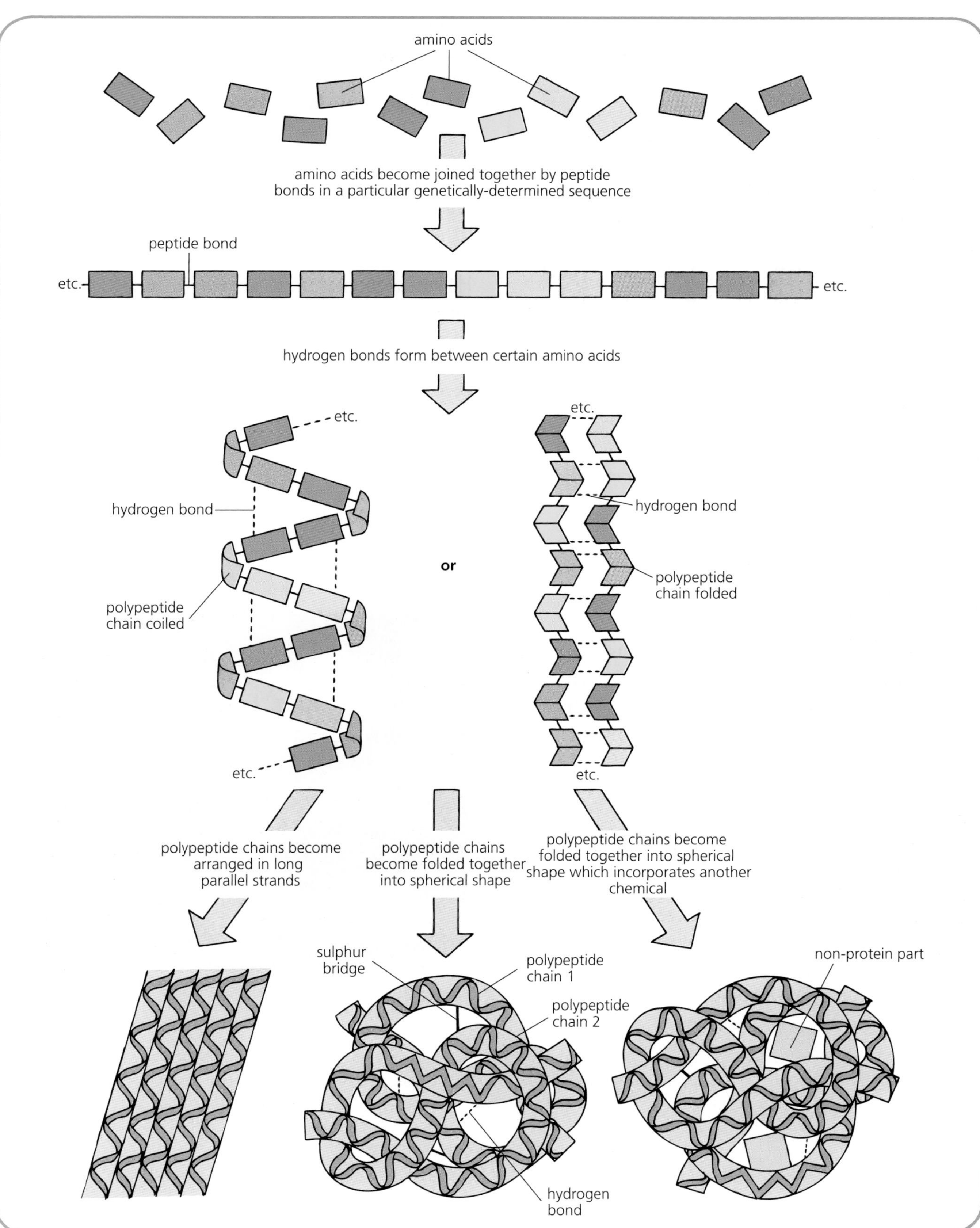

Figure 3.13 Structure of proteins

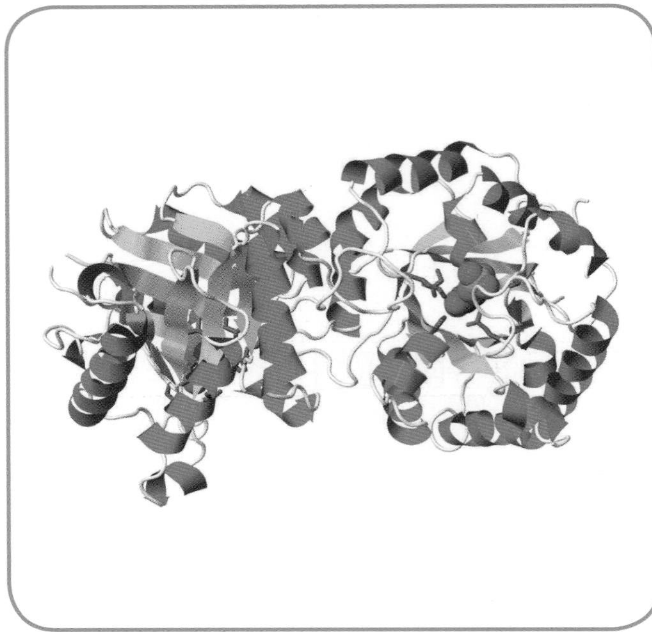

Figure 3.14 Protein molecule as visualised by Jmol software

Functions of proteins

A vast variety of structures and shapes exists among proteins and as a result they are able to perform a wider range of functions than any other type of molecule in the body. Some are found in connective tissue, bone and muscle where their strong fibres provide support and allow movement. Other proteins are vital components of all living cells and play a variety of roles as follows.

Enzymes

Each molecule of **enzyme** is made of protein and folded in a particular way to expose an active surface which combines readily with a specific substrate (also see Chapter 8). Since intracellular enzymes speed up the rate of biochemical processes such as photosynthesis, respiration and protein synthesis, they are essential for the maintenance of life.

Structural protein

Protein is one of two components which make up the **membrane** surrounding a living cell. Similarly it forms an essential part of all membranes possessed by a cell's organelles. Therefore this type of protein plays a vital structural role in every living cell.

Hormones

These are **chemical messengers** transported in an animal's blood to 'target' tissues where they exert a specific effect. Some hormones are made of protein and exert a regulatory effect on the animal's growth and metabolism.

Antibodies

Although Y-shaped rather than spherical, **antibodies** are also composed of protein (see Figure 3.15). They are made by white blood cells to defend the body against antigens.

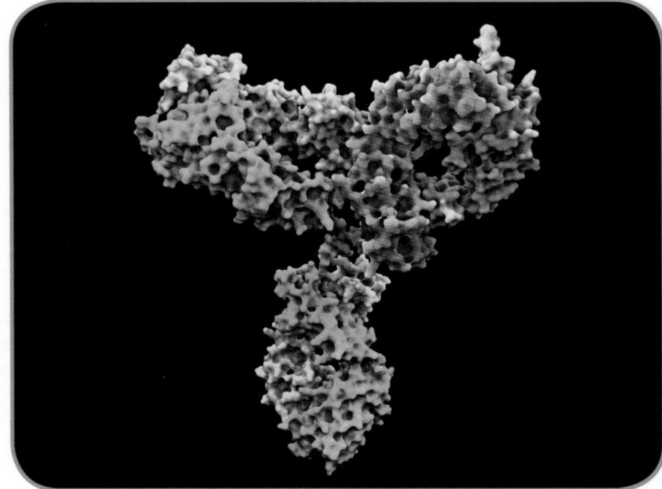

Figure 3.15 Antibody molecule

Related Activity

Separation and identification of amino acids using paper chromatography

Chromatography

Chromatography is a technique used to separate the components of a mixture which differ in their degree of solubility in the solvent. During ascending paper chromatography, the mixture is applied to absorbent paper. As the solvent passes up through the paper, it carries the components of the mixture up to different levels depending on their degree of solubility in the solvent (and the extent to which they are absorbed by the paper).

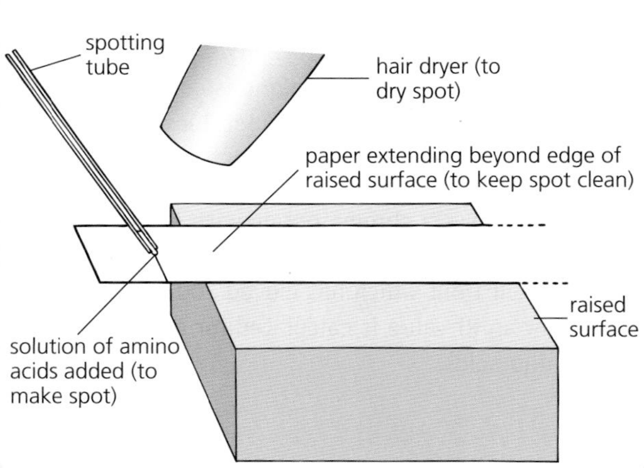

Figure 3.17 Spotting and drying

A strip of chromatography paper is prepared so that it fits the gas jar and the split stopper as shown in Figure 3.16. Plastic gloves are worn when handling the paper to prevent its contamination by amino acids from the skin. The solution of amino acids is spotted and dried repeatedly (see Figure 3.17). A small volume of solvent is added to the gas jar and the end of the paper is dipped into the solvent (see Figure 3.18).

The **chromatogram** is allowed to run until the solvent has almost reached the top of the paper. A second pencil line is drawn at the highest point reached by the solvent (called the **solvent front**) and then the strip is hung up to dry. The chromatogram is developed in a fume cupboard using ninhydrin spray. The amino acids show up as purple spots. The solvent has carried the most soluble amino acids to the highest position and so on down the paper to the least soluble one.

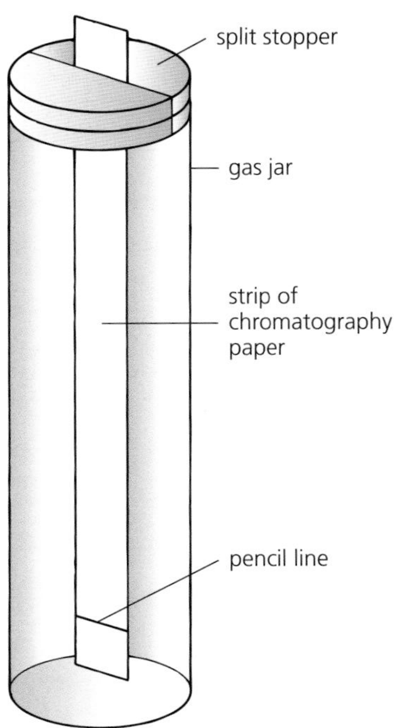

Figure 3.16 Preparing a strip of chromatography paper

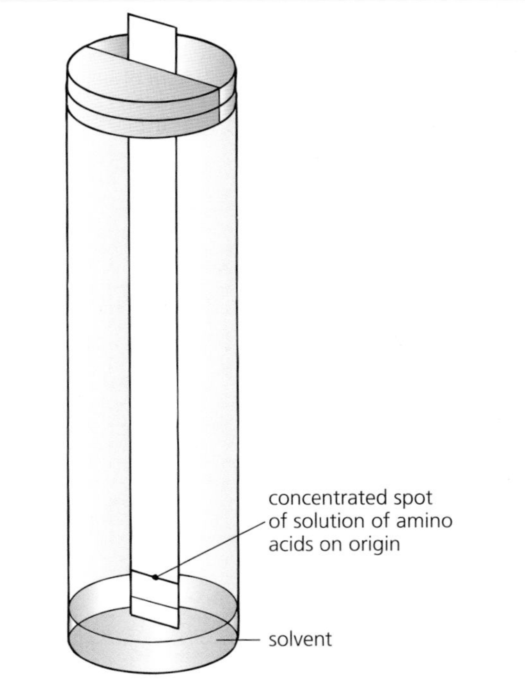

Figure 3.18 Running the chromatogram

→

Rf (relative front) value

Each amino acid separated by chromatography has an **Rf value**. This is the ratio of the distance moved by the amino acid front from the origin to the distance moved by the solvent front from the origin. It is normally expressed as a decimal fraction and is constant for each amino acid provided that the solvent and the type of paper strip used remain the same each time.

Figure 3.19 shows a chromatogram with three spots each indicating the position of an amino acid relative to the solvent front which has moved 200 mm. Table 3.3 shows how an Rf value is calculated for each of the three amino acids.

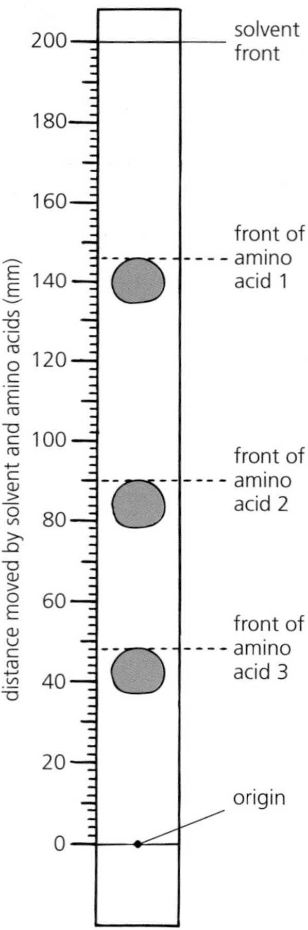

Figure 3.19 Chromatogram

Amino acid	Distance moved by amino acid's front (mm)	Rf value
1	146	146/200 = 0.73
2	90	90/200 = 0.45
3	48	48/200 = 0.24

Table 3.3 Rf calculations

Identification

The three amino acids can be identified by referring to a table of known Rf values for amino acids in the same solvent (see Table 3.4). Amino acid 1 is leucine, amino acid 2 is tyrosine and amino acid 3 is aspartic acid.

Amino acid	Abbreviation	Rf
histidine	his	0.11
glutamine	gln	0.13
lysine	lys	0.14
arginine	arg	0.20
aspartic acid	asp	0.24
glycine	gly	0.26
serine	ser	0.27
glutamic acid	glu	0.30
threonine	thr	0.35
alanine	ala	0.38
cysteine	cys	0.40
proline	pro	0.43
tyrosine	tyr	0.45
asparagine	asn	0.50
methionine	met	0.55
valine	val	0.60
tryptophan	trp	0.66
phenylalanine	phe	0.68
isoleucine	ile	0.72
leucine	leu	0.73

Table 3.4 Rf values

Related Activity

Separation of fish proteins by agarose gel electrophoresis

Gel electrophoresis (also see page 23) is a technique used to separate electrically charged molecules by subjecting them to an electric current which forces them to move through a sheet of gel.

Eliminating two variable factors

The behaviour of a protein molecule during gel electrophoresis is affected by three variable factors: its shape, its size and its net electric charge. In this experiment variation in shape of the protein molecule and variation in electric charge (in other words, positive or negative) are eliminated by subjecting each protein sample to a negatively charged detergent, a session of heat treatment and a reducing agent. These processes give the molecule a uniformly negative charge and disrupt its hydrogen bonding and sulphur bridges. All the protein molecules become converted to one or more negatively charged linear polypeptides and therefore only vary in **size** (molecular weight).

The technique illustrated in Figure 2.13 on page 22 is used to separate the proteins present in extracts from four species of fish (W, X, Y and Z) and those in a standard sample of known proteins.

Identification of fish proteins

Figure 3.20 shows a gel with five lanes resulting from the electrophoresis process. The standard sample is known to contain the seven fish proteins listed in Table 3.5. The band nearest to the well in lane 1 represents the largest protein molecule which has moved the shortest distance (in other words, protein a). Similarly band b represents the next largest and so on to band g that has moved the furthest distance and must be the smallest. Comparison of each of lanes 2–5 with the standard reveals the identity of the proteins present in the extract from a particular species of fish.

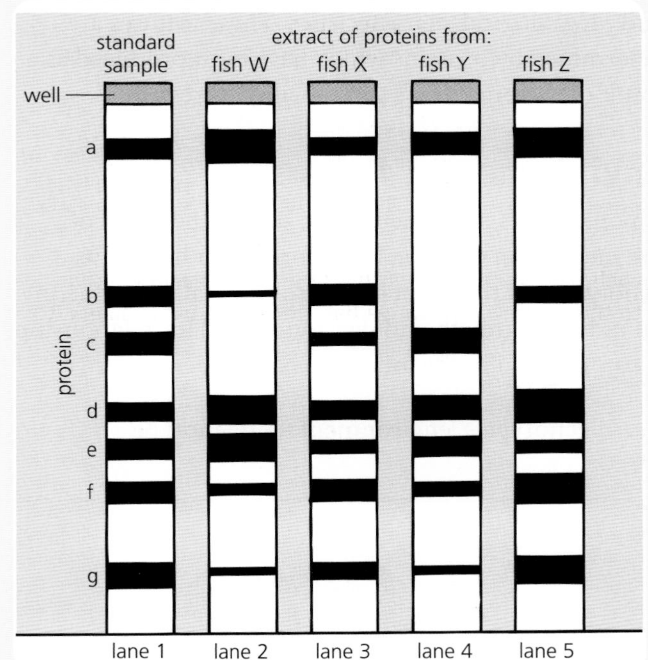

Figure 3.20 Banding results of gel electrophoresis

Protein	Molecular weight (kDa)
a	210
b	107
c	90
d	42
e	35
f	30
g	5

Table 3.5 Proteins in standard sample

Proteins and phenotype

An organism's phenotype is the sum of all its physical and physiological characteristics. These features are determined by proteins which have been produced as a result of the organism's genes being expressed.

Environmental factors

An organism's final phenotypic state is the result of the interaction between the information held in its genotype and the effect of environmental factors acting on it during growth and development. This relationship is summarised in the following equation:

$$\text{genotype} + \text{environment} \rightarrow \text{phenotype}$$

The phenotypic expression of some inherited traits remains unaffected by environmental factors. Tongue-rolling and blood group in humans, for example, are determined solely by genotype.

The phenotype of some other characteristics is influenced in part by environmental factors. If, for example, a person inherits the genetic information to become tall but consumes a poor diet during childhood then they will not reach their full potential height. Therefore it is possible for individuals such as identical twins to have the same genotype but to have different phenotypes. For example, one could become tall and the other one remain of medium height because they have been raised separately in different environments (see Figure 3.21).

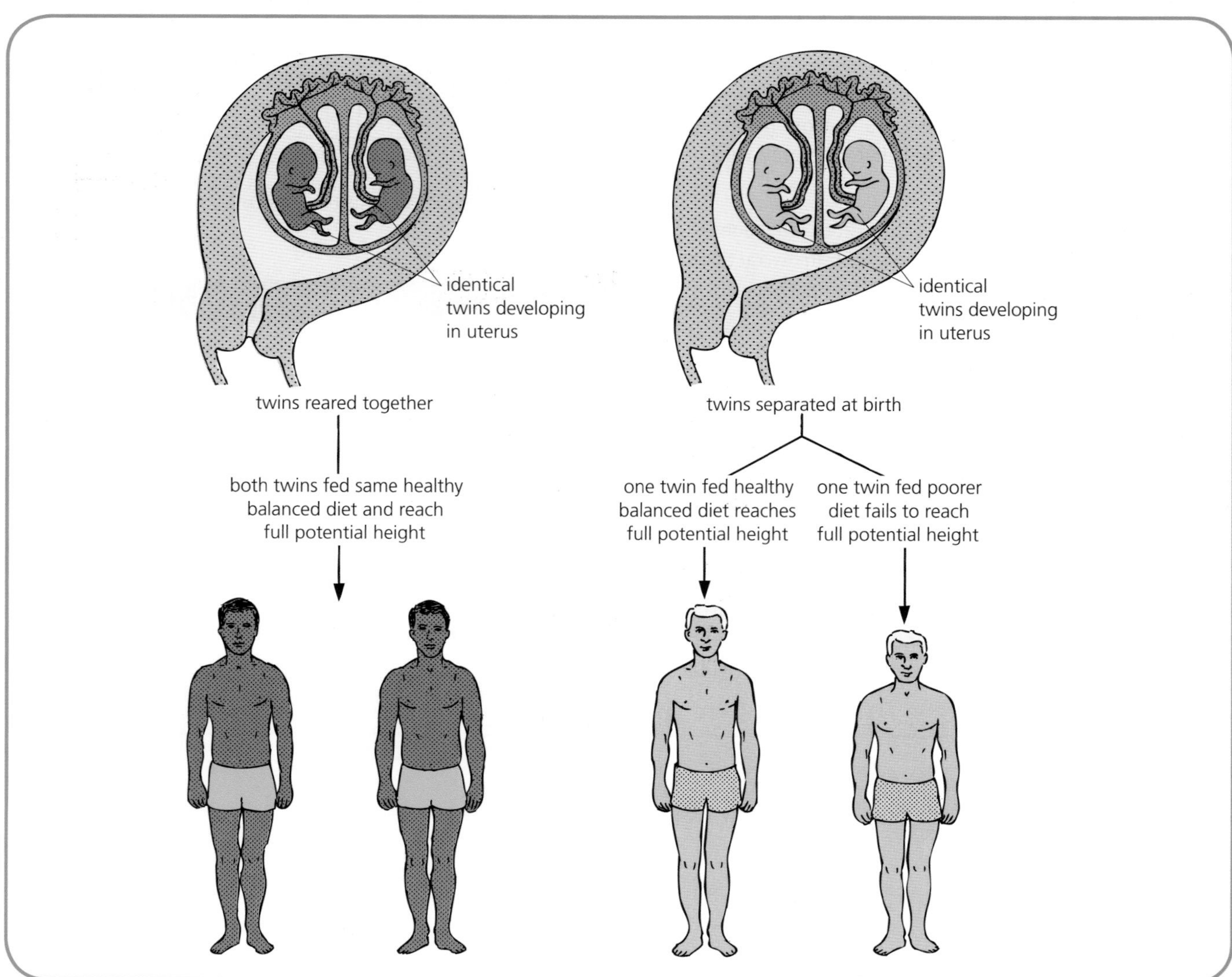

Figure 3.21 Twin study

Testing Your Knowledge 2

1 a) How many anticodons in a molecule of tRNA are exposed? (1)

b) Each molecule of tRNA has a site of attachment at one end. What becomes attached to this site? (1)

2 a) What is a *ribosome*? (1)

b) i) How many tRNA binding sites are present on a ribosome?

ii) To what does a tRNA's anticodon become bound at one of these sites? (2)

c) What type of bond forms between adjacent amino acids attached to tRNA molecules? (1)

d) What is the fate of a tRNA molecule once its amino acid has been joined to the polypeptide chain? (2)

3 a) Copy and complete Table 3.6. (2)

b) Name the third type of RNA. (1)

Stage of synthesis	Site in cell
formation of primary transcript of mRNA	
modification of primary transcript of mRNA	
collection of amino acid by tRNA	
formation of codon–anticodon links	

Table 3.6

4 Choose the correct answer from the underlined choice for each of the following statements. (6)

a) The basic units of the genetic code present on mRNA are called <u>anticodons/codons</u>.

b) The synthesis of mRNA from DNA is called <u>transcription/translation</u>.

c) A non-coding region of mRNA is called an <u>intron/exon</u>.

d) The basic units of the genetic code present on molecules of tRNA are called <u>anticodons/codons</u>.

e) Protein synthesis occurs at a <u>nucleus/ribosome</u>.

5 a) How many different types of amino acid are known to occur in proteins? (1)

b) What name is given to the chain formed when several amino acids become linked together? (1)

c) What determines the order in which amino acids are joined together into a chain? (1)

d) Describe TWO ways in which chains of amino acids can become arranged to form a protein. (2)

Applying Your Knowledge and Skills

1 Figure 3.22 shows the method by which the genetic code is transmitted during protein synthesis. Table 3.7 gives some of the triplets which correspond to certain amino acids.

a) Identify bases 1–9. (2)

b) Name processes P and Q. (1)

c) Copy and complete Table 3.7. (2)

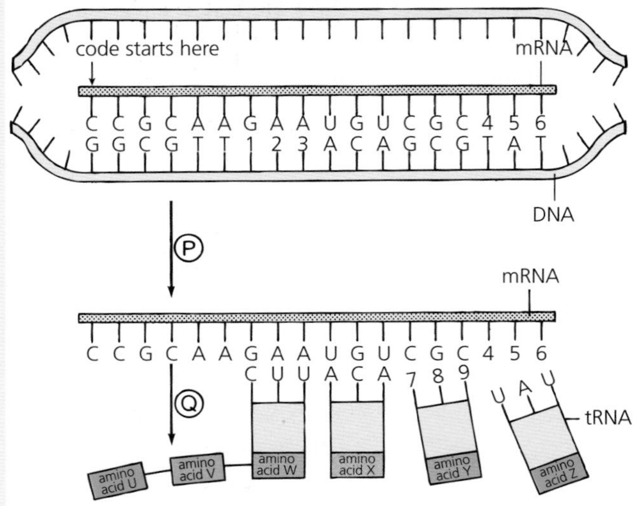

Figure 3.22

Amino acid	Codon	Anticodon
alanine		CGC
arginine	CGC	
cysteine		ACA
glutamic acid	GAA	
glutamine		GUU
glycine	GGC	
isoleucine		UAU
leucine	CUU	
proline		GGC
threonine	ACA	
tyrosine		AUA
valine	GUU	

Table 3.7

d) Give the triplet of bases that would be exposed on a molecule of tRNA to which valine would become attached. (1)

e) Use your table to identify amino acids U, V, W, X, Y and Z. (2)

f) i) Work out the mRNA code for part of a polypeptide chain with the amino acid sequence:

threonine–leucine–alanine–glycine.

ii) State the genetic code on the DNA strand from which this mRNA would be formed. (2)

2 The information in Table 3.8 refers to the relative numbers of ribosomes present in the cells of a new leaf developing at a shoot tip.

a) Construct an hypothesis to account for the trend shown by these data. (2)

b) Explain why the relative numbers of ribosomes in the cells of a fully grown leaf do not drop to zero. (1)

Age of new leaf (days)	Relative number of ribosomes in cells
1	300
3	500
5	650
7	450
9	210
11	200
13	200
15	200

Table 3.8

3 Give an account of translation of mRNA into a polypeptide. (9)

4 Figure 3.23 shows a chromatogram of three amino acids (X, Y and Z).

a) Identify the three amino acids with the aid of Table 3.9. (3)

b) Which of these amino acids was
i) most soluble in the solvent?
ii) least soluble in the solvent?
iii) Explain your answer. (2)

→

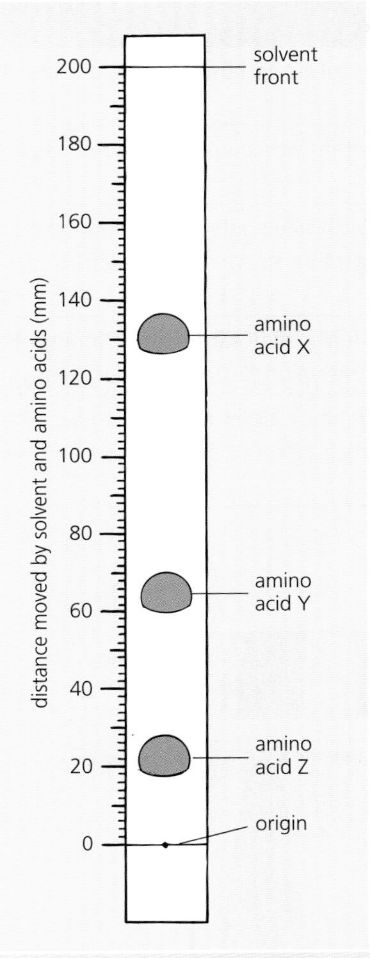

Figure 3.23

5 A particular polypeptide chain was known to be ten amino acids in length. When enzymes were used to break down several molecules of it at three different places along its length, the fragments shown in Figure 3.24 were obtained.

(Note: AA = amino acid; N = one end of the polypeptide chain.)

Draw a diagram of the complete polypeptide chain. (1)

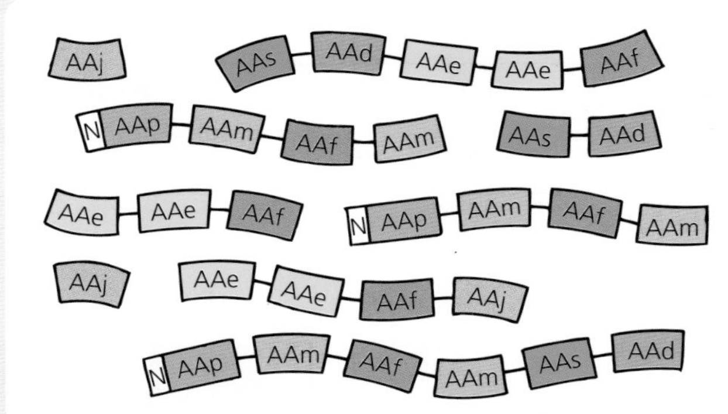

Figure 3.24

6 Human blood serum contains two major groups of protein: albumin and globulins. Some examples of these proteins are given in Table 3.10. Figure 3.25 shows a separation of serum proteins by electrophoresis and the results presented as a graph.

Amino acid	Abbreviation	Rf	Amino acid	Abbreviation	Rf
histidine	his	0.11	cysteine	cys	0.40
glutamine	gln	0.13	proline	pro	0.43
lysine	lys	0.14	tyrosine	tyr	0.45
arginine	arg	0.20	asparagine	asn	0.50
aspartic acid	asp	0.24	methionine	met	0.55
glycine	gly	0.26	valine	val	0.60
serine	ser	0.27	tryptophan	trp	0.66
glutamic acid	glu	0.30	phenylalanine	phe	0.68
threonine	thr	0.35	isoleucine	ile	0.72
alanine	ala	0.38	leucine	leu	0.73

Table 3.9

<table>
<tr><th colspan="2">Blood protein group</th><th>Details</th></tr>
<tr><td colspan="2">albumin</td><td>It makes up more than half of the protein in blood serum and prevents blood from leaking out of vessels.</td></tr>
<tr><td rowspan="4">globulins</td><td>alpha-1-globulins ($\alpha1$)</td><td>They include a high-density lipoprotein which contains 'good' cholesterol not taken into the artery wall.</td></tr>
<tr><td>alpha-2-globulins ($\alpha2$)</td><td>They include a protein that binds with haemoglobin. Some of the proteins in this group are increased in concentration in conditions such as diabetes and cirrhosis of the liver.</td></tr>
<tr><td>beta-globulins (β)</td><td>They include a protein called transferrin that carries iron through the bloodstream and increases in concentration during iron-deficiency anaemia.</td></tr>
<tr><td>gamma-globulins (γ)</td><td>Many are antibodies whose numbers increase in response to viral invasion and some cancers such as myeloma (which affects bone marrow) and lymphatic leukaemia.</td></tr>
</table>

Table 3.10

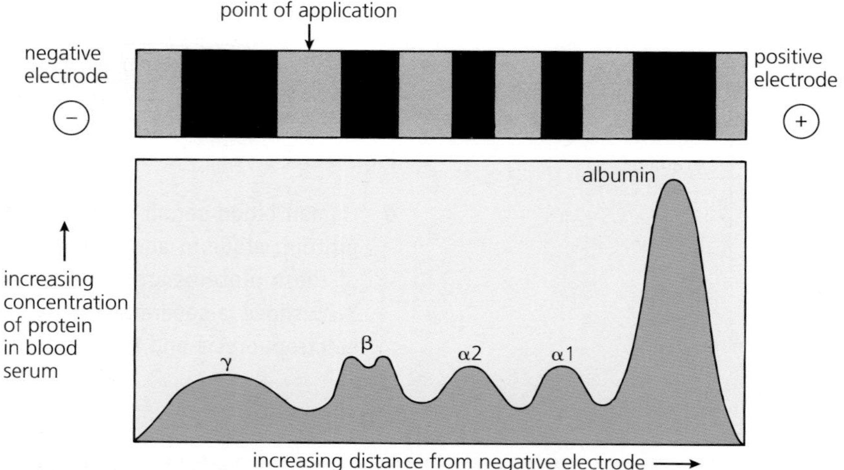

Figure 3.25

a) i) Compared with the globulins as a group, do albumin proteins have a higher or a lower molecular weight?

ii) Explain how you arrived at your answer. (2)

b) i) Based only on the information given in Table 3.10, identify the specific group of blood proteins that could indicate liver disease if its concentration increased greatly.

ii) What condition might a person have if the concentration of beta-globulins in their bloodstream increased to an abnormal level? (2)

c) i) The graph in Figure 3.26 shows a patient's results from a serum protein electrophoresis test. Which of the following could this person have?

(Your answer should be based only on the information in Table 3.10.)

A leakage of blood from vessels

B cirrhosis of the liver

C iron-deficiency anaemia

D cancer of cells in bone marrow.

ii) Why is this diagnosis ***not*** conclusive? (2)

→

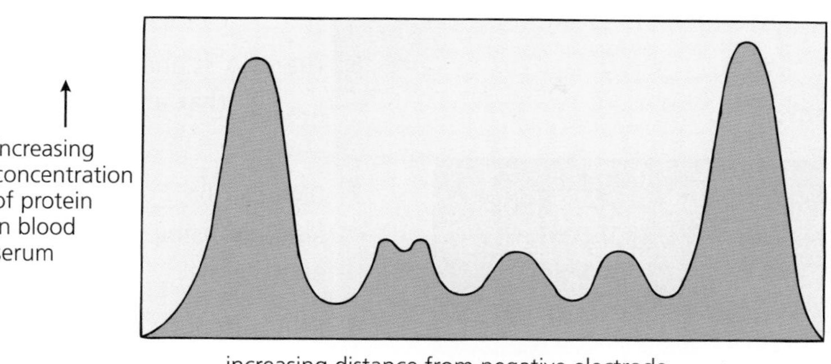

Figure 3.26

7 Some amino acids can be synthesised by the body from simple compounds; others cannot be synthesised and must be supplied in the diet. The latter type are called the **essential amino acids**. The graph in Figure 3.27 shows the results of an experiment using rats where group 1 was fed zein (maize protein), group 2 was fed casein (milk protein) and group 3 was fed a diet which was changed at day 6.

a) One of the proteins contains all of the essential amino acids whereas the other lacks two of them.
  i) Identify each protein.
  ii) Explain how you arrived at your answer. (4)

b) i) State which protein was given to the rats in group 3 during the first six days of the experiment.
  ii) Suggest TWO different ways in which their diet could have been altered from day 6 onwards to account for the results shown in the graph. (3)

c) By how many grams did the mean body weight of the rats in group 2 increase over the 20-day period? (1)

d) Calculate the percentage decrease in mean body weight shown by the rats in group 1 over the 20-day period. (1)

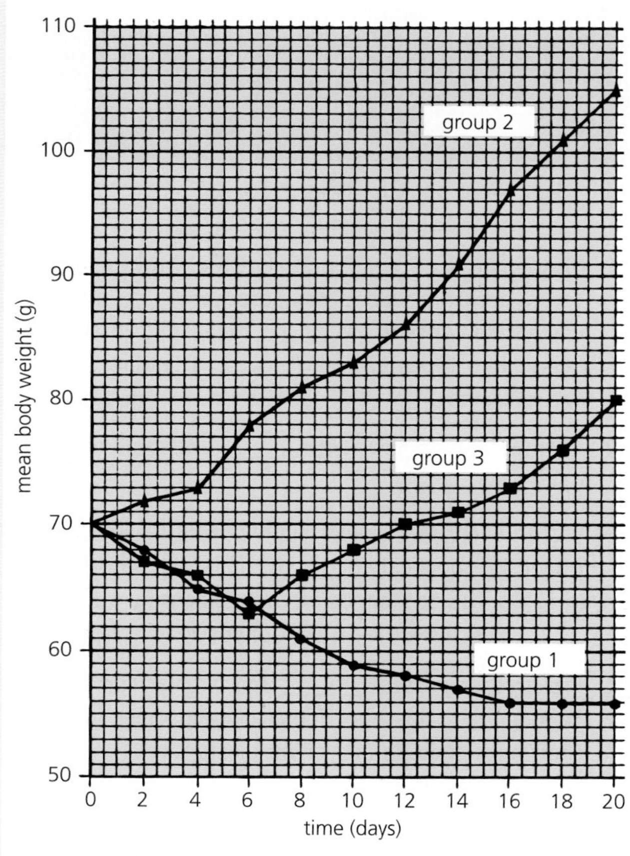

Figure 3.27

What You Should Know

Chapters 1–3

(See Table 3.11 for word bank.)

adenine	deoxyribose	plasmids
amino	DNA polymerase	polypeptide
amplified	environmental	primary
anticodons	eukaryotes	primer
antiparallel	exons	protein
backbone	folded	replication
bonds	fragments	ribose
chain	genetic	ribosomes
codons	guanine	RNA polymerase
coiled	helix	splicing
complementary	introns	thymine
cross-connections	ligase	transcription
cycling	linear	translation
cytosine	nucleotides	twenty
	peptide	uracil

Table 3.11 Word bank for chapters 1–3

1 DNA consists of two strands twisted into a double ______. Each strand is composed of ______. Each nucleotide consists of ______ sugar, phosphate and one of four types of base (______, thymine, ______ and cytosine).

2 Adenine always pairs with ______; guanine always pairs with ______.

3 Within each DNA strand neighbouring nucleotides are joined by chemical ______ into a sugar–phosphate ______. The backbones of complementary strands are ______ because they run in opposite directions.

4 Prokaryotes possess ______ and circular chromosomal DNA. ______ possess circular chromosomes in their mitochondria and chloroplasts, and ______ chromosomes made of tightly ______ DNA in their nucleus.

5 DNA is unique because it can direct its own ______. This begins by DNA unwinding and its two strands separating at a starting point. A ______ forms beside the DNA strand with the 3′ end. Individual nucleotides aligned with ______ nucleotides on the DNA strand become joined into a new DNA strand by the enzyme ______.

6 The DNA strand with the 5′ end is replicated in ______ that are joined together by the enzyme ______.

7 DNA can be ______ by the polymerase ______ reaction using primers, heat-tolerant DNA polymerase and thermal ______.

8 RNA differs from DNA in that it is single-stranded, contains ______ (not deoxyribose) and has the base ______ in place of thymine.

9 DNA contains a species' ______ information as a coded language determined by the sequence of its bases arranged in triplets called ______. Expression of this information through ______ synthesis occurs in two stages when a gene is switched on.

10 The first stage, ______, begins when the enzyme ______ becomes attached to and moves along the DNA bringing about the synthesis of a ______ transcript of mRNA from individual RNA nucleotides. Primary RNA is cut and spliced to remove non-coding regions called ______ and bind together coding regions called ______.

11 The second stage, ______, occurs at ______ where codons on the mRNA strand match up with the ______ on tRNA molecules carrying amino acids. These become joined together by peptide bonds to form a ______ chain whose amino acid sequence reflects the code on the mRNA.

12 Alternative ______ of primary mRNA enables a gene to be expressed as several proteins.

13 Proteins consist of subunits called ______ acids of which there are ______ different types.

14 Amino acid molecules are joined together by ______ bonds to form polypeptides. Polypeptides are coiled and ______ to form protein molecules whose three-dimensional shape, which is maintained by ______ between amino acids, is directly related to its function.

15 An organism's phenotype is affected by both its genotype and ______ factors.

4 Cellular differentiation

A multicellular organism consists of a large number of cells. Rather than each cell carrying out every function for the maintenance of life, a **division of labour** occurs and most of the cells become differentiated. **Differentiation** is the process by which an unspecialised cell becomes altered and adapted to perform a special function as part of a permanent tissue.

Selective gene expression

Since all the cells in a multicellular organism have arisen from the zygote by repeated cell division, every cell possesses all the genes for constructing the whole organism. These genes all have the potential to become switched on and some of them are already switched on at this stage.

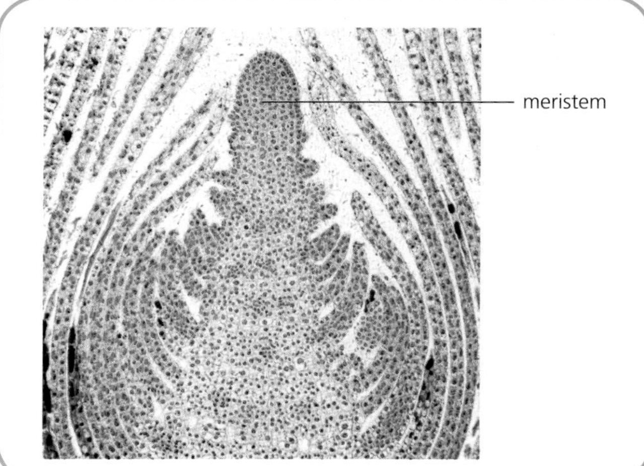

Figure 4.1 Apical meristem at shoot tip

During differentiation:

- many genes remain switched on
- some genes that control characteristic features of this type of differentiated cell now become switched on
- some unnecessary genes now become switched off (otherwise they would be coding for proteins not required by this type of differentiated cell).

Growth and differentiation in multicellular plants

Meristems

In multicellular plants, growth is restricted to regions called **meristems** (see Figures 4.1 and 4.2). A meristem is a group of unspecialised plant cells capable of dividing repeatedly throughout the life of the plant. Some of the cells formed remain meristematic and go on to produce new cells while others become differentiated.

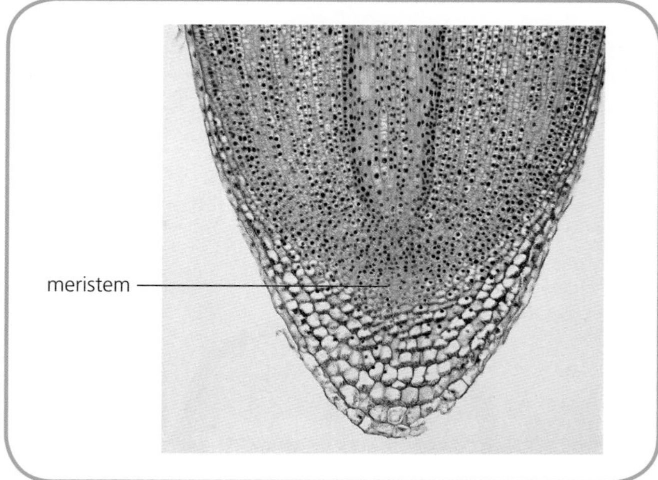

Figure 4.2 Apical meristem at root tip

Growth and differentiation in multicellular animals

A multicellular animal such as a human being begins life as a fertilised egg (zygote) as shown in Figure 4.3. The zygote divides repeatedly by mitosis and cell division to form an embryo. Like the cells in an adult, each **embryonic cell** possesses all the genes for constructing the whole organism. However, unlike the adult, all the genes in cells at this early stage are still switched on or have the potential to become switched on.

As development proceeds, the cells undergo **differentiation** and become **specialised** in structure, making them perfectly adapted for carrying out a particular function. For example, a motor neuron (see Figure 4.3) is a type of nerve cell which possesses an axon (a long insulated cytoplasmic extension). This structure is ideally suited to the transmission of nerve impulses. Similarly, the cells of the epithelial lining of the windpipe are perfectly suited to their job of sweeping dirty mucus up and away from the lungs.

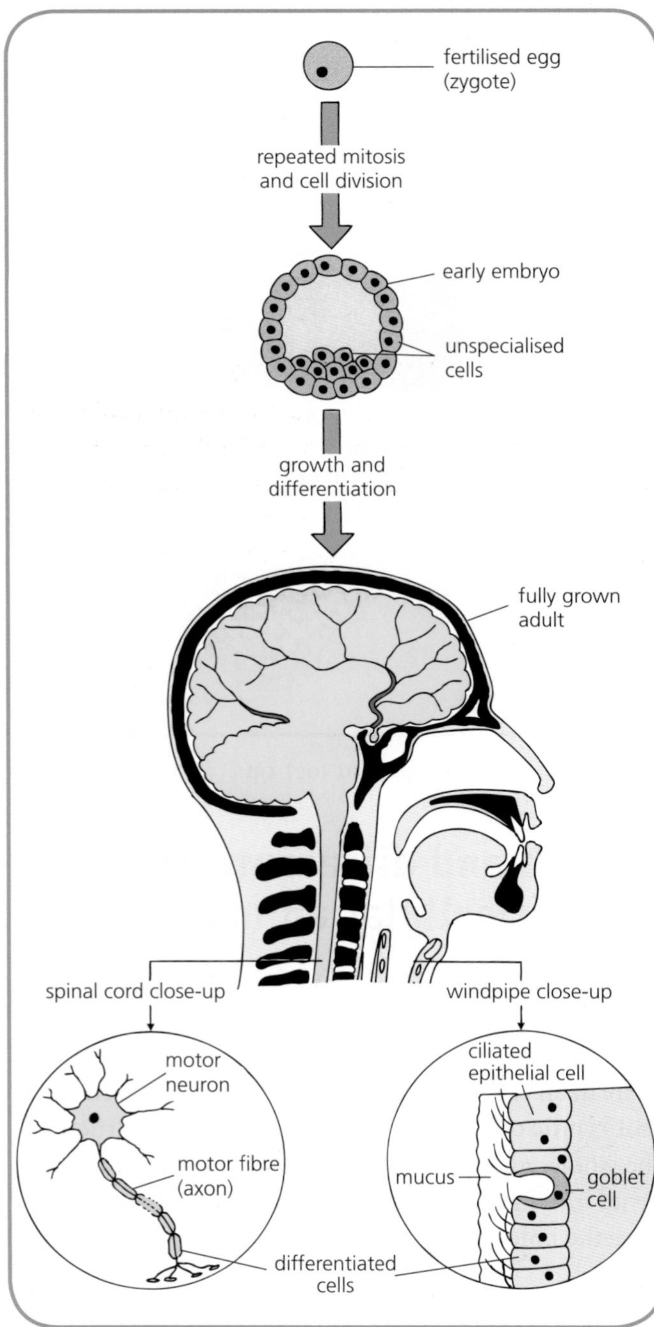

Figure 4.3 Differentiation

Once a cell becomes differentiated, it only expresses the genes that code for the proteins **specific** to the workings of that particular type of cell. For example, in a nerve cell, genes that code for the formation of neurotransmitter substances are switched on and continue to operate but those for mucus are switched off. The reverse is true of the genes in a goblet cell in the lining of the windpipe. Only a fraction of the genes in a specialised cell are expressed (for example, 3–5% in a typical human cell).

Stem cells

Stem cells are unspecialised cells that can:

- **reproduce** (self-renew) themselves by repeated mitosis and cell division while remaining undifferentiated
- **differentiate** into specialised cells when required to do so by the multicellular organism that possesses them.

Embryonic stem cells

A very early embryo consists of a ball of **embryonic stem cells** (see Figures 4.4 and 4.5). All of the genes in an embryonic cell have the potential to be switched on, therefore the cell is capable of differentiating into all of the cell types (more than 200) found in the human body. Such embryonic cells are described as being **pluripotent** (also see page 51).

Tissue stem cells

Tissue stem cells are found in locations such as skin and red bone marrow (see Figure 4.4). They have a much **narrower differentiation potential** than embryonic stem cells because many of their genes are already switched off. Therefore they only give rise to a limited range of cell types and are described as being **multipotent**. However, they are able to replenish continuously the supply of certain differentiated cells needed by the organism for growth, repair and renewal of tissues. Tissue stem cells in bone marrow, for example, give rise to red blood cells, platelets and various types of white blood cells.

Research value of stem cells

Much of the research to date has been carried out using stem cells from embryos of mice and humans. Human stem cells will renew themselves when grown in optimal

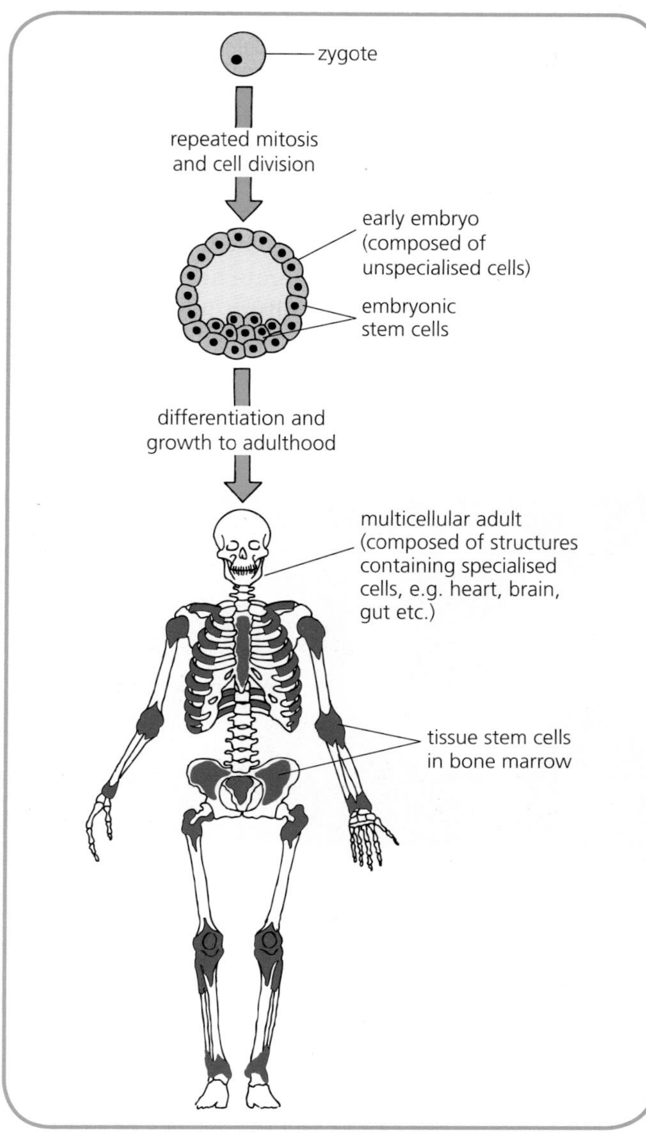

Figure 4.4 Two types of stem cells

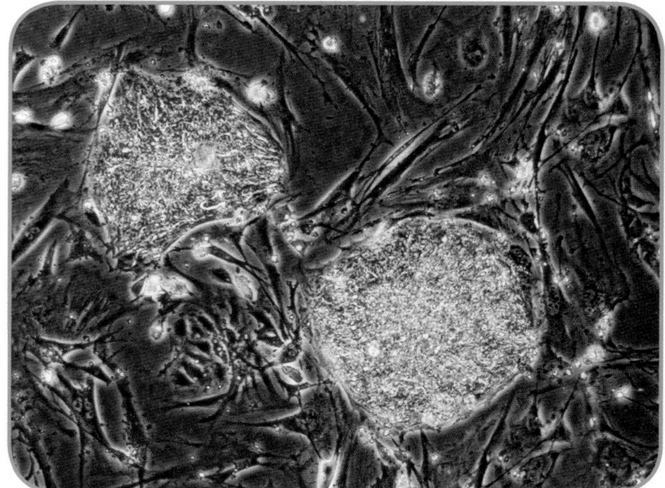

Figure 4.5 Embryonic stem cells

culture conditions provided that certain growth factors are present. In the absence of these growth factors, the stem cells rapidly differentiate.

By investigating why stem cells continue to multiply in the presence of a certain chemical yet undergo differentiation in its absence, scientists are attempting to obtain a fuller understanding of cell processes such as growth and differentiation. It is hoped that this will lead in turn to a better understanding of gene regulation (including the molecular biology of cancer). Since stem cells are genetically identical to differentiated cells, they can be used in research as **model cells** to investigate

- the means by which certain diseases and disorders develop
- the responses of cells to new pharmaceutical drugs.

Therapeutic uses of stem cells

Corneal repair

In recent years scientists have shown that **corneal damage** by chemical burning can be treated successfully using stem cell tissue. This is grown from the patient's own stem cells located at the edge of the cornea. In most cases the person's eyesight is restored following grafting of the stem cell tissue of the healthy eye to the surface of the damaged eye.

Skin graft

In a traditional skin graft, a relatively large section of skin is removed from a region of the person's body and grafted to the site of injury. This means that the person has two bodily areas that need careful treatment and time to heal.

A skin graft using stem cells only requires a **small sample** of skin to be taken to obtain stem cells. Therefore the site needs much less healing time and results in minimum scarring. The sample is normally taken from an area close to and similar in structure to the site of injury. Enzymes are used to isolate and loosen the stem cells which are then cultured. Once a **suspension of new stem cells** has developed, they are sprayed over the damaged area to bring about regeneration of missing skin.

Bone marrow transplant

Tissue stem cells present in bone marrow and blood are routinely used in bone marrow transplantation to treat cancers of the blood.

Case Study **Future potential therapeutic uses of stem cells**

Treatment of burns

Recently, human embryonic stem cells grown on synthetic scaffolds have been used to treat burn victims. The stem cell tissue provides a source of **temporary skin** while the patient is waiting for grafts of their own skin to develop. This potential use of stem cells is expected to be developed further in the future.

Treatment of degenerative conditions

Embryonic stem cells are able to differentiate into any type of cell in the body. Therefore they are believed to have the potential to provide treatments in the future for a wide range of disorders and degenerative conditions such as **diabetes**, **Parkinson's disease** and **Alzheimer's disease** (see Figure 4.6) that traditional medicine has been unable to cure. Already scientists have managed to generate nerve cells from embryonic stem cells in culture. It is hoped that this work will eventually be translated into effective therapies to treat neurological disorders such as multiple sclerosis. However, the use of embryonic stem cells raises questions of **ethics** (see below).

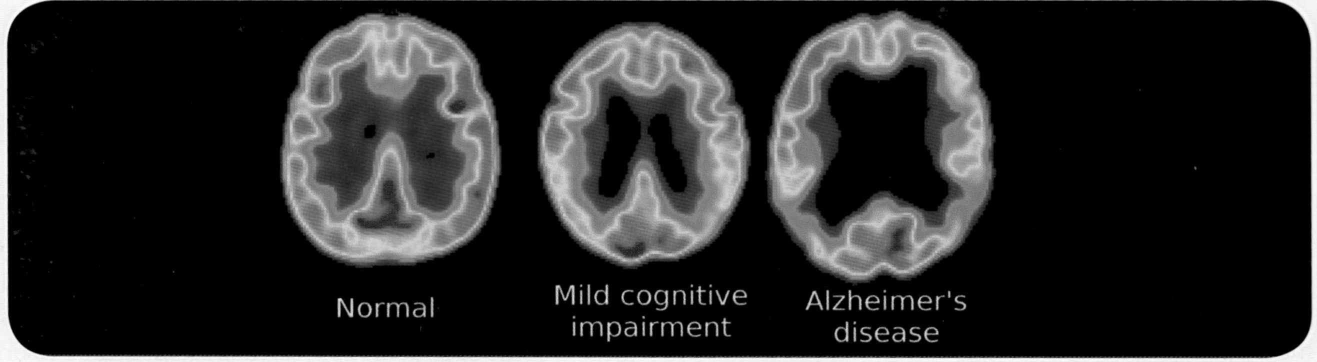

Figure 4.6 Brain scans

Ethical issues

Ethics refers to the moral values and rules that ought to govern human conduct. The use of stem cells raises several **ethical issues**. For example, the extraction of human embryonic stem cells to create a stem cell line (a continuous culture) for research purposes results in the destruction of the human embryo. Many people believe strongly that this practice is unethical. (See Case Study – Embryonic stem cell debate.)

Ethical issues are also raised by the use of **induced pluripotent stem cells** (see Case Study – Sources of stem cells) and by the use of **nuclear transfer techniques** (see Case Study – Nuclear transfer technique).

Case Study **Embryonic stem cell debate**

At present the creation of a human embryonic stem cell line using cells from a human embryo (of no more than 14 days) results in the destruction of the embryo. The ethical debate about the use of embryonic stem cells most commonly rests on the controversial question: 'Is a human embryo of less than two weeks a human *person*?'

People on one side of the debate believe that the embryo is definitely a human person and argue that fatally extracting stem cells from it constitutes murder. People on the other side of the debate feel certain that the embryo is not yet a person and believe that removing stem cells from it is morally acceptable.

The people who are against stem cell research using human embryos often support their case with the following claims:

- A human life begins when a sperm cell fuses with an egg cell and it is inviolable (in other words it is sacred and must not be harmed).
- A unique version of human DNA is created at conception.

→

- A fertilised egg is a human being with a soul.
- Stem cell research violates the sanctity of life.

The people who are in favour of stem cell research using human embryos often support their case with the following arguments:

- An embryo is not a person although it has the potential to develop into a person.
- At 14 days or less an embryo is not sentient (in other words, it does not have a brain, a nervous system, consciousness or powers of sensation).
- The death of a very young embryo is not of serious moral concern when it has the potential to benefit humanity (particularly people whose daily lives are compromised by debilitating medical conditions).
- Abortion is legal in many countries including the UK. Destroying a 14-day-old embryo is far less objectionable to most people than terminating a fetus at 20 weeks.
- Stem cell research uses embryos that were generated for IVF (*in vitro* fertilisation) but were not used and would be destroyed as a matter of course.

Possible solutions to the problem

In the future, the ethical issues raised by the use of embryonic stem cells may become less heated if advances in **induced pluripotent stem cell** technology (see below) and increased use of stem cells from **amniotic fluid** (see below) reduce the need for their use.

Case Study Sources of stem cells

Donated embryos

At present, patients undergoing infertility treatment may agree to donate any **extra embryos** that are not required for their treatment to medical science. These very early embryos proved an immediate source of embryonic stem cells for research. In addition, **long-term cultures** originally set up using cells isolated from donated embryos provide a further source of embryonic stem cells.

However, the number of human embryonic cells available for research remains limited and this restricts the ability of scientists to carry out research work in this important area.

Amniotic fluid

Scientists continue to search for new sources of stem cells. One of these is **amniotic fluid**. The stem cells which it contains can be harvested from the fluid removed from pregnant women for amniocentesis tests (see Figure 4.7). The stem cells obtained are capable of differentiating into many types of specialised cells such as bone, muscle, nerve and liver. One advantage of using stem cells from amniotic fluid is that it does not involve the destruction of a human embryo.

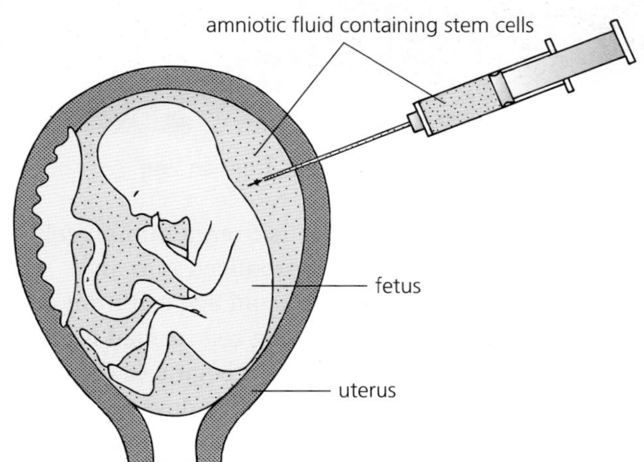

Figure 4.7 Stem cells in amniotic fluid

Induced pluripotent stem cells

A **totipotent** stem cell is one that is able to differentiate into any cell type and is capable of giving rise to the complete organism. A **pluripotent** stem cell is a descendent of a totipotent stem cell and is capable of differentiating into all the cell types that make up a human being except the placenta.

Induced pluripotent stem cells are, strictly speaking, not true stem cells. They are differentiated cells (for example, from human skin) that have been genetically reprogrammed using transcription factors to switch some of their turned-off genes back on again. As a result they act as stem cells and can be used for research.

Case Study Nuclear transfer technique

This technique involves removing the nucleus from an egg (see Figure 4.8) and then replacing it with a nucleus from a donor cell. Some cells constructed in this way divide normally, producing undifferentiated stem cells.

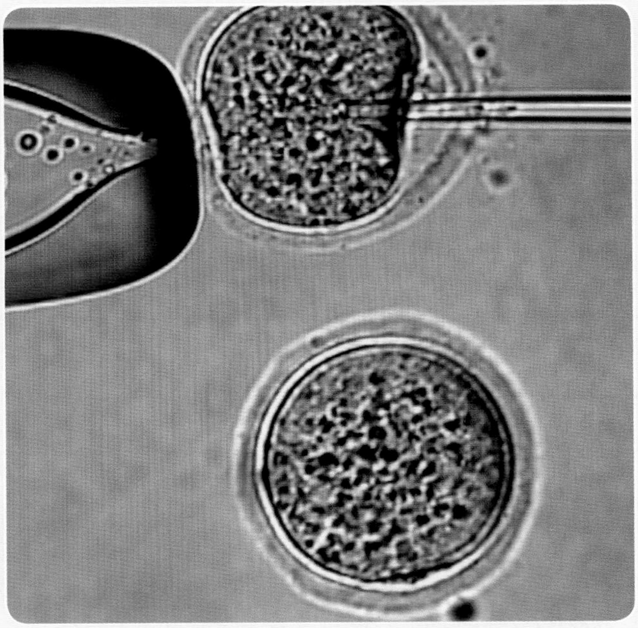

Figure 4.8 Removing the nucleus from an egg

Using this technique, a nucleus from a human cell (such as a skin cell) can be introduced into an **enucleated** animal cell (such as an egg cell from a cow), as shown in Figure 4.9. The cell formed is called a **cytoplasmic hybrid cell**. Once it begins to divide, stem cells can be extracted after five days and used for research. However, they are not 100% human and must not be used for therapeutic procedures.

Some people feel that it is unethical to mix materials from human cells with those of another species even if the hybrid cells formed are used strictly for research purposes only. Other people support the production of cytoplasmic hybrid cells because it helps to relieve the shortage of human embryonic stem cells available for research. In addition, they point out that the practice allows the nucleus from a diseased human cell (for example, from a patient with a degenerative disease or cancer) to be introduced into the enucleated animal egg. This may allow scientists to study the gene expression in these cells, observe how the disease develops and eventually develop new treatments that disrupt the disease process.

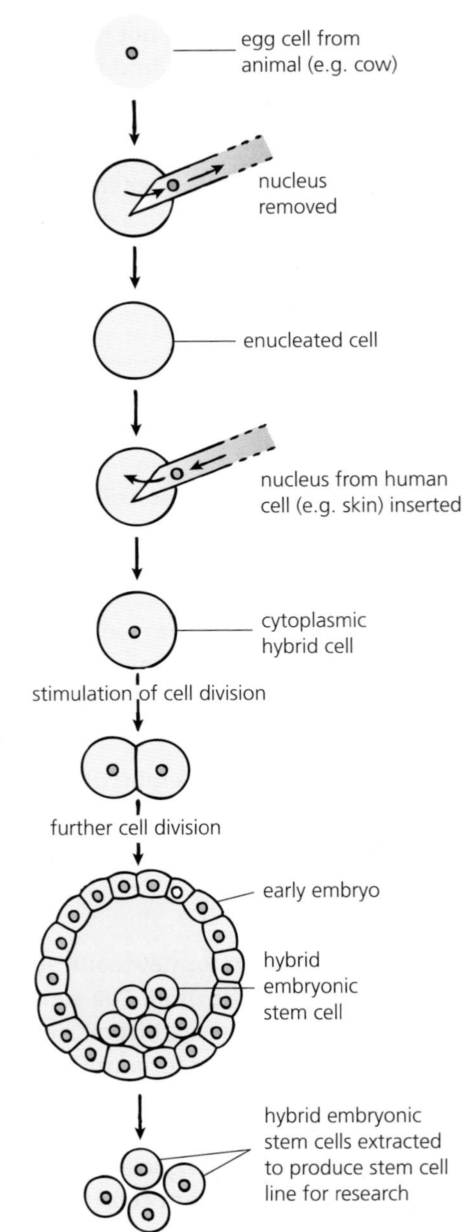

Figure 4.9 Nuclear transfer technique

Testing Your Knowledge

1 a) Define the term *differentiation*. (1)
b) What is a *meristem*? (1)

2 a) In what way is a ciliated epithelial cell a good example of a specialised cell? (1)
b) A goblet cell in the lining of the windpipe produces mucus but not insulin. Explain briefly how this specialisation is brought about with reference to genes. (2)

3 a) Give TWO characteristics of stem cells. (2)
b) i) Name TWO types of stem cell found in humans.
ii) For each type, identify ONE location where these cells could be found. (4)
c) Which type of stem cell is capable of differentiating into all the types of cell that make up the organism to which it belongs? (1)

4 a) i) Name ONE medical condition that is routinely treated using tissue stem cells.
ii) From where in the human body are these cells obtained? (2)
b) Give an example of a medical condition that may be treated in the future using stem cells. (1)
c) Why can the stem cells used to treat the medical condition you gave as your answer to **a)** not be used to treat patients with the condition you gave as your answer to **b)**? (2)

5 One definition of the word *ethical* is 'in accordance with principles that are morally correct'. Briefly explain why stem cell research using human embryos raises ethical issues. (2)

Applying Your Knowledge and Skills

1 Figure 4.10 shows a possible use of stem cells in the future.

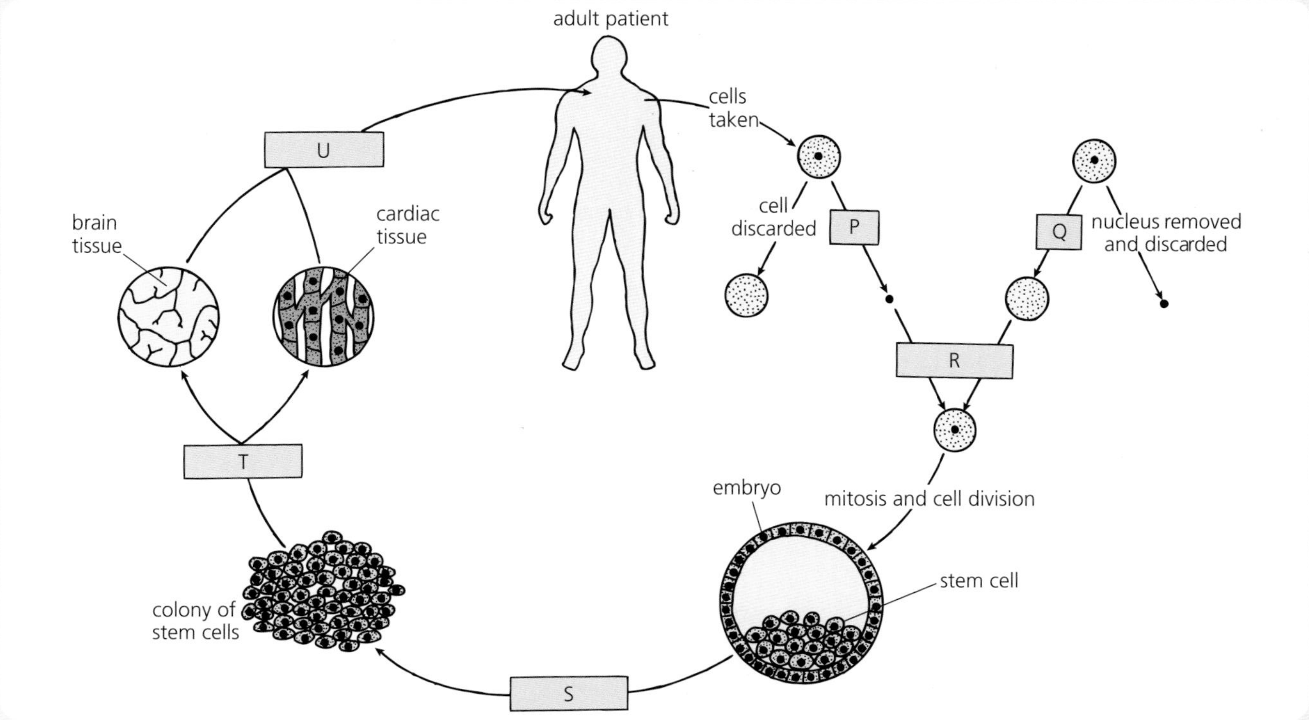

Figure 4.10

→

a) Match blank boxes P, Q, R, S, T and U with the following possible answers. (6)
   i) stem cells induced to differentiate
   ii) nucleus removed and retained
   iii) stem cells removed and cultured in laboratory
   iv) egg lacking nucleus retained
   v) matching tissue transplanted to patient without fear of rejection
   vi) nucleus inserted into egg

b) Which of the following is NOT represented in Figure 4.10? (1)

A cytoplasmic hybrid cell

B amniotic stem cell line

C undifferentiated stem cells

D nuclear transfer technique

c) i) Name a source of the donor egg cells used at present in this line of research.
   ii) Why does this prevent the series of events shown in the diagram being put into practice? (2)

2 The procedure that was adopted to produce 'Dolly the sheep' is shown in Figure 4.11.

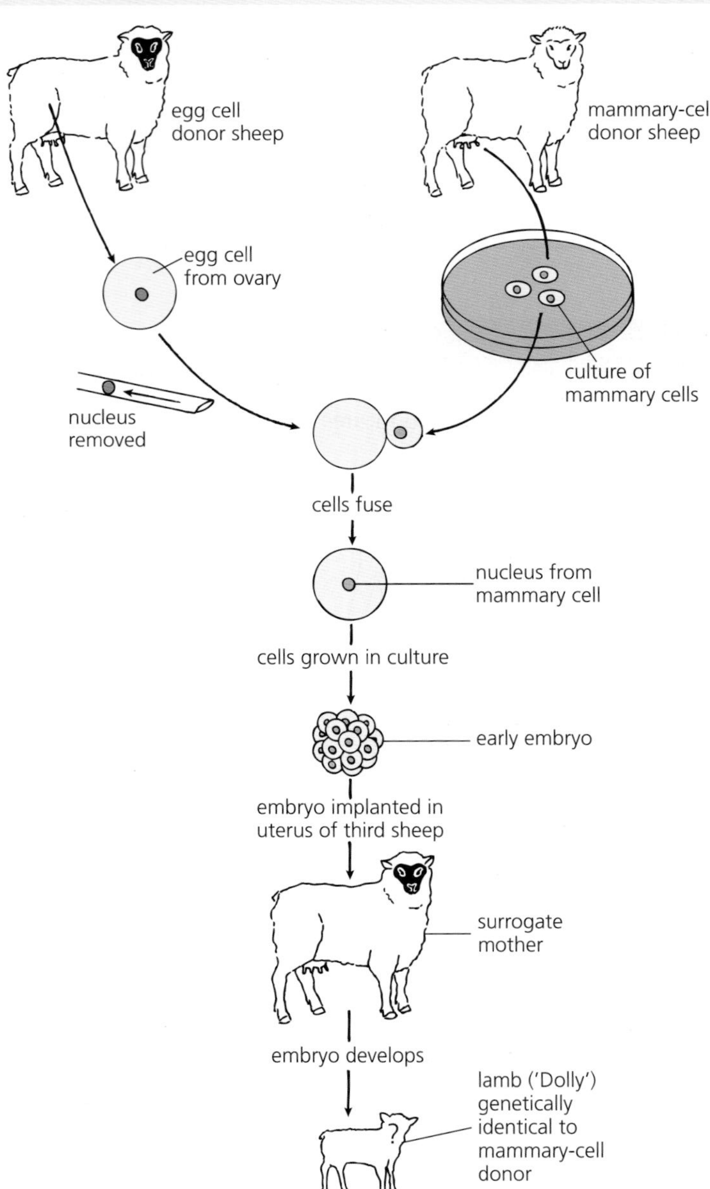

Figure 4.11

a) What name is given to the technique employed to create the original cell which gave rise to Dolly? (1)

b) Why is Dolly said to be the result of a ***cloning*** procedure? (1)

c) i) Did Dolly develop a black face or a white face?

ii) Explain your answer. (2)

d) i) What was the chance of Dolly being a ram?
A 0 **B** 1 in 1 **C** 1 in 2

ii) Explain your choice of answer. (2)

3 The four people shown in Figure 4.12 all support the use of human embryos in stem cell research.

a) Who is making a statement based on fact rather than expressing an opinion? (1)

b) i) Do you consider this person's statement to be a convincing or an unconvincing argument in support of the use of embryonic stem cells for research?

ii) Justify your answer. (2)

4 Give an account of the two types of stem cell and their research and therapeutic value. (9)

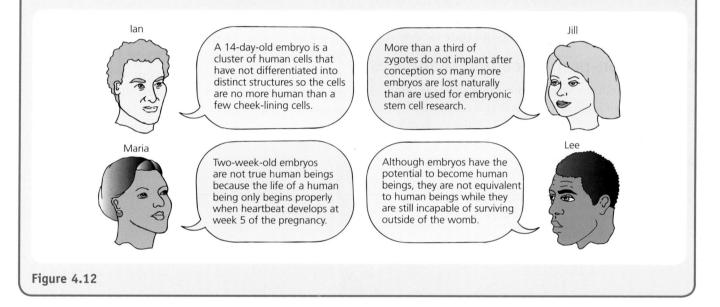

Maria
Two-week-old embryos are not true human beings because the life of a human being only begins properly when heartbeat develops at week 5 of the pregnancy.

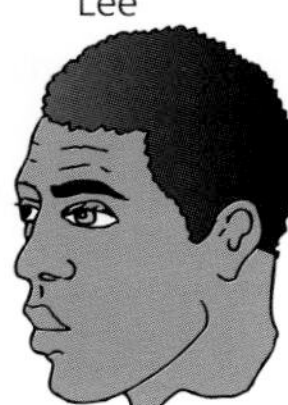

Figure 4.12

5 Genome and mutations

Structure of the genome

An organism's **genome** consists of all the genetic information encoded in the DNA of a complete set of its chromosomes. **Genes** are sequences of DNA bases that **code** for protein. However, a genome is not made up exclusively of genes. In eukaryotes (but not prokaryotes) only a tiny proportion of the genome actually consists of genes. The remainder of the genome is composed of many lengthy sequences of DNA that do not code for protein.

Role of non-protein-coding sequences

The **non-protein-coding regions** of the genome often take the form of DNA sequences that are repeated thousands of times over. Although the functions of many of these still remain unknown, the roles played by some sequences are now known and are described below.

Regulation of transcription

RNA polymerase, the enzyme that drives transcription (see page 33), is unable to initiate the process on its own. It needs the assistance of **transcription factors**. Some of these are called **activators** and they are bound to non-coding **regulator** sequences of DNA, often at some distance from the coding genes (see Figure 5.1).

Bending of the DNA strand brings the regulator bearing its activators into contact with other transcription factors close to the promoter of the gene to be transcribed. A molecular complex now forms at the gene's promoter site and the process of transcription begins. By this means a sequence of DNA bases that does not code for protein **regulates** the transcription of a gene.

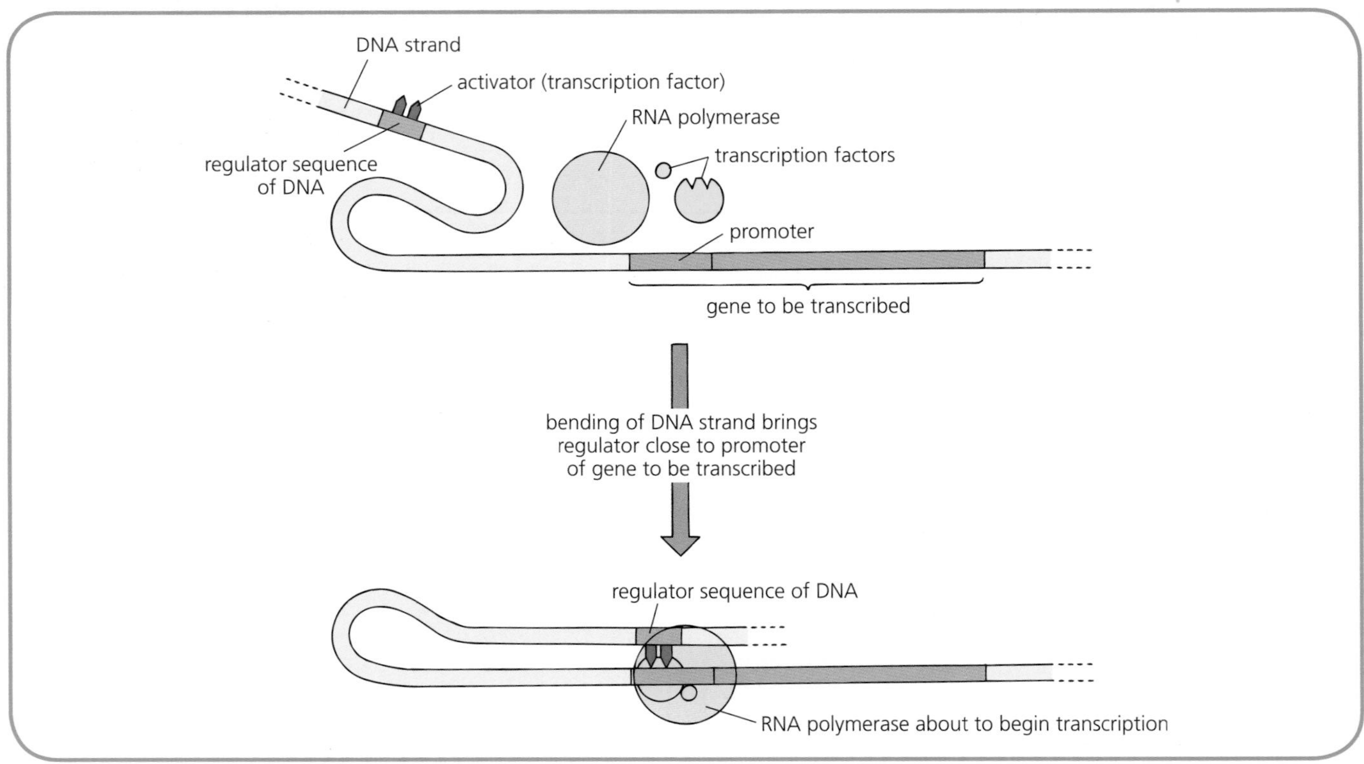

Figure 5.1 Regulation of transcription

Protection

Some of a genome's repetitive DNA sequences make up a protective structure called a **telomere** at each end of a chromosome. It prevents the chromosome from becoming damaged by 'fraying' at its ends.

Transcription of non-translated RNA

Genes code for mRNA whose sequence of bases is translated into protein (see page 30). However, some non-protein-coding sequences of DNA do code for forms of RNA other than mRNA. These types of RNA have specific functions but they are not translated into protein. (See Related Topic – Non-translated forms of RNA.)

Related Topic

Non-translated forms of RNA

Two types of RNA transcribed from DNA but not translated into protein are as follows.

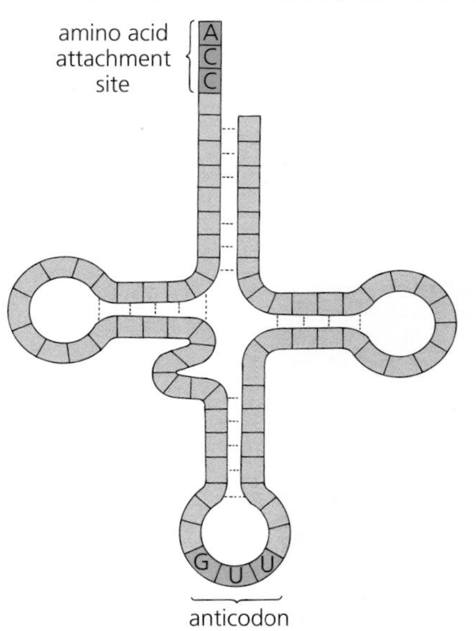

Figure 5.2 tRNA

tRNA

The structure of a molecule of **transfer RNA (tRNA)** is shown in Figure 3.9 on page 31. Each single-stranded molecule folds up on itself following some pairing of bases. This gives a molecule with some regions of double-stranded RNA separated by some regions of unpaired RNA. Sometimes this is depicted as the clover leaf model of tRNA (see Figure 5.2).

Each tRNA molecule transports a specific amino acid molecule to a ribosome to be used during the translation process. However, the tRNA itself is not translated into protein.

rRNA

A **ribosome** consists of a large and a small subunit (see Figure 5.3). Each subunit consists of a complex of protein and **ribosomal RNA (rRNA)** molecules. The rRNA molecules are transcribed from non-protein-coding sequences of DNA and combine with protein to form a subunit in the nucleus. This is exported via a nuclear pore to be assembled with the other subunit into a ribosome in the cytoplasm.

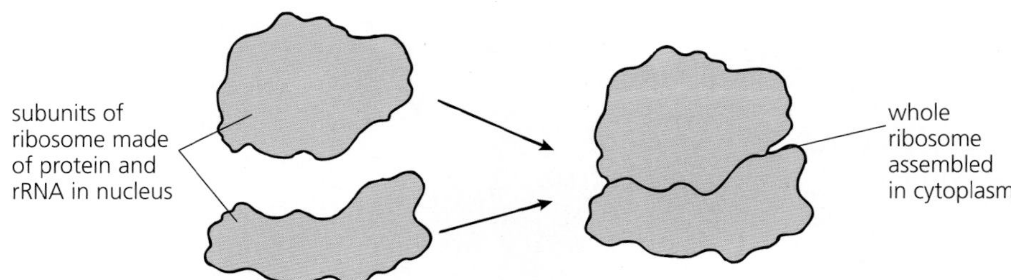

Figure 5.3 Ribosome structure

Mutation

A **mutation** is a change in the structure or amount of an organism's genome. It varies in form from a tiny change in the DNA structure of a gene to a large-scale alteration in chromosome structure or number. A mutation may result in an altered protein or no protein being synthesised. When a mutation produces a change in phenotype, the individual affected, such as the albino ivy plant shown in Figure 5.4, is called a **mutant**.

Frequency of mutation

In the absence of outside influences, gene mutations arise **spontaneously** and at **random** but only occur **rarely**. The mutation frequency of a gene is expressed as the number of mutations that occur at that gene site per million gametes. Mutation frequency varies from gene to gene and species to species as shown in Table 5.1.

Organism	Mutant characteristic	Mutation frequency (mutations at gene site/million gametes)	Chance of new mutation occurring
fruit fly	ebony body white eye	20 40	1 in 50 000 1 in 25 000
mouse	albino coat	10	1 in 100 000
human	haemophilia muscular dystrophy	5 80	1 in 200 000 1 in 12 500

Table 5.1 Mutation frequency

Figure 5.4 Mutant ivy plant

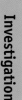

Investigation

Effect of UV radiation on yeast

Investigating the effect of UV radiation on UV-sensitive yeast cells

Normal yeast cells have genes which code for enzymes that repair damage done to their DNA by UV radiation. **UV-sensitive yeast** is a strain that has had these genes 'knocked out' by genetic engineering. Therefore it is unable to repair damaged DNA resulting from exposure to UV radiation.

The experiment is carried out as shown in Figure 5.5. After two days of incubation, plate X is found to

sterile inoculating loop
culture of UV-sensitive yeast
loop used to inoculate plate with yeast
sterile yeast extract dextrose (YED agar)
multiple streaking carried out
plate wrapped in metal foil and incubated at 30 °C for 2 days
UV-sensitive yeast
single colony transferred using a sterile loop
1 ml transferred after mixing, every time
A
B
C
D
E
tube containing 10 ml of sterile water at start
tubes containing 9 ml of sterile water at the start
sample of yeast suspension added to plates X and Y and spread thoroughly

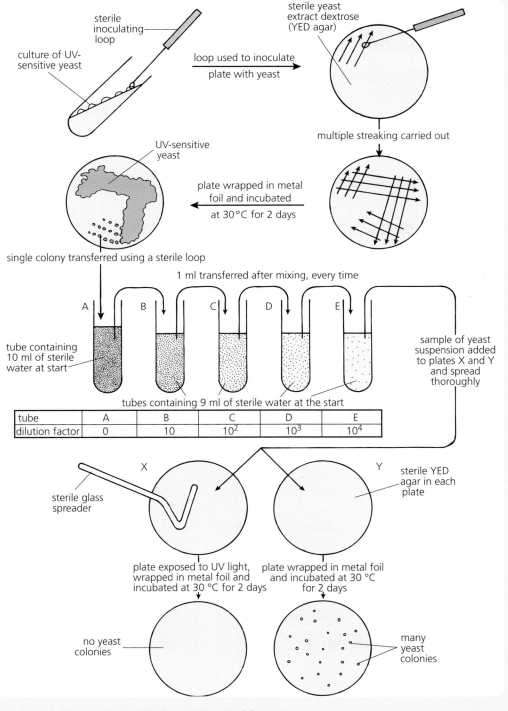

tube	A	B	C	D	E
dilution factor	0	10	10^2	10^3	10^4

X
Y
sterile glass spreader
sterile YED agar in each plate
plate exposed to UV light, wrapped in metal foil and incubated at 30 °C for 2 days
plate wrapped in metal foil and incubated at 30 °C for 2 days
no yeast colonies
many yeast colonies

Figure 5.5 Effect of UV radiation on UV-sensitive yeast

lack yeast colonies whereas plate Y, the control, has many colonies. Therefore it is concluded that exposure to UV radiation has had a lethal effect on UV-sensitive yeast.

Investigating the effect of UV radiation on 'protected' UV-sensitive yeast

The previous experiment is repeated and extended to include two further Petri dishes V and W wrapped in cling film. Their top surfaces are smeared each with a different sun barrier cream (for example, protection factors 6 and 20) and then exposed to the UV light source as before. If more yeast colonies grow on the plate with the higher protection factor then this result suggests that the higher factor has given them more protection from the harmful UV rays than the lower factor.

Single gene mutation

This type of mutation involves an alteration of nucleotide sequence in a gene's DNA.

Point mutation

A point mutation is a type of single-gene mutation that involves a **change in one nucleotide** in the DNA

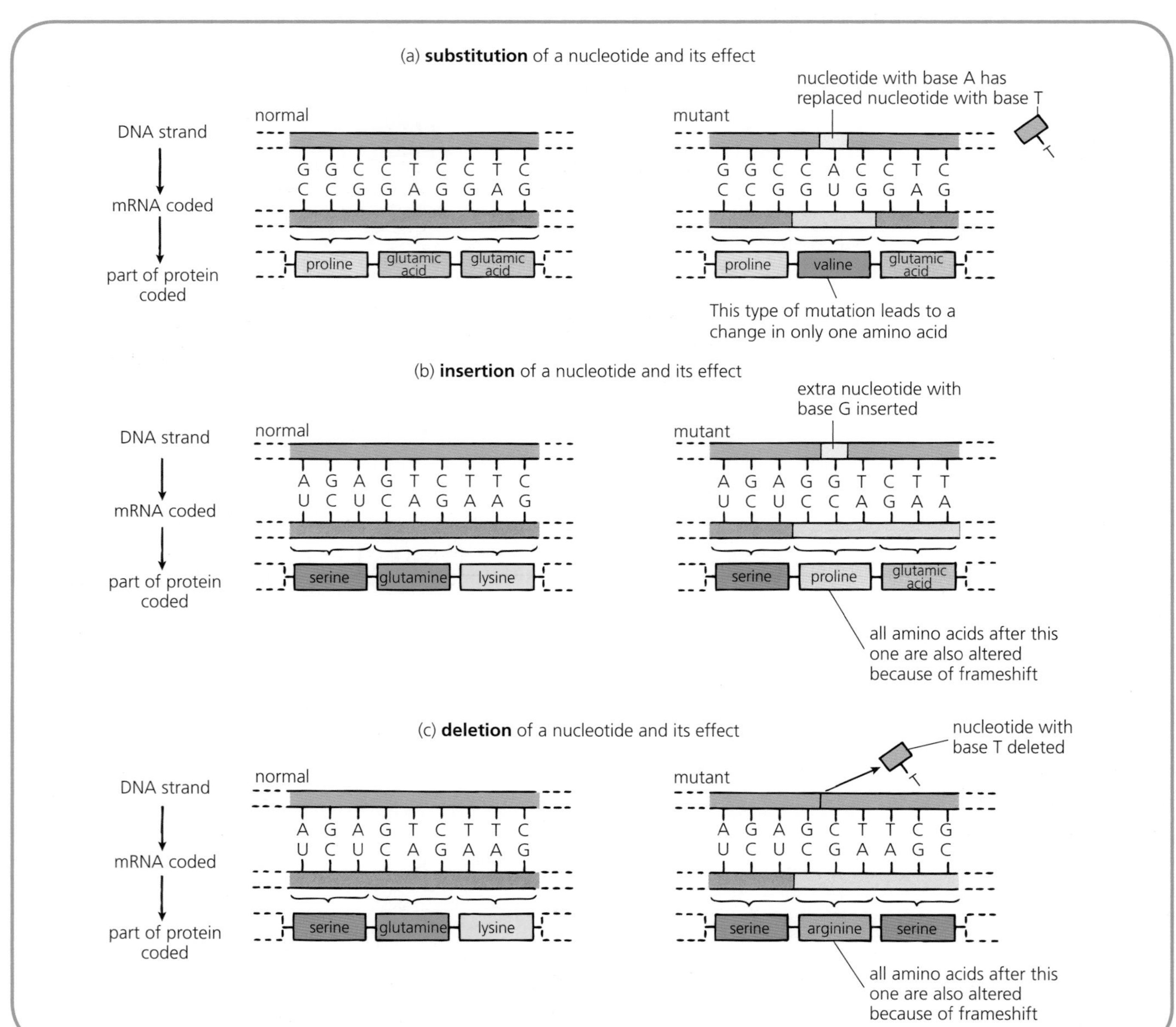

Figure 5.6 Types of point mutation

sequence of the gene. Three types of point mutation are shown in Figure 5.6. A single nucleotide is either **substituted, inserted** or **deleted.** In each case this results in one or more codons for one or more amino acids becoming altered.

Splice-site mutation

Before mRNA leaves the nucleus, introns (non-coding regions) are removed and exons (coding regions) are joined together. This process of post-transcriptional processing of mRNA is called **splicing** (see Figure 3.6 on page 29). Splicing is controlled by specific nucleotide sequences at splice sites on those parts of introns that flank exons. If a mutation occurs at one of these splice sites, the codon for an **intron–exon splice** may be altered. Then an intron may be retained in error by the mature transcript of mRNA (see Figure 5.7) or an essential exon may not be retained by the modified mRNA.

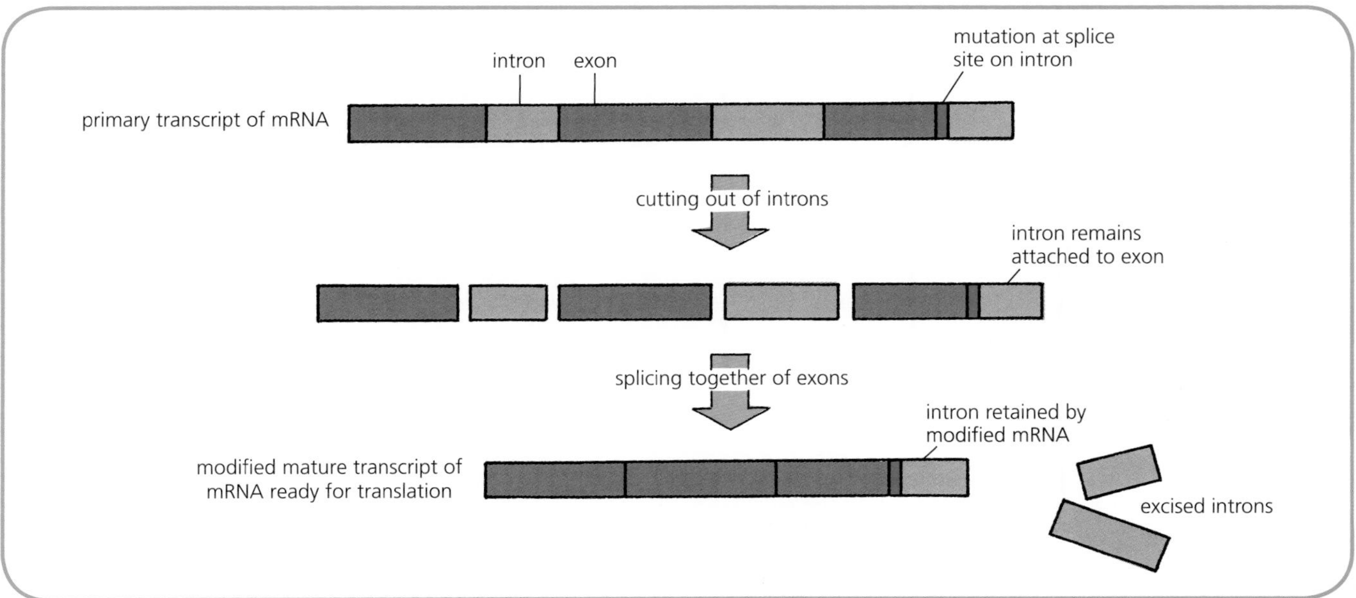

Figure 5.7 Effect of splice-site mutation

Impact on protein structure

Missense

Following a **substitution**, the altered codon codes for an amino acid that still makes sense but not the original sense (see Figure 5.8). This change in genome is called a **missense** mutation. The organism may be affected only slightly or not at all. However, if the substituted amino acid occurs at a critical position in the protein, then a major defect may arise (for example, the formation of haemoglobin S and sickle cell anaemia as described on pages 63–4).

Nonsense

As a result of a **substitution**, a codon that used to code for an amino acid is exchanged for one that acts as a premature **stop codon** (UAG, UAA or UGA). It causes protein synthesis to be halted prematurely (see Figure 5.8) and results in the formation of a polypeptide chain that is shorter than the normal one and unable to function. This change in genome is called a **nonsense** mutation.

Splice-site mutation

If one or more introns have been retained by the modified mature transcript of mRNA, they may in turn be translated into an **altered protein** that does not function properly. Thalassemia, a type of anaemia that results from a defect in the synthesis of haemoglobin, is caused by a mutation at a splice site.

Frameshift

mRNA is read as a series of triplets (codons) during translation. Therefore, if one base pair is inserted or deleted (see Figure 5.6) this affects the **reading frame** (triplet grouping) of the genetic code. It becomes shifted

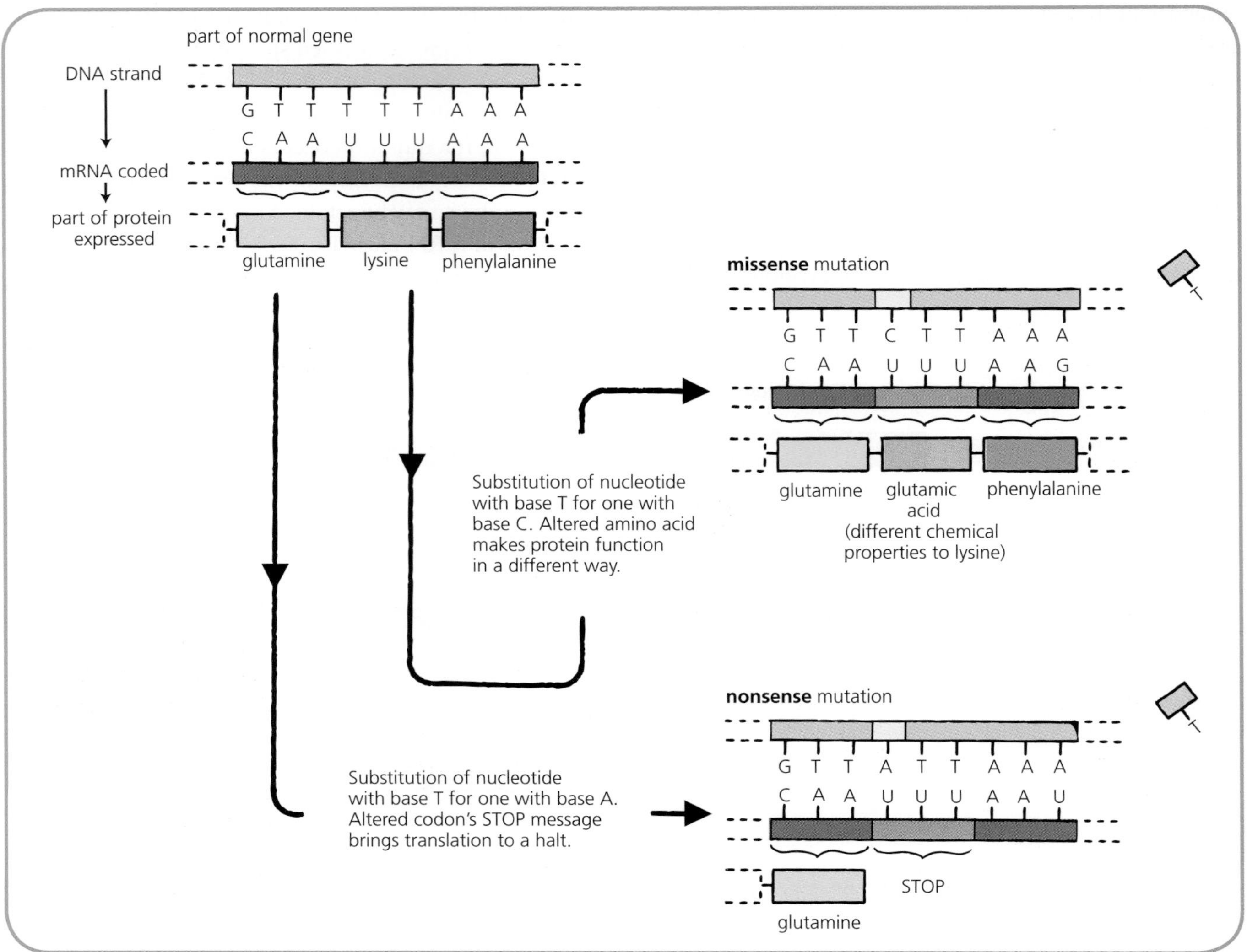

Figure 5.8 Possible effects of a base-pair substitution on sequence of amino acids

in a way that alters every subsequent codon and amino acid all along the remaining length of the gene. The protein formed is almost certain to be non-functional. This change in genome is called a **frameshift** mutation.

Inferior phenotype

Since most proteins are indispensable to the organism, most gene mutations produce an **inferior version** of the phenotype (see Table 5.2). If this results in the death of the mutant (an albino plant, for example, cannot photosynthesise) then the altered gene is said to be **lethal**.

Evolutionary importance of mutation

Mutation is the only source of **new variation**. Point mutations bring about changes in genes' DNA sequences and are therefore the means by which new alleles of genes arise. Without mutations, all organisms would be homozygous for all genes and no variation would exist.

Most mutations are harmful or even lethal. However, on very rare occasions there occurs by mutation a mutant allele which confers some **advantage** on the organism that receives it. Such mutant alleles, which are better than the originals, provide the alternative choices upon which natural selection can act. They are therefore the raw material of **evolution** (see Chapter 6).

Organism	Characteristic controlled by normal gene	Mutant characteristic resulting from gene mutation
fruit fly	long wing grey body red eye	vestigial wing yellow body white eye
human	normal blood clotting secretion of normal mucus in lung normal haemoglobin and biconcave red blood cells	haemophilia secretion of abnormally thick mucus which blocks bronchioles (cystic fibrosis) haemoglobin S and sickle-shaped red blood cells (see Figure 5.10)
mouse	brown coat	white coat (albino) lacking melanin pigment
ivy	green leaves containing chlorophyll	albino leaves lacking chlorophyll (see Figure 5.4)

Table 5.2 Mutant characteristics

Case Study Sickle-cell disease

When one of the genes on chromosome 11 that codes for haemoglobin undergoes a **substitution** (see Figure 5.9), it becomes expressed as an unusual form of haemoglobin called **haemoglobin S**. This is an example of **missense**. Although haemoglobin S differs from normal haemoglobin by only one amino acid, that one tiny alteration leads to profound changes in the folding and ultimate shape of the haemoglobin S molecule, making it a very inefficient carrier of oxygen.

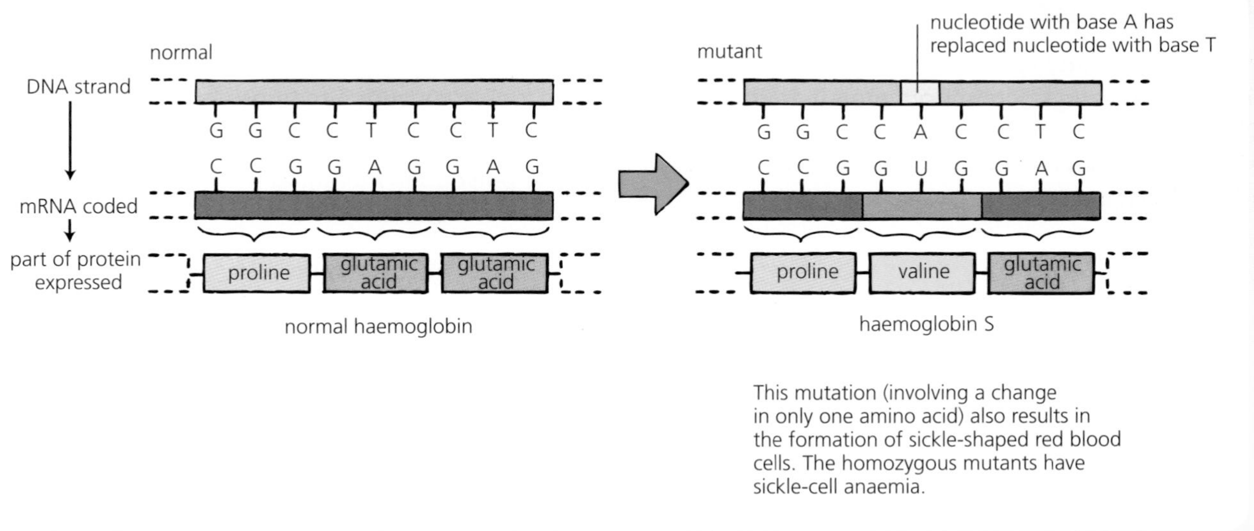

Figure 5.9 Mutation causing sickle-cell disease

People who are homozygous for the mutant allele face drastic consequences. In addition to all of their haemoglobin being type S, which fails to perform the normal function properly, they also possess distorted, sickle-shaped red blood cells (see Figure 5.10). These are less flexible than the normal type and tend to stick together and interfere with blood circulation. The result of these problems is severe shortage of oxygen followed by damage to vital organs and, in many cases, death. This disorder is called **sickle-cell anaemia**.

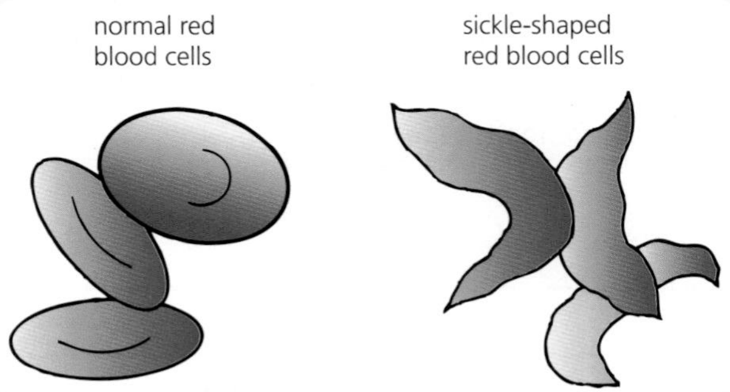

Figure 5.10 Two types of red blood cell

Sickle-cell trait

People who are heterozygous for the mutant allele do not have sickle-cell anaemia. Instead they are found to have a milder condition called **sickle-cell trait**. Their red blood cells contain both forms of haemoglobin but do not show 'sickling'. The slight anaemia that they tend to experience does not prevent moderate activity.

Resistance to malaria

The sickle-cell mutant allele is rare in most populations. However, in some parts of Africa up to 40% of the population have the heterozygous genotype. This is because people with sickle-cell trait are **resistant to malaria**. The parasite cannot make use of the red blood cells containing haemoglobin S. This situation, where a genetic disorder confers an advantage on people with the condition, is very unusual.

Case Study Phenylketonuria (PKU)

Phenylalanine and tyrosine are two amino acids that human beings obtain from protein in their diet. During normal metabolism, excess phenylalanine is acted on by an enzyme (enzyme 1 in the pathway shown in Figure 5.11).

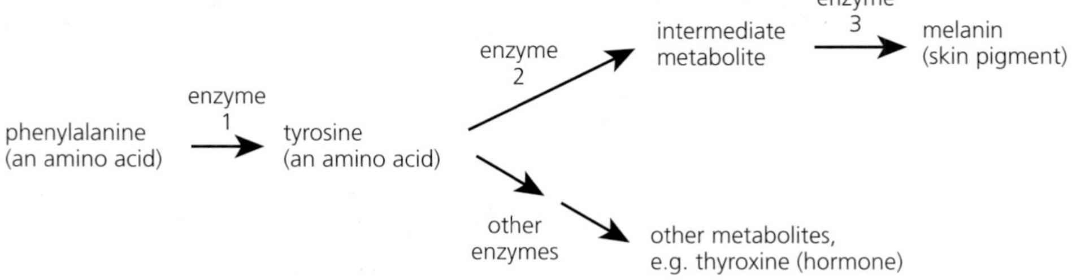

Figure 5.11 Normal fate of phenylalanine

Phenylketonuria is a genetic disorder caused by a mutation to a gene on chromosome 12 that normally codes for enzyme 1 in the pathway. Most commonly, the mutated gene has undergone a **substitution** of a nucleotide and **missense** occurs. The altered form of the protein expressed contains a copy of tryptophan in place of arginine and is non-functional. As a result of this **inborn error of metabolism**, phenylalanine is no longer converted to tyrosine. Instead it accumulates and some of it is converted to toxins.

These poisonous metabolites inhibit one or more of the enzymes that control biochemical pathways in brain cells. The brain fails to develop properly, resulting in the person having severe learning difficulties. In Britain, newborn babies are screened for PKU and any found to be affected are put on a diet containing low phenylalanine. By this means, the worst effects of PKU are reduced to a minimum.

Case Study Duchenne muscular dystrophy (DMD)

Duchenne muscular dystrophy (DMD) is caused by any one of several types of mutation to a particular gene on chromosome X, such as a **deletion** or a **nonsense** mutation. The affected gene fails to code for a protein called dystrophin which is essential for the functioning of muscles. In skeletal and cardiac muscle, for example, dystrophin is part of a group of proteins that strengthen muscle fibres and protect them from injury during contraction and relaxation.

DMD is the most common form of muscular dystrophy (muscle-wasting disease). In the absence of dystrophin, skeletal muscles become weak and lose their normal structure. This condition is accompanied by progressive loss of co-ordination. People with this condition are severely disabled from an early age and normally die young without passing the mutant allele on to the next generation. DMD is a sex-linked recessive trait and is almost entirely restricted to males, being passed on by carrier mothers to their sons.

Case Study Beta (β) thalassemia

A molecule of haemoglobin is composed of two alpha-globin and two beta-globin polypeptide chains. These polypeptides are encoded by genes.

Beta (β) thalassemia is a genetic disorder caused by any one of several types of mutation that affect a gene on chromosome 11 that codes for beta-globin. One of the most common of these mutations is a **substitution** that occurs at a **splice site** on an intron and causes base G to be replaced by base A.

There are several forms of β thalassemia, some more severe than others. One type, for example, is characterised by the complete lack of production of beta-globin; another by the production of an altered version of the protein. In either case, the person has a relative excess of alpha-globin in their bloodstream which tends to bind to, and damage, red blood cells. Patients with severe β thalassemia require medical treatment such as blood transfusions.

Case Study Tay-Sachs disease

Tay-Sachs disease is a genetic disorder resulting from a mutation to a gene on chromosome 15. Under normal circumstances the gene is responsible for encoding an enzyme that controls an essential biochemical reaction in nerve cells.

Changes to the gene take the form of point mutations such as **insertions** and **deletions** which result in the **frameshift** effect. The protein expressed is so different from the normal one that it is non-functional. As a result, the enzyme's unprocessed substrate accumulates in brain cells. This leads to neurological degeneration, generalised paralysis and death at about 4 years of age.

Case Study Cystic fibrosis

Cystic fibrosis is a genetic disorder caused by a three-base-pair **deletion** to a gene on chromosome 7. This type of mutation removes a codon for phenylalanine and causes the coded message to be seriously altered by the **frameshift** effect producing a non-functional protein.

The normal allele for the gene codes for a **membrane protein** that assists in the transport of chloride ions into and out of cells. In the absence of this protein, an abnormally high concentration of chloride gathers outside cells. Those regions of the body that coat their cells with mucus become affected because the high concentration of chloride causes mucus to become **thicker** and **stickier**. Organs such as the lungs, pancreas and alimentary canal become congested and blocked. Regular pounding on the chest to clear thick mucus (see Figure 5.12) and daily use of antibiotics can extend the person's life into their thirties and beyond. Untreated people with cystic fibrosis may only live to 4 or 5 years.

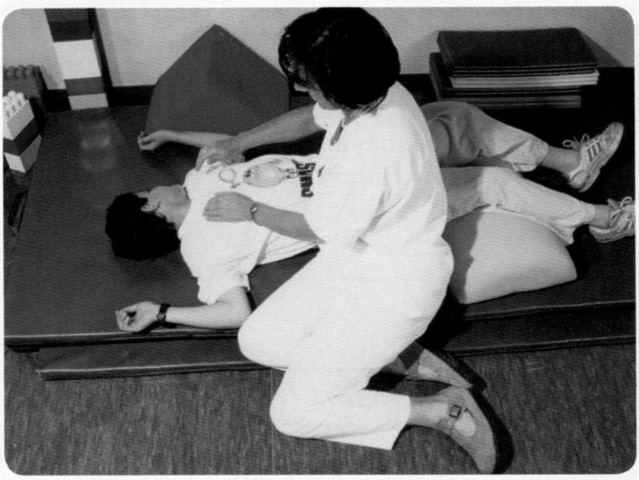

Figure 5.12 Easing symptoms of cystic fibrosis

Testing Your Knowledge 1

1 With the aid of an example, distinguish between the terms *mutant* and *mutation*. (2)

2 **a)** State TWO characteristics of mutant alleles with respect to their occurrence and frequency. (2)

b) i) Are all mutant alleles an inferior version of the original allele?

ii) Explain your answer. (2)

3 **a)** Identify THREE types of point mutation. (3)

b) i) Which of these is *least* likely to have a major impact on gene expression?

ii) Explain your answer. (2)

4 Decide whether each of the following statements is true or false and then use T or F to indicate your choice. Where a statement is false, give the word that should have been used in place of the word in bold print. (6)

a) The sum of the hereditary information encoded in an organism's DNA is called its **ribosome**.

b) A **gene** is a DNA sequence that codes via mRNA for a protein.

c) Most of the genome of a **prokaryote** is composed of DNA sequences that do not code for protein.

d) Some non-coding sequences regulate the process of **transcription**.

e) The function of many **coding** sequences of DNA remains unknown.

f) A mutation to a splice site may alter **pre**-transcriptional processing of mRNA.

Chromosome structure mutations

This type of mutation involves the breakage of one or more chromosomes. A broken end of a chromosome is 'sticky' and it can join to another broken end. There are four different ways that this can happen and bring about a change in the **number** or **sequence** of the genes in a chromosome.

Deletion

The chromosome breaks in two places and the segment in between becomes detached (see Figures 5.13 and 5.14). The two ends then join up giving a shorter chromosome which **lacks** certain genes. **Deletion** normally has a drastic effect on the organism involved.

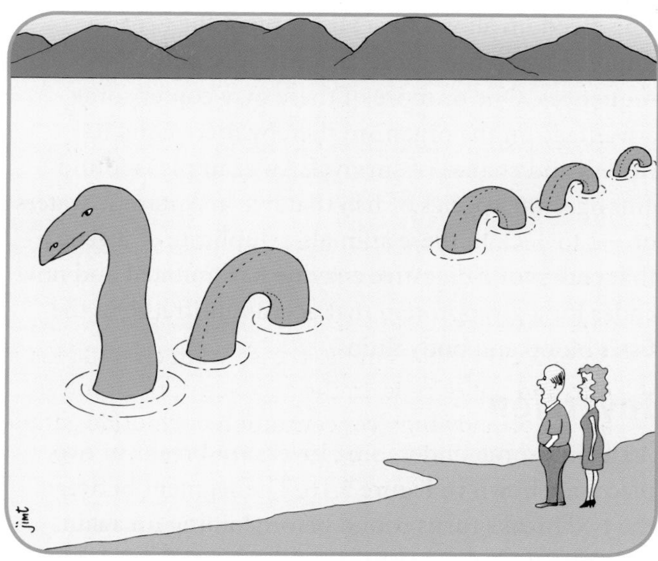

Figure 5.14 'Look! Nessie's had a deletion.'

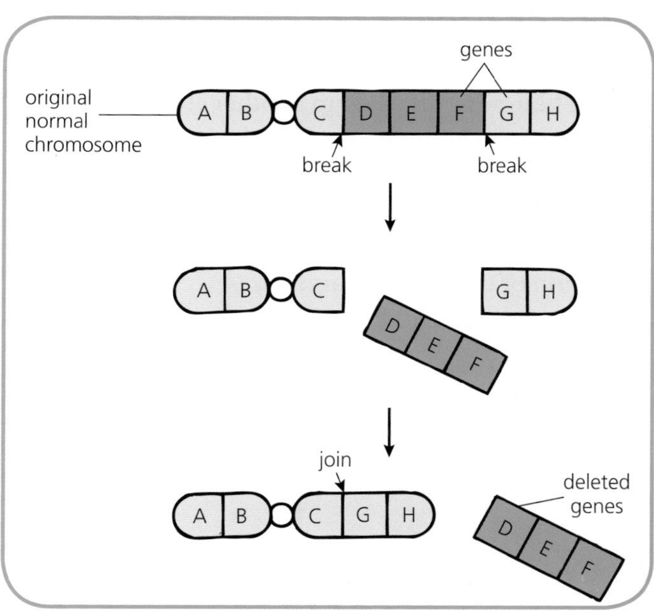

Figure 5.13 Deletion

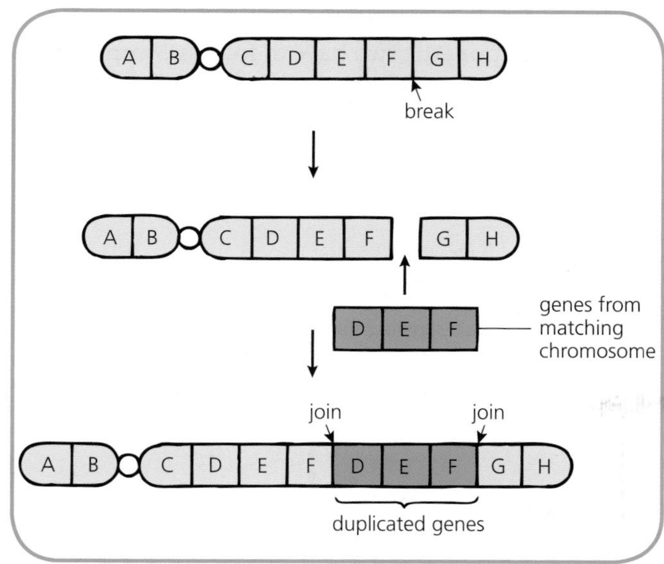

Figure 5.15 Duplication

Duplication

A chromosome undergoes this type of change when a segment of genes (such as deleted genes from its homologous partner) becomes attached to one end of the first chromosome or becomes inserted somewhere along its length as shown in Figure 5.15. This results in a set of genes being **repeated**.

Some **duplications** of genes may have a detrimental effect on the organism. For example, the duplication of certain genes (called oncogenes) is a common cause of cancer. Other gene duplications may be of advantage to the species as follows.

Importance of gene duplication in evolution

Duplication of a gene produces a second copy that is free from selection pressure. It is thought that this extra copy can become altered without interfering with the original gene's function or affecting the organism in some harmful way.

This freedom allows the extra copy of the gene to undergo point mutations that produce **new DNA sequences**. One or more of these may confer some advantage on the organism thereby **increasing its fitness** and chance of survival. An example is found among some species of fish that live in Antarctic waters of −2 to 4 °C. In these animals a duplicate of a gene that codes for a digestive enzyme has mutated and now codes for a glycoprotein that acts as antifreeze in the fish's blood and body fluids.

Inversion

A chromosome undergoing **inversion** breaks in two places as shown in Figure 5.16. The segment between the two breaks turns round before joining up again. This brings about a **reversal** of the normal sequence of genes in the affected section of chromosome.

When a chromosome which has undergone inversion meets its normal, non-mutated matching partner at gamete formation, the two have to form a complicated loop in order to pair up. This often results in the formation of **non-viable** gametes.

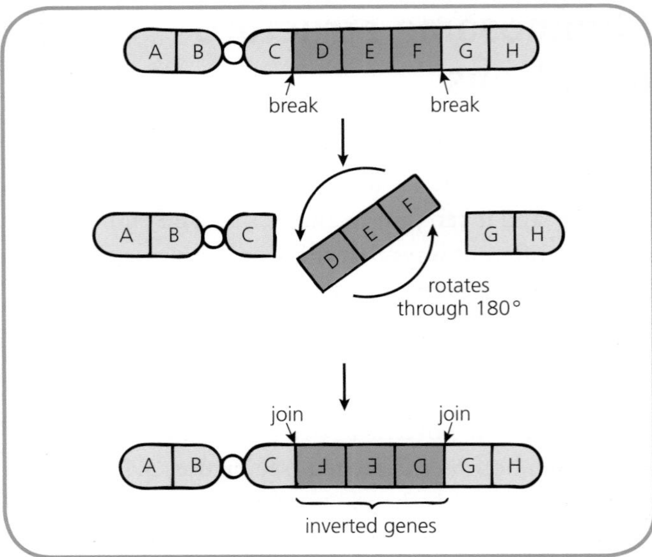

Figure 5.16 Inversion

Translocation

This involves a section of one chromosome breaking off and becoming attached to another chromosome which is not its homologous (matching) partner. Figure 5.17 shows two ways in which this may occur. **Translocation** usually leads to problems during pairing of chromosomes at gamete formation and results in the formation of **non-viable** gametes.

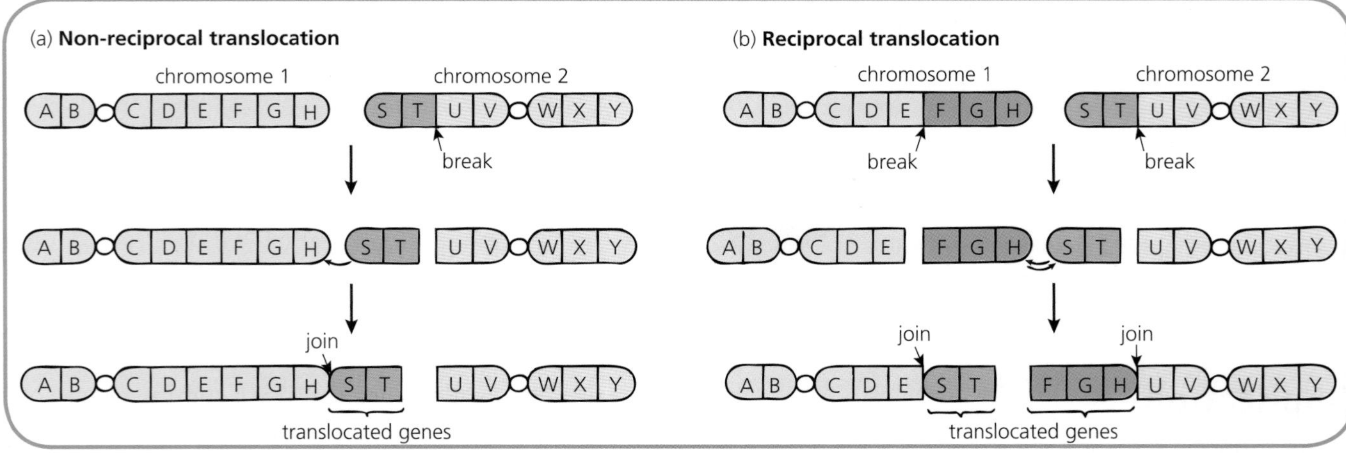

Figure 5.17 Translocation

Case Study Cri-du-chat syndrome

Cri-du-chat syndrome is caused by a **deletion** of part of chromosome 5. Children born with this genetic disorder have severe learning difficulties. They develop a small head with unusual facial features and widely spaced eyes. The condition is so-called because the infant's crying resembles that of a distressed cat. Affected individuals usually die early in childhood.

Case Study Haemophilia A

Haemophilia A is a genetic disorder caused by several types of mutation. One of the most common of these is an **inversion** within the gene that produces **blood-clotting factor VIII**.

People with this condition fail to make normal factor VIII. In the absence of this protein, an untreated person experiences prolonged external bleeding following injury and prolonged internal bleeding both post-operatively and spontaneously into soft tissues and joints. The condition is successfully treated with regular intravenous infusions. Since haemophilia A is a sex-linked recessive trait, it occurs in males and homozygous females.

Case Study Chronic myeloid leukaemia (CML)

Chronic myeloid leukaemia is a form of cancer that affects some of the stem cells that give rise to white blood cells. These stem cells are affected by a **reciprocal translocation** involving genetic material on chromosomes 9 and 22 as shown in Figure 5.18. This translocation results in the formation of an **oncogene**. An oncogene encodes a protein that promotes uncontrolled cell growth (i.e. cancer). In CML the encoded protein is called tyrosine kinase.

CML is treated by using drugs that inhibit the effect of tyrosine kinase and reduce the number of white blood cells produced in the bone marrow. CML occurs most commonly in middle-aged and elderly people. Its incidence is increased by exposure to ionising radiation. The atomic bombing of Hiroshima and Nagasaki in Japan at the end of the Second World War resulted in greatly increased rates of CML among the population. The condition is lethal if left untreated.

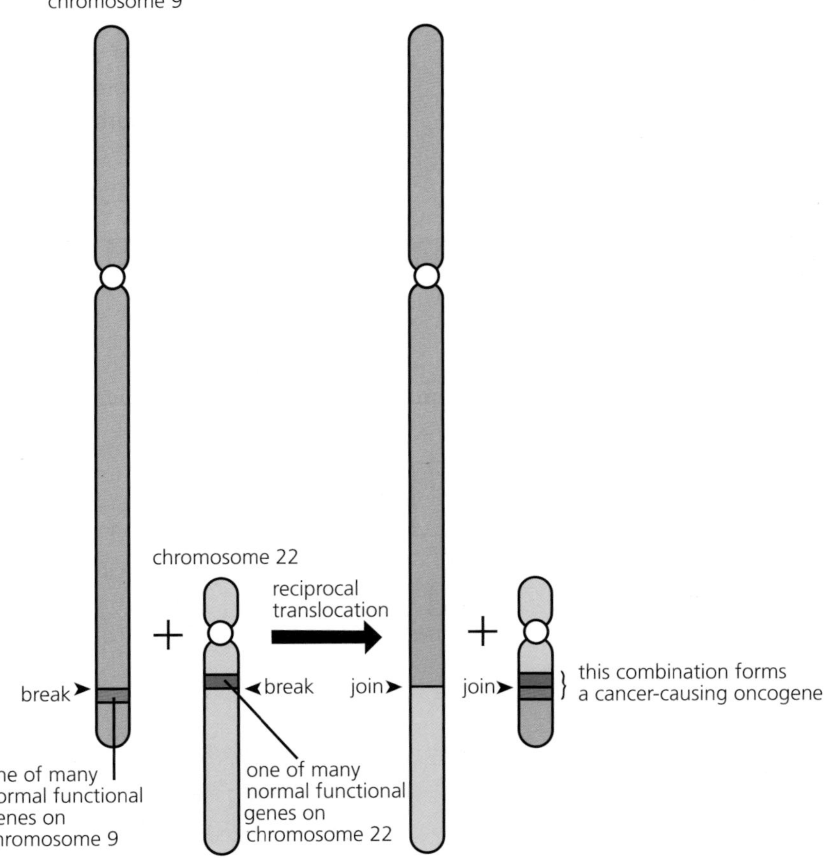

Figure 5.18 Mutation causing chronic myeloid leukaemia

Testing Your Knowledge 2

1 a) i) What name is given to a change which involves a chromosome breaking in two places and a segment of genes dropping out?
ii) Is this type of mutation likely to be beneficial or harmful to the organism affected?
iii) Explain why. (3)

b) i) Name the type of change which involves a chromosome breaking in two places and the affected length of genes becoming rotated through 180° before becoming reunited with the chromosome.
ii) What effect does this change have on the sequence of the genes in the affected segment? (2)

2 a) i) What name is given to the type of chromosomal change which involves a segment of genes from one chromosome becoming inserted somewhere along the length of its matching partner?
ii) Why might this mutation be of benefit to the organism affected? (2)

b) i) Name the type of change which involves a section of one chromosome breaking off and joining onto another non-matching chromosome.
ii) What effect does this change have on the number of genes present on each of the affected chromosomes? (2)

Applying Your Knowledge and Skills

1 Figure 5.19 shows a DNA strand and three types of RNA that are transcribed from it.

a) Name the types of RNA shown at A, B and C. (3)

b) Which of DNA sequences W, X and Y is coding for protein? (1)

c) Match blank boxes 1, 2 and 3 with the following answers:
i) combines with proteins to form complexes
ii) a sequence of codons becomes translated
iii) folding of molecule by some bases pairing with one another. (2)

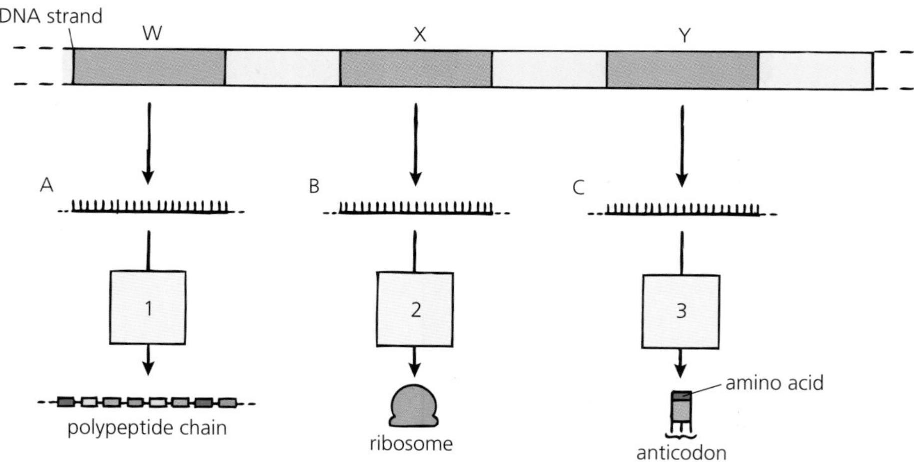

Figure 5.19

2 In the following three sentences, a small error alters the sense of the message. To which type of point mutation is each of these equivalent? (3)

a) Intended: She ordered boiled rice.
Actual: She ordered boiled ice.

b) Intended: He walked to the pillar box.
Actual: He talked to the pillar box.

c) Intended: He put a quid in his pocket.
Actual: He put a squid in his pocket.

3 Figure 5.20 shows the base sequence on a region of DNA undergoing a type of point mutation.

a) i) Identify the type of point mutation that occurred.
   ii) Describe the way in which the DNA has been altered. (2)

b) Refer back to Table 3.2 (and Table 3.4) and work out the amino acid sequence for
   i) the original DNA
   ii) the mutant DNA. (2)

c) Suggest why this mutation is described as being *silent*. (1)

d) i) Refer back to Figure 5.6 and state whether the point mutation that results in the formation of haemoglobin S is missensical or nonsensical.
   ii) Explain your choice. (2)

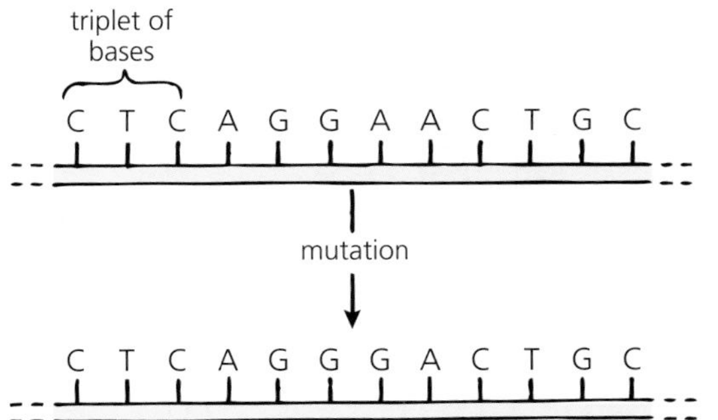

Figure 5.20

4 The data in Table 5.3 refer to the results of an experiment to investigate the effect of increasing doses of radiation on root length in germinating chick peas.

Radiation dose (units)	Mean root length (mm)
0	180
100	152
200	140
300	106
400	90
500	74
600	44

Table 5.3

a) In this experiment, which factor is the
   i) dependent variable?
   ii) independent variable? (1)

b) Plot the points on a sheet of graph paper and draw the line of best fit. (4)

c) i) What relationship exists between radiation dose and mean root length?
   ii) Suggest why. (2)

d) From your graph, extrapolate the radiation dose that would result in no root growth. (1)

e) In this experiment, ten seeds were planted per pot and four replicates of each pot were set up. Explain why. (1)

f) Although gene mutations in the absence of mutagenic agents are rare, the gene controlling grain colour in maize mutates as often as one in 2000 gametes on average. Express this as a mutation frequency. (1)

5 a) Show diagrammatically the proportion of offspring with sickle-cell trait that would result from a cross between
   i) a normal parent and a parent with sickle-cell trait
   ii) two parents with sickle-cell trait. (4)

b) Figure 5.21 shows a region in Africa. Each circle on the map represents the percentage of local people with sickle-cell trait. Explain the variation in incidence of sickle-cell trait in this region. (2)

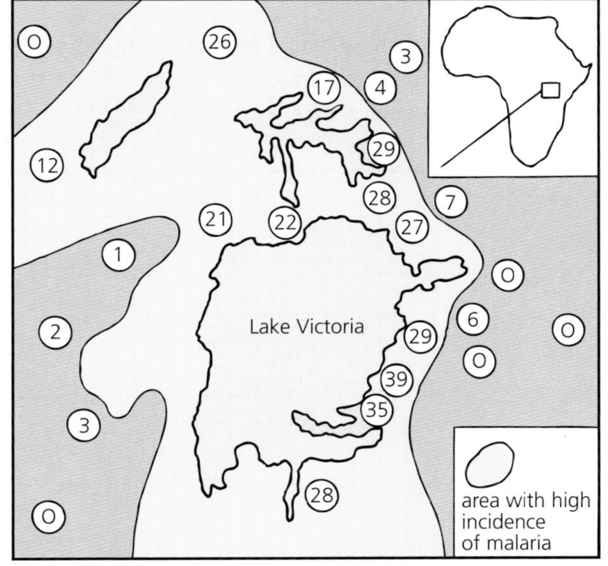

Figure 5.21

→

6 During gamete formation, chromosomes normally form pairs that match one another, gene for gene, all along their length. The members of each pair of chromosomes shown in Figure 5.22 do not match properly because of a mutation.

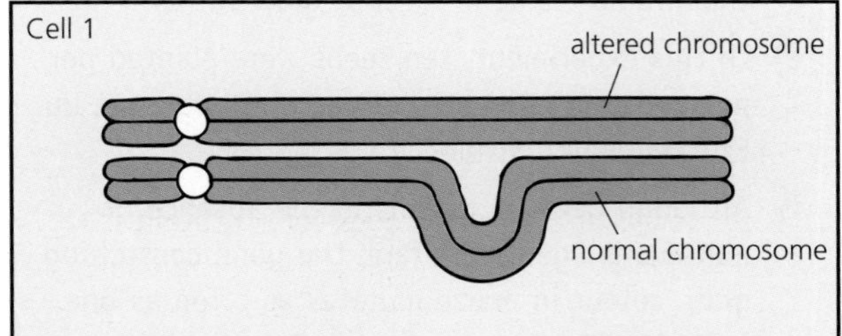

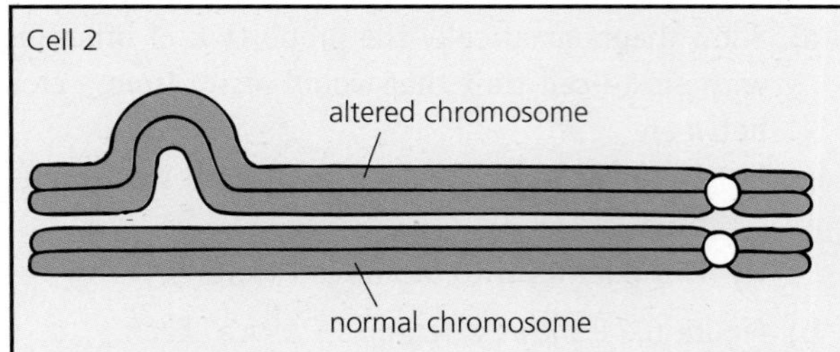

Figure 5.22

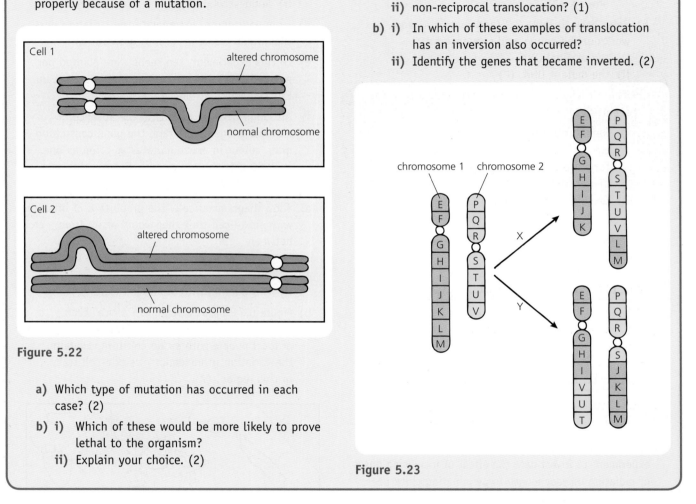

a) Which type of mutation has occurred in each case? (2)

b) i) Which of these would be more likely to prove lethal to the organism?
ii) Explain your choice. (2)

7 Figure 5.23 shows two chromosomes undergoing mutations.

a) Which arrow represents
i) reciprocal translocation?
ii) non-reciprocal translocation? (1)

b) i) In which of these examples of translocation has an inversion also occurred?
ii) Identify the genes that became inverted. (2)

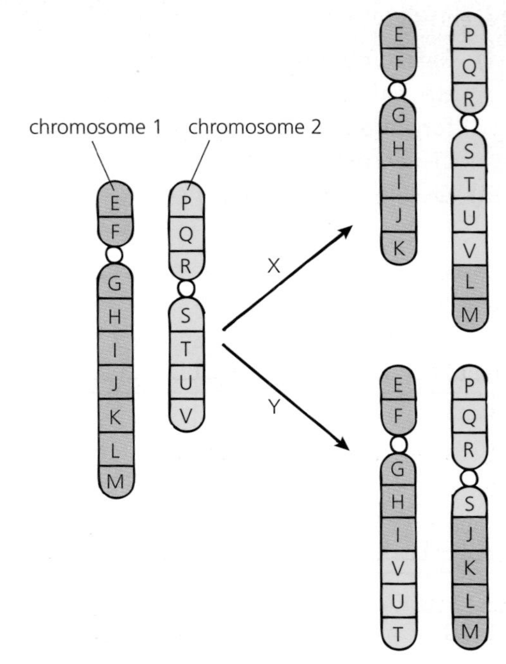

Figure 5.23

What You Should Know

Chapters 4–5

(See Table 5.4 for word bank)

adult	genes	proteins
advantage	genome	random
altered	insertion	range
chromosome	inverted	research
deleted	meristems	selection
differentiate	multicellular	specialised
differentiation	mutation	therapy
embryonic	nucleotide	transcription
evolution	phenotype	unspecialised
gene	point	variation

Table 5.4 Word bank for chapters 4–5

1 Regions of unspecialised cells in plants that are capable of cell division are called ______. The process by which an unspecialised cell becomes altered and adapted to perform a special function is called ______.

2 Cells produced in meristems differentiate into ______ cells. A specialised cell only expresses the ______ that code for proteins needed for the workings of that type of cell.

3 Stem cells are ______ cells that can reproduce themselves and ______ into specialised cells.

4 ______ stem cells are able to differentiate into all the cell types that make up a ______ organism. Tissue (______) stem cells are only able to generate a limited ______ of cell types.

5 Stem cells are used in ______ to gain a better understanding of cell growth and gene regulation. In the future several debilitating conditions may be treated successfully using stem cell ______.

6 The sum of all the genetic information encoded in an organism's DNA is called its ______.

7 A sequence of DNA that codes for a protein is defined as a ______. However, most of a eukaryote's genome is composed of DNA sequences that do not code for ______. Some of these regulate ______.

8 A ______ is a change in the structure or amount of an organism's genome. Spontaneous mutations occur infrequently and at ______. They can affect gene expression and lead to the synthesis of an ______ protein that changes the organism's ______.

9 A ______ mutation involves the substitution, ______ or deletion of a ______ in the DNA chain.

10 Point mutations are the source of the new ______ that provides the raw material for evolution. Very rarely a mutant allele arises that confers an ______ on an organism which is then favoured by natural ______.

11 A ______ may undergo a structural mutation if one or more of its genes becomes duplicated, ______, translocated or ______. If a duplicated gene is further changed by a point mutation it may provide more raw material for ______.

6 Evolution

Evolution is the process of gradual change in the characteristics of a population of organisms that occurs over successive generations as a result of variations in the population's genome. These variations take the form of **changes** in the **frequencies** of certain **genetic sequences** (in other words, alleles of genes).

Evolution accounts for the origin of existing species from ancestors that lived long ago and were often very different from present-day species. It involves the processes of inheritance, selection and speciation.

Inheritance

Vertical transfer of genetic material

Genetic sequences of DNA such as protein-coding genes are transferred **vertically** from parent down to offspring. This vertical inheritance may be the result of sexual or asexual reproduction.

In **sexual** reproduction the parents normally differ from one another genetically and produce offspring that vary further in their genetic make-up (see Figure 6.1). Vertical inheritance in humans is often represented as a family tree (see Figure 6.2).

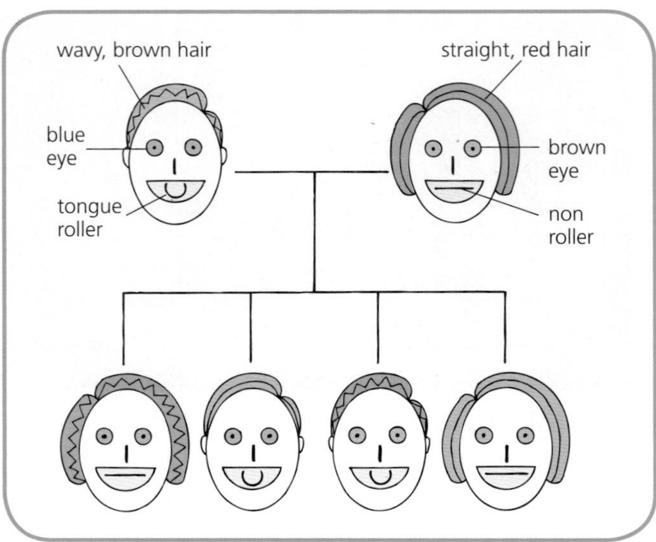

Figure 6.2 Vertical inheritance in a family tree

In **asexual** reproduction a single parent with a certain genome produces offspring with exactly the same genome and no variation results among successive generations (see Figure 6.3). Vertical inheritance occurs among eukaryotes and prokaryotes.

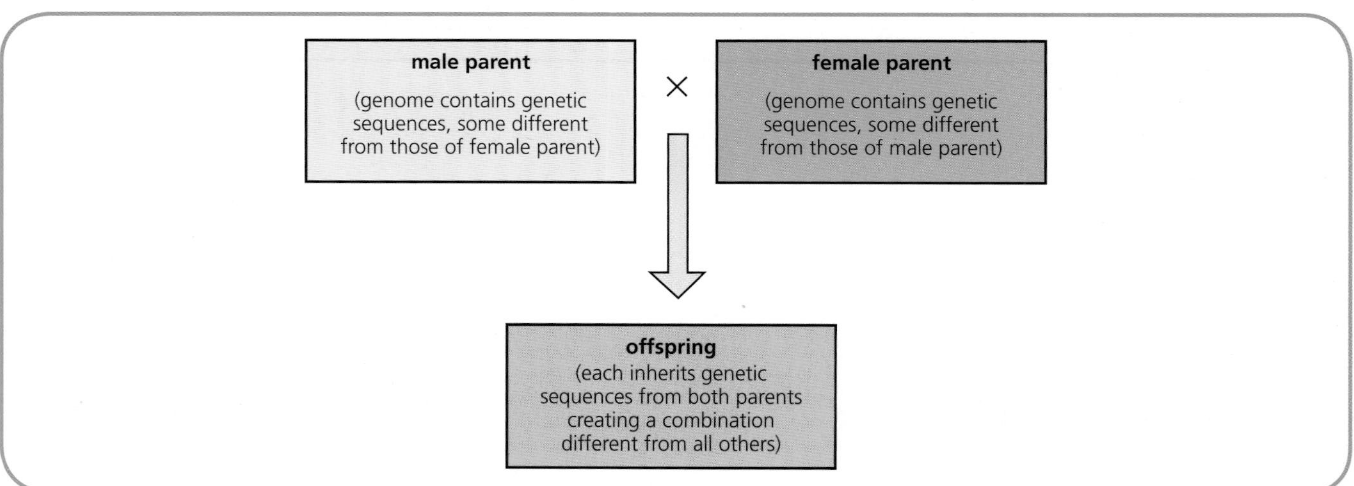

Figure 6.1 Vertical inheritance by sexual reproduction

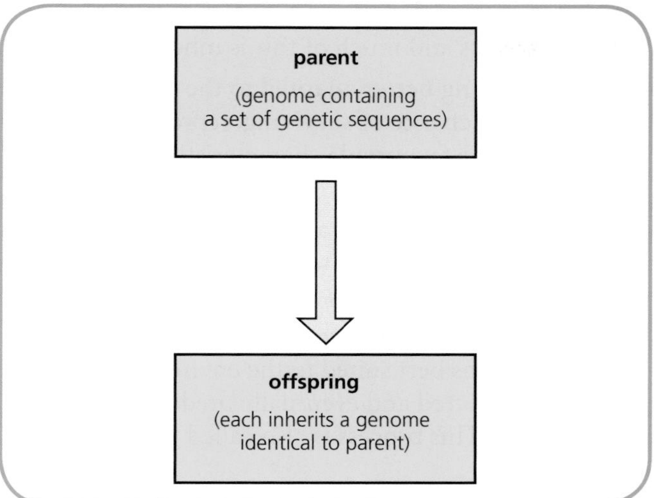

Figure 6.3 Vertical inheritance by asexual reproduction

Horizontal transfer of genetic material

In prokaryotes, genetic material can pass across from one cell to another **horizontally**. Therefore genetic sequences are not only handed down from one generation to the next by reproduction but are also exchanged among and between contemporary members of the population. The cells involved may not even belong to the same species.

Rapid evolutionary change in prokaryotes

It is thought that the rate of gene loss during genome replication and vertical inheritance was high during the very early stages of prokaryotic evolution. Experts believe that at the time this loss was compensated for by a high rate of **horizontal gene transfer** (HGT) which led to a rapid spread of new genetic sequences. These promoted the build-up of larger genomes and allowed rapid natural selection (see page 76) and **faster evolutionary change** to occur among early prokaryotes. After all, the acquisition of a beneficial gene from a neighbouring cell is a much faster method of obtaining it than waiting for it to evolve (if ever) by natural means.

However, there is no guarantee that a genetic sequence gained horizontally will confer an advantage on the recipient. In fact it may turn out to be useless or even harmful. For this reason HGT is a **risky** evolutionary strategy. By comparison, vertical inheritance is much **safer** because the genes that are passed on to successive generations have been tested and tried by the parent.

It is thought, therefore, that as organisms reached a certain level of multicellular complexity, the importance of gaining sequences by HGT decreased. Gradually, as the role of vertical inheritance increased in importance, distinct lineages began to emerge with their own sets of specific genes, eventually giving rise to the 'tree of life' (see Appendix 4).

Related Topic

MRSA

A significant amount of HGT still takes place in modern prokaryotes. An important example occurs among certain **bacteria** that exchange plasmids carrying the genes that confer **resistance to antibiotics** from one bacterial species to another.

This has resulted in the emergence of strains of bacteria resistant to several antibiotics. One of these, called **MRSA** (methicillin-resistant *Staphylococcus aureus*) is shown in Figure 6.4. It causes infections which are extremely difficult to cure and which sometimes prove to be fatal.

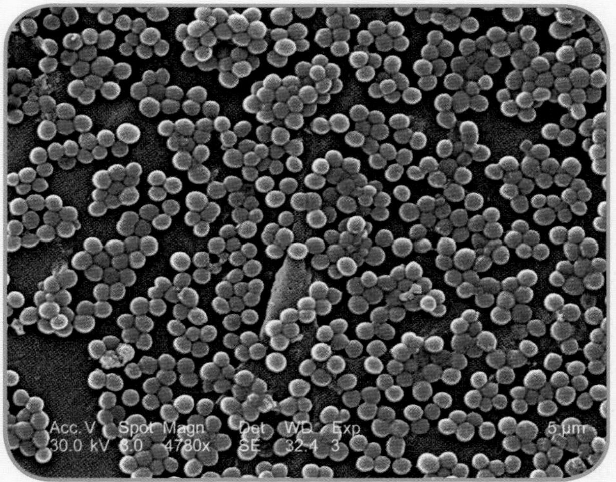

Figure 6.4 MRSA (methicillin-resistant *Staphylococcus aureus*)

Selection

Selection is the process by which the frequencies of some DNA sequences within a population increase because they are selected **for** and others decrease because they are selected **against**, in a non-random manner.

Natural selection

In 1858 Charles Darwin and Alfred Wallace presented a joint paper suggesting that the main factor producing evolutionary change is **natural selection**. In *On the Origin of Species*, Darwin amplified his ideas as follows:

- Organisms tend to produce **more offspring** than the environment can support.
- A **struggle** for existence follows and many offspring die before reaching reproductive age because of factors such as lack of food, overcrowding and lack of resistance to disease.
- Members of a species show **variation** in all characteristics and much of this is inherited.
- Those offspring **better adapted** to the environment have a better chance of surviving, reproducing and passing on the favourable characteristics to their offspring.
- Those offspring **less well adapted** to the immediate environment die and fail to pass on the less favourable characteristics.
- This process is repeated generation after generation. The organisms best suited to the environment are naturally **selected** and eventually predominate in the population. This process is also called the **survival of the fittest**.

Thus **natural selection** is a non-random process that results in the **increase in frequency** among a population of organisms of those **genetic sequences** (such as certain alleles of genes) that confer an advantage on members of the population and help them to survive.

Related Topic

Evolution of resistant insects

DDT is a poisonous chemical which has been widely used against many insects. These include mosquitoes, which carry malaria and yellow fever, and insect pests which destroy crops. Within a few years of use, many **mutant** forms of insects **resistant** to the insecticide 'appeared'. These mutants are able to make an enzyme that renders the chemical harmless. They had not arisen in response to DDT. A tiny number of resistant mutants just happened to be present within the natural insect populations or arose later by a chance mutation.

When the spray was applied, the vast majority of non-resistant insects died and the resistant mutants suddenly enjoyed a **selective advantage** and multiplied (see Figure 6.5). Under such conditions natural selection may enable the mutants to eventually replace their wild-type relatives. This non-random increase in frequency of the gene for DDT resistance in the population increased the mosquito's chance of survival. Many new pesticides have been developed in recent years but strains of pests resistant to these chemicals are now emerging.

→

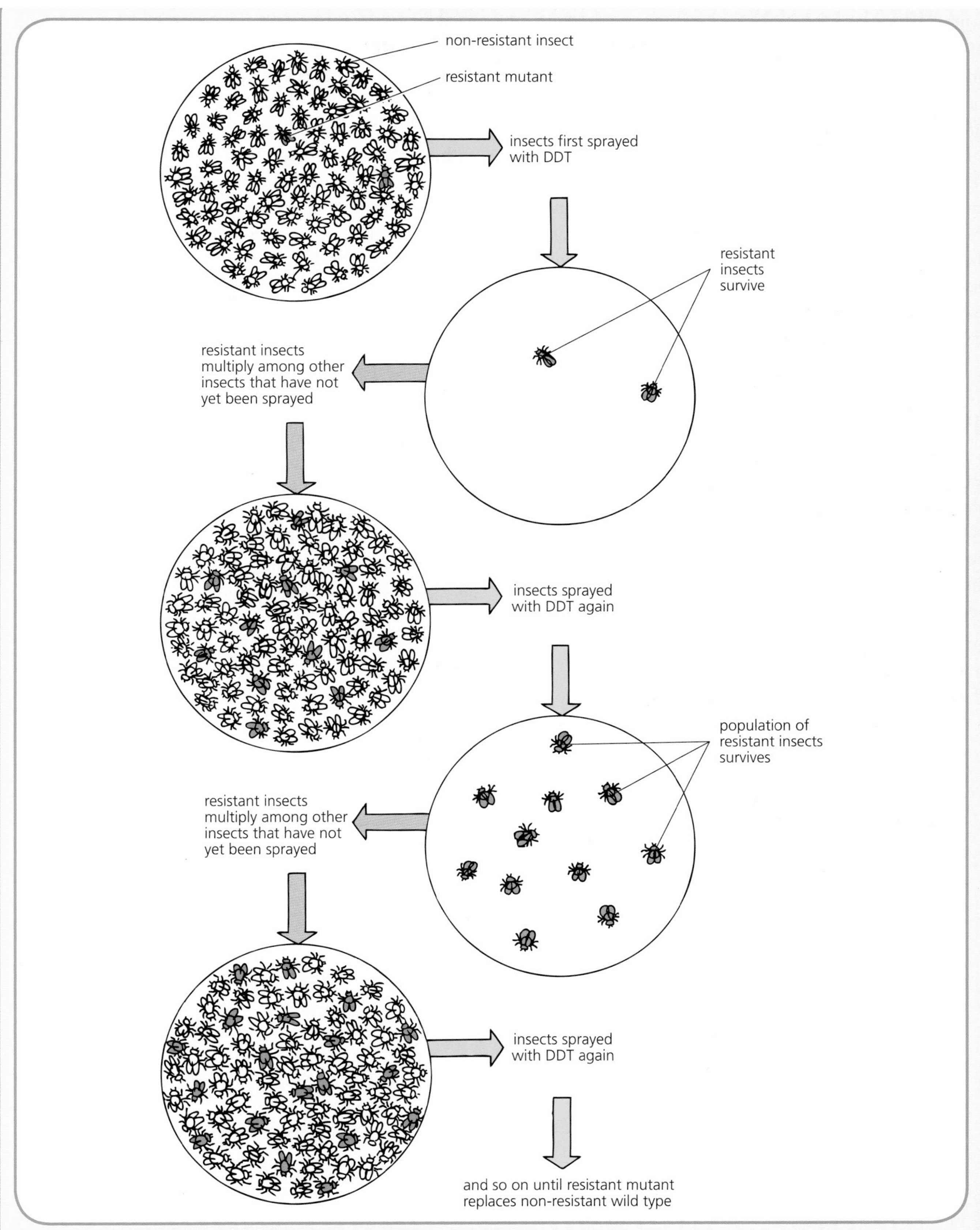

Figure 6.5 Selection of insects resistant to DDT

Selection against deleterious sequences

Where a deleterious genetic sequence codes for an inferior version of a characteristic which leaves the individual poorly adapted to the environment, an affected individual will leave fewer than the average number of offspring. Since fewer and fewer copies of the deleterious sequence will be passed on to successive generations, a non-random **reduction** in its frequency will occur. It may even be eliminated from the population in time. A sequence that is **lethal** will disappear much more quickly than one that is less harmful since the selection pressure acting against a lethal allele will be of maximum strength.

Types of selection for a quantitative trait

A polygenic trait is controlled by the interaction of several genes. A characteristic of this type, such as seed mass, is quantitative. When the data for a large population are graphed, the result is a bell-shaped **normal distribution**. Natural selection can affect the frequency of a quantitative trait within a large population in any one of the following three ways.

Stabilising selection

This form of selection exerts its pressure against the extreme variants of the phenotypic range and favours the intermediate versions of the trait as shown in Figure 6.6. It leads to a **reduction** in genetic diversity without a change in the mean value. It operates in an unchanging environment and **maintains the status quo** for the best-adapted genotypes in the population. For example, stabilising selection keeps natural human birth mass, almost without exception, in the 3–4 kg range. Babies of a very low body mass are more susceptible to fatal diseases and those of a very high mass encounter difficulties at birth passing through the mother's pelvis.

Directional selection

This type of selection is most common during a period of environmental change. It favours an extreme version of the characteristic that was initially a less common form and results in a **progressive shift** in the population's mean value for the trait (see Figure 6.6). For example, fossil evidence shows that European black bears increased, on average, in body size during each ice age. Since a larger body loses relatively less heat than a

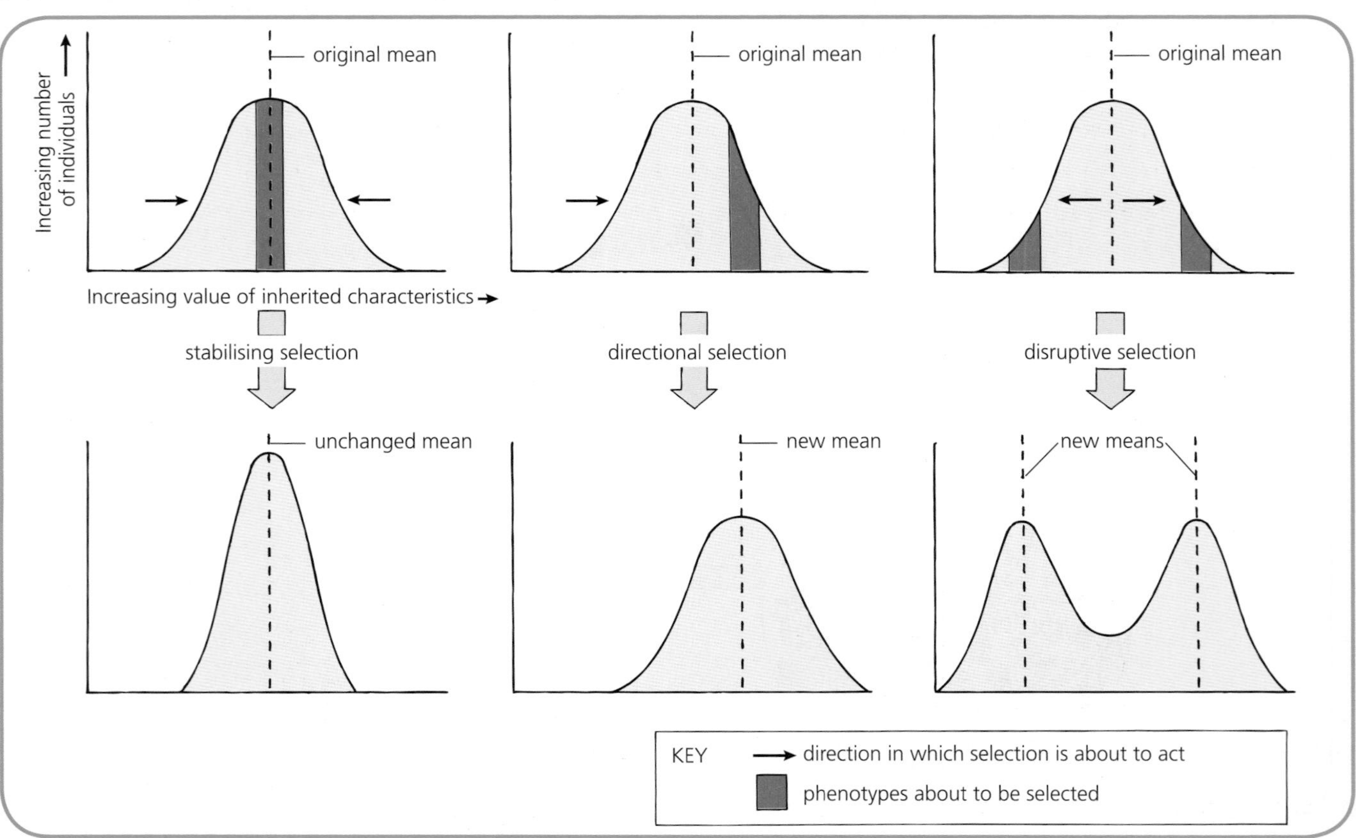

Figure 6.6 Three types of selection

smaller one, a larger body is of survival value in a cold climate. **Artificial selection** practised by humans to improve strains of domesticated plants and animals is also a form of directional selection.

Disruptive selection

This is a form of selection in which two or more extreme versions of a trait are favoured at the expense of the intermediates. It results in the population becoming split into **two distinct groups** each with its own mean value as shown in Figure 6.6. Under natural conditions it occurs when two different habitats or types of resource become available. It is considered to be the driving force behind **sympatric speciation** (see page 81). Plant and animal breeding involving artificial selection of extremely large and extremely small varieties (for example, breeds of dog) is also a form of disruptive selection.

Testing Your Knowledge 1

1 Figure 6.7 shows different directions of transfer of genetic material where P = prokaryotes and E = eukaryotes. Which of them is NOT a natural form of inheritance? (1)

2 a) What are thought to have been TWO benefits of a high rate of horizontal gene transfer during the early stages of prokaryotic evolution? (2)
  b) Briefly explain why horizontal gene transfer might prove to be a risky strategy. (1)

3 a) Within a breeding population, some organisms fail to reach reproductive age. Give TWO possible factors that could be responsible for their early death. (2)
  b) i) Other offspring do survive and reproduce on reaching adulthood. What name did Darwin give to this 'weeding out' process that promotes the survival of the fittest?
    ii) Define this process including in your answer the terms: *advantage*, *frequency*, *genetic sequences*, *population*. (3)

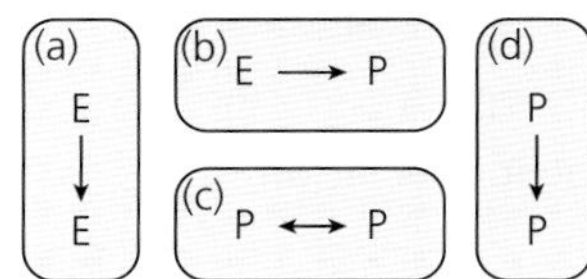

Figure 6.7

4 Briefly explain the difference between *stabilising* selection and *disruptive* selection. (2)

Species and speciation

Species

A **species** is a group of organisms that are able to interbreed with one another and produce fertile offspring. They are genetically isolated and cannot produce viable, fertile offspring with members of other groups (in other words, different species). It is thought that there may be about 5–20 million different species on Earth at present. However, species are not constant, immutable units. Their number and kinds are always changing. At any given moment some will be enjoying a **stable relationship** with the environment, some will be moving towards **extinction** and others will be undergoing **speciation**.

Speciation

Speciation is the formation of new biological species. It is brought about by **evolution** as a result of isolation, mutation and selection. It occurs when circumstances arise that interrupt gene flow between two populations causing their gene pools to diverge. Two types of speciation are allopatric speciation and sympatric speciation.

Allopatric speciation

In **allopatric** speciation, gene flow between two (or more) populations is prevented by a **geographical barrier**. Some examples are given in Figure 6.8.

A simplified version of allopatric speciation is shown in Figure 6.9.

1 The members of a large population of a species occupy an environment. They share the same gene pool and interbreed freely.

2 The population becomes split into two completely isolated sub-populations by a **geographical barrier** which prevents interbreeding and gene exchange.

3 **Mutations** occur at random. Therefore most of the mutations that occur within one sub-population are different from those that occur within the other sub-population. This results in **new variation** arising within each group which is not shared by both groups.

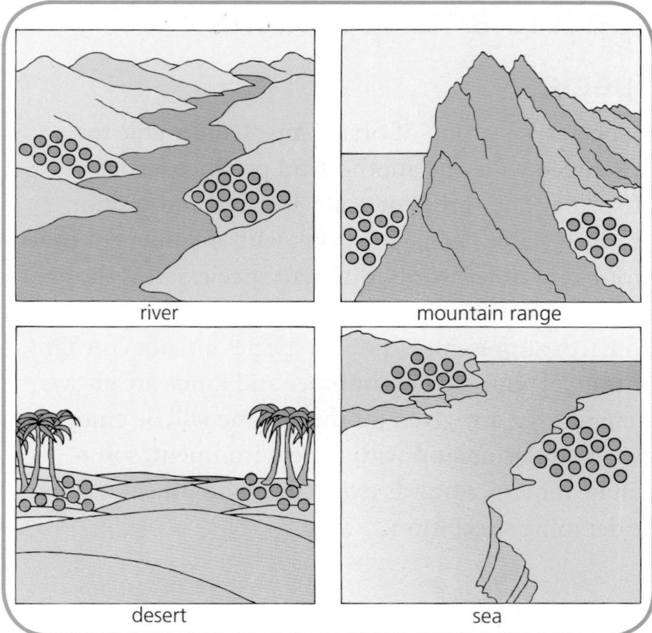

Figure 6.8 Geographical barriers

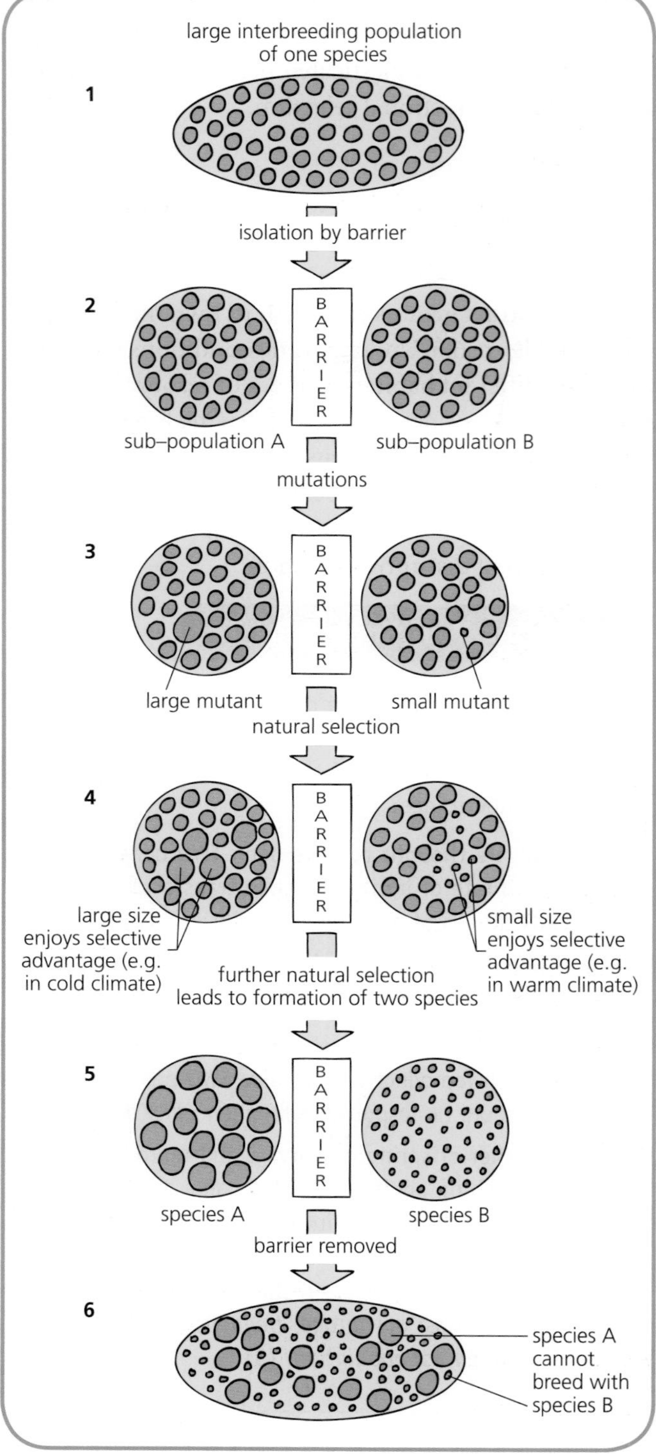

Figure 6.9 Allopatric speciation

4 The selection pressures acting on each sub-population are different depending on local conditions such as climate, predators and disease. **Natural selection** affects each sub-group in a different way by favouring those alleles which make the members of that sub-population best at exploiting their environment.

5 Over a very, very long period of time, stages 3 and 4 cause the two gene pools to become so altered that the groups become genetically distinct and **isolated**.

6 If the barrier is removed, they are no longer able to interbreed since their chromosomes cannot make matching pairs. **Speciation** has occurred and two separate distinct species have evolved.

European wren

Figure 6.10 refers to three subspecies of the European wren. Geographical isolation from the original mainland population and from one another has caused each island's wren population to take its own course of evolution. Although the populations have not been isolated for long enough to allow distinct species to arise, it is interesting to note that the subspecies on St Kilda (which is further from the mainland than the Hebrides) is already more different. Similarly four different subspecies of Orkney vole are found on four different islands in the Orkney group.

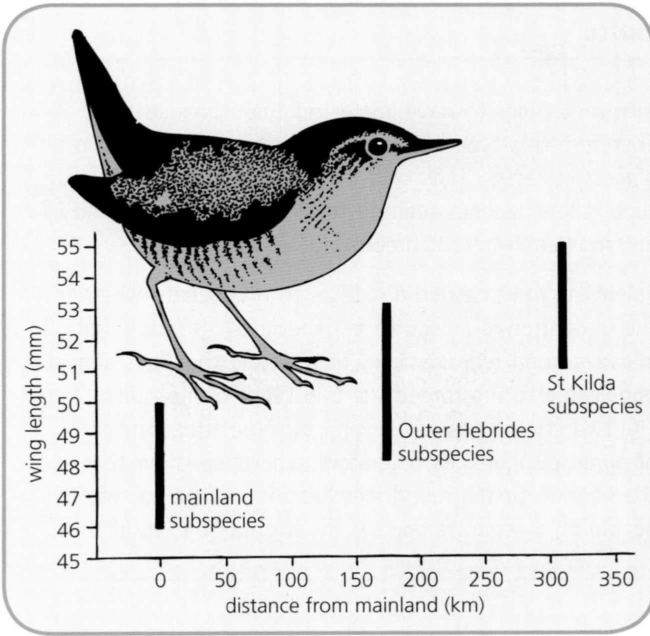

Figure 6.10 Subspecies of wren

Sympatric speciation

In **sympatric** speciation, the two (or more) populations live in close proximity to one another in the same environment yet they become genetically **isolated**. This happens because gene flow between them is prevented by the presence of a **behavioural** or **ecological barrier**. Sympatric speciation is promoted by disruptive selection (see page 79). A simplified version of sympatric speciation is shown in Figure 6.11.

Fruit flies

A species of fruit fly is found in North America. It has evolved over a very long period of time. It lives on hawthorn trees and its larvae feed on the berries. About 200 years ago, settlers introduced a type of apple tree to North America. Some of the fruit flies soon began to exploit this new ecological niche and made use of the apples to feed their larvae.

The type of fruit tree where males search for mates and where females lay their eggs is normally the same as the type of fruit upon which the flies developed as larvae. This difference in behaviour has created an ecological barrier to gene flow between 'hawthorn' fruit flies and 'apple' fruit flies. Already some genetic differences have developed between the two groups and they are thought to have begun the process of **sympatric speciation**. Similarly many parasites that are specific to a certain host have arisen by sympatric speciation.

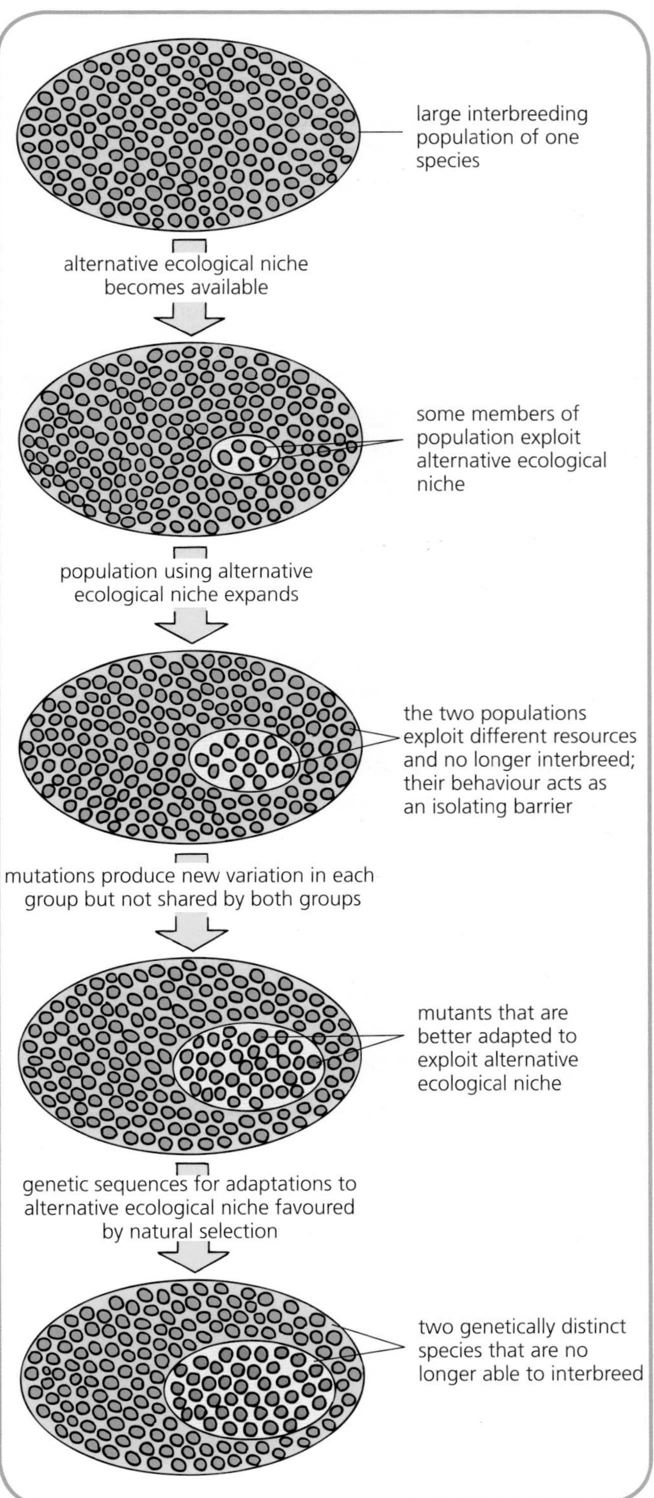

Figure 6.11 Sympatric speciation

Research Topic: London Underground mosquito

Culex pipiens is a species of mosquito that lives above ground in Europe and in many other parts of the world. It feeds by biting birds and sucking their blood. It is able to tolerate cold conditions in winter by 'hibernating'.

The London Underground mosquito is thought to have evolved from members of the above-ground species that moved below ground as a splinter group over 100 years ago when the tunnels for the Tube were being dug. Over the past century these mosquitoes have exploited their new ecological niche by breeding in pools of stagnant water in the tunnels and by feeding on rats and mice. They are also known to have bitten humans sheltering from air raids during the Second World War Blitz and in recent times to have 'molested' maintenance staff. Therefore this type of mosquito has been named *Culex pipiens molestus*. It is cold intolerant because over time it has become adapted to the warm underground environment where it breeds all the year round.

Scientists have carried out tests to investigate whether the underground mosquito is genetically different from its above-ground relative. They have found that it is almost impossible to interbreed the two types of mosquito. The two strains possess several different frequencies of genetic sequences. Therefore experts consider that the London Underground mosquito is well on its way to becoming a separate species by **allopatric speciation**.

Testing Your Knowledge 2

1 When a horse is crossed with a donkey, the result is a sterile animal called a mule.
   a) Do a mule's parents belong to the same species? (1)
   b) Explain your answer. (1)

2 a) Define the term *speciation*. (1)
   b) Arrange the following stages into the correct order in which they would occur during allopatric speciation. (1)
      A isolation of gene pools
      B formation of new species
      C mutation
      D occupation of territory by one species with one gene pool
      E natural selection

3 a) i) Identify the type of speciation that involves geographical barriers.
      ii) Give TWO examples of geographical barriers. (2)
   b) Identify the type of speciation that involves behavioural or ecological barriers. (1)

Applying Your Knowledge and Skills

1 The parents in the family tree shown in Figure 6.12 have eight children, none of whom is identical to any other with respect to the three genetically controlled traits shown.

a) Is this form of inheritance horizontal or vertical? (1)

b) Make a simple diagram to show the possible phenotypes of sons and daughters B–G. (3)

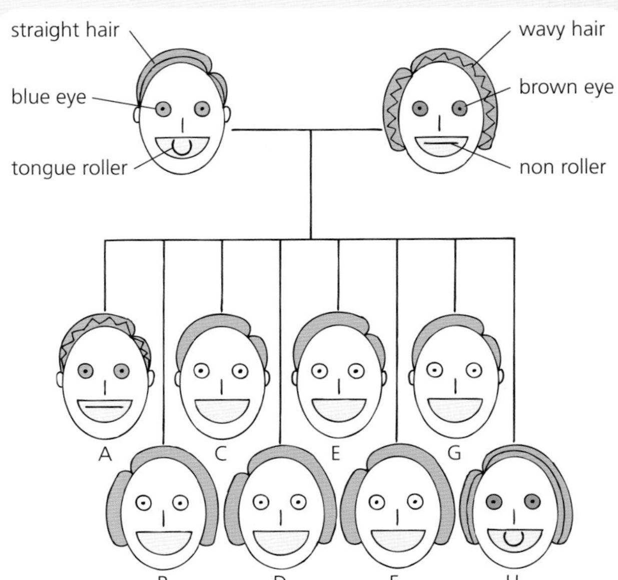

Figure 6.12

2 Figure 6.13 shows how a mutant form of a species that possesses some advantageous characteristic could spread through a population. Imagine that before dying, each mutant form leaves, on average, two offspring as part of the next generation whereas each wild type leaves only one.

a) Which symbol represents the mutant form? (1)

b) i) Draw and complete a box to represent the F_2.

ii) State the ratio that applies to the members of its population. (2)

c) Continue the series of diagrams until you can state the generation in which:

i) mutants outnumber wild type for the first time

ii) the ratio of wild type to mutants is 1:8. (2)

3 For many years, rats were successfully controlled by a poison called warfarin which interferes with the way that vitamin K is used in the biochemical pathway shown in Figure 6.14. However, in 1958, strains of rat appeared in Scotland which were resistant to warfarin. This resistance is an inherited characteristic as shown by the information in Table 6.1.

Rat's genotype	Rat's phenotype
W^sW^s	sensitive to warfarin
W^sW^r	resistant to warfarin (but needs some extra vitamin K in diet)
W^rW^r	resistant to warfarin (but needs 20 times normal amount of vitamin K in diet)

Table 6.1

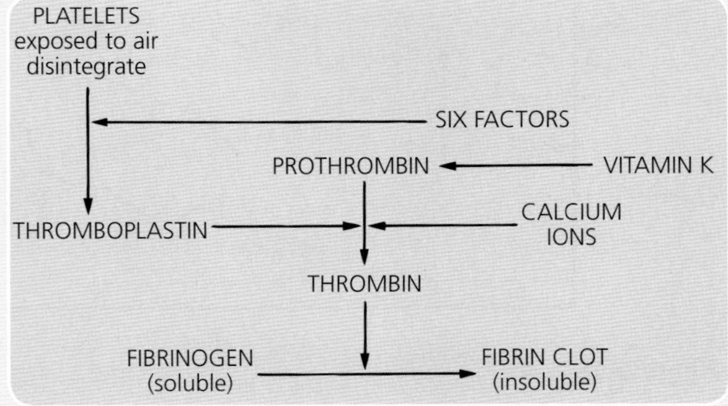

Figure 6.14

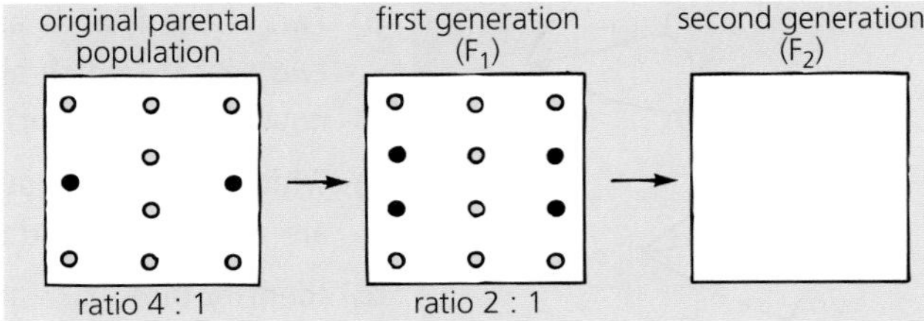

Figure 6.13

a) With reference to the diagram, explain why warfarin is lethal to normal rats. (1)

b) Why are people who get thrombosis (internal blood clotting) given small amounts of warfarin to take? (1)

c) Suggest how the resistant strain of rat arose. (1)

d) The resistant strain has increased greatly in number over the years. Explain this success in terms of natural selection. (1)

e) If the use of warfarin is continued, predict the fate of the normal wild type rat. (1)

f) Construct a hypothesis to account for the fact that most of the rats resistant to warfarin are heterozygotes. (2)

4 Figure 6.15 shows a map of the Galapagos Islands. Each figure in brackets refers to the number of different species of finch found on the island. Table 6.2 gives further information about six of the islands. Table 6.3 and Figure 6.16 refer to ground-living finches on two of the islands.

Name of island	Percentage number of finch species only found on this island
Culpepper	75.0
Pinta	33.3
Espanola	66.7
Isabella	22.2
Santa Cruz	0.0
Santa Fe	14.3

Table 6.2

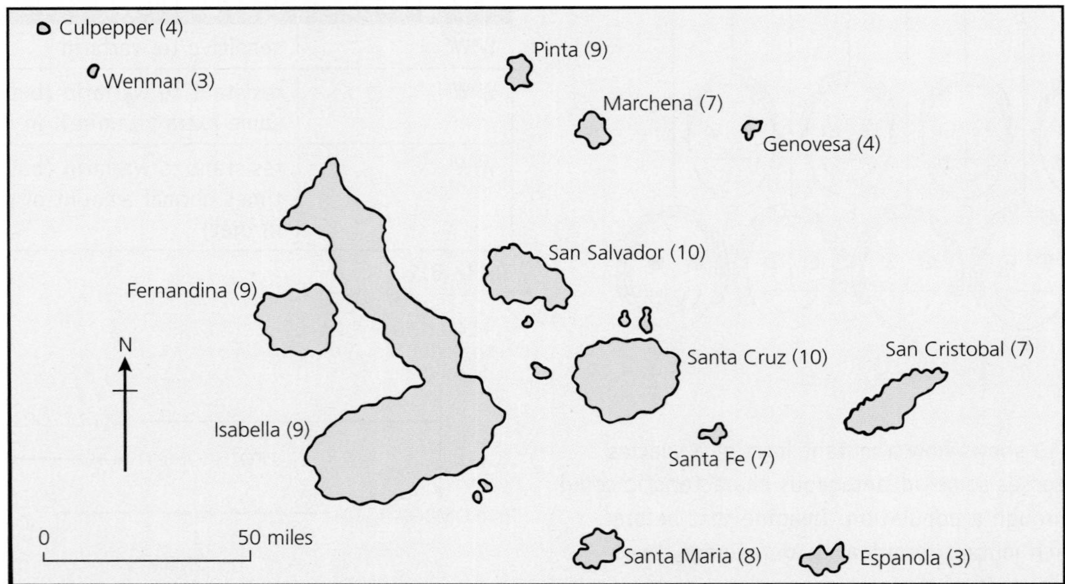

Figure 6.15

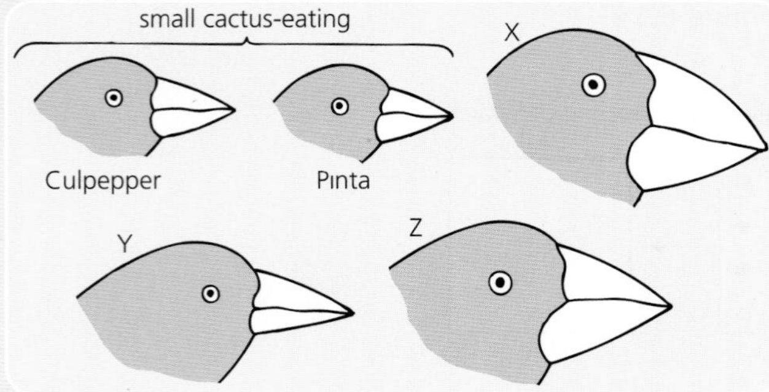

Figure 6.16

a) Calculate the actual number of finch species which are found on Culpepper and on no other island. (1)

b) Two of the islands each have an actual number of two species which are found on that island and nowhere else. Identify the islands. (2)

c) Which island is populated by finch species which are all found on other islands? (1)

d) Identify birds X, Y and Z using the information in Table 6.3. (2)

Species of ground finch		Island	
Body size	**Food**	**Culpepper**	**Pinta**
small	cactus	present (11.3 × 9.0)	present (9.7 × 8.5)
large	cactus	absent	present (14.6 × 9.7)
medium	seeds	present (15.0 × 16.5)	absent
large	seeds	absent	present (16.0 × 20.0)

Table 6.3 (numbers in brackets = length × depth of beak in mm)

e) **i)** If a large number of finch type Z were transported from Culpepper to Pinta, which native type on Pinta would face most competition?

ii) Predict the possible outcome over a long period of time. (2)

5 With respect to the effect that natural selection can have on the frequency of a quantitative trait within a large population, give an account of:

a) stabilising selection (3)

b) directional selection (3)

c) disruptive selection (3).

7 Genomic sequencing

Genomics is the study of genomes. It involves determining the sequence of the nucleotide base molecules all the way along an organism's DNA (**genomic sequencing**) and then relating this genetic information about genes to their functions. Progress in this area has been accelerated by bioinformatics, making genomics one of the major scientific advances of recent years.

Bioinformatics is the name given to the fusion of molecular biology, statistical analysis and computer technology. This ever-advancing area now enables scientists to carry out rapid mapping and analysis of DNA sequences on a huge scale and then compare them. Information about genetic sequences that used to take years to unravel is now obtained in days or even hours. It can be used to investigate evolutionary biology, inheritance and personalised medicine.

Genomic sequencing

Use of restriction endonucleases

A **restriction endonuclease** is a type of enzyme that recognises a specific short sequence of DNA nucleotides called a restriction site on a DNA strand (see page 185). It 'cuts' the DNA at this exact site all the way along the DNA strand. The restriction site recognised by one restriction endonuclease is different to that recognised by any other endonuclease.

Genome shotgun approach

The genome to be investigated is cut up into **fragments of DNA** using a restriction endonuclease. A further copy of the genome is cut up using a different endonuclease. Since these enzymes recognise different restriction sites on the DNA, they cut the genome at different points as shown in Figure 7.1.

Each DNA fragment produced is sequenced to establish the **order of its bases** (see Research Topic – Sequencing DNA). This information is entered into a computer. Because of the way in which the DNA has been cut, many of the fragments overlap with one another. For example Figure 7.2 shows, in a simplified way, how the 'end' of fragment p overlaps with the 'start' of fragment q and the 'end' of fragment q overlaps with the 'start' of fragment r.

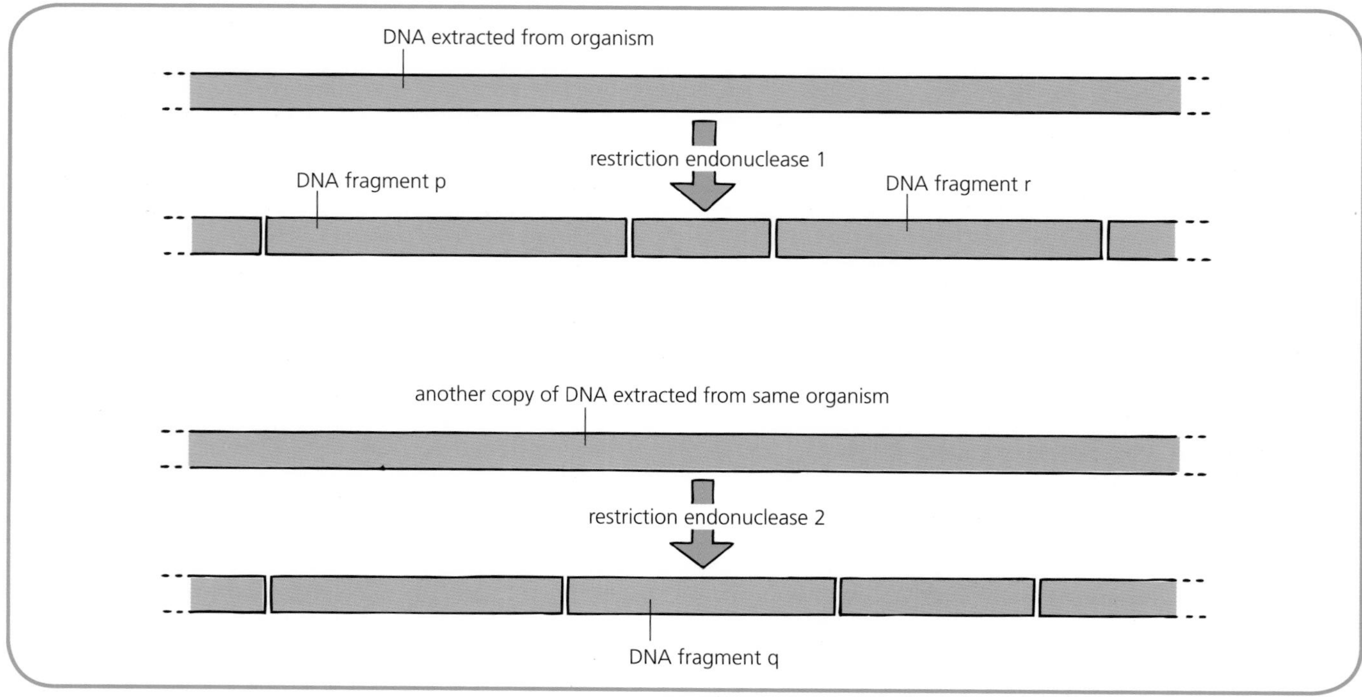

Figure 7.1 Cutting a genome using two different restriction endonucleases

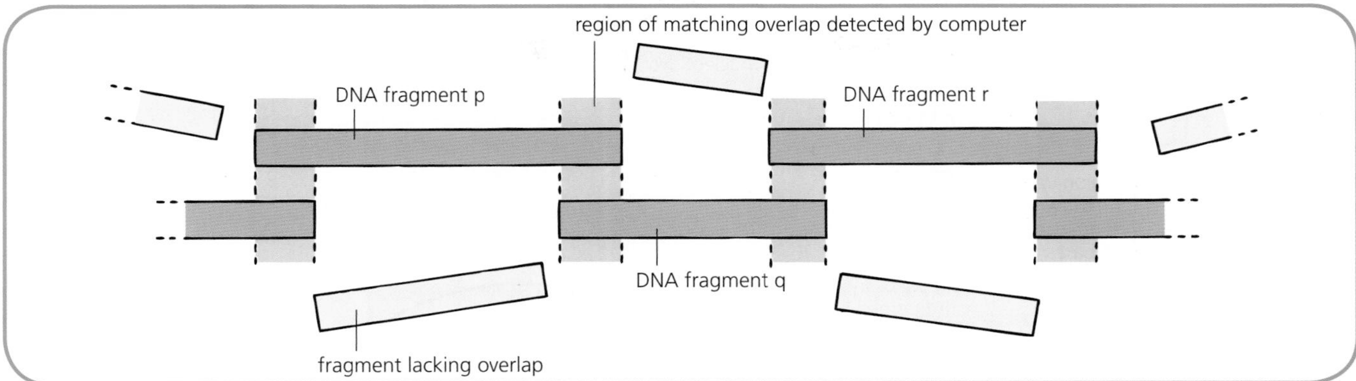

Figure 7.2 Overlapping DNA fragments

In this case the computer would recognise that the sequence of nucleotide bases at the 'end' of fragment p matches the sequence of nucleotide bases at the 'start' of fragment q and that the sequence of bases at the 'end' of q matches the sequence at the 'start' of r. It would therefore deduce that the fragments (and the order of their bases) run in the sequence p–q–r.

The computer analyses all the areas of overlap between the DNA fragments in a sample and is able to compile a **complete genome** based on these overlaps. Since each fragment has had its DNA sequenced, this procedure enables scientists to determine the sequence of bases for individual genes and for entire genomes.

Research Topic Sequencing DNA

A portion of DNA with an unknown base sequence is chosen to be sequenced. Many copies of one of this DNA's strands (the template) are synthesised. Then, in order to make DNA strands that are complementary to these template strands, all the ingredients needed for synthesis are added to the preparation. These include DNA polymerase, primer and the four types of DNA nucleotide as shown in Figure 7.3. In addition the preparation receives a supply of **modified nucleotides** (ddA, ddT, ddG and ddC) each tagged with a different **fluorescent dye**.

Every so often during the synthesis process, a molecule of modified nucleotide just happens to be taken up instead of a normal one. However, when a modified nucleotide is incorporated into the new DNA strand, it brings the synthesis of that strand to a halt because a modified nucleotide does not allow any subsequent nucleotide to become bonded to it. Provided that the process is carried out on a large enough scale, the synthesis of a complementary strand will have been **stopped at every possible nucleotide position** along the DNA template.

The resultant mixture of DNA fragments of various lengths (each with its modified nucleotide and its unique fluorescent tag) is separated using **electrophoresis**. In this process the smallest (shortest) fragments travel the furthest distance. The identity and sequence of nucleotides (as indicated by their fluorescent dyes) is then read for the complementary DNA using this separation. From this information the sequence of the bases in the original DNA can be deduced.

This process has been automated by linking the detection of the four fluorescent dyes to a computer. As these are monitored, the computer, working as an **automated sequence analyser**, processes the information and rapidly displays the sequence of bases in the DNA sample as a series of peaks (see Figure 7.4).

→

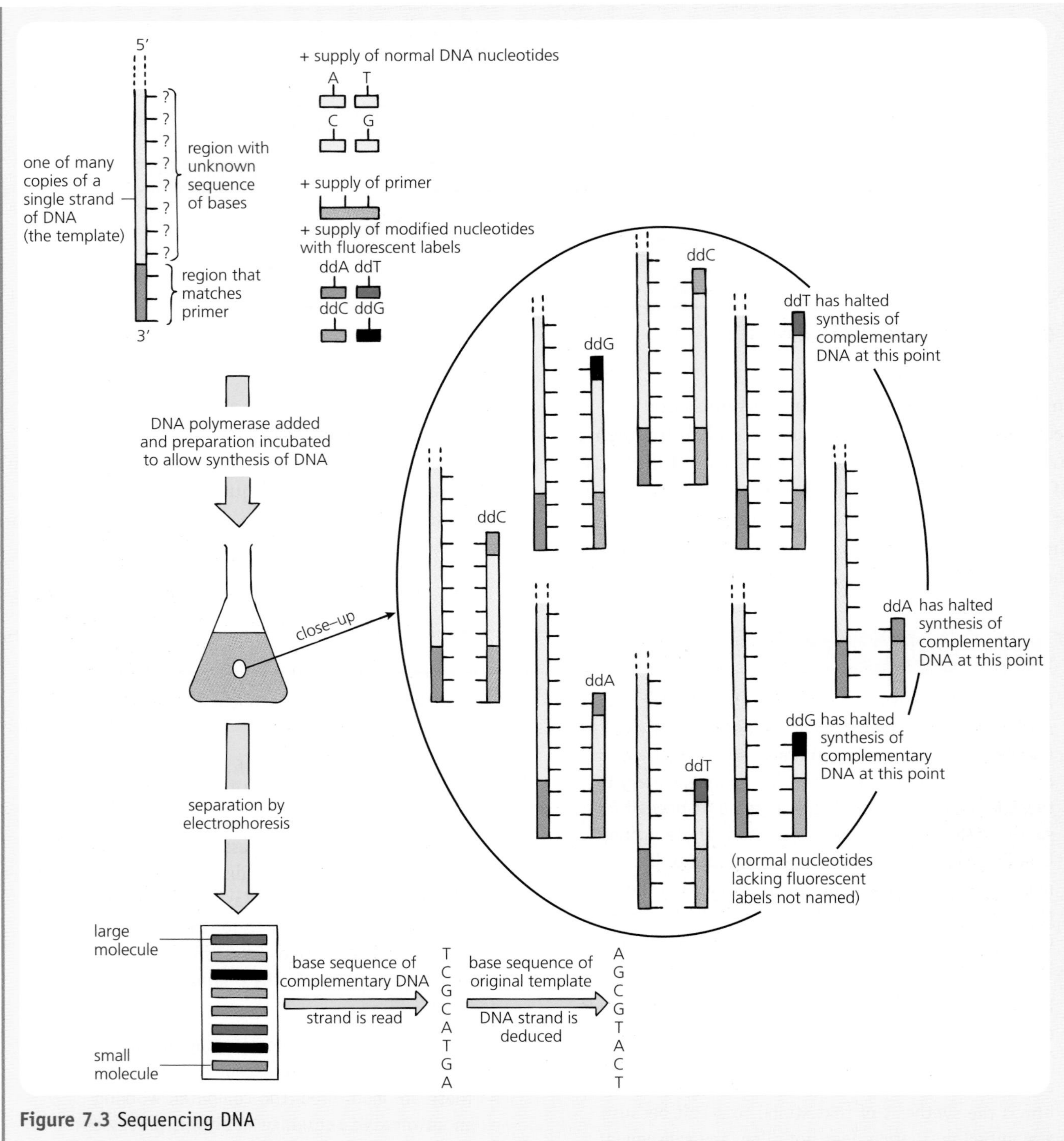

Figure 7.3 Sequencing DNA

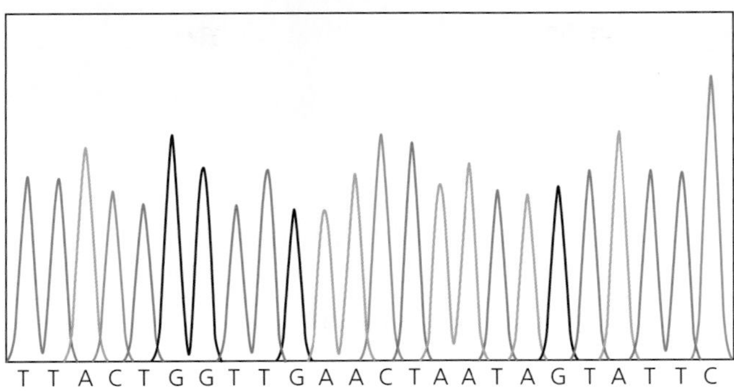

Figure 7.4 Printout from DNA sequence analyser

Genomics

A milestone in human history was reached in 2003 when the DNA sequence of the **human genome** was completed. It is based on the combined genome of a small number of donors and is regarded as the reference genome. It was sequenced using the **shotgun method**.

In addition to sequencing the human genome, part of the project's remit was to determine the nucleotide sequence of the genome of a range of other species. Many have now been sequenced and they include the following.

- A large number of **viruses** and many **bacteria**, most of which are disease-causing agents. A few of these species of pathogenic bacteria are given in Table 7.1.
- Many **pest species** such as the type of mosquito that acts as a vector for malaria and the unicellular organism (*Plasmodium*) that causes malaria.
- Species called **model organisms** (see Table 7.2) that are important for research because they possess genes equivalent to genes in humans responsible for inherited diseases and disorders. Therefore they may provide understanding of the malfunctioning of these genes and even lead the way to the development of new treatments.

Species of pathogenic bacterium	Number in single (haploid) genome		Disease or disorder caused
	Base pairs	Genes	
Chlamydia trachomatis	1 042 519	936	sexually transmitted disease
Rickettsia prowazekii	1 111 523	834	typhus
Treponema pallidum	1 138 011	1 039	syphilis
Campylobacter jejuni	1 641 481	1 708	food poisoning
Helicobacter pylori	1 667 867	1 589	stomach ulcers
Neisseria meningitidis	2 272 351	2 221	meningitis
Propionibacterium acnes	2 560 265	2 333	acne
Vibrio cholerae	4 033 460	3 890	cholera
Mycobacterium tuberculosis	4 411 532	3 959	tuberculosis

Table 7.1 Genomes of pathogenic bacteria

Model organism	Number in single (haploid) genome		Notes
	Base pairs	Genes	
Escherichia coli	4 639 221	4 377	Bacterium that is an important model for molecular biology, genetics and biotechnology.
Saccharomyces cerevisiae	12 495 682	5 770	Unicellular yeast that has many genes in common with humans therefore it has potential for research as a model for eukaryotes.
Caenorhabditis elegans	100 258 171	c. 20 500	Tiny multicellular worm that acts as a simple model for multicellular organisms in the study of genetics and molecular aspects of development, nerve functioning and ageing.
Arabidopsis thaliana	115 409 949	c. 28 000	Small flowering plant that acts as a model for other plants in the study of genetics and molecular aspects of plant development.
Drosophila melanogaster	122 653 977	c. 17 000	Fruit fly that has many genes in common with those that cause disease in humans therefore it is useful in research programmes.
Mus musculus	3.4×10^9	c. 23 000	Mouse that possesses many genes present in humans. Useful information obtained using 'knock-out' mice where a gene is deleted or replaced with a mutant allele to investigate the effect.

(c. = approximately)

Table 7.2 Genomes of model organisms

Research Topic Importance of *Fugu* genome

The genome of *Fugu rubripes*, the pufferfish (see Figure 7.5), has been sequenced completely. Its genome is one of the smallest found among the vertebrates. It is more than seven times smaller than that of humans yet its gene number is higher than that of humans (see Table 7.3).

The *Fugu* genome has a higher density of genes because it possesses **fewer introns** and very **little repetitive DNA**. By comparison the human genome is gene-sparse. The difference between the two is the result of a high rate of **deletion** in the chromosomes of *Fugu* which is largely confined to the intron regions of DNA.

Figure 7.5 Pufferfish

Organism		Genome size (Mb)	Approximate number of protein-coding genes
Scientific name	Common name		
Fugu rubripes	pufferfish	393	31 000
Rattus norvegicus	rat	2 750	30 200
Homo sapiens	human	2 900	20 000

Note: 1 megabase (Mb) = 1×10^6 bases

Table 7.3 Genomes of three vertebrates

➜

The genome of *Fugu* is important because:

- many of its protein-coding genes and regulatory gene sequences are equivalent to those in the human genome
- the compact nature of its DNA makes it easier for scientists to locate the genes and their regulatory sequences for study.

It is likely that decoding the functions of *Fugu*'s genes will in turn increase the understanding of the functions of many human genes.

Comparative genomics

This branch of genomics compares the sequenced genomes of:

- members of different species – for example, disease-causing micro-organisms – to investigate whether they have important genetic sequences in common
- members of the same species – for example, the harmless strain of *E. coli* with the strain that causes serious food poisoning to discover which genetic sequences cause illness
- cancerous cells and normal cells from the same individual to try to discover the specific 'driver' mutations that cause a healthy cell to divide uncontrollably and form a tumour.

Differences in genome

A variation in DNA sequence that affects a single base pair in a DNA chain is called a **single nucleotide polymorphism** (SNP). SNPs are one of the ways in which genomes are found to differ from one another. For example, the DNA of two people might differ by the SNP shown in Figure 7.6. This difference has arisen as a result of a point mutation where one base pair has been **substituted** for another.

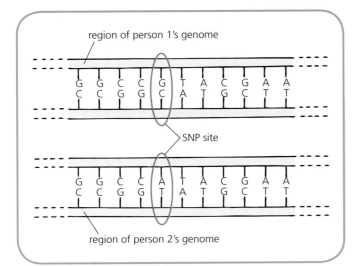

Figure 7.6 Single nucleotide polymorphism (SNP)

The use of bioinformatics has enabled scientists to catalogue more than a million SNPs and specify their exact locations in the human genome. They believe that this SNP map will help them to identify and understand the workings of genes associated with diseases. Therefore SNPs are regarded as a valuable tool in research and may aid the development of future treatments for genetic disorders.

Similarities in genome

In addition to displaying significant differences, a close comparison of genomes often reveals important similarities. For example, they may show a high level of **conservation**. This means that the same or very similar DNA sequences are present in the genomes. Much of the genome is found to be highly conserved in a wide variety of organisms and across species boundaries. Among the most **highly conserved genetic sequences** are those that code for the active sites of essential enzymes and have been positively selected over time.

Humans and whales are very distant relatives. Their evolutionary paths diverged about a hundred million years ago. However, when their genomes are compared, the base sequences of many of their genes are found to be very similar. Figure 7.7 compares a section of the male-determining gene in both species. Apart from four point mutations, the sequence of DNA bases that they received from their common ancestor has been accurately conserved over evolutionary time.

Highly conserved DNA sequences can be used in comparisons of genomes of two groups to find out how **close** or **distant** their relationship is. The greater the number of conserved DNA sequences that their genomes have in common, the more closely related the two groups that possess them.

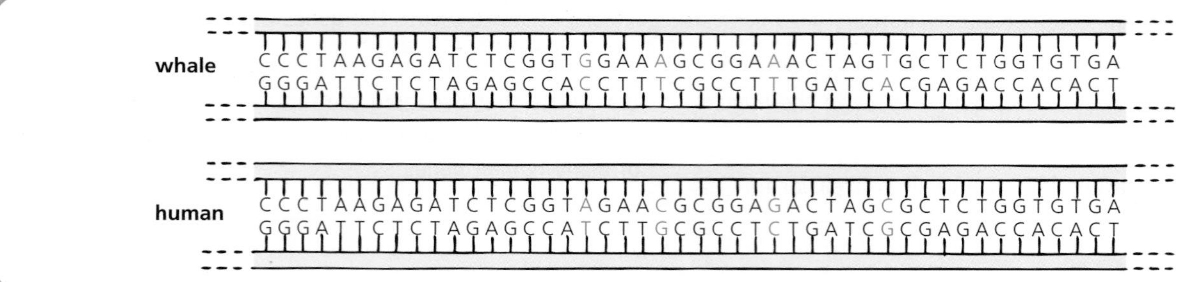

Figure 7.7 Part of male sex-determining gene in two species

Related Topic

Shared genes

Certain biochemical processes, such as respiration, occur in all living things. These processes require **similar genes** to code for the functional enzymes that operate the basic metabolic pathways involved. Therefore, although two organisms such as yeast and human may appear very different in structure, they have much in common at a cellular level and have many shared genes. Around 33% of yeast's genes have survived millions of years of evolutionary change and are conserved in the human genome.

Caenorhabditis elegans is a tiny transparent nematode worm. An adult's body consists of 959 cells of which 302 are nerve cells. *C. elegans* was the first multicellular eukaryote to have its entire genome sequenced. It possesses about 20 500 genes of which about 19 000 are protein-coding. Humans possess about 20 000 protein-coding genes and share approximately 7000 of these with *C. elegans*.

In humans about 300 genes are implicated in diseases. Of these about 42% are found in an equivalent form in *C. elegans*. These genes include the ones closely associated with Alzheimer's disease, colon cancer, spinal muscular atrophy and many other diseases and disorders. The effect of the absence of the product normally coded for by one of these genes can be investigated by inactivating and **silencing** the equivalent gene in *C. elegans* and finding out what happens. By using *C. elegans* as a model organism in this way, it is hoped that a greater understanding of the biological functioning of these genes will be obtained and that it will in turn lead to the development of treatments.

Drosophila melanogaster, the fruit fly, has about 17 000 genes in its genome. Although fewer in number than *C. elegans*, they show a similar diversity of function. The fruit fly has many genes in common with vertebrates. It has been used by scientists to determine developmental and neurological pathways conserved through evolutionary time from invertebrates to humans.

The fruit fly also possesses equivalent versions of about 75% of the 300 genes in the human genome known to be involved in diseases and disorders. By disrupting these genes in fruit flies (and the proteins that they code for), it is possible to investigate some of the events that may also occur in humans. Therefore *Drosophila melanogaster* is also a valuable model organism in the study of diseases that affect humans and in the development of new treatments.

Related Topic

Comparison of human and chimp genomes

The genomes of humans and chimpanzees (our closest living relatives) are very similar. A chimpanzee has **24** chromosomes in a single copy of its genome; a human has **23** (see Figure 7.8). This difference is thought to be the result of the fusion of two ancestral ape chromosomes during evolution.

Analysis of the difference in base sequence of the two genomes reveals that humans and apes have about **98.5%** of their DNA in common. Fossil evidence shows that they shared a common ancestor until about six million years ago when the two groups diverged.

Around 600 genes, common to both chimps and humans, have been identified that are thought to have been conserved by **positive selection**. Many of these genes are known to be involved in defence of the body by the immune system against pathogenic micro-organisms. For example, one of them codes for granulysin, a membrane-disrupting protein that destroys microbes such as *Mycobacterium tuberculosis*.

Despite the very close similarities in genomes, some important differences exist between the two species. Several key genes that give resistance to the parasite that causes sleeping sickness are completely deleted from the chimpanzee genome. Humans, on the other hand, appear to have lost the function of a gene that produces an enzyme thought to give protection against Alzheimer's disease. Most of the differences in genome between humans and chimps take the form of gene duplications and single base-pair substitutions.

Of all the genetic differences that have arisen, one of the most striking involves a gene called HAR1. In the chimpanzee and the chicken, this gene is an exact match for 116 out of 118 bases, amounting to a total of two changes over a period of 310 million years. However, in the chimpanzee and the human genomes, this gene is a match in only 100 out of 118 bases, amounting to 18 changes over 6 million years.

The HAR1 gene encodes a form of RNA needed by a region of the brain for its proper development. The same cells with the active HAR1 gene also produce a protein that is essential to regulate proper **neural development** of the brain's cerebral cortex. It is thought that the genetic differences between the chimp and the human versions of the HAR1 gene account in part for the development of the more **advanced brain** present in humans. These differences in genome tend to support the idea that humans are 'naked apes with big brains'.

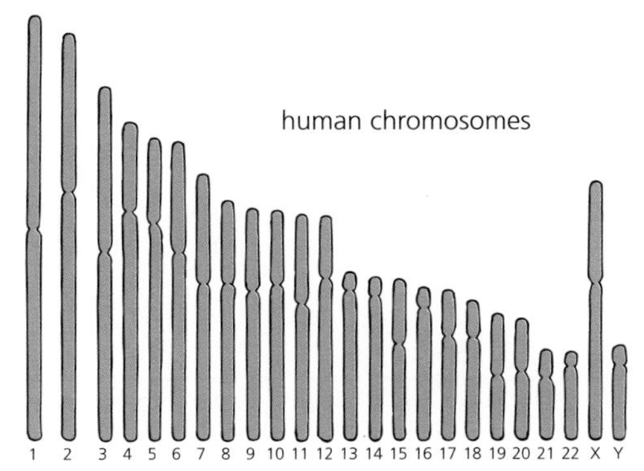

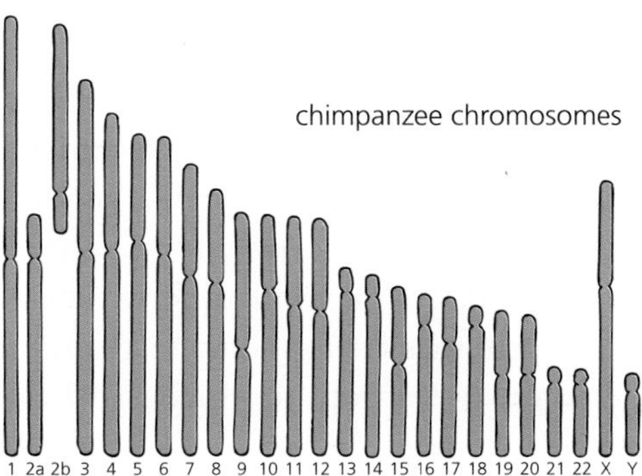

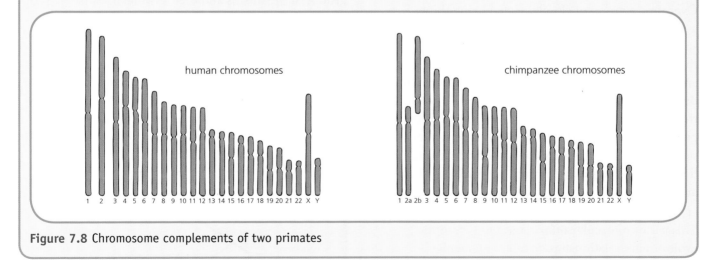

Figure 7.8 Chromosome complements of two primates

Testing Your Knowledge 1

1 a) What information is obtained by the process of genomic sequencing? (1)
b) i) What is meant by the term *bioinformatics*?
ii) What use do biologists make of bioinformatics? (2)

2 a) Name TWO examples of *model* organisms whose genomes have been sequenced. (2)
b) Explain why model organisms are important for research. (2)

3 What is the purpose of comparing the genome of a harmless strain of *E. coli* with that of a pathogenic strain? (2)

4 a) 'Much of the genome is highly conserved across different organisms.' Explain what this statement means. (1)
b) Suggest why a genetic sequence responsible for an enzyme used in photosynthesis is highly conserved among plant species. (1)

Phylogenetics

The study of evolutionary history and relatedness among different groups of organisms is called **phylogenetics**. It makes use of information obtained from comparisons of genome sequence data to deduce phylogenies (sequences of events involved in a group's evolution) and to construct phylogenetic trees (diagrams that show evolutionary relationships). Closely related species are found to have genomes that are very similar in the sequence of their nucleotide bases.

Divergence

Over time a group of closely related living things acquires its own set of **mutations** (such as nucleotide substitutions) which gradually alter its genome. If the group gives rise to two groups that become more and more different from one another and eventually **diverge**, then changes occur in each group's genome that are distinct from those occurring in the other group's genome. Therefore the more different the base sequences of two genomes are found to be, the more **distantly related** the two groups to which they belong and vice versa.

This use of molecular information to determine evolutionary relationships is called **molecular phylogenetics**. It has the advantage over the sole use of structural features for this purpose in that it can be used in a wider range of situations. These vary from the comparison of two groups that are not structurally alike (but may be related genetically) to those that are physically indistinguishable (including many species of bacteria).

When the base sequences for a selection of genes from two different groups are compared and found to differ by only a few bases, this suggests that the groups share a **common ancestor** and that they diverged fairly recently. The greater the number of differences, the longer the time since the point of divergence.

The number of differences per unit length of DNA sequence (in other words, quantity of genetic change) between the two genomes is regarded as a measure of **evolutionary distance** between the two groups. These distances can then be used to construct a **phylogenetic tree** which shows the probable evolution of related groups of organisms and their phylogenetic patterns of divergence.

An example of a phylogenetic tree is shown in Figure 7.9. It is based on differences between several genetic sequences possessed by five related species. In this tree the length of a branch represents the number of changes that have occurred in that species' genome

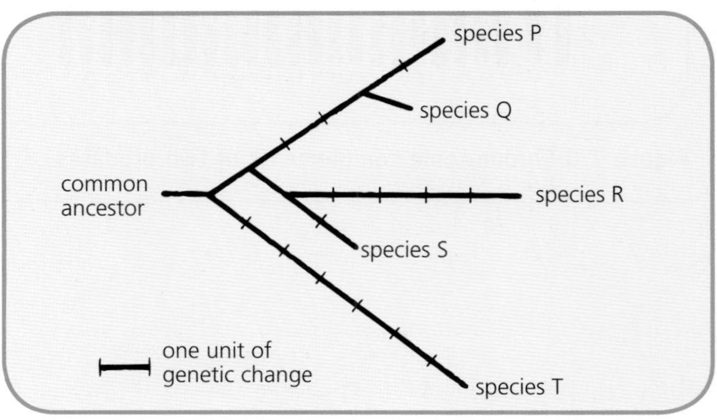

Figure 7.9 Phylogenetic tree

compared with the others. Species P and Q are the most closely related and species T is the most distantly related to the others.

Molecular clocks

Molecules of nucleic acid (and the proteins that they code for) gradually change over time as they are affected by mutations such as nucleotide substitutions. When equivalent genetic sequences for two related groups of organisms (that are known from fossil evidence to have diverged at a certain point in geological time) are compared, the **number of nucleotide substitutions** by which they differ is regarded as being proportional to the **length of time** that has elapsed since the groups diverged. In other words, the quantity of molecular difference that exists between the two groups is a measure of how long ago it is since they shared a common ancestor.

Therefore a molecule of nucleic acid (or a protein coded for by the nucleic acid) can be regarded as a **molecular clock**. It can be given an actual time scale by graphing the number of molecular differences it has evolved against a time scale based on fossil records. Molecular clocks assume a constant mutation rate. They are used as tools to try to date the **origins** of groups of living things and to determine the **sequence** in which they evolved.

Figure 7.10 shows the use of **α-globin** (a polypeptide present in haemoglobin) as a molecular clock. Each lettered point on the graph refers to a comparison of two groups of organisms. For example, the groups compared at point A have nine molecular differences between their versions of α-globin and diverged from one another about 450 million years ago. On the other hand, the groups compared at point E have only two differences and shared a common ancestor until about 100 million years ago when they diverged.

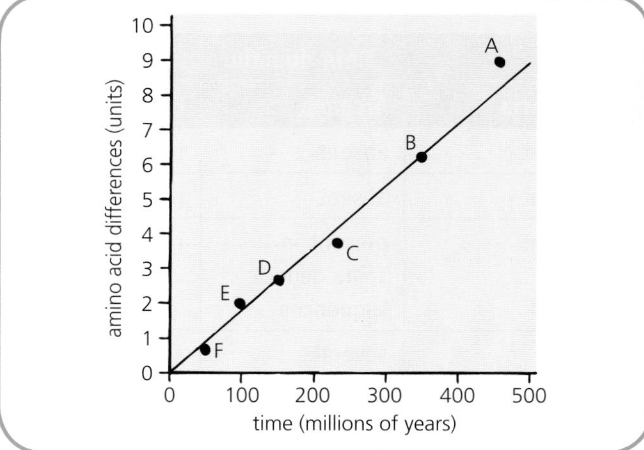

Figure 7.10 α-globin as a molecular clock

Molecular versus structural

Comparison of dolphins and bats using molecular clocks shows the two groups to be more closely related than sharks and tuna fish. This agrees with fossil evidence that suggests that sharks and tuna have been evolving along separate lineages for longer than dolphins and bats. In this case, changes in the molecules that have acted as molecular clocks give a more accurate indication of underlying evolutionary relationships than the changes that have occurred to the animals' body structure.

Shortcomings

Use of a molecular clock assumes that the mutation rate affecting that type of molecule has been relatively constant over time. Although this is often the case for closely related species, it may not be true for groups that diverged early in the evolution of life on Earth. Therefore molecular clocks are less reliable for use in dating the origins of distantly related groups.

Three domains of living things

RNA and DNA sequences have been used to trace the primary evolutionary lineages of all living things. This work was based largely on comparisons of nucleotide sequences of **ribosomal RNA** (rRNA) from many organisms. rRNA is used as a molecular clock for constructing phylogenies because the genes that code for rRNA are ancient, have experienced little or no horizontal gene transfer and are possessed by all living things. Molecular evidence obtained in this way has been used to construct a phylogenetic tree as shown in Figure 7.11. It supports the idea that living things are made up of three main **domains**:

- the **bacteria** (traditional prokaryotes)
- the **archaea** (mostly prokaryotes that inhabit extreme environments such as hot springs and salt lakes)
- the **eukaryotes** (fungi, plants and animals).

The three domains are compared in Table 7.4. Bacteria and archaea are found to be as different genetically as bacteria and eukaryotes. Sequencing of genomes has

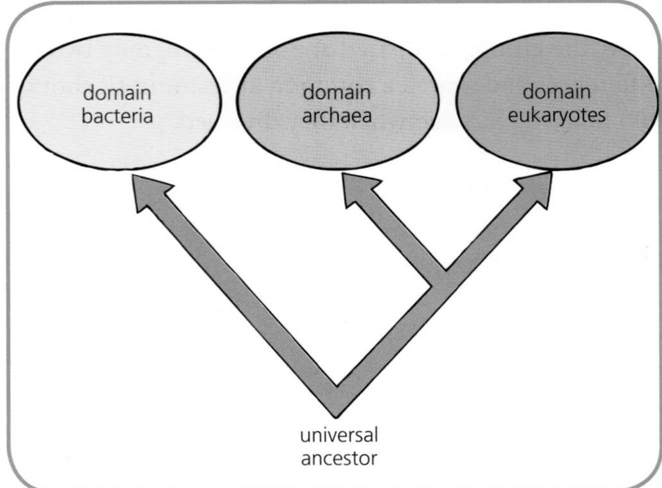

Figure 7.11 Three main domains

shown the archaean genes involved in DNA replication, transcription and translation to resemble much more closely those of eukaryotes than those of bacteria.

The deep evolutionary divisions that separate bacteria and archaea were not obvious from their phenotypes and only came to light following comparison of their rRNA. This work also revealed that the genetic material present in mitochondria and chloroplasts originated in prokaryotes.

Sequence of evolutionary events

Fossils

Fossilisation normally involves the conversion of hard parts of the body such as bone, teeth or shells into rock. The age of a fossil can be determined by estimating the age of the rock of which it is composed. The older the rock, the less radioactivity it emits.

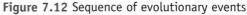

Combined evidence

Scientists have used a combination of genome sequence data and fossil evidence to work out the **sequence** in which key events in evolution have taken place. The evidence strongly supports the theory that living things have undergone a series of **modifications** from the first emergence of life on Earth through to the present day, gradually becoming more and more **complex** as evolution has progressed. Some of the important events are shown in Figure 7.12.

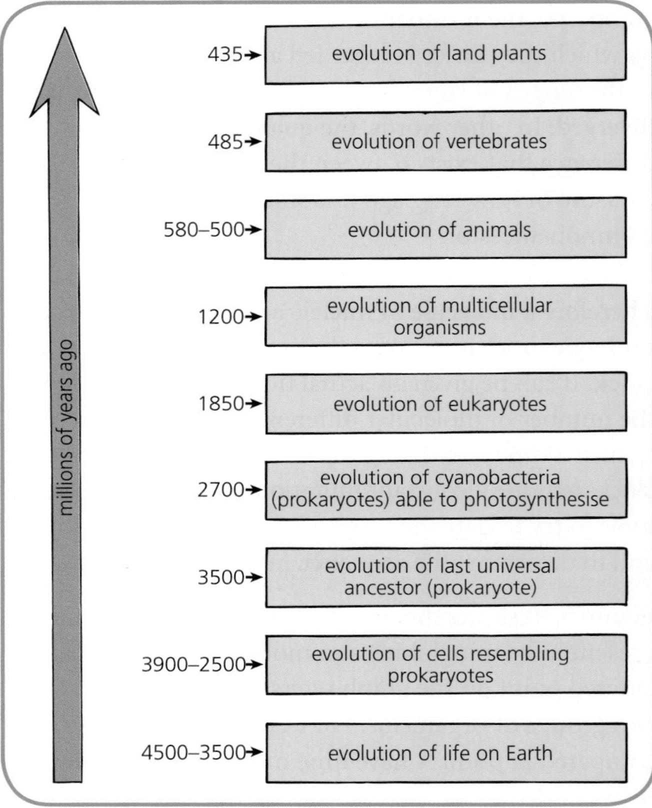

Figure 7.12 Sequence of evolutionary events

Characteristic	Main domain		
	Bacteria	**Archaea**	**Eukaryotes**
true nucleus bound by double membrane	absent	absent	present
membrane-enclosed organelles (such as mitochondria)	absent	absent	present
introns	absent	present in some genetic sequences	present
number of types of RNA polymerase	one	several	several
normal response to streptomycin (an antibiotic)	growth inhibited	growth not inhibited	growth not inhibited

Table 7.4 Comparison of three main domains

Personal genomics

The branch of genomics involved in sequencing an individual's genome and analysing it using bioinformatic tools is called **personal genomics**. As a result of advances in computer technology, the process of sequencing DNA is rapidly becoming **faster** and **cheaper**. Routine sequencing of an individual's DNA for medical reasons will soon become a real possibility. In years to come, a person's entire genome may be sequenced early in life and stored as an electronic medical record available for future consultation by doctors when required.

Genetic variation

Many different types of **variation in genome** are found to occur among the members of a human population. These differences are largely the result of **mutations** and **rearrangements** of parts of the sequence of bases. They range from completely missing or extra chromosomes to single nucleotide changes (such as single nucleotide polymorphisms – see page 91) in a protein-coding gene or non-coding regulatory sequence.

Harmful and neutral mutations

Having located the mutant variants present in the genome, it is often difficult to distinguish between those altered genetic sequences that are genuinely **harmful** (for example, those which fail to code for an essential protein) and those that are **neutral** (in other words have no negative effect).

Genetic disorders

A **genetic disorder** or **disease** is the result of a variation in genomic DNA sequence. The challenge for scientists is to establish a **causal link** between a particular mutant variant in a genomic sequence and a specific genetic disease or disorder. When this is achieved, the disease or disorder is said to be **molecularly characterised**.

The causal genetic sequence has been identified, at least in part, for around 2200 genetic disorders and diseases in humans. However, this does not mean that it is a simple matter to produce treatments for these disorders.

The nature of disease is highly complex. Most medical disorders depend on both **genetic** and **environmental** factors for their expression, though the specific effects of these are not fully understood.

Pharmacogenetics and personalised medicine

Pharmacogenetics is the use of genome information in the choice of drugs. Already it is known that one in ten drugs (including the blood thinner warfarin) varies in effect depending on differences such as SNPs in the person's DNA profile. In the future it may be possible to customise medical treatment to suit an individual's exact metabolic requirements. The most **suitable drug** and the **correct dosage** would be prescribed as indicated by personal genomic sequencing (and *not* as shown in Figure 7.13!). Ideally this advance would increase drug effectiveness while reducing side effects and the 'one-size-fits-all' approach would be consigned to history.

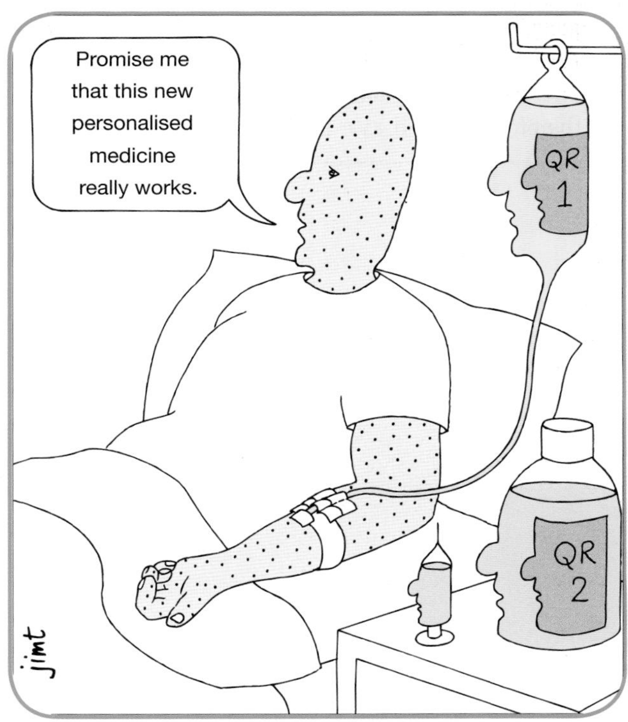

Figure 7.13 'Personalised' medicine

Risk prediction

Already variations in DNA have been linked to conditions such as diabetes, heart disease, schizophrenia and cancer. In the future, when the location in the human genome of many more markers for common diseases and disorders have been established, it should become possible to scan an individual's genome for **predisposition** to a disease and **predict risk** early enough to allow suitable action to be taken. Eventually reduction of risk may be achieved through appropriate drug treatment combined with a healthy lifestyle.

Testing Your Knowledge 2

1 **a)** What is meant by the term *phylogenetics*? (1)

b) Identify the advantage of using molecular phylogenetics rather than structural features to determine the evolutionary relatedness of two groups of bacteria. (2)

2 Rewrite the following paragraph and complete the blanks using the answers given below it. (3)

Two related groups of organisms known from fossil records to have ______ at a certain point in ______ time are chosen. Many genetic ______ for the two groups are compared and the number of nucleotide ______ by which they differ is determined. The quantity of molecular change in their ______ that has occurred is a measure of how long ago the groups diverged from a common ______. The DNA can therefore be used as a molecular ______.

ancestor, clock, diverged, DNA, geological, sequences, substitutions

3 **a) i)** Name the THREE main domains of living things.

ii) Identify TWO features possessed by members of the most recently evolved domain that are absent in members of the other two domains. (3)

b) Draw a flow chart to show the evolution of plants using the following terms:

photosynthetic eukaryotes, last universal ancestor, photosynthetic land plants, photosynthetic prokaryotes, multicellular green plants. (2)

4 **a)** What is meant by the term *personal genomics*? (2)

b) Give TWO possible benefits of personalised medicine to patients in the future. (2)

Applying Your Knowledge and Skills

1 Figure 7.14 shows the DNA fragments which resulted from two copies of part of a genome, each cut by a different restriction endonuclease. The computer found that the four larger fragments possessed overlaps.

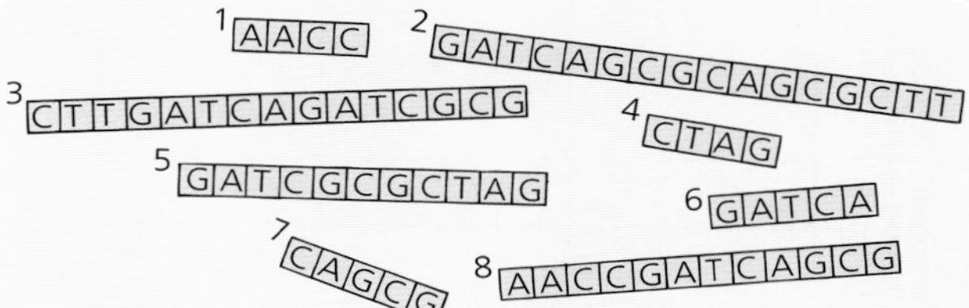

Figure 7.14

Draw a copy of these four larger fragments on squared paper, cut them out and use them to construct this part of the person's genome. (1)

2 The DNA fragments shown in Figure 7.15 were formed during the type of sequencing technology illustrated in Figure 7.3. Each fluorescent tag indicates the point on the strand where replication of complementary DNA was brought to a halt by a modified nucleotide.

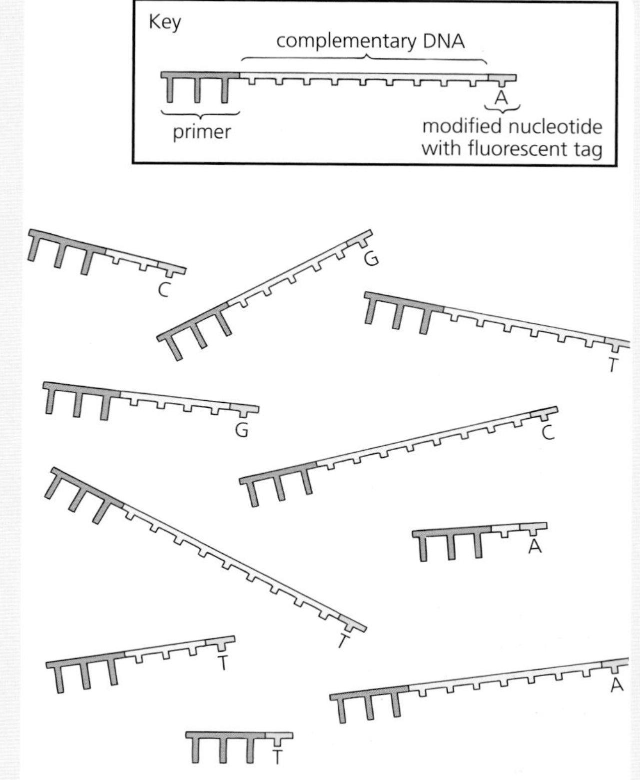

Figure 7.15

Group	Ethnic origin	Number of individuals in group	Set of bases at six sites in genome as a result of SNPs					
			Site 1	Site 2	Site 3	Site 4	Site 5	Site 6
1	W	1	C	C	T	A	T	G
2	W	17	T	C	C	A	C	A
3	W	63	T	T	C	A	C	A
4	W	21	C	T	T	A	T	G
5	X	44	T	C	C	A	C	A
6	X	36	C	T	T	A	T	G
7	X	1	C	C	T	A	T	G
8	X	1	T	T	C	A	C	A
9	Y	47	T	C	C	A	C	A
10	Y	1	T	C	C	G	C	A
11	Y	87	C	T	T	A	T	G
12	Y	1	C	C	T	A	T	G

Table 7.5

→

a) Work out the sequence of the bases in the complementary DNA strand. (1)

b) Deduce the sequence of bases in the original DNA strand. (1)

3 Genetic material from a sample of volunteers of differing ethnic origin was collected and analysed. Table 7.5 summarises the results and shows the SNPs (single nucleotide polymorphisms) that occur at six closely located sites on the genome of these people. The results refer to a single strand of DNA.

a) i) What is a single nucleotide polymorphism?
ii) Which site in the table appears to have been least affected by SNPs? (2)

b) By how many bases at sites 1–6 do the genomes of groups 7 and 10 differ? (1)

c) Which group(s) has the same set of bases at these six sites in their genome as:
i) group 1?
ii) group 2?
iii) group 3? (3)

d) How many people of ethnic origin W have the same genotype as people in group 9? (1)

e) i) Which set of six bases occurs most frequently among the total sample group?
ii) What percentage of the total sample group possesses this set of bases in their genome? (2)

f) Which group has the least common set of bases in its genome? (1)

g) If the set of bases in the genome possessed by group 6 is strongly associated with a fatal disease, which other groups are at equal risk? (1)

h) What TWO things could be done to increase the reliability of the results? (2)

4 The graph in Figure 7.16 shows the use of cytochrome c as a molecular clock.

a) Which groups had a common ancestor until about 400 million years ago? (1)

b) i) Which groups are the most different from one another according to a comparison of their cytochrome c?
ii) When does the fossil record suggest that they diverged? (2)

c) i) Which is thought to have evolved first from a common reptilian ancestor, the birds or the mammals?

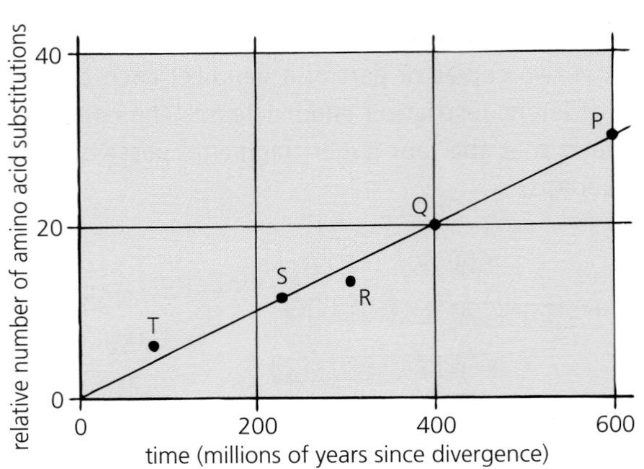

Key

comparison	groups being compared
P	vertebrates and insects
Q	fish and reptiles
R	reptiles and mammals
S	reptiles and birds
T	two mammalian groups

Figure 7.16

ii) Justify your answer with molecular evidence from the graph. (2)

d) Amphibians share a common ancestor with fish. In the absence of fossil evidence, what investigation could be carried out to attempt to estimate their time of divergence using the molecular clock in Figure 7.16? (2)

5 The data in Table 7.6 refer to three micro-organisms.

a) Calculate the values that should have been entered in boxes **(i)**, **(ii)** and **(iii)**. (3)

b) Which micro-organism has the lowest ratio of genome length to gene number? (1)

c) i) Which micro-organism is most likely to have the highest amount of repetitive DNA in its genome?
ii) Explain your answer. (2)

d) To which of the three main domains of living things does each of the three micro-organisms belong? Give ONE reason to support your choice in each case. (3)

→

Micro-organism	Characteristics				
	Length of genome (kb)	Number of genes that code for protein	Ratio of genome length to gene number	Relative number of introns	Nucleus with double membrane
1 *Neurospora crassa*	**(i)**	10 100	3.96 : 1	many	present
2 *Bradyrhizobium japonicum*	9 154.2	**(ii)**	1.1 : 1	none	absent
3 *Methanosarcina acetivorans*	5 750	4 662	**(iii)**	a few	absent

Note: 1 kilobase (kb) = 1×10^3 bases

Table 7.6

6 Read the passage and answer the questions that follow it.

Debrisoquine hydroxylase is an enzyme made by cells in the liver. It is responsible for the breakdown of drugs used to treat a variety of disorders such as nausea, depression and heart disorders once the drugs have brought about their desired effect.

Several alleles of the gene that codes for this enzyme occur among the members of the human population. These alleles code for different versions of the enzyme which, in turn, vary in their ability to metabolise drugs. Depending on their particular genotype, a person may produce no functional enzyme and be a poor metaboliser because both of their alleles are null and void.

If the person has one null allele and one inferior allele that code for a partly functional version of debrisoquine hydroxylase, they are said to be an intermediate metaboliser. An extensive metaboliser has one or two normal alleles which code for the fully functional form of the enzyme. Some people possess more than two copies of the normal allele and their metabolic profile is described as ultra-rapid.

a) Copy and complete Table 7.7 which summarises the passage. (4)

b) What type of mutation could account for an ultra-rapid metaboliser having more than two copies of the allele of the gene that codes for debrisoquine hydroxylase? (1)

c) i) Which group of people are most likely to be at risk of harmful side effects if given a standard dose of a drug normally broken down by debrisoquine hydroxylase?

ii) Explain your answer. (2)

d) i) For which group of people would a standard dose of such a drug probably be ineffective?

ii) Explain your answer. (2)

e) In what way might personalised medicine (pharmacogenetics) solve the problems referred to in questions **c)** and **d)**? (2)

Alleles of gene present in genome	State of enzyme	Person's metabolic profile
	non-functional	
one null allele and one inferior allele		
		extensive
	highly functional	

Table 7.7

What You Should Know

Chapters 6–7
(See Table 7.8 for word bank)

allopatric	diverged	mutations
archaea	domains	natural
asexual	environmental	non-random
bioinformatics	evolution	nucleotide
change	fertile	pharmacogenetics
clocks	frequency	phylogenetics
closely	genomics	risk
conservation	geological	sequencing
customised	horizontally	speciation
decrease	interbreed	stabilising
directional	life	sympatric
disruptive	model	vertical

Table 7.8 Word bank for chapters 6–7

1 The gradual ______ in the characteristics of a population from generation to generation as a result of variations in its genome is called ______.

2 Genetic sequences of DNA are passed down from parent to offspring by ______ inheritance. This occurs as a result of both sexual and ______ reproduction. In prokaryotes, genetic material may also be transferred ______ between one another.

3 Within a population, ______ selection brings about both the non-random increase in ______ of genetic sequences that increase the chance of survival and the ______ decrease in frequency of deleterious sequences that ______ the chance of survival.

4 ______ selection favours the intermediate versions of a trait and acts against the extreme variants. ______ selection favours a version of the trait that was less common and results in a progressive shift of the population's mean. ______ selection favours extreme versions of a trait at the expense of the intermediates.

5 A group of organisms that can ______ and produce ______ offspring belong to the same species.

6 The formation of new species by evolution is called ______. The barriers involved in ______ speciation are geographical whereas those involved in ______ speciation are normally behavioural or ecological.

7 Determining the sequence of ______ bases along an organism's genes or entire genome is called genomic ______. Use is made of ______, involving computing and statistics, to compare sequence data.

8 The study of genomes is called ______. Many genomes have been sequenced including those of ______ organisms important for research. In addition to many differences, comparative genomics shows that a high degree of ______ exists among different organisms.

9 ______ is the study of evolutionary relatedness found among different groups of organisms. The use of sequence data indicates that distantly related groups ______ at an earlier point in ______ time and that more ______ related groups diverged more recently.

10 Molecules such as DNA are affected by ______ and gradually change over time therefore they can be used as molecular ______.

11 Comparison of nucleic acid sequences provides evidence to support the idea that living things fall into three main ______: bacteria, ______ and eukaryotes. A combination of sequence data and fossil evidence has enabled scientists to determine the main sequence of events in the evolution of ______ on Earth.

12 In the future, routine sequencing of an individual's genome may lead to personalised medicine (______). This could involve predicting ______ of disease through knowledge of the person's genome and administering ______ drugs in appropriate dosages. Diseases are complex and are often affected by both genetic and ______ factors.

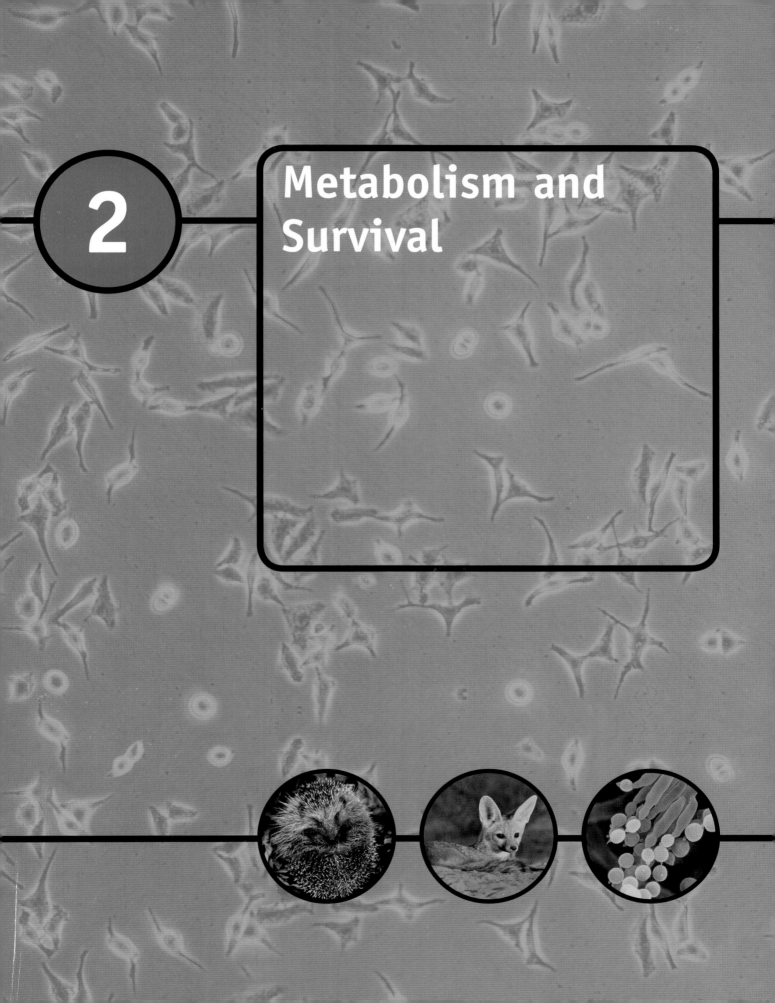

2

Metabolism and Survival

8 Metabolic pathways

Cell metabolism

Cell metabolism is the collective term for the thousands of biochemical reactions that occur within a living cell. The vast majority of these are steps in a complex network of connected and integrated pathways that are catalysed by enzymes.

Metabolic pathways

The biochemical processes upon which life depends take the form of **metabolic pathways** which fall into two categories:

- **Catabolic** pathways which bring about the breakdown of complex molecules to simpler ones usually releasing energy and often providing building blocks.
- **Anabolic** pathways which bring about the biosynthesis of complex molecules from simpler building blocks and require energy to do so.

Such pathways are closely integrated and one often depends upon the other. For example, aerobic respiration in living cells is an example of **catabolism** which releases the energy needed for the synthesis of protein from amino acids (an example of **anabolism**). This close relationship is shown in Figure 8.1. An important chemical called ATP (see Chapter 9) plays a key role in the transfer of energy between catabolic and anabolic reactions.

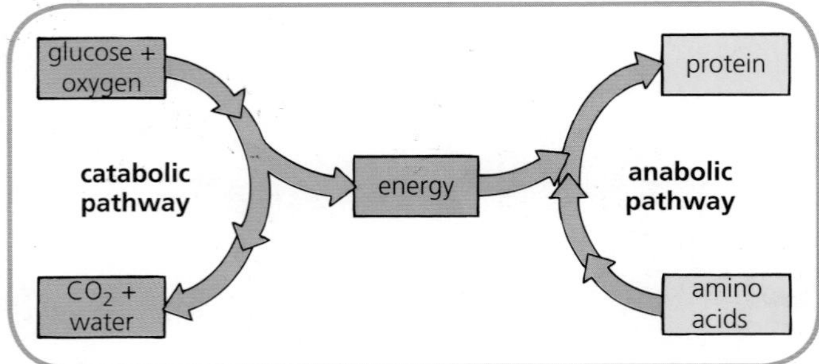

Figure 8.1 Two types of metabolic pathway

Reversible and irreversible steps

Metabolic pathways are regulated by enzymes which catalyse specific reactions. A pathway often contains both **reversible** and **irreversible** steps which allow the process to be kept under precise control. **Glycolysis** (see page 126) is the metabolic pathway that converts **glucose** to an intermediate metabolite called **pyruvate** at the start of respiration. Figure 8.2 shows the first three enzyme-controlled steps in a long pathway.

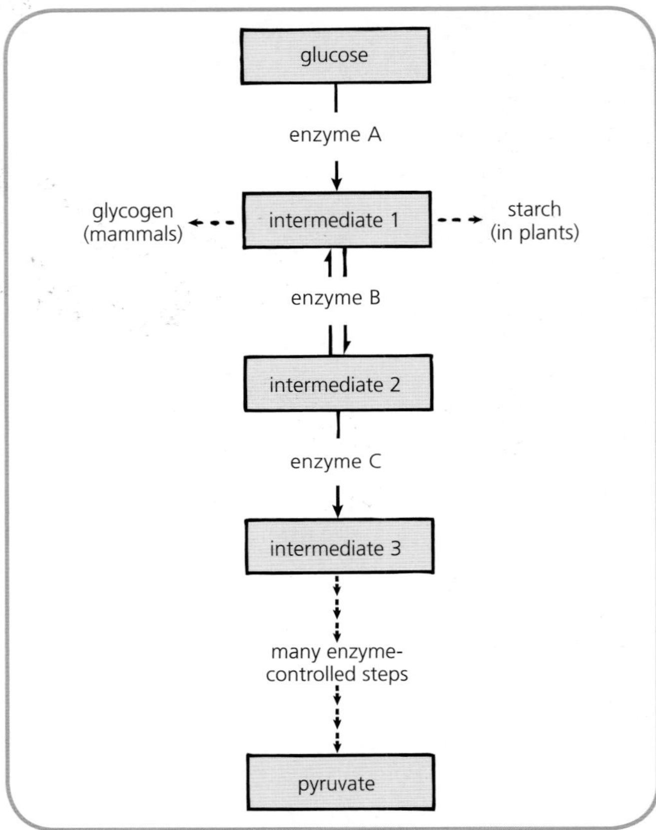

Figure 8.2 Metabolic pathway

Glucose diffusing into a cell from a high concentration outside to a low concentration inside is irreversibly converted to intermediate 1 by enzyme A. This process is of advantage to the cell because it maintains a low concentration of glucose inside the cell and therefore promotes continuous diffusion of glucose into the cell from the high concentration outside.

The conversion of intermediate 1 to intermediate 2 by enzyme B is **reversible**. If more intermediate 2 is formed than the cell requires for the next step then some can be converted back to intermediate 1 and used in an alternative pathway (for example, to build glycogen in animal cells or starch in plant cells). The conversion of intermediate 2 to intermediate 3 by enzyme C is **irreversible** and is a **key regulatory point** in the pathway. There is no going back for the substrate now. It is committed to following glycolysis through all the steps to pyruvate.

Alternative routes

Metabolic pathways may also contain **alternative routes** that allow steps in the pathway to be bypassed. Figure 8.3 shows a pathway from glucose via an intermediate (called sorbitol) that bypasses the steps controlled by enzymes A, B and C but returns to glycolysis later in the pathway. This bypass is used when the cell has a plentiful supply of sugar.

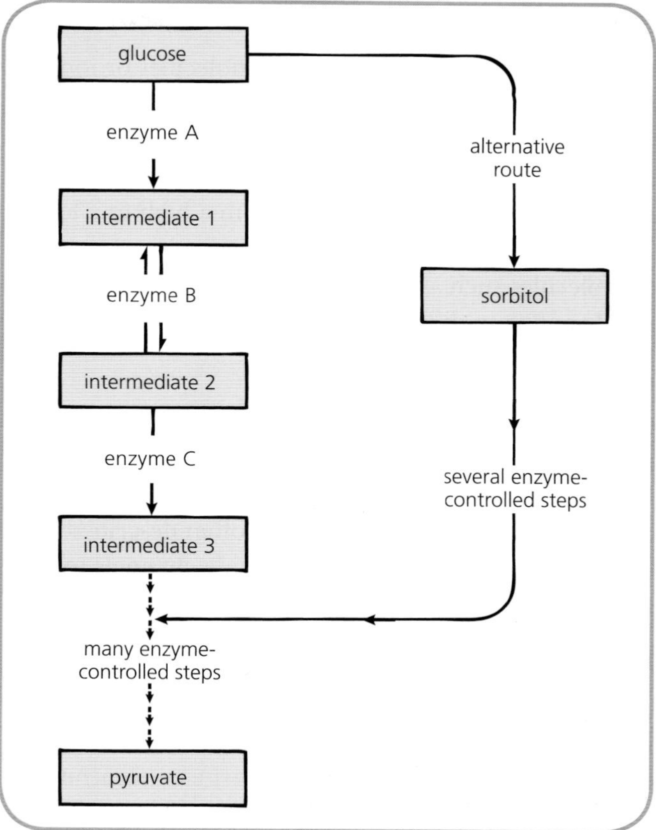

Figure 8.3 Alternative route

Membranes

The cell **membrane** is the boundary that separates the internal living contents of a cell from its external surroundings. The membrane regulates the flow of materials into and out of the cell by allowing selective communication between intracellular and extracellular environments. Cell organelles are also bounded by membranes.

Structure of cell membrane

The cell membrane consists of **protein** and **phospholipid** molecules. Most evidence supports the **fluid mosaic model** of membrane structure (see Figure 8.4). This proposes that the cell membrane consists of a fluid bilayer of constantly moving phospholipid molecules that forms a stable boundary. It contains a patchy mosaic of protein molecules that vary in size and structure.

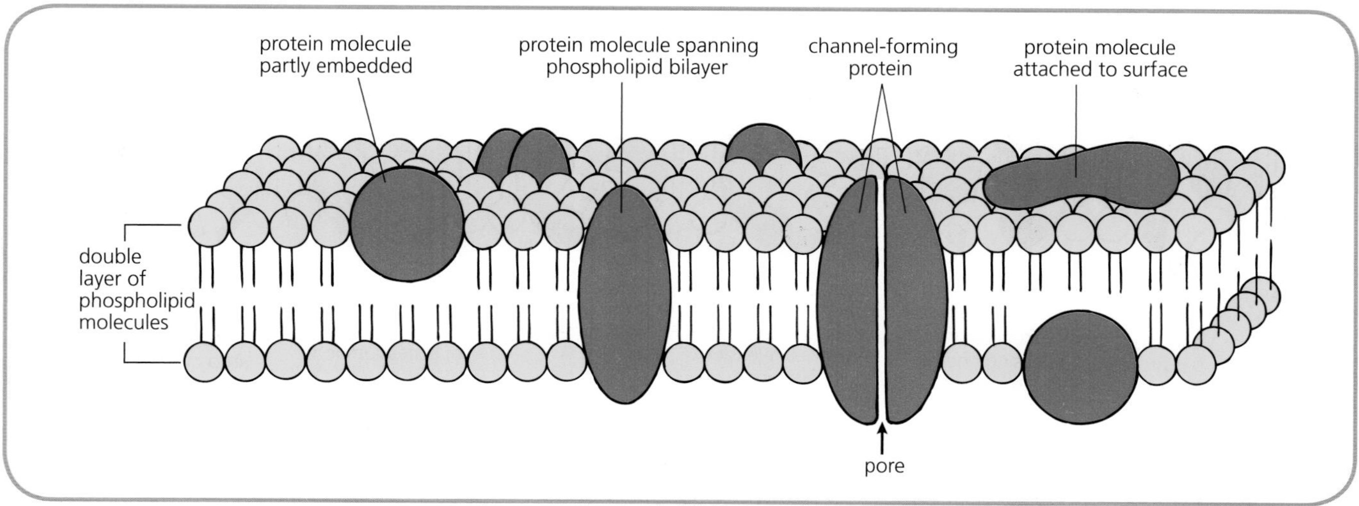

Figure 8.4 Fluid mosaic model of cell membrane structure

Molecular transport

Diffusion is the net movement of molecules or ions from a region of high concentration to a region of low concentration of that type of molecule or ion. During diffusion, molecules or ions always move down a **concentration gradient** from a high to a low concentration. The cell membrane is freely permeable to molecules such as oxygen and carbon dioxide that are small enough to diffuse through the phospholipid bilayer.

Role of protein pores

Larger molecules to be transported depend on certain protein molecules to let them move across the membrane into or out of the cell. These transport molecules contain **pores**. They are often described as channel-forming because they provide channels for specific substances to diffuse across the membrane.

Active transport

Active transport is the movement of molecules and ions across the cell membrane from a low to a high concentration, in other words **against** a concentration gradient. Active transport works in the opposite direction to diffusion and always requires energy.

Protein pumps

Certain protein molecules present in the cell membrane act as **carrier** molecules which recognise specific ions and transfer them across the cell membrane. These active transport carriers are often called **pumps**.

Conditions required by protein pumps

A pump requires energy. Therefore factors such as temperature and availability of oxygen and food, which directly affect a cell's respiratory rate, also affect the rate of active transport.

Enzymes in membrane

Some protein molecules embedded in a membrane of phospholipids are **enzymes** which catalyse the steps in a metabolic process essential to the cell. For example, the enzyme ATP synthase, which catalyses the synthesis of ATP, is a membrane protein present in mitochondria, chloroplasts and prokaryotes.

Related Topic

Sodium/potassium pump

Some protein pumps play a dual role in that they exchange one type of ion for another. An example is the **sodium/potassium pump** where the same carrier molecule actively pumps sodium ions out of the cell and potassium ions into the cell each against its own concentration gradient (as shown in Figure 8.5). Maintenance of the difference in ionic concentration by this pump is particularly important for the proper functioning of muscle and nerve cells. It accounts for around 30% of a typical animal cell's energy expenditure.

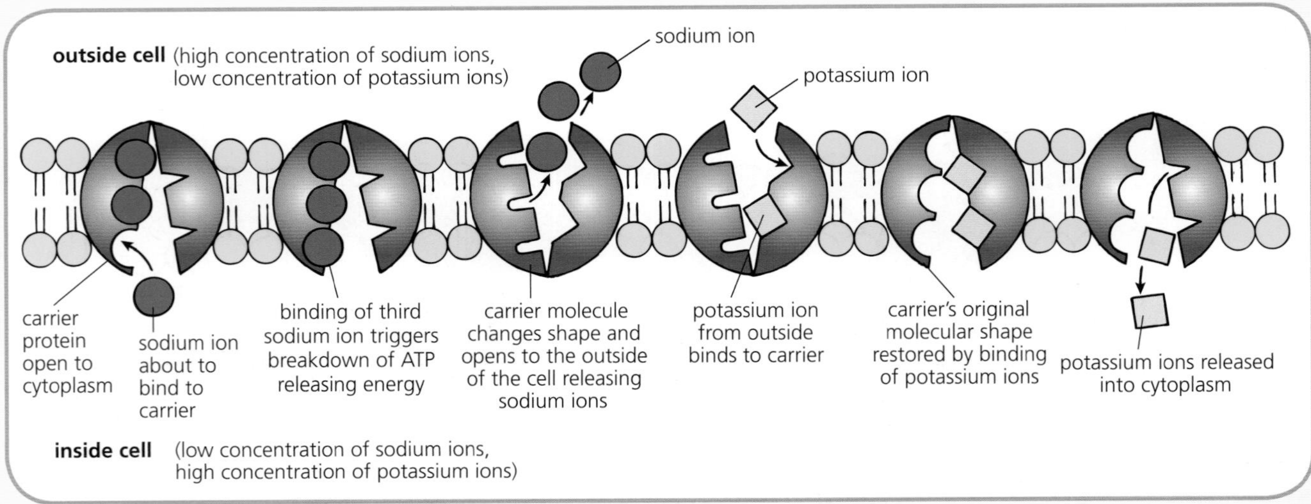

Figure 8.5 Active transport by the sodium/potassium pump

Testing Your Knowledge 1

1 a) Define the term *metabolism*. (2)
   b) Describe TWO ways in which the two types of metabolic pathway differ from one another. (2)
2 Briefly describe the role played by a) protein pores and b) protein pumps present in the membrane that bounds a cell. (4)

Activation energy and enzyme action

The rate of a chemical reaction is indicated by the amount of chemical change that occurs per unit time. Such a change may involve the joining together of simple molecules into more complex ones or the splitting of complex molecules into simpler ones. In either case the energy needed to break chemical bonds in the reactant chemicals is called the **activation energy**.

The bonds break when the molecules of reactant have absorbed enough energy to make them unstable. They are now in the **transition state** and the reaction can occur. This energy input often takes the form of heat energy and the reaction only proceeds at a high rate if the chemicals are raised to a high temperature (see Figure 8.6).

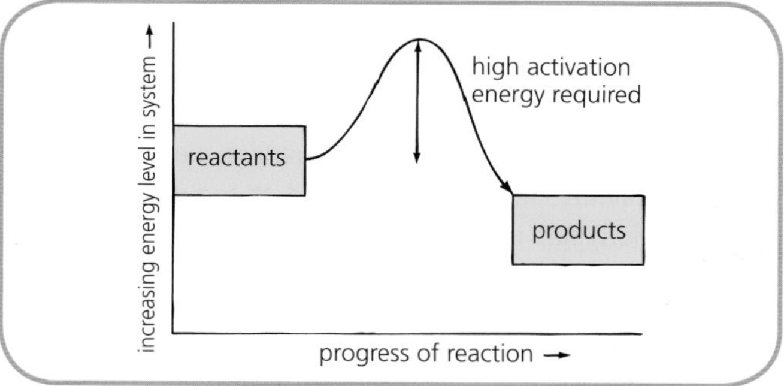

Figure 8.6 Uncatalysed reaction

Related Activity

Investigating the effect of heat on the breakdown of hydrogen peroxide

Hydrogen peroxide is a chemical that breaks down into water and oxygen as shown in the following equation:

hydrogen peroxide	→	water	+	oxygen
$(2H_2O_2)$		$(2H_2O)$		(O_2)

In the experiment shown in Figure 8.7, test tubes containing hydrogen peroxide and drops of detergent are placed in five water baths at different temperatures. The detergent is used to sustain any oxygen bubbles that are released as a froth.

After 30 minutes the tubes are inspected for the presence of a froth of oxygen bubbles which indicates the breakdown of hydrogen peroxide. The diagram shows a typical set of results where the volume of froth is found to increase with an increase in temperature.

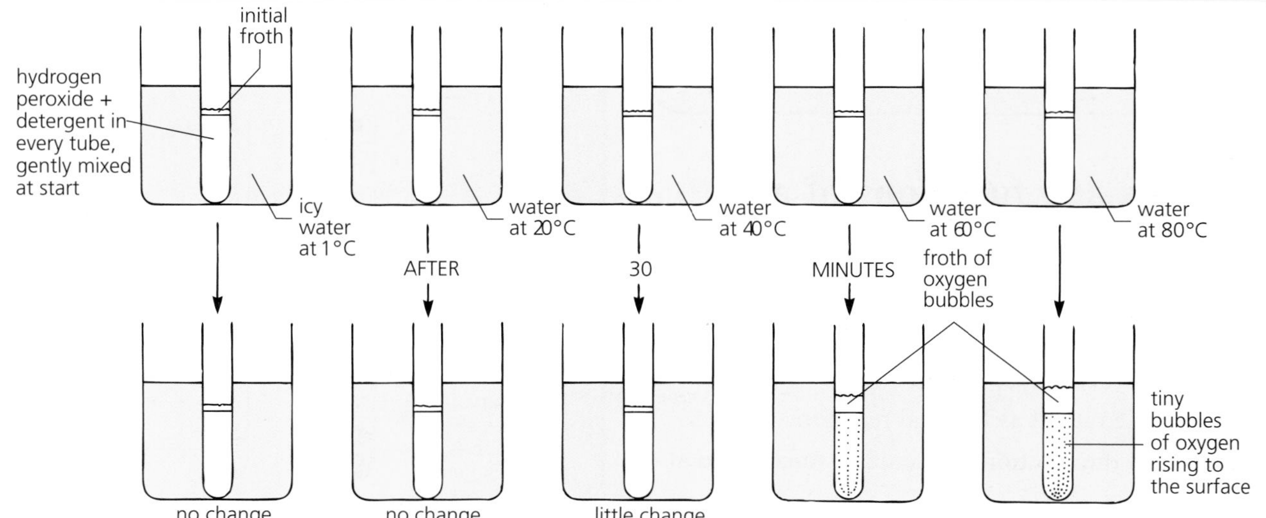

Figure 8.7 Investigating the effect of heat on the breakdown of hydrogen peroxide

Related Activity

Investigating the effect of manganese dioxide on the breakdown of hydrogen peroxide

In the experiment shown in Figure 8.8, the bubbles forming the froth in tube A are found to relight a glowing splint. This shows that oxygen is being released during the breakdown of hydrogen peroxide. In tube B, the control, the breakdown process is so slow that no oxygen can be detected.

It is concluded therefore that manganese dioxide (which remains chemically unaltered at the end of the reaction) has increased the rate of this chemical reaction which would otherwise have only proceeded very slowly. A substance that has this effect on a chemical reaction is called a **catalyst**.

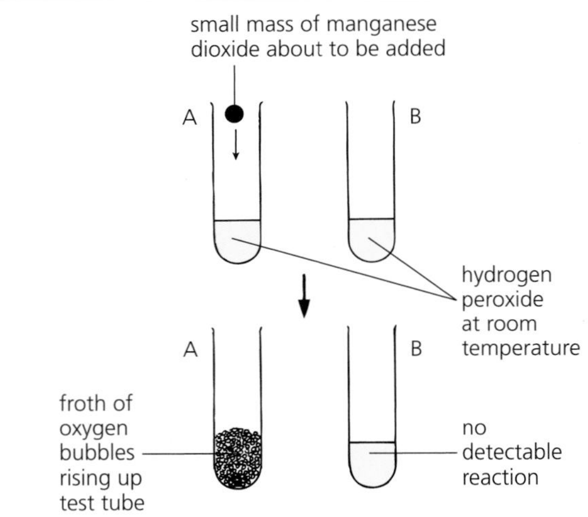

Figure 8.8 Effect of a catalyst

Properties and functions of a catalyst

A catalyst is a substance that:

- lowers the activation energy required for a chemical reaction to proceed (see Figure 8.9)
- speeds up the rate of a chemical reaction
- takes part in the reaction but remains unchanged at the end of it.

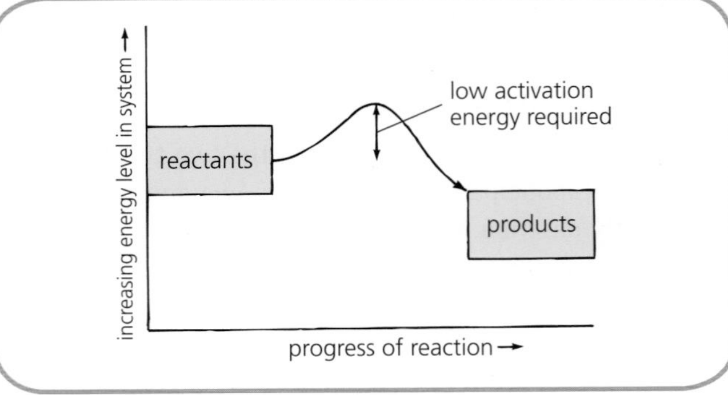

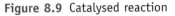

Figure 8.9 Catalysed reaction

Related Activity

Investigating the effect of catalase on the breakdown of hydrogen peroxide

Catalase is an enzyme made by living cells. It is especially abundant in fresh liver cells. In the experiment shown in Figure 8.10 the bubbles produced in tube C are found to relight a glowing splint. This shows that oxygen is being released during the breakdown of hydrogen peroxide as follows:

$$\underset{\text{(substrate)}}{\text{hydrogen peroxide}} \xrightarrow[\text{(enzyme)}]{\text{catalase}} \underset{\text{(end products)}}{\text{water + oxygen}}$$

In tube D, the control, the breakdown process is so slow that no oxygen can be detected. It is concluded that the enzyme catalase has increased the rate of this chemical reaction which would otherwise have proceeded only very slowly.

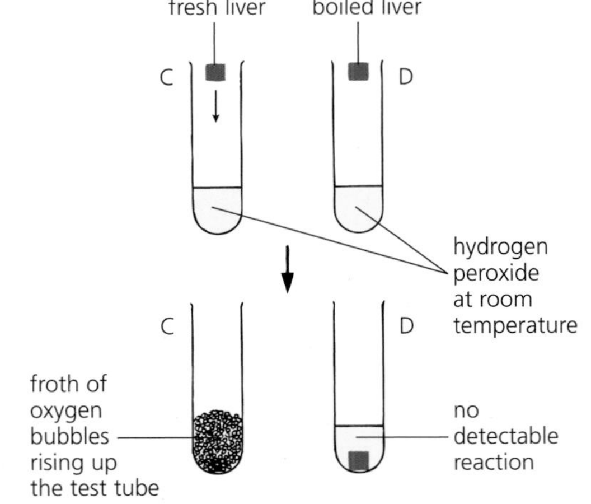

Figure 8.10 Effect of catalase

Importance of enzymes

Living cells cannot tolerate the high temperatures needed to make chemical reactions proceed at a rapid rate. Therefore they make use of **biological catalysts** called **enzymes**.

Enzymes speed up the rate of the reactions in a metabolic pathway by **lowering the activation energy** needed by the reactant(s) to form the transition state. It is from this unstable state that the end products of the reaction are produced.

By this means biochemical reactions are able to proceed rapidly at the relatively low temperatures (such as 5–40 °C) needed by living cells to function properly. In the absence of enzymes, biochemical pathways such as respiration and photosynthesis would proceed so slowly that life as we know it would cease to exist.

Enzyme action

Enzyme molecules are made of **protein**. Somewhere on an enzyme's surface there is a groove or hollow where its **active site** is located. This site has a particular shape which is determined by the chemical structure of, and bonding between, the amino acids in the polypeptide chains that make up the enzyme molecule.

Specificity

An enzyme acts on one type of substance (its **substrate**) whose molecules exactly fit the enzyme's active site. The enzyme is **specific** to its substrate and the molecules of substrate are complementary to the enzyme's active site for which they show a **high affinity** (chemical attraction).

Induced fit

The active site is not a rigid structure. It is **flexible** and **dynamic**. When a molecule of substrate enters the active site, the shape of the enzyme molecule and the active site change slightly making the active site fit very closely round the substrate molecule. This is called **induced fit** (see Figure 8.11). The process is like a rubber glove, slightly too small, exerting a very tight fit round a hand. Induced fit ensures that the active site comes into very close contact with the molecules of substrate and increases the chance of the reaction taking place.

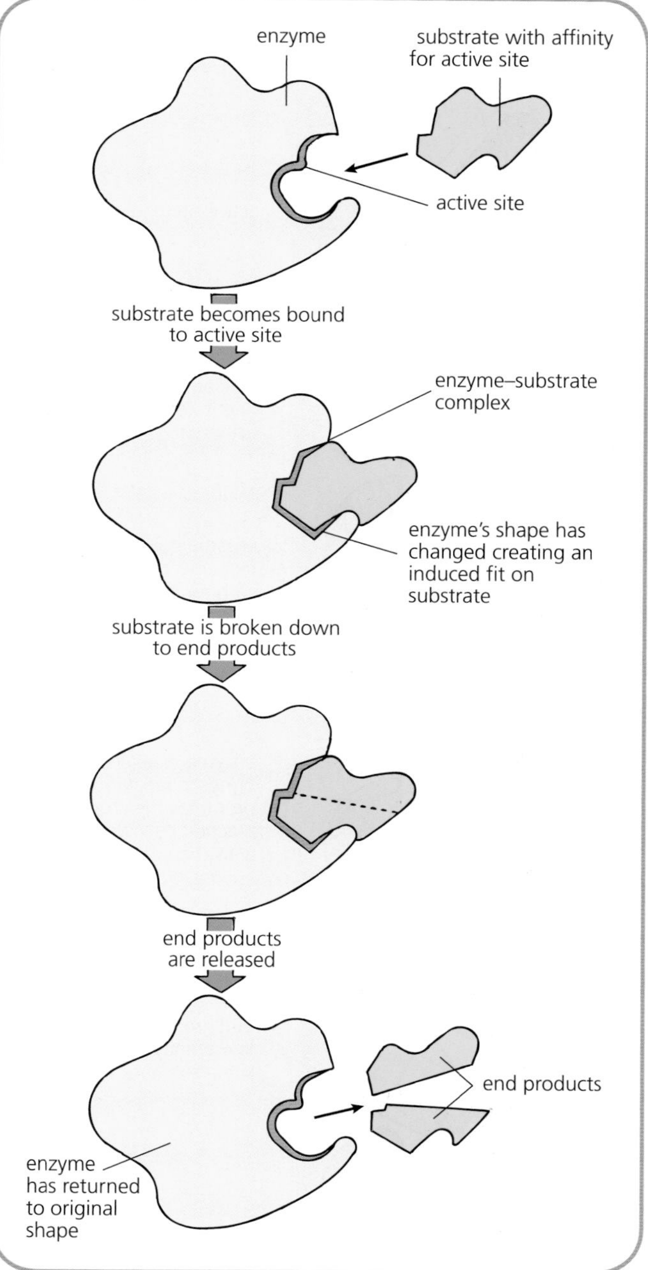

Figure 8.11 Induced fit during an enzyme-catalysed reaction

Orientation of reactants

When the reaction involves two (or more) substrates (see Figure 8.12), the shape of the active site determines the **orientation** of the reactants. This ensures that they are held together in such a way that the reaction between them can take place.

First the active site holds the two reactants closely together in an induced fit. Then it acts on them to weaken chemical bonds that must be broken during the

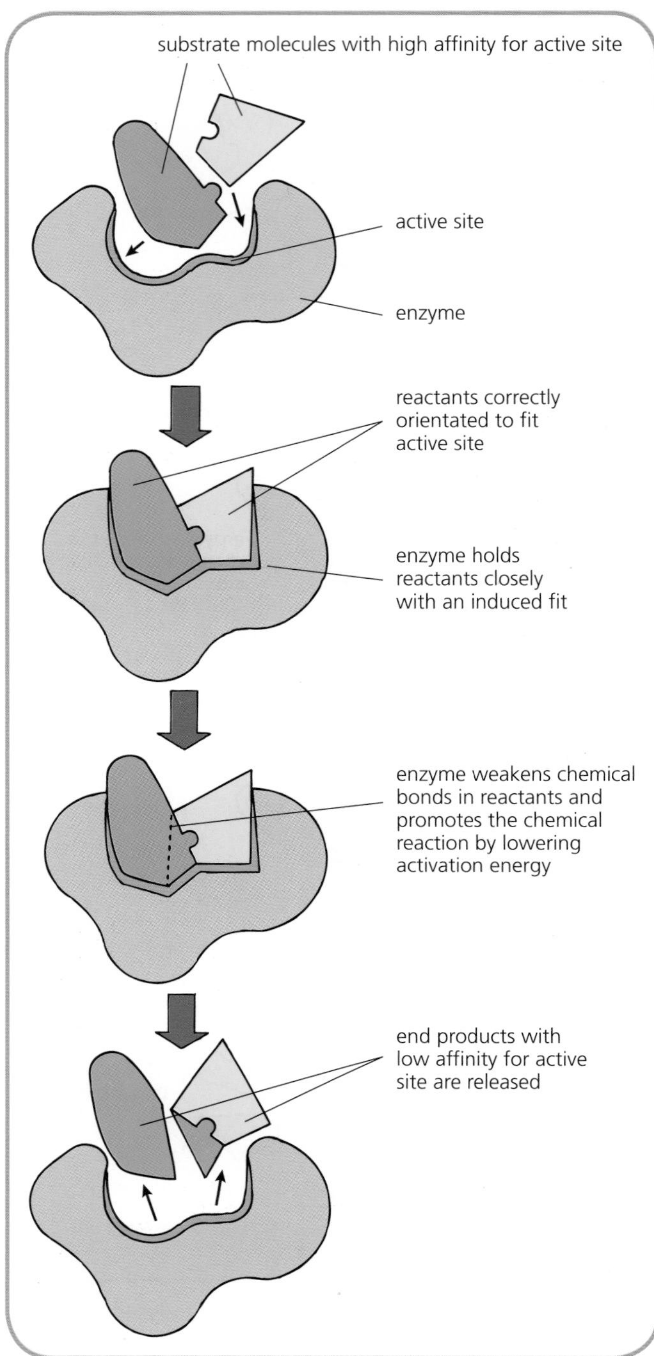

Figure 8.12 Orientation of reactants during an enzyme-catalysed reaction

reaction. This process **reduces the activation energy** needed by the reactants to reach the **transition state** that allows the reaction to take place.

Once the reaction has occurred, the products have a **low affinity** for the active site and are released. This leaves the enzyme free to repeat the process with new molecules of substrate.

Factors affecting enzyme action

To function efficiently, an enzyme requires a suitable temperature, an appropriate pH and an adequate supply of substrate. Inhibitors (see page 113) may slow down the rate of an enzyme-controlled reaction or bring it to a halt.

Effect of substrate concentration on enzyme activity

The graph in Figure 8.13 shows the effect of increasing substrate concentration on the rate of an enzyme-controlled reaction for a limited concentration of enzyme. At low concentrations of substrate, the reaction rate is low since there are too few substrate molecules present to make maximum use of all the active sites on the enzyme molecules. An increase in substrate concentration results in an increase in reaction rate since more and more active sites become involved.

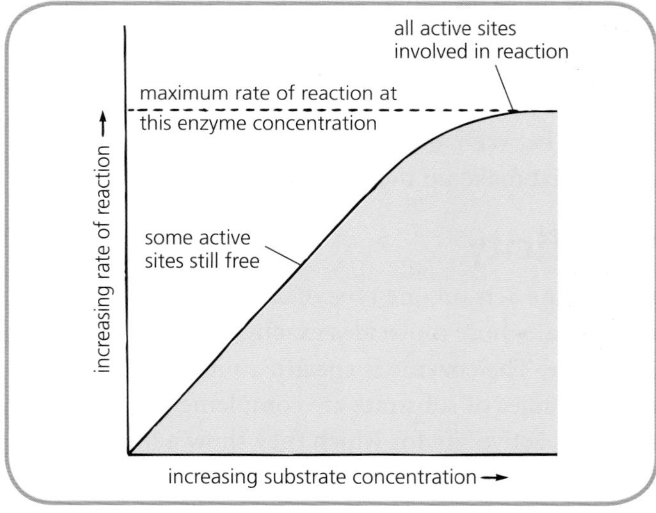

Figure 8.13 Effect of increasing substrate concentration

This upward trend in the graph continues as a straight line until a point is reached where a further increase in substrate concentration fails to make the reaction go any faster. At this point all the active sites are occupied (the enzyme concentration has become the **limiting factor**). The graph levels off since there are now more substrate molecules present than there are free active sites with which to combine. The effect of increasing substrate concentration is summarised at a molecular level in a simplified way in Figure 8.14.

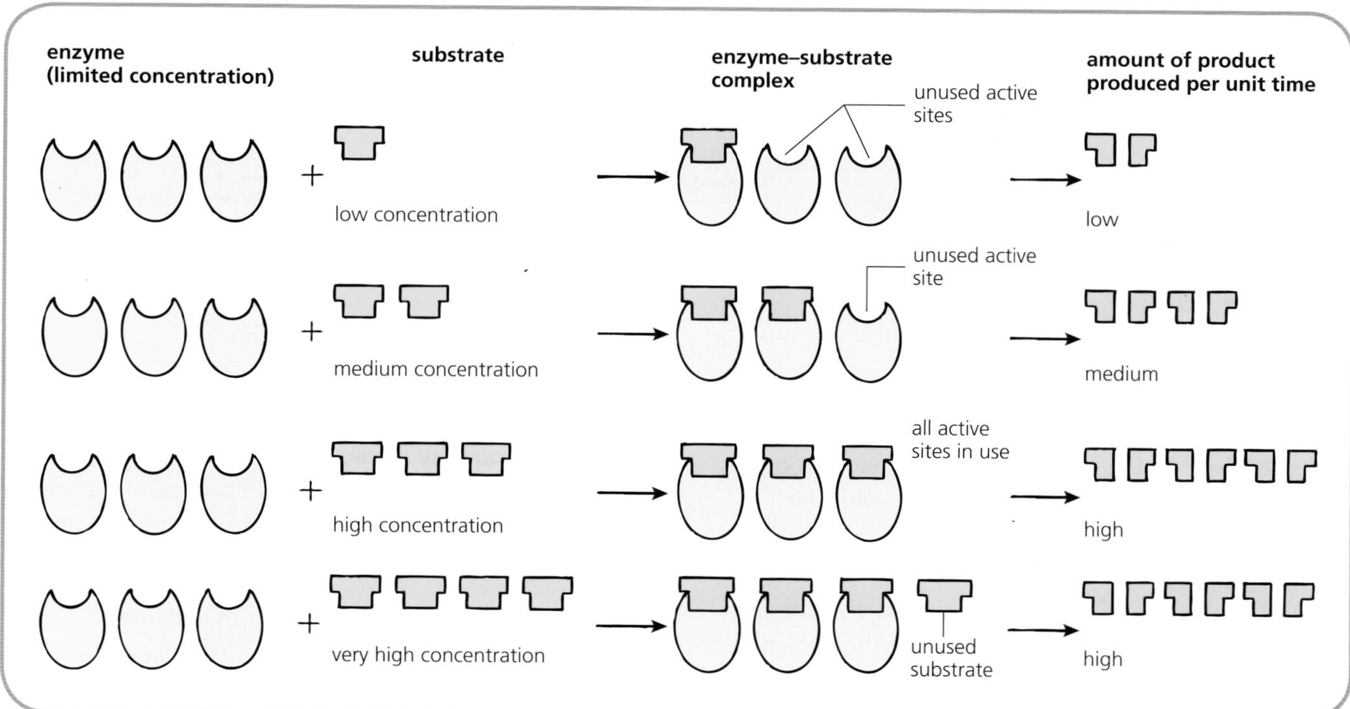

Figure 8.14 Effect of increasing substrate concentration at a molecular level

Related Activity

Investigating the effect of increasing substrate concentration

Liver cells contain the enzyme catalase which catalyses the breakdown of hydrogen peroxide to water and oxygen. In the experiment shown in Figure 8.15, the one variable factor is the concentration of the substrate (hydrogen peroxide). When an equal mass of fresh liver is added to each cylinder, the results shown in the diagram are produced. The height of the froth of oxygen bubbles indicates the activity of the enzyme at each concentration of substrate.

From the experiment it is concluded that an increase in substrate concentration results in increased enzyme activity until a point is reached (in cylinder G) where some factor other than substrate concentration has become the limiting factor.

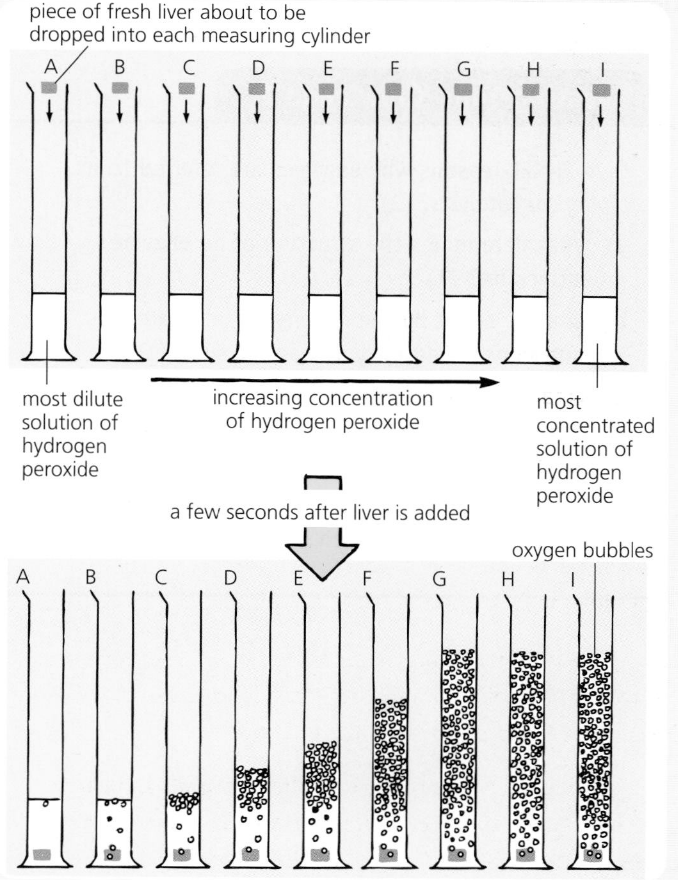

Figure 8.15 Effect of substrate concentration on enzyme activity

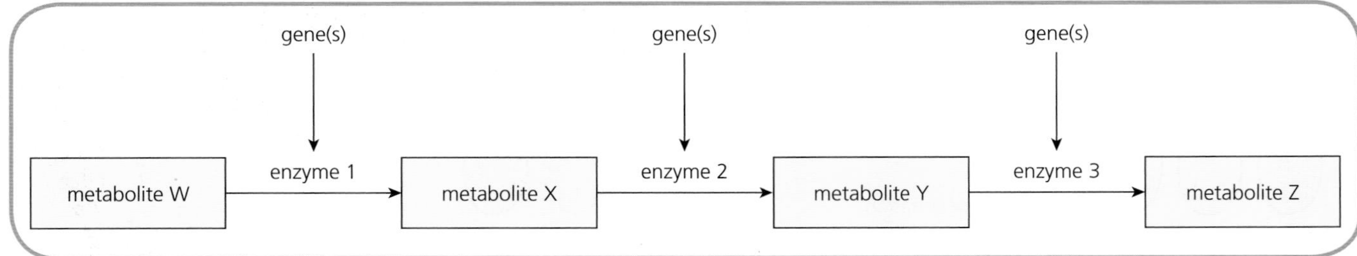

Figure 8.16 Action of a group of enzymes

Direction of enzyme action

A **metabolic pathway** normally consists of several stages, each of which involves the conversion of one metabolite to another. Each stage in a metabolic pathway is driven by a specific enzyme as shown in Figure 8.16. Each enzyme is coded for by one or more genes. As substrate W becomes available, enzyme 1 becomes active and converts W to X. In the presence of metabolite X, enzyme 2 becomes active and converts X to Y and so on. A continuous supply of W entering the system drives the sequence of reactions in the direction W to Z with the product of one reaction acting as the substrate of the next.

Reversibility

Most metabolic reactions are **reversible**. Often an enzyme can catalyse a reaction in both a forward and a reverse direction. The actual direction taken depends on the relative concentrations of the reactant(s) and product(s).

A metabolic pathway rarely occurs in isolation. If, as a result of related biochemical pathways, the concentration of metabolite Y in Figure 8.16 were to increase to an unusually high level and that of X were to decrease, then enzyme 2 could go into reverse and convert some of Y back to X until a balanced state (equilibrium) was restored once more.

Testing Your Knowledge 2

1 Give THREE reasons why enzymes are referred to as *biological catalysts*. (3)

2 **a)** What determines the structure of an enzyme's active site? (1)

b) What is meant by the *affinity* of substrate molecules for an enzyme's active site? (1)

c) What term means 'the change in shape of an active site to enable it to bind more snugly to the substrate'? (1)

d) Rewrite the following sentences choosing the correct answer from each underlined choice. (3)

The shape of the active site ensures that the reactants are correctly orientated/denatured so that the reaction can take place. This is made possible by the fact that the enzyme increases/decreases the activation energy needed by the reactants to reach the transitory/transition state.

3 **a)** What is meant by the term *rate of reaction*? (See page 107 for help.) (1)

b) i) What effect does an increase in concentration of substrate have on the reaction rate when a limited amount of enzyme is present?

ii) Explain why. (4)

Control of metabolic pathways

Regulation by switching genes on or off

Some metabolic pathways are only required to operate under certain circumstances. To prevent resources being wasted, the genes that code for the enzymes controlling certain stages in the pathway can be switched on (by an **inducer**) or off (by a **repressor**) as required. (Also see Related Activity – Investigating enzyme induction.)

Related Activity

Investigating enzyme induction

ONPG is a colourless synthetic chemical which can be broken down by the enzyme β-galactosidase as follows:

$$\text{ONPG} \xrightarrow{\beta\text{-galactosidase}} \text{galactose} + \text{yellow compound}$$

The presence of the yellow colour indicates activity by β-galactosidase. The experiment is set up as shown in Figure 8.17.

From the results it is concluded that:

- In tube 1, lactose has acted as an inducer and switched on the gene in *E. coli* that codes for β-galactosidase. This enzyme has acted on the ONPG forming the yellow colour.
- In tubes 2 and 4, no yellow colour was produced because ONPG was absent.
- In tube 3, β-galactosidase has acted on ONPG forming the yellow compound.

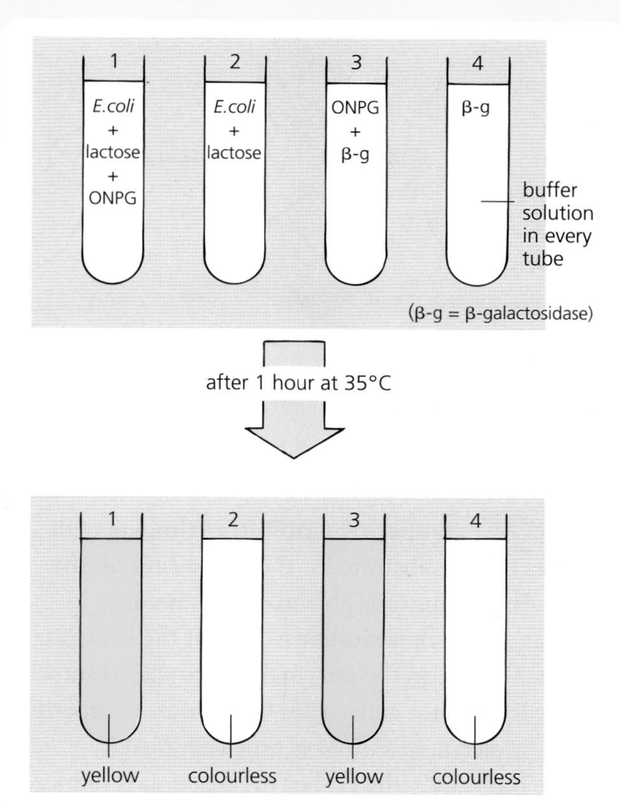

Figure 8.17 Investigating enzyme induction

Regulation by inhibition of enzyme action

Some metabolic pathways (for example, glycolysis) are required to operate continuously. The genes that code for their enzymes are always switched on and the enzymes which they code for are always present in the cell. Control of these metabolic pathways can be achieved by means of inhibitors. An **inhibitor** is a substance that decreases the rate of an enzyme-controlled reaction.

Regulation of the action of an enzyme controlling a stage in a metabolic pathway can be brought about by one of the following processes:

- competitive inhibition
- non-competitive inhibition
- feedback inhibition.

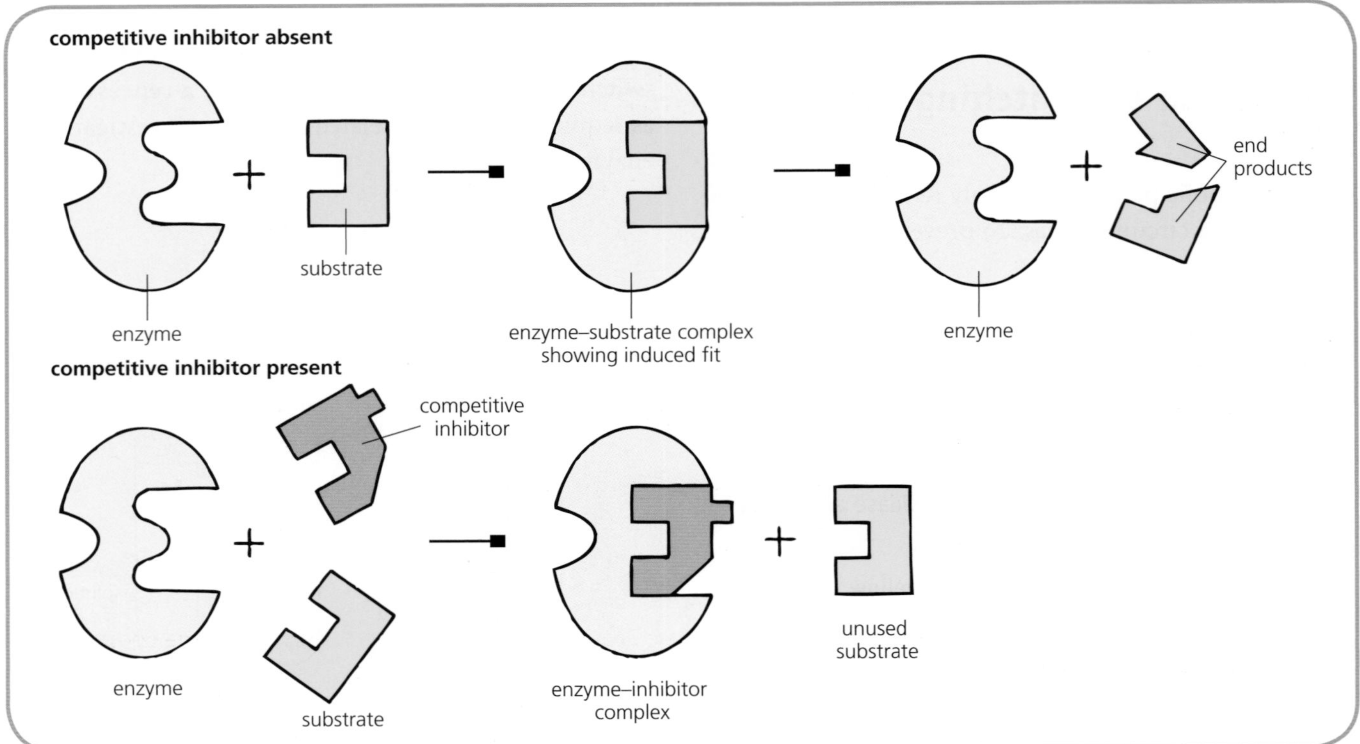

Figure 8.18 Effect of the competitive inhibitor

Competitive inhibitors

Molecules of a **competitive inhibitor** compete with molecules of the substrate for the active sites on the enzyme. The inhibitor is able to do this because its molecular structure is **similar** to that of the substrate and it can attach itself to the enzyme's active site, as shown in Figure 8.18. Since active sites **blocked** by competitive inhibitor molecules cannot become occupied by substrate molecules, the rate of the reaction is reduced.

Effect of increasing substrate concentration

The graph in Figure 8.19 compares the effect of increasing substrate concentration on rate of reaction for a limited amount of enzyme affected by a limited amount of inhibitor. In graph line 1 (the control), an increase in substrate concentration brings about an increase in reaction rate until a point is reached where all active sites on the enzyme molecules are occupied and then the graph levels off.

In graph line 2, an increase in substrate concentration brings about a gradual increase in the reaction rate. Although the competitive inhibitor is competing for and occupying some of the enzyme's active sites, the true substrate is also occupying some of the sites. As substrate molecules increase in concentration and outnumber those of the competitive inhibitor, more and more active sites become occupied by true substrate rather than inhibitor molecules. The reaction rate continues to increase until all the active sites are occupied (almost all of them by substrate).

This experiment shows that competitive inhibition is reversed by increasing substrate concentration.

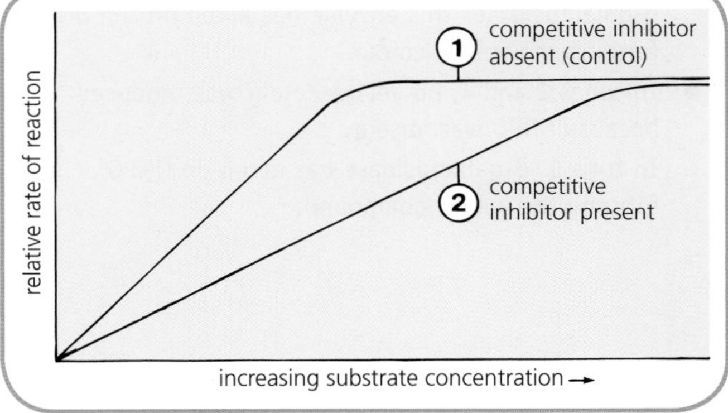

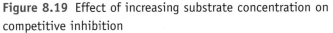

Figure 8.19 Effect of increasing substrate concentration on competitive inhibition

Investigation

Inhibition of β-galactosidase by galactose

Normally the enzyme β-galactosidase catalyses the reaction:

$$\text{lactose} \xrightarrow{\beta\text{-galactosidase}} \text{glucose} + \text{galactose}$$

However, it is also able to break down a colourless, synthetic compound called ONPG as follows:

$$\text{ONPG} \xrightarrow{\beta\text{-galactosidase}} \text{galactose} + \text{yellow compound}$$

The experiment shown in Figure 8.20 is set up to investigate the inhibitory effect of galactose on the action of β-galactosidase as the concentration of the substrate, ONPG, is increased. The **independent variable** in this experiment is substrate concentration.

At the end of the experiment, an increasing intensity of yellow colour (indicating products of enzyme activity) is found to be present in the tubes, with tube 1 the least yellow and tube 4 the most yellow. The intensity of colour can be measured quantitatively using a **colorimeter**. This allows the results to be displayed as a graph.

A possible explanation for these results is that galactose acts as a **competitive inhibitor**, having the most effect at low concentrations of substrate. As the concentration of substrate increases, more and more active sites on the enzyme become occupied by substrate, not inhibitor, and the reaction rate increases.

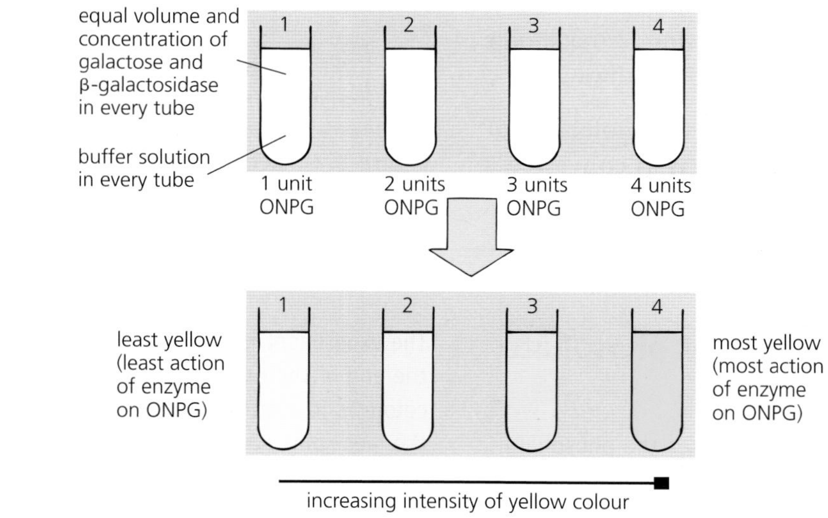

Figure 8.20 Investigating the inhibitory effect of galactose

Non-competitive inhibitors

A **non-competitive inhibitor** does not combine directly with an enzyme's active site. Instead it becomes attached to a non-active site and **changes the shape** of the enzyme molecule. This results in the active site becoming **altered indirectly** and being unable to combine with the substrate, as shown in Figure 8.21.

The larger the number of enzyme molecules affected in this way, the slower the enzyme-controlled reaction. Therefore the non-competitive inhibitor acts as a type of regulator.

Non-competitive inhibition is not reversed by increasing substrate concentration.

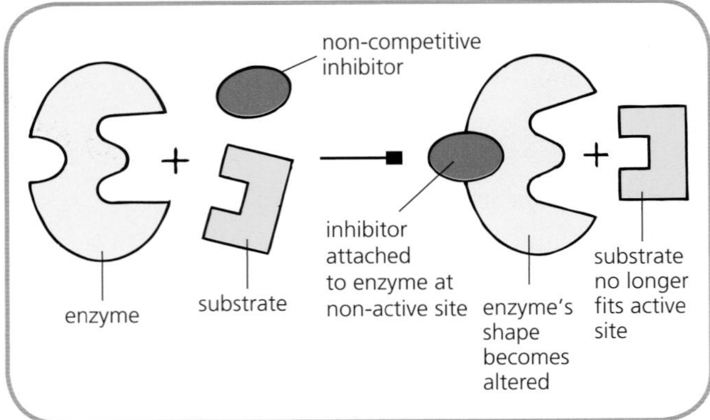

Figure 8.21 Effect of the non-competitive inhibitor

Feedback inhibition by an end product

End-product inhibition (see Figure 8.22) is a further way in which a metabolic pathway can be regulated. As the concentration of end product (metabolite Z) builds up and reaches a critical concentration, some of it binds to and inhibits molecules of enzyme 1 in the pathway. This slows down the conversion of metabolite W to X and in turn regulates the whole pathway.

As the concentration of Z drops, fewer molecules of enzyme 1 are affected and more of W is converted to X and so on. The pathway is kept under **finely-tuned control** by this means (called **negative feedback control**) and wasteful conversion and accumulation of intermediates and final products are avoided.

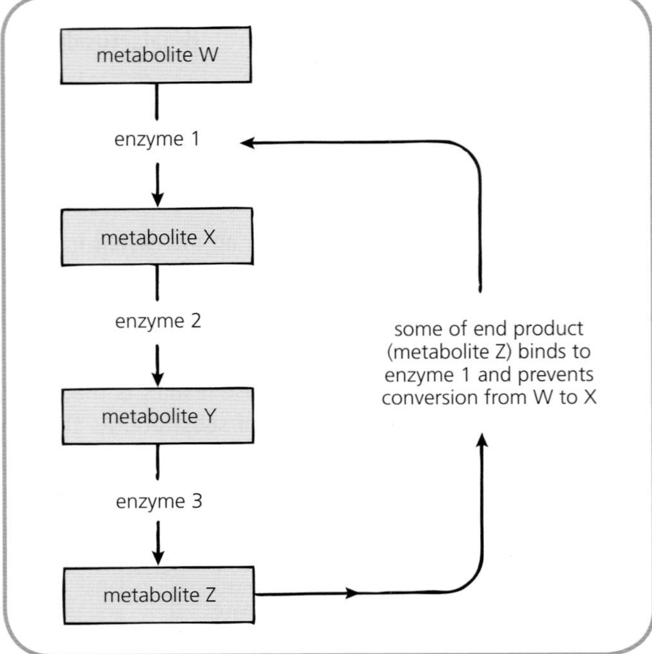

Figure 8.22 Regulation by feedback inhibition

Related Activity

Investigating the effect of phosphate on phosphatase

Phosphatase is an enzyme that releases the phosphate group from its substrate for use in cell metabolism. Phosphatase is present in the extract obtained from ground-up mung bean sprouts. **Phenolphthalein phosphate** is a chemical that can be broken down by phosphatase as follows:

The experiment is set up as shown in Figure 8.23. At the end of the experiment a decreasing intensity of pink colour is found to be present in the tubes. Tube 1 is the most pink and tube 5 is the least pink. From these results it is concluded that tube 1 contains most free phenolphthalein as a result of most enzyme activity and that tube 5 contains least free phenolphthalein as a result of least enzyme activity. In other words, as phosphate concentration increases, the activity of the enzyme phosphatase decreases. A possible explanation for this effect is that phosphate acts as an **end-product inhibitor** of the enzyme phosphatase.

$$\text{phenolphthalein phosphate} \xrightarrow{\text{phosphatase}} \text{phenolphthalein (pink in alkaline conditions)} + \text{phosphate}$$

→

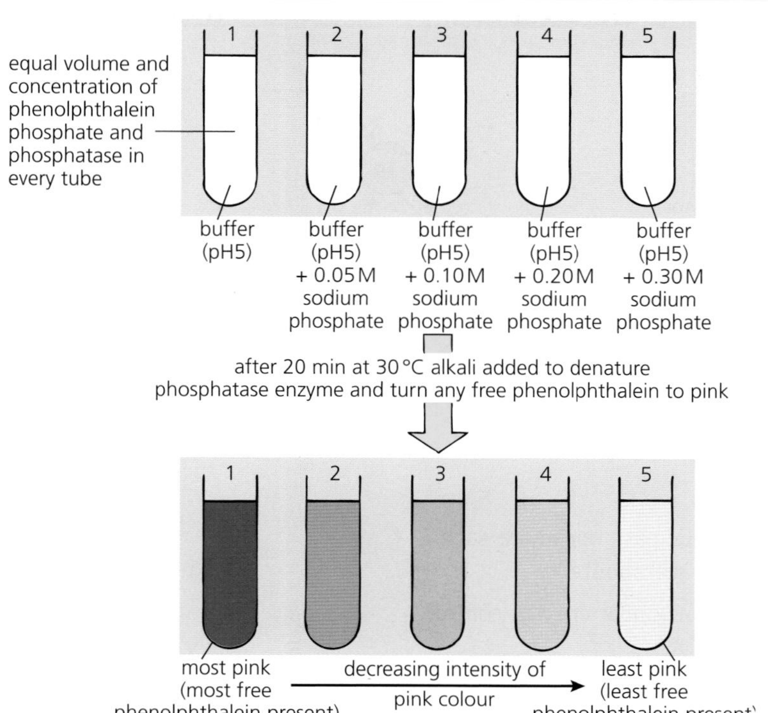

Figure 8.23 Investigating the effect of phosphate on phosphatase

Testing Your Knowledge 3

1 a) What property of a competitive inhibitor enables it to compete with the substrate? (1)

b) i) What effect does an increase in concentration of a substrate have on the rate of a reaction when a limited amount of competitive inhibitor and enzyme are present?

ii) Explain why. (3)

2 Which form of enzyme inhibition cannot be reversed by increasing substrate concentration? (1)

3 Figure 8.24 shows a metabolic pathway where metabolites P, Q and R are present in equal quantities at the start.

a) Name enzyme X's

i) substrate

ii) product. (2)

b) Name enzyme Y's

i) substrate

ii) product. (2)

c) In which direction will the pathway proceed if more of metabolite P is added to the system? (1)

d) i) Metabolite R can act as an end-product inhibitor. Describe how this would work.

ii) What is the benefit of end-product inhibition? (3)

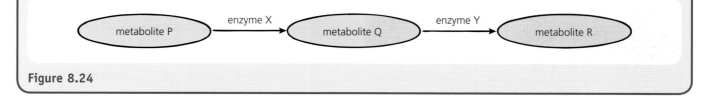

Figure 8.24

Applying Your Knowledge and Skills

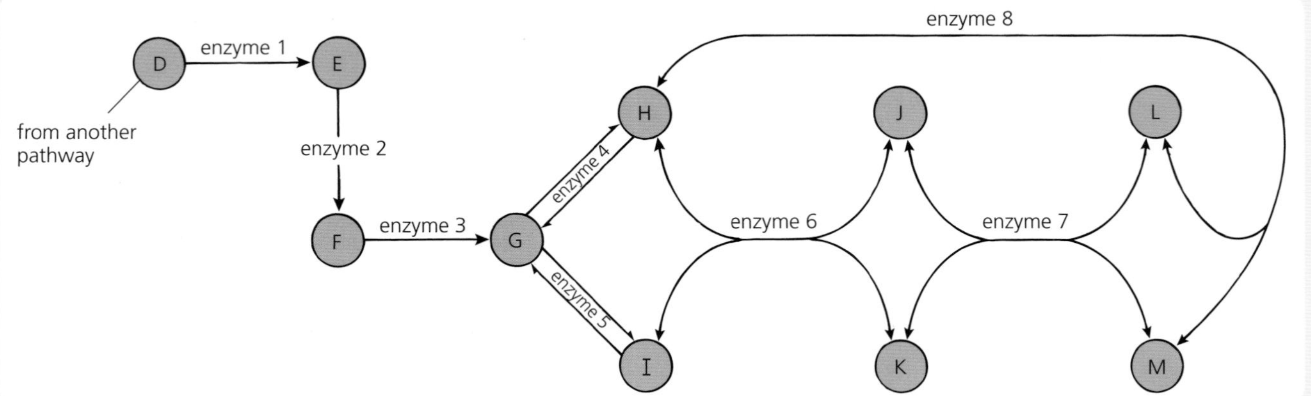

Figure 8.25

1 Figure 8.25 shows a metabolic pathway where each encircled letter represents a metabolite.

a) How many of the reactions under enzyme control in this pathway are
   i) reversible? ii) irreversible? (2)

b) Predict what would happen if metabolite I built up to a concentration far in excess of that of metabolite H. (2)

c) i) By what alternative route could a supply of intermediates J and K be obtained if enzyme 6 becomes inactive?
   ii) By what alternative route could a supply of intermediates L and M be obtained if enzyme 8 becomes inactive?
   iii) By what alternative route could a supply of metabolite I be obtained if enzyme 5 becomes inactive? (3)

d) Suggest a benefit to a living organism of its metabolic pathways possessing alternative routes. (1)

2 The data in Table 8.1 refer to ions present inside and outside muscle cells in the body of an amphibian. The graph in Figure 8.26 represents the results of an experiment set up to investigate the effect of oxygen concentration on uptake of potassium ions and consumption of sugar by muscle cells in a tissue culture.

Ion	Concentration (mmol/l)	
	Intracellular	Extracellular
potassium	137.5	2.5
sodium	13.0	104.0

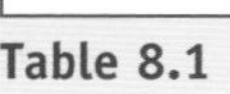
Table 8.1

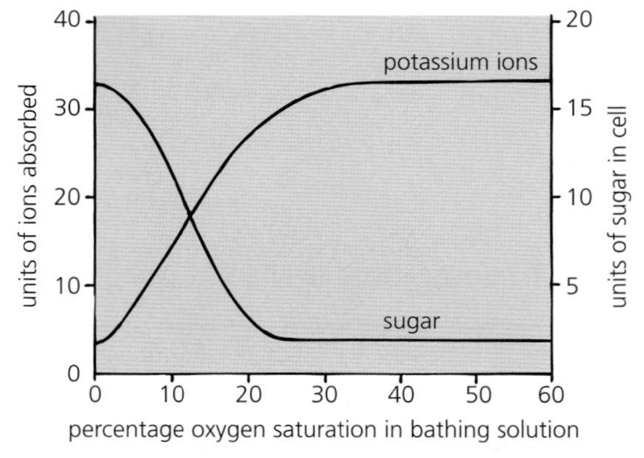

Figure 8.26

a) Identify the region of:
   i) high sodium concentration
   ii) low sodium concentration
   iii) high potassium concentration
   iv) low potassium concentration.
   v) Explain how these differences in ionic concentration are maintained. (3)

b) By how many times is the concentration of potassium inside a cell greater than that outside? (1)

c) i) From the graph in Figure 8.26, state the effect that an increase in oxygen concentration from 0–30% has on the rate of ion uptake.
   ii) Suggest why ion uptake levels off beyond 30% oxygen.
   iii) What relationship exists between units of ions absorbed and units of sugar remaining in the cell? Suggest why. (4)

→

3 Figure 8.27 shows the stages that occur during an enzyme-controlled reaction.

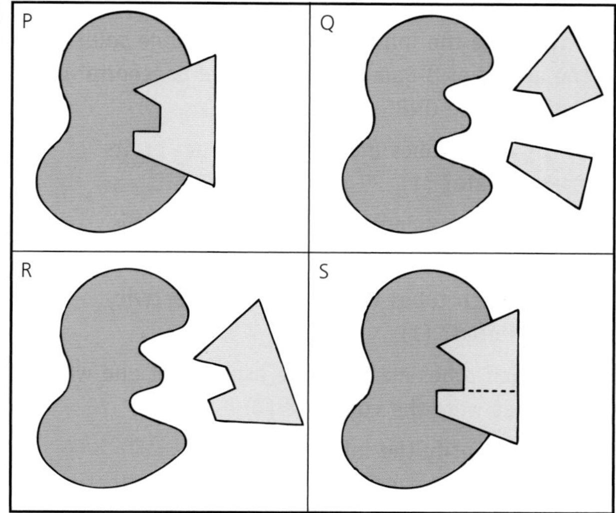

Figure 8.27

a) Which of these stages illustrates induced fit? (1)

b) Using the letters given, indicate the correct sequence in which the four stages would occur if the enzyme were promoting:

i) the build-up of a molecule from smaller components

ii) the breakdown of a molecule into smaller constituents. (2)

4 The graph shown in Figure 8.28 summarises the results from an experiment involving an enzyme-controlled reaction.

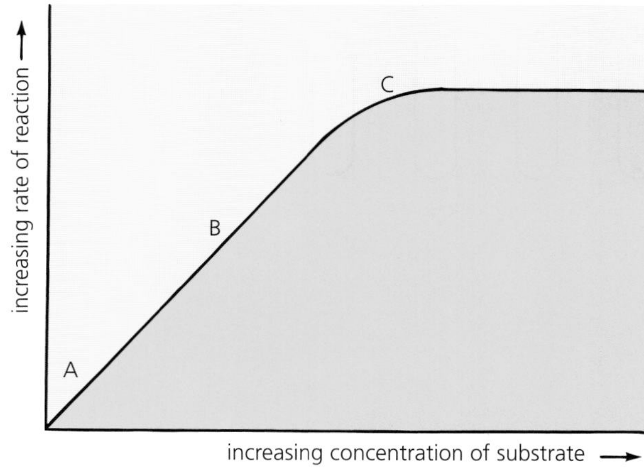

Figure 8.28

a) i) In this experiment, the enzyme concentration was kept constant. From the graph, identify the factor that was varied by the experimenter.

ii) Is this factor called the dependent or the independent variable?

iii) What effect did an increase in this factor have over region AB of the graph? (3)

b) Suggest which factor became limiting at point C on the graph. (1)

c) Which letter on the graph represents the situation where

i) almost all of the active sites

ii) none of the active sites

iii) about half of the active sites

on enzyme molecules are freely available for attachment to substrate molecules? (3)

d) Suggest what could be done to increase the rate of the reaction beyond the level it has reached at C. (1)

5 Tables 8.2 and 8.3 give the results from an experiment set up to compare the activity of an enzyme (alkaline phosphatase) with its substrate (para-nitrophenol phosphate) in the presence and absence of a competitive inhibitor.

Concentration of substrate (n moles l^{-1})	Enzyme activity (units)
0	0.0
10	1.8
20	2.6
30	3.3
40	3.6
50	3.8
60	4.0
70	4.0

Table 8.2

Concentration of substrate (n moles l^{-1}) + inhibitor	Enzyme activity (units)
0	0.0
10	0.6
20	1.0
30	1.5
40	2.1
50	2.7
60	3.0
70	3.6

Table 8.3

a) i) Draw a curve of best fit for the results in Table 8.2.
ii) On the same graph, draw a line of best fit for the results in Table 8.3.
iii) Mark 'presence of inhibitor' and 'absence of inhibitor' on your graph to identify the lines. (4)

b) At which of the following ranges of substrate concentration (in n moles l^{-1}) did the enzyme activity increase at the fastest rate in the absence of inhibitor? (1)

A 0–19 **B** 20–39 **C** 40–59

c) By how many times was enzyme activity at substrate concentration of 10 n moles l^{-1} greater when the inhibitor was absent? (1)

d) Calculate the percentage decrease in enzyme activity caused by the inhibitor at a substrate concentration of:
i) 20 n moles l^{-1}
ii) 70 n moles l^{-1}. (2)

e) Why would the two lines on the graph fail to meet even if higher concentrations of substrate were used? (1)

6 The experiment shown in Figure 8.29 was set up to investigate the inhibitory effect of iodine solution on the action of β-galactosidase as the concentration of the substrate ONPG was increased.

a) In which tubes did the enzyme act on its substrate? (1)

b) Identify the independent variable in this experiment. (1)

c) In which tubes was the enzyme's activity inhibited? (1)

d) Which tubes made up the experiment and which tubes were the controls? (2)

e) i) Identify the inhibitor.
ii) Did it act competitively or non-competitively?
iii) Explain your choice of answer. (4)

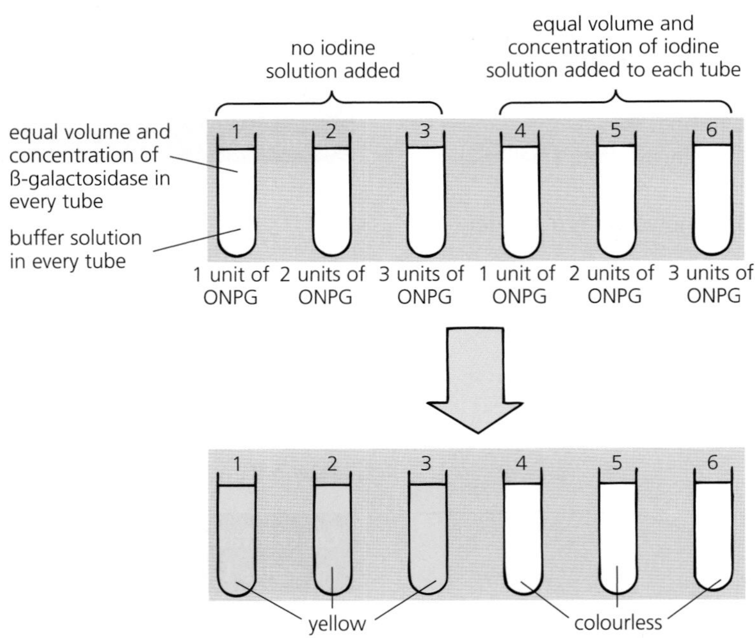

Figure 8.29

→

7 Figure 8.30 shows a metabolic pathway that occurs in cells of *E. coli*.

a) Identify enzyme P's
 i) substrates
 ii) products
 iii) end-product inhibitor. (3)

b) i) If there is little or no demand for cytidylic acid for use in other metabolic pathways, what effect will this have on the concentration of carbamyl phosphate?
 ii) Explain your answer. (2)

c) i) If there is a high demand for cytidylic acid in other metabolic pathways, will the negative feedback process be increased or decreased?
 ii) Explain your answer. (2)

d) Which of the following statements is/are true? (1)
 1 The end product inhibits an early step in its own synthesis.
 2 The negative feedback mechanism regulates the rate of synthesis of metabolic intermediates.
 3 End-product inhibition prevents the build-up of intermediates which would be wasteful to the cell.

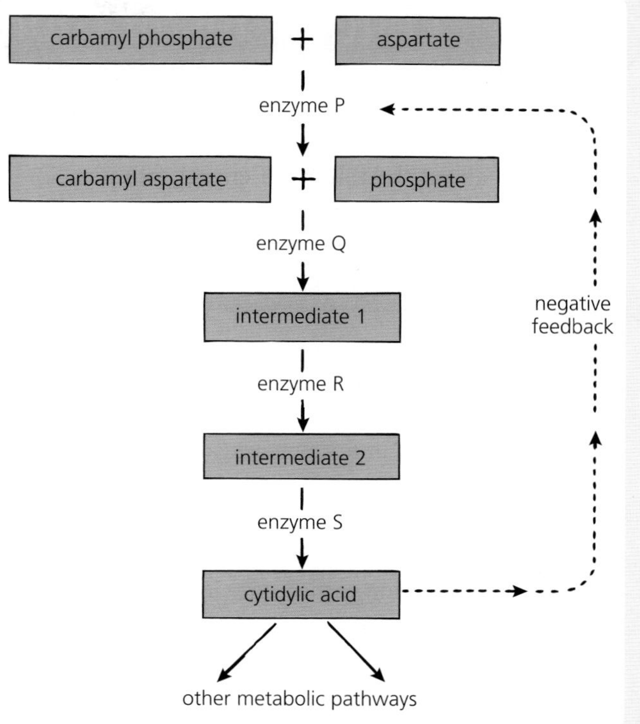

Figure 8.30

9 Cellular respiration

Cellular respiration is a series of metabolic pathways which brings about the release of energy from a foodstuff and the regeneration of the high-energy compound **ATP**. It occurs in the cells of all members of the three domains of life.

Investigation and Report

Investigating the use of three different sugars as respiratory substrates by yeast

Background biology

- Figure 9.1 shows, in a simple way, the molecular structure of three types of sugar and the digestive enzymes needed to break down maltose and sucrose.
- Strictly speaking, this activity is really three investigations being carried out simultaneously.
- In each case the independent variable is time.
- The dependent variable that you are going to measure is the volume of carbon dioxide released as a result of yeast using a particular type of sugar as its respiratory substrate.

You need

three graduated tubes

three large beakers (such as 500 ml) of coloured tap water

three clamp stands

one container of glucose solution (10 g in 90 ml of water)

one container of maltose solution (10 g in 90 ml of water)

one container of sucrose solution (10 g in 90 ml of water)

three conical flasks (250 ml) each with a rubber stopper and delivery tube

three labels

three portions of dried yeast, each 1 g

a stopclock

What to do

1 Read all of the instructions in this section and prepare your results table before carrying out the experiment.

2 Fill each graduated tube with coloured tap water and clamp it in an inverted position in a beaker of coloured water as shown in Figure 9.2.

3 Label the conical flasks 'glucose', 'maltose' and 'sucrose', respectively, and add your initials.

4 Pour the appropriate sugar solution into each conical flask and add a portion of dried yeast.

5 Assemble the stoppers and delivery tubes as shown in Figure 9.2.

6 Start the clock and record, at 5-minute intervals, the total volume of carbon dioxide that has been released for each flask over a period of 2 hours.

7 If other students have carried out the same experiment, pool the results.

Reporting

Write up your report by doing the following:

1 Rewrite the title given at the start of this activity.

2 Put the subheading '**Aim**' and state the aim of your experiment.

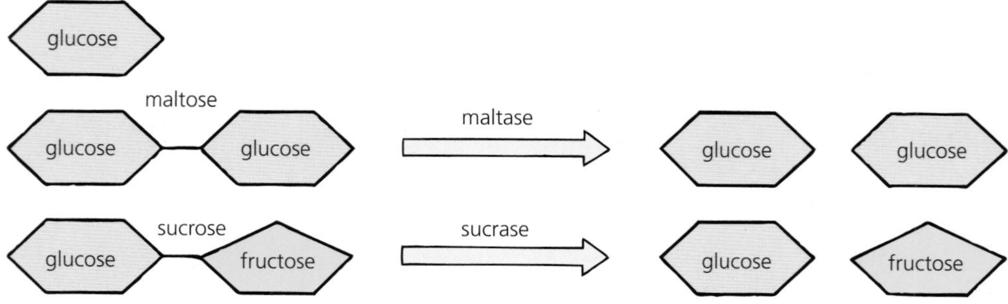

Figure 9.1 Relationship between three sugars

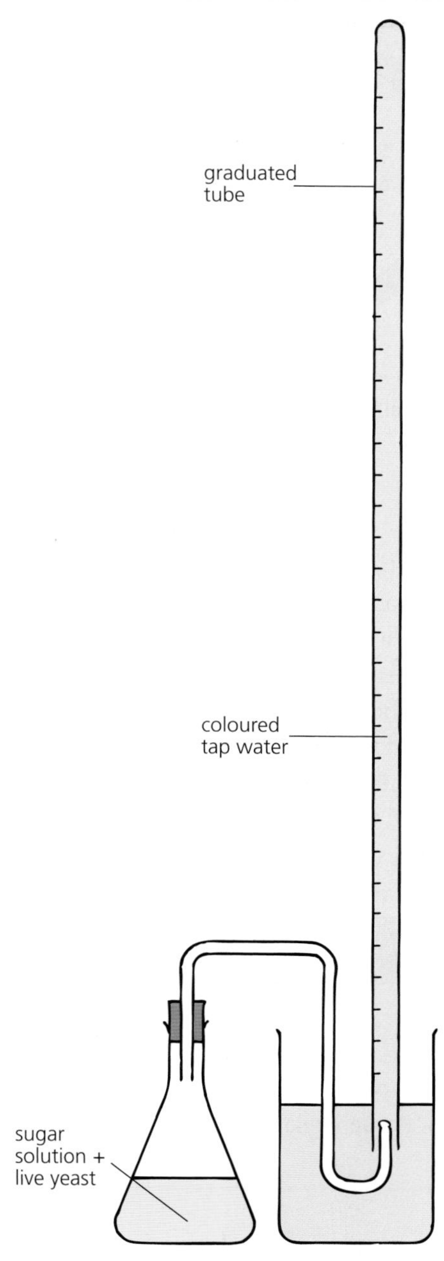

Figure 9.2 Yeast investigation set-up

3 a) Put the subheading '**Method**'.

b) Draw a diagram of your apparatus set-up at the start of the experiment after the yeast has been added and bubbles of carbon dioxide are being released.

c) Briefly describe the experimental procedure that you followed using the impersonal passive voice. (Note: The impersonal passive voice avoids the use of 'I' and 'we'. Instead it makes the apparatus the subject of the sentence. In this experiment, for example, you could begin your report by saying 'Three graduated tubes were filled with coloured water ...' etc. *not* 'I filled three graduated tubes with coloured water ...' etc.)

d) Continuing in the impersonal passive voice, state how your results were obtained.

4 Put the subheading '**Results**' and draw a final version of your table of results.

5 Put a subheading '**Analysis and Presentation of Results**'. Present your results as three line graphs with shared axes on the same sheet of graph paper.

6 Put the subheading '**Conclusion**' and write a short paragraph to state what you have found out from a study of your results. This should include answers to the following questions:

a) Which respiratory substrate(s) was yeast able to use effectively?

b) Which enzyme (see Figure 9.1) is probably produced by yeast cells in adequate quantities to digest its substrate before its use in cellular respiration?

c) Which respiratory substrate(s) was yeast not able to use effectively?

d) Which enzyme is probably not produced by yeast in adequate quantities within the 2-hour timescale to digest its substrate and make it suitable for use in cellular respiration?

7 Put a final subheading '**Evaluation of Experimental Procedure**'. Give an evaluation of your experiment (keeping in mind that you may comment on any stage of the experiment that you wish).

Try to incorporate answers to the following questions in your evaluation. Make sure that at least one of your answers includes a supporting statement.

a) Why is the same mass of yeast and the same mass of sugar used in every flask?

b) Why is the same genetic strain of yeast used in each flask?

c) Why must the rubber stoppers be tightly fitting?

d) Why should a control flask containing distilled water and yeast have been included in this investigation?

e) What is the purpose of pooling results with other groups?

Adenosine triphosphate

A molecule of **adenosine triphosphate (ATP)** is composed of adenosine and three inorganic phosphate (P_i) groups as shown in Figure 9.3. Energy held in an ATP molecule is released when the bond attaching the terminal phosphate is broken by enzyme action. This results in the formation of **adenosine diphosphate (ADP)** and inorganic **phosphate**. On the other hand, energy is required to regenerate ATP from ADP and inorganic phosphate. This relationship is summarised in Figure 9.4.

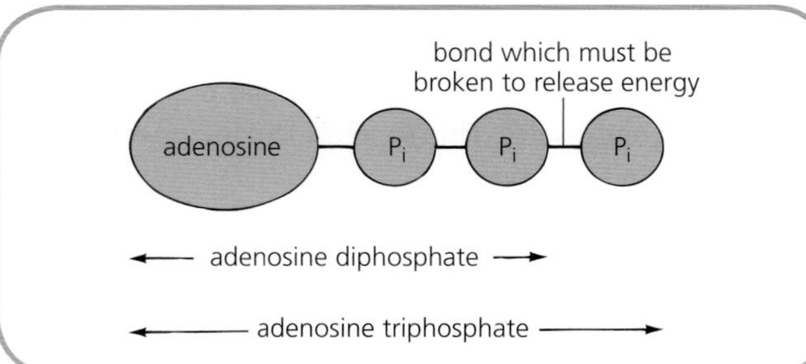

Figure 9.3 Structure of ATP

Phosphorylation

Phosphorylation is an enzyme-controlled process by which a phosphate group is added to a molecule. Phosphorylation occurs, for example, when the reaction shown in Figure 9.4 goes from the bottom to the top and P_i combines with low-energy ADP to form high-energy ATP.

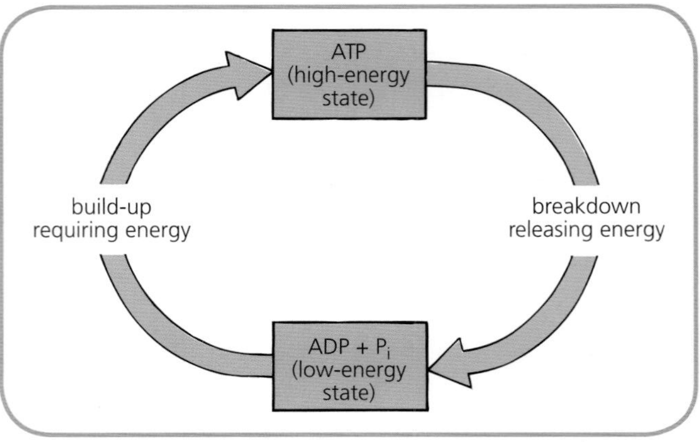

Figure 9.4 Relationship between ATP and ADP + P_i

Phosphorylation of a reactant in a pathway

Phosphorylation also occurs when phosphate and energy are transferred from ATP to the molecules of a reactant in a metabolic pathway making them **more reactive**. Often a step in a pathway can proceed only if a reactant becomes **phosphorylated** and energised. In the early stages of cellular respiration, for example, some reactants must undergo phosphorylation during what is called the energy investment phase. One of these steps is shown in Figure 9.5.

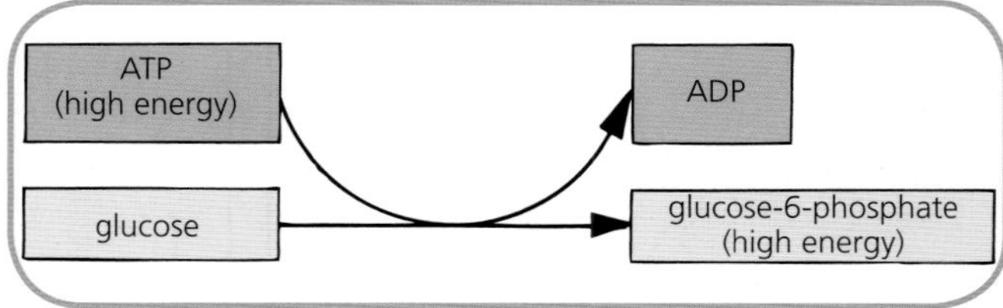

Figure 9.5 Phosphorylation of glucose

Investigation

A phosphorylated substrate

Background

- **Glucose-1-phosphate** is a phosphorylated form of glucose.
- A molecule of starch is composed of many glucose molecules linked together in a long chain.
- Potato tuber cells contain **phosphorylase**, an enzyme that promotes the synthesis of starch.
- Potato extract containing phosphorylase is prepared by liquidising a mixture of potato tuber and water and then centrifuging the mixture until the potato extract (see Figure 9.6) is **starch free**.

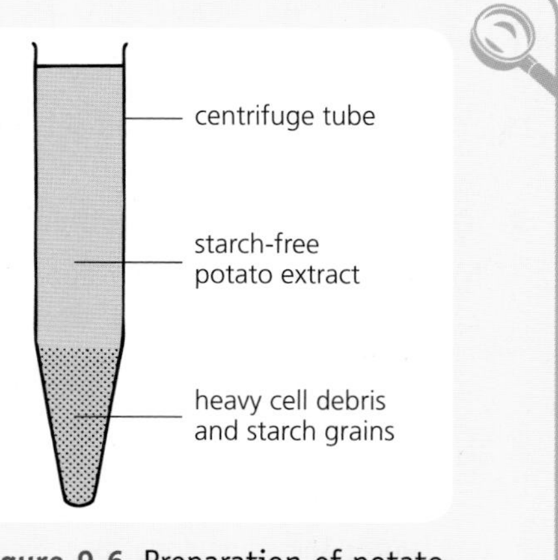

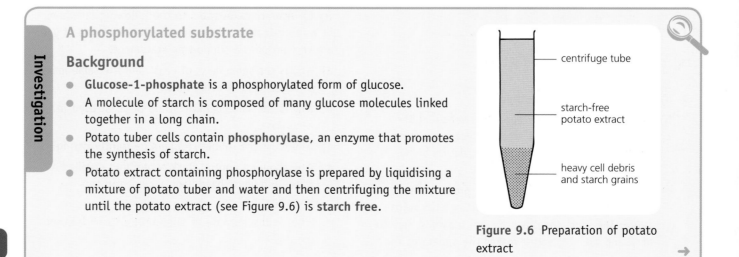

Figure 9.6 Preparation of potato extract

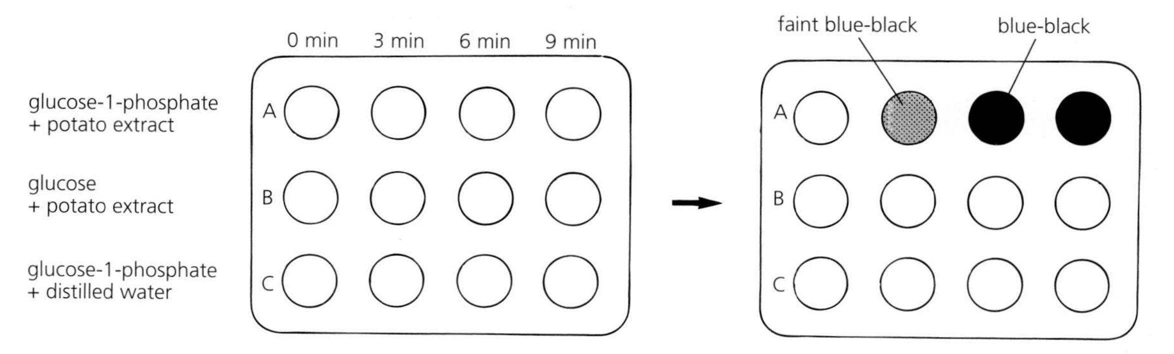

Figure 9.7 Investigating a phosphorylated substrate

The experiment is set up on a dimple tile as shown in Figure 9.7. One dimple from each row is tested at 3-minute intervals with iodine solution. Starch is found to be formed in row A only. It is concluded that in row A, phosphorylase has promoted the conversion of the **phosphorylated** (and more reactive) form of the substrate, glucose-1-phosphate, to starch as in the following equation:

$$\underset{\text{(phosphorylated substrate)}}{\text{glucose-1-phosphate}} \quad \xrightarrow[\text{(enzyme)}]{\text{phosphorylase}} \quad \underset{\text{(end product)}}{\text{starch}}$$

In row B (a control), phosphorylase has failed to convert the more stable (and less reactive) form of the substrate, glucose, to starch.

In row C (a control), the molecules of glucose-1-phosphate have failed to become bonded together into starch without the aid of phosphorylase.

Positive and negative controls

A **positive control** is set up to assess the validity of a testing procedure or design and ensure that the equipment and materials being used are in working order and appropriate for use in the experiment being carried out. For example, a positive control for the above experiment could be set up as row D, which would contain starch in every dimple. If the addition of iodine solution at each 3-minute interval gave a blue-black colour, this would confirm that:

- the iodine solution being used was working properly as a testing reagent for starch
- the experiment was not adversely affected in some way, for example by the contamination of the spotting tile or by changes in room temperature.

If a positive control does not produce the expected result, then this indicates that there is something wrong with the design of the testing procedure or with the materials being used.

A **negative control** is one that should not work. It is a copy of the experiment in which all factors are kept exactly the same except the one being investigated. When the results are compared, any difference found between the experiment and a negative control must be due to the factor being investigated.

In the above investigation, starch is not synthesised in row B, showing that the glucose must be in a phosphorylated state to become converted to starch. If row B had not been set up, it would be valid to suggest that starch would have been formed whether or not the glucose was phosphorylated. Similarly, starch was not formed in row C showing that phosphorylase (in potato extract) must be present for phosphorylated glucose to be converted to starch. If row C had not been included, it would be valid to suggest that phosphorylated glucose would have become starch whether or not phosphorylase was present. Therefore rows B and C in this investigation are negative controls.

Metabolic pathways of cellular respiration

Glycolysis

The process of cellular respiration begins in the cytoplasm of a living cell with a molecule of **glucose** being broken down to form **pyruvate**. This process of 'glucose-splitting' is called **glycolysis**. It consists of a series of enzyme-controlled steps. Those in the first half of the chain make up the **energy investment phase** (where two ATP are used up per molecule of glucose); those in the second half of the chain make up the **energy payoff phase** (where four ATP are produced per molecule of glucose) as shown in Figure 9.8.

Phosphorylation of intermediates occurs twice during the first phase:

- at step 1 where an intermediate is formed that can connect with other metabolic pathways
- at step 3 which is an irreversible reaction leading only to the rest of the glycolytic pathway.

The generation of four ATP that occurs during the second half of the pathway gives a **net gain of two ATP** per molecule of glucose during glycolysis. In addition, during the energy payoff phase, H ions are released from the substrate by a dehydrogenase enzyme. These H ions are passed to a coenzyme molecule called **NAD** forming **NADH**.

The process of glycolysis does not require oxygen. However, NADH only leads to the production of further molecules of ATP at a later stage in the respiratory process if oxygen is present. In the absence of oxygen, fermentation occurs (see page 131).

Citric acid cycle

If oxygen is present, aerobic respiration proceeds and pyruvate is broken down into carbon dioxide and

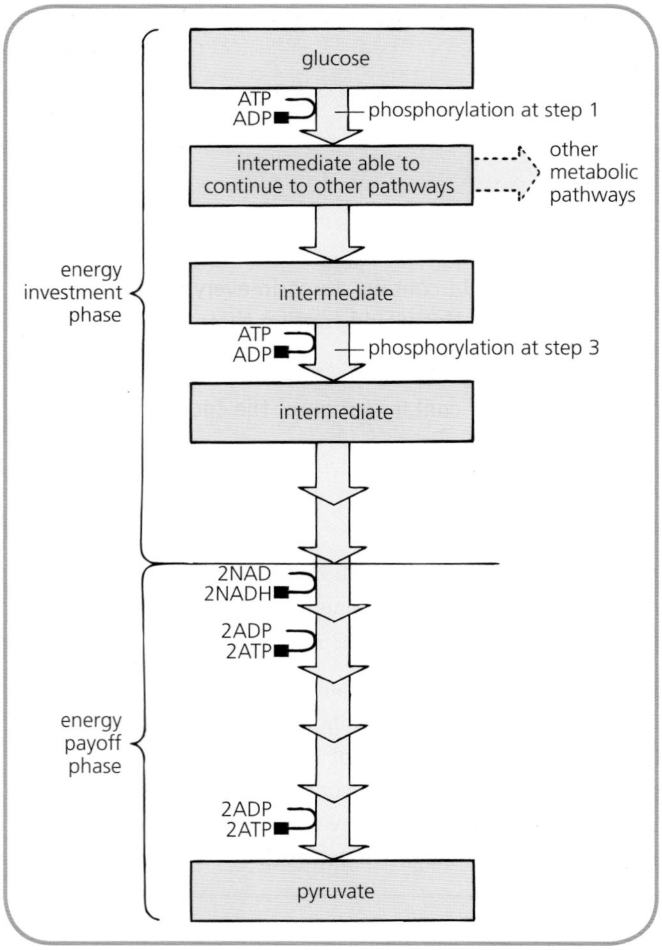

Figure 9.8 Glycolysis

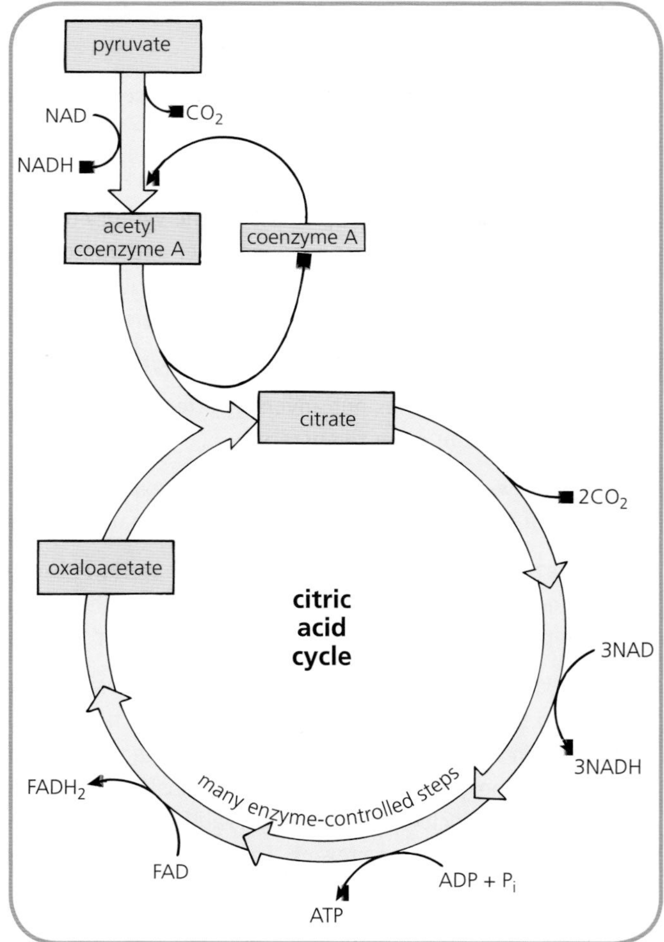

Figure 9.9 Citric acid cycle

an **acetyl group**. Each acetyl group combines with **coenzyme A** to form **acetyl coenzyme A**. During this process, further H ions are released and become bound to NAD forming NADH. A simplified version of the metabolic pathway is shown in Figure 9.9.

The acetyl group of acetyl coenzyme A combines with **oxaloacetate** to form **citrate** and enter the **citric acid cycle**. This cycle consists of several enzyme-mediated stages which occur in the central matrix of mitochondria and result finally in the regeneration of oxaloacetate.

At three steps in the cycle, dehydrogenase enzymes remove **H ions** and electrons from the respiratory substrate and pass them to the coenzyme NAD to form NADH. In addition, ATP is produced at one of the steps and carbon dioxide is released at two of the steps.

Research Topic: Discovery of citric acid cycle by Hans Krebs

Many people refer to the citric acid cycle as the **Krebs cycle** in recognition of the contribution to its discovery made by **Hans Krebs**, a German biochemist. Prior to the work done by Krebs, scientists had established that a cell extract from respiring animal tissue in the presence of oxygen is able to rapidly break down chemicals such as citrate, fumarate, malate and succinate and release carbon dioxide. Then they found that the following sequence of reactions occurs:

succinate → fumarate → malate → oxaloacetate

And later they established that the following sequence also takes place:

citrate → isocitrate → α-ketoglutarate → succinyl-CoA → succinate

Therefore, at the time of Krebs' discovery, this part of the biochemistry of respiration was thought to occur as the pathway shown in Figure 9.10.

However, Krebs, using extract from respiring muscle tissue, found that citrate could be formed if oxaloacetate and pyruvate were added to it. He therefore deduced that the process occurred as a **cycle** (see Figure 9.11) and *not* as a linear pathway. Further evidence to support this conclusion came from the fact that the addition of any of the intermediate reactants resulted in the generation of all the others. Experimental work involving **competitive inhibition** of the enzymes that convert one intermediate to another also lent support to what has now become accepted as fact.

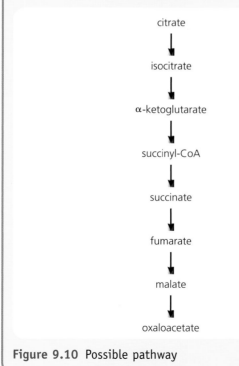

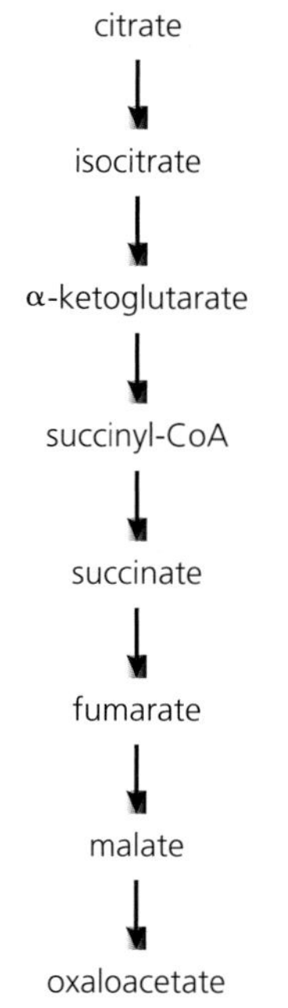

citrate
↓
isocitrate
↓
α-ketoglutarate
↓
succinyl-CoA
↓
succinate
↓
fumarate
↓
malate
↓
oxaloacetate

Figure 9.10 Possible pathway

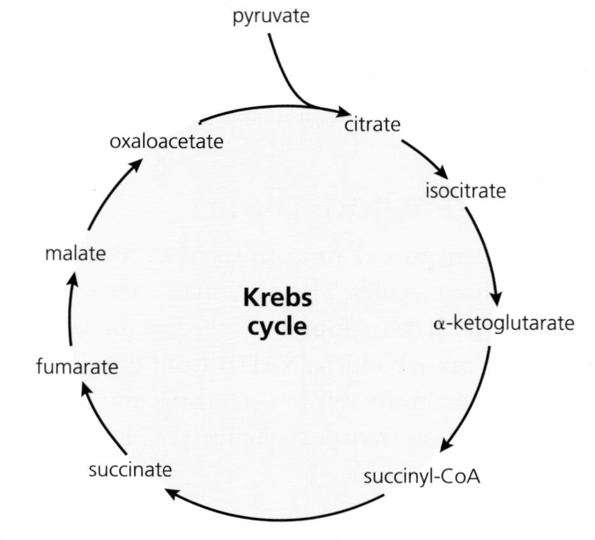

Figure 9.11 Krebs cycle

Related Activity

Demonstrating the effect of malonic acid

Background

- Succinate and fumarate are two of the intermediates in the citric acid cycle.
- When succinate is converted to fumarate during respiration in a living cell, hydrogen is released and passed to the coenzyme FAD. The reaction is catalysed by the enzyme **succinic dehydrogenase** as follows:

$$\text{succinate} + \text{FAD} \xrightarrow{\text{succinic dehydrogenase}} \text{fumarate} + \text{FADH}_2$$

- **Malonic acid** is a chemical that inhibits the action of succinic dehydrogenase.
- In this investigation a chemical called DCPIP is used as the hydrogen acceptor. DCPIP changes colour upon gaining hydrogen as follows:

dark blue (lacks hydrogen) → colourless (has gained hydrogen)

The experiment is carried out as shown in Figure 9.12. From the results it is concluded that in tube A, succinic dehydrogenase in respiring mung bean cells has converted succinate to fumarate and that the hydrogen released has been accepted by DCPIP, turning it colourless. It is concluded that in tube B, the respiratory pathway has been blocked and no hydrogen has been released for DCPIP to accept because malonic acid has inhibited the action of succinic dehydrogenase.

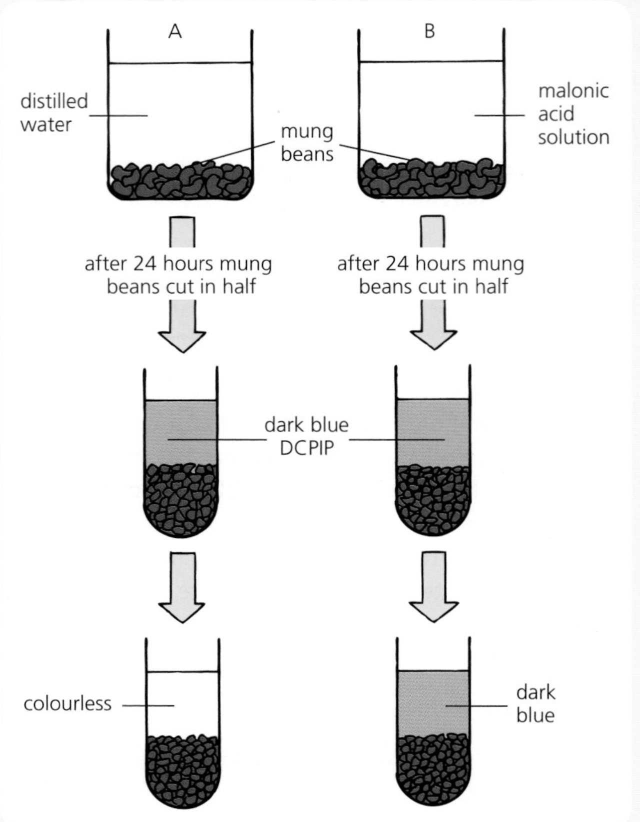

Figure 9.12 Effect of malonic acid

Electron transport chain

An **electron transport chain** consists of a group of carrier protein molecules. There are many of these chains in a cell. They are found attached to the **inner membrane** of mitochondria. NADH from the glycolytic and citric acid pathways releases **electrons** and passes them to the electron transport chains (see Figure 9.13).

ATP synthesis

As the electrons flow along a chain of electron acceptors, they release energy. This is used to pump **hydrogen ions** across the membrane from the inner cavity (matrix) side to the intermembrane space where a higher concentration of hydrogen ions is maintained. The return flow of hydrogen ions to the matrix (the region of lower H^+ concentration) via molecules of **ATP synthase** drives this enzyme to synthesise ATP from ADP and P_i. Most of the ATP generated by cellular respiration is produced in mitochondria in this way.

When the electrons come to the end of the electron transport chain, they combine with **oxygen**, the final electron acceptor. At the same time, the oxygen combines with a pair of hydrogen ions to form **water**. In the absence of oxygen, the electron transport chains do not operate and this major source of ATP becomes unavailable to the cell.

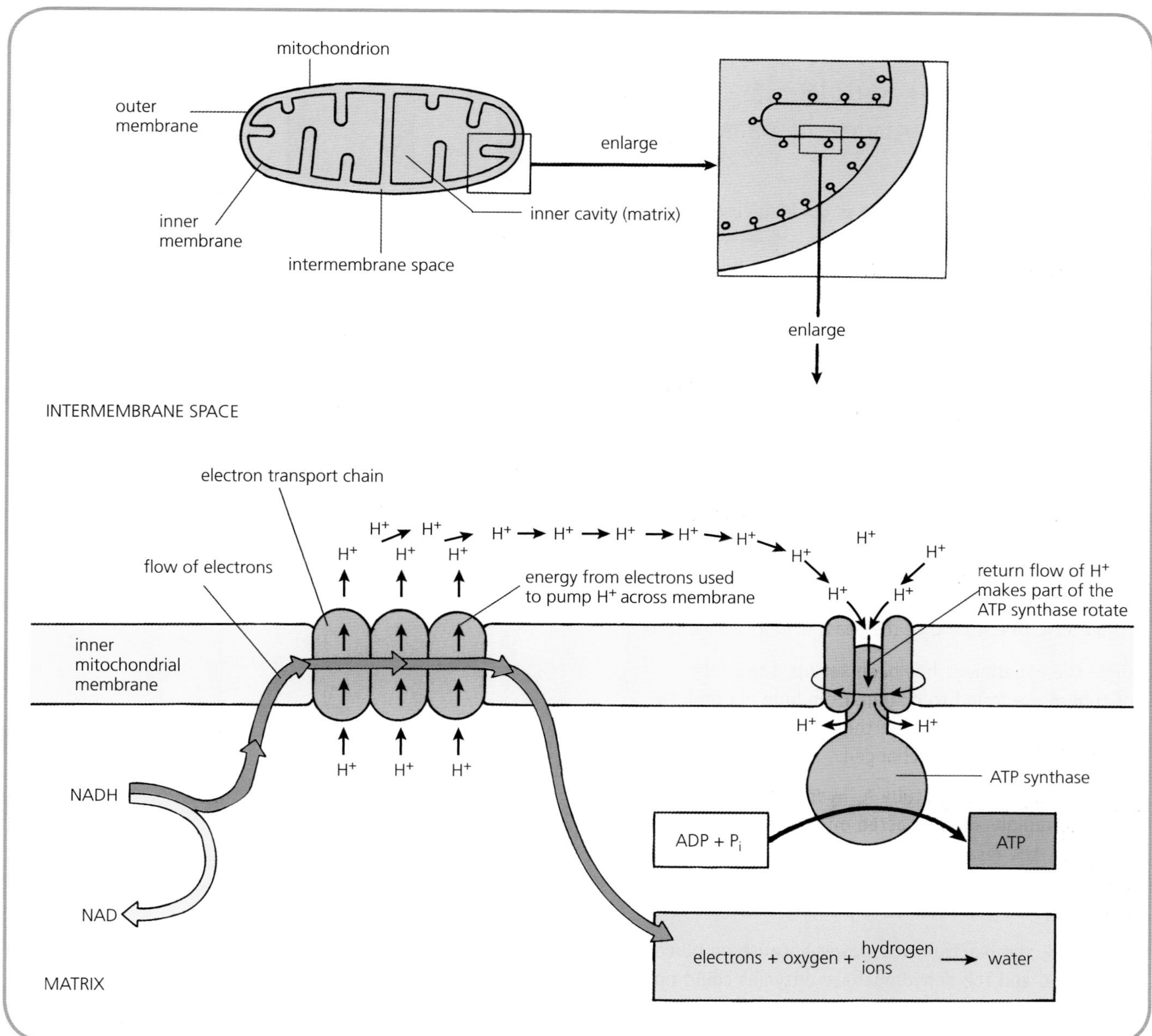

Figure 9.13 Electron transport chain

Related Activity

Investigating the activity of dehydrogenase enzyme in yeast

Background

- During respiration, glucose is gradually broken down and hydrogen released at various stages along the pathway. Each of these stages is controlled by an enzyme called a **dehydrogenase.**
- Yeast cells contain small quantities of stored food which can be used as a respiratory substrate.
- Resazurin dye is a chemical which changes colour upon gaining hydrogen as follows:

blue (lacks hydrogen)	→	pink (some hydrogen gained)	→	colourless (much hydrogen gained)

- Before setting up the experiment shown in Figure 9.14, dried yeast is added to water and aerated for an hour at 35 °C to ensure that the yeast is in an active state.

Once the experiment has been set up, the contents of tube A are found to change from blue via pink to colourless much faster than those in tube B. Tube C, the control, remains unchanged.

It is concluded that in tube A, hydrogen has been released rapidly and has acted on, and changed the colour of, the resazurin dye. For this to be possible, dehydrogenase enzymes present in the yeast cells must have acted on glucose, the respiratory substrate.

In tube B, the reaction was slower because no glucose was added and the dehydrogenase enzymes could only act on any small amount of respiratory substrate already present in the yeast cells.

In tube C, boiling has killed the cells and denatured the dehydrogenase enzymes.

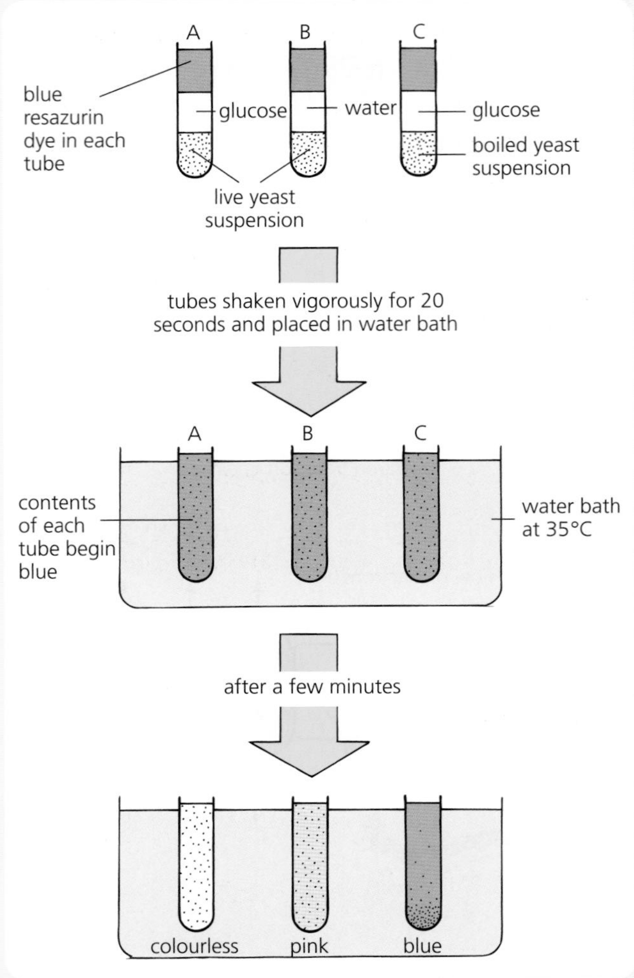

Figure 9.14 Dehydrogenase activity

Fermentation

This is the process by which a little energy is derived from the **partial** breakdown of sugar in the absence of oxygen (anaerobic conditions). Since oxygen is unavailable to the cell, the citric acid cycle and electron transport chain cannot operate. Only glycolysis can occur. Each glucose molecule is converted to pyruvate and yields a net gain of two ATP. Then the pyruvate continues along an alternative metabolic pathway. The form that this takes depends on the type of organism involved.

Plants

The equation below summarises fermentation in plant cells such as yeast deprived of oxygen and cells of roots in water-logged soil:

glucose → pyruvate → alcohol (ethanol) + carbon dioxide

Animals (and some bacteria)

The equation below summarises fermentation in animal cells such as skeletal muscle tissue:

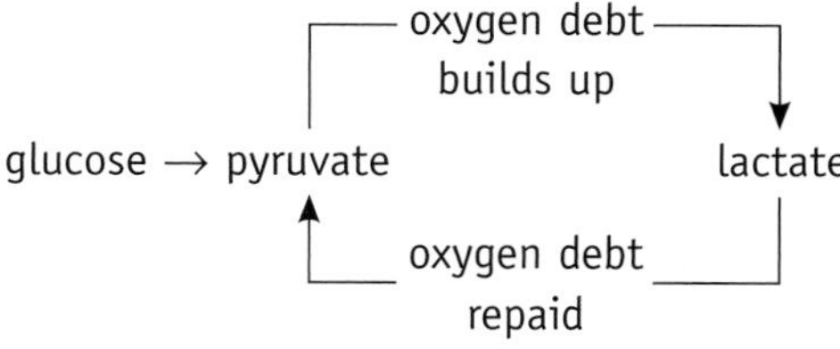

During the formation of lactate (lactic acid), the body accumulates an **oxygen debt**. This is repaid when oxygen becomes available (see Figure 9.15) and lactate is converted back to pyruvate which enters the aerobic pathway. Fortunately the biochemical conversion suggested in Figure 9.16 never really takes place.

Figure 9.15 Repayment of oxygen debt

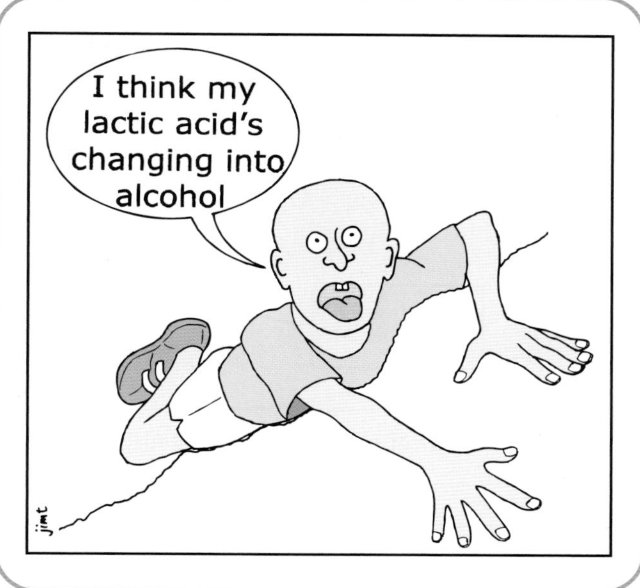

Figure 9.16 Anaerobic nightmare

ATP totals

Compared with aerobic respiration, fermentation is a less efficient process. It produces only two ATP per molecule of glucose whereas respiration in the presence of oxygen produces many ATP per molecule of glucose. The majority of living things thrive in oxygen and respire aerobically. They only resort to fermentation to obtain a little energy for survival while oxygen is absent.

Role of ATP

Transfer of energy via ATP

ATP is important because it acts as the **link** between catabolic energy-releasing reactions (such as respiration) and anabolic energy-consuming reactions (such as the synthesis of proteins, transmission of nerve impulses and replication of DNA). It provides the means by which chemical energy is transferred from one type of reaction to the other in a living cell (see Figures 9.17 and 9.18).

Turnover of ATP molecules

It has been estimated that an active cell (for example, a bacterium undergoing cell division) requires approximately two million molecules of ATP per second to satisfy its energy requirements. This is

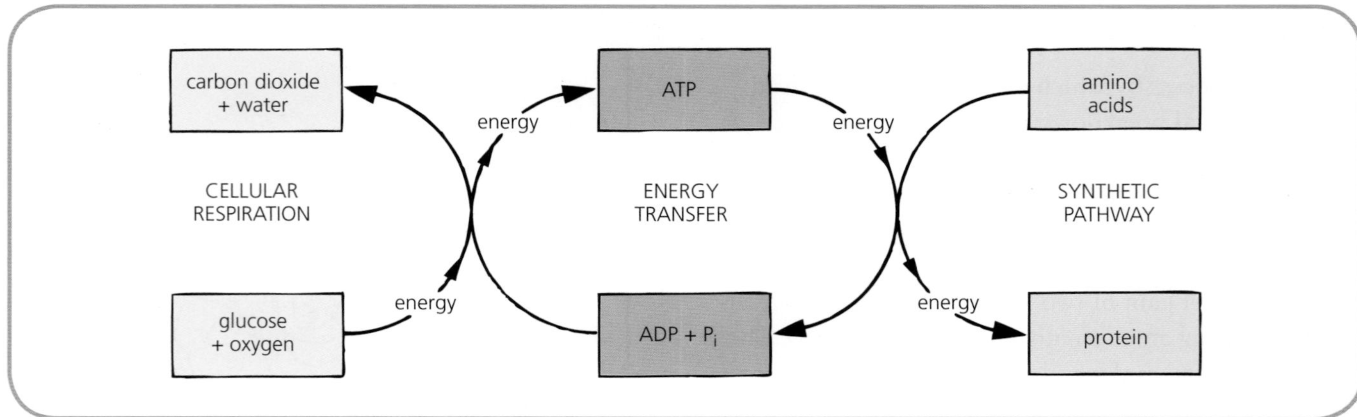

Figure 9.17 Transfer of chemical energy by ATP

Figure 9.18 'What do you mean, I need **ATP**? I thought you said **a teepee**.'

made possible by the fact that a **rapid turnover** of ATP molecules occurs constantly in a cell. At any given moment some ATP molecules are undergoing breakdown and releasing the energy needed for cellular processes while others are being regenerated from ADP and P_i using energy released during cell respiration.

Fixed quantity of ATP

Since ATP is manufactured at the same time as it is used up, there is no need for a living organism to possess a vast store of ATP. The quantity of ATP present in the human body, for example, is found to remain fairly **constant** at around 50 g despite the fact that the body may be using up *and* regenerating ATP at a rate of about 400 g h^{-1}.

Related Activity

Measuring ATP using luciferase

Background

- **Luciferase** is an enzyme present in the cells of fireflies. It is involved in the process of bioluminescence (the production of light by a living organism).
- Luciferase catalyses the following reaction:

$$\text{luciferin} + \text{ATP} \xrightarrow{\text{luciferase}} \text{end products} + \text{light energy}$$

- The presence of **ATP** is essential for the production of light energy and the reaction does not proceed in its absence.

→

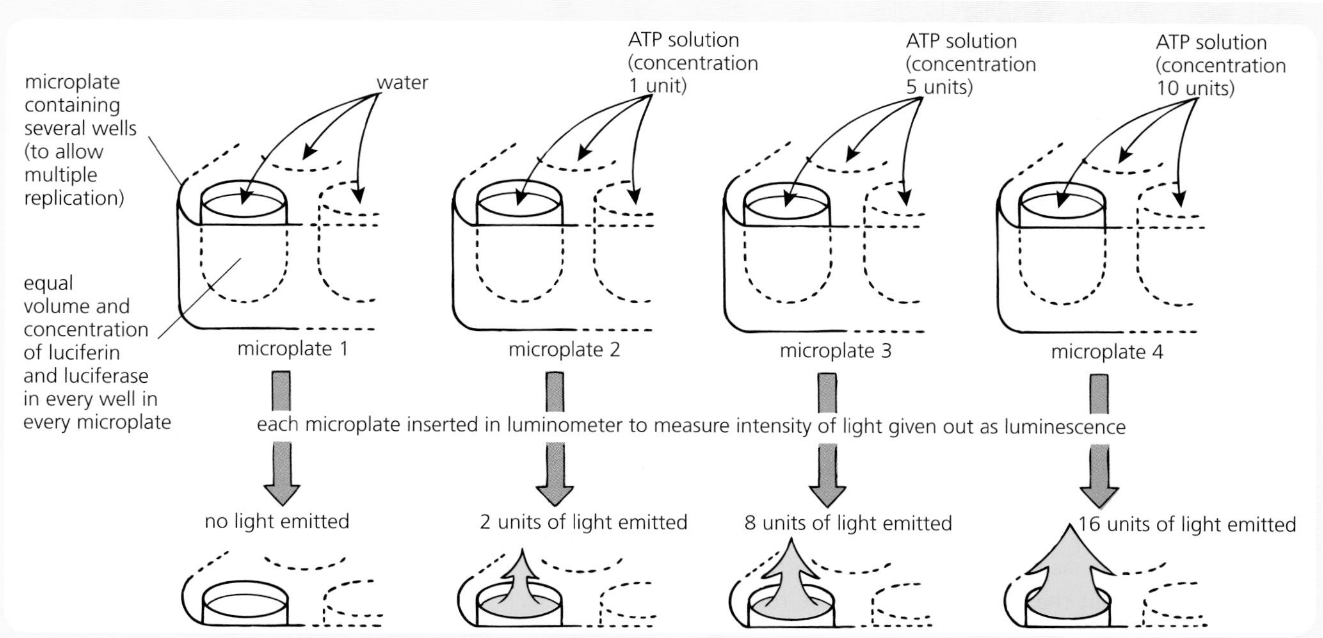

Figure 9.19 Measuring light emitted from known concentrations of ATP

- When luciferin and luciferase are plentiful and ATP is the limiting factor, the intensity of light emitted is proportional to the concentration of ATP present.

The experiment is carried out as shown in Figure 9.19 and the results used to draw a graph of known values of ATP concentration (see Figure 9.20). When the experiment is repeated using material of unknown ATP content, the ATP concentration can be determined from the graph. For example, the sample shown in Figure 9.21 would contain 7.5 units of ATP.

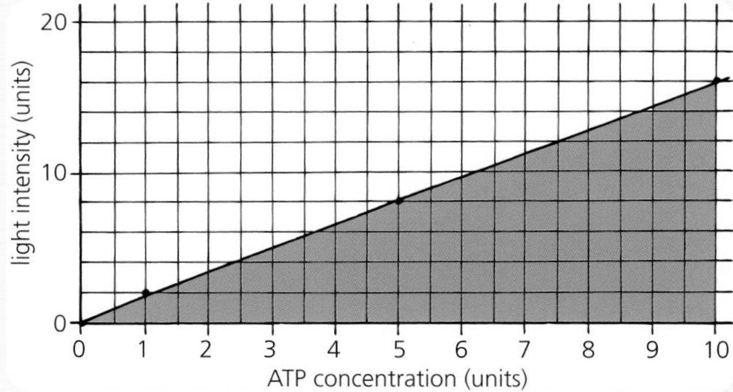

Figure 9.20 Graph of luciferase results

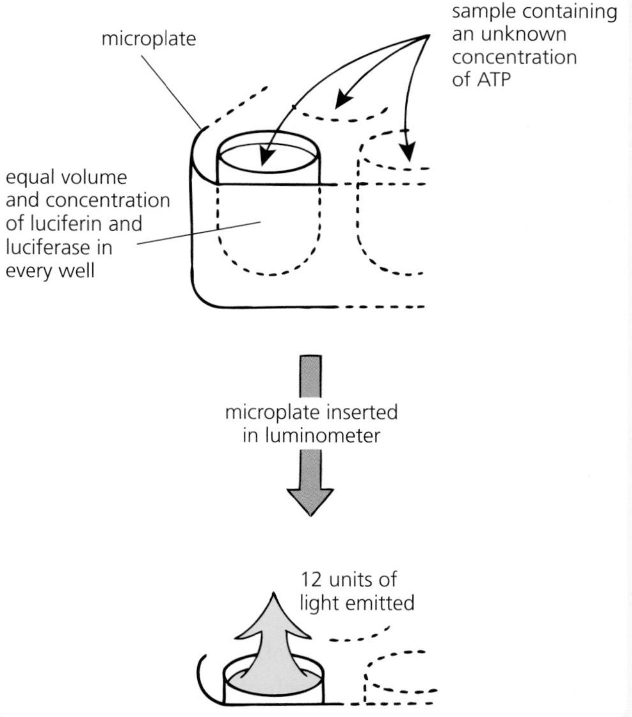

Figure 9.21 Measuring light emitted from an unknown concentration of ATP

Testing Your Knowledge

1 a) What compound is represented by the letters ATP? (1)
b) What is the structural difference between ATP and ADP? (1)
c) Give a word equation to indicate how ATP is regenerated in a cell. (2)

2 Explain each of the following:
a) During the glycolysis of one molecule of glucose, the net gain is two and not four molecules of ATP. (1)
b) Living organisms have only small quantities of oxaloacetate in their cells. (1)

3 Using the letters G, C and E, indicate whether each of the following statements refers to glycolysis (G), citric acid cycle (C) or electron transport chain (E). (Some statements may need more than one letter.) (8)
a) It brings about the breakdown of glucose to pyruvate.
b) It ends with the production of water.
c) It begins with acetyl from acetyl coenzyme A combining with oxaloacetate.
d) It involves a cascade of electrons which are finally accepted by oxygen.
e) It has an energy investment phase and an energy payoff phase.
f) It results in the production of NADH.
g) It involves the release of carbon dioxide.
h) It results in the production of ATP.

4 a) Give the word equation of fermentation in
i) a plant cell
ii) an animal cell. (2)
b) Which of these forms of respiration is quickly reversed when oxygen becomes available? (1)

5 Rewrite the following sentence to include the correct answer from each underlined choice. (4)

<u>ADP/ATP</u> is used to transfer energy to cellular processes such as protein <u>synthesis/digestion</u> and transmission of nerve <u>impulses/responses</u> that <u>release/require</u> energy.

Applying Your Knowledge and Skills

1 Figure 9.22 shows, in a simple way, the molecular structure of three types of sugar.

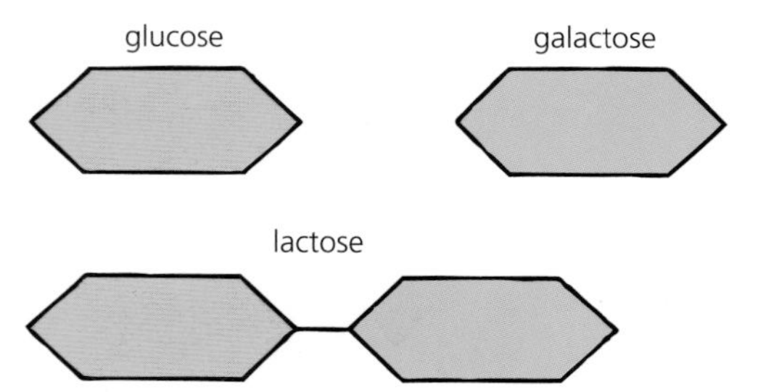

Figure 9.22

Table 9.1 shows the results of an investigation into the use by yeast of each of these sugars as its respiratory substrate.

Time (min)	Total volume of carbon dioxide released (ml)		
	Glucose	Galactose	Lactose
0	0.0	0.0	0.0
10	0.0	0.0	0.0
20	0.5	0.0	0.0
30	3.0	0.5	0.5
40	6.0	1.0	0.5
50	11.0	1.5	1.5
60	16.0	1.5	1.5
70	18.0	2.0	2.0
80	32.5	2.0	2.5
90	39.5	2.0	3.0

Table 9.1

a) Draw a line graph of the results on the same sheet of graph paper using three different colours. (4)
b) i) Identify the glucose result that was least reliable.
ii) Justify your choice. (2)
c) What percentage increase in total volume of carbon dioxide released occurred for glucose between 20 minutes and 80 minutes? (1)
d) What conclusion can be drawn about yeast's ability to make use of each of the sugars as its respiratory substrate? (1)

→

e) i) What conclusion can be drawn about yeast's ability to break lactose down into its component sugars within the given time span?
ii) Explain your answer. (2)

f) What could be done to improve the reliability of the results? (1)

2 Table 9.2 refers to the process of cellular respiration in the presence of oxygen.

a) Copy the table and complete the blanks indicated by brackets. (5)

b) Which stage consists of an energy investment phase followed by an energy pay-off phase? (1)

c) At which stage is *most* ATP produced per molecule of glucose? (1)

d) Which TWO stages would fail to occur in the absence of oxygen? (2)

3 Refer back to the investigation shown in Figure 9.12 on page 128. Based only on these results, it could be argued that malonic acid has simply killed the cells.

a) How could the experiment be adapted to investigate whether malonic acid really does act as a competitive inhibitor? (1) (Hint: see Chapter 8, page 114.)

b) Explain how you would know from the results of your redesigned experiment whether or not malonic acid had acted as a competitive inhibitor. (2)

4 Figure 9.23 shows a small region of an inner mitochondrial membrane.

a) i) Which side of the membrane has the higher concentration of H^+ ions?
ii) Explain how this higher concentration of H^+ ions is maintained. (3)

b) i) Name molecule X.
ii) Briefly describe how it works. (3)

c) Cyanide is a chemical that binds with the electron transport chains and brings the flow of high-energy electrons to a halt. Explain why cyanide is poisonous. (2)

5 Metabolism falls into two parts:

- **anabolism** consisting of energy-requiring reactions which involve synthesis of complex molecules
- **catabolism** consisting of energy-yielding reactions in which complex molecules are broken down.

Transfer of energy from catabolic reactions to anabolic reactions is brought about by **ATP**. Figure 9.24 is a summary of this information.

Stage of respiratory pathway	Principal reaction or process that occurs	Products
glycolysis	splitting of glucose into [_____]	[_____], NADH and pyruvate
[_____] acid cycle	removal of [_____] ions from molecules of respiratory [_____]	[_____], [_____] and ATP
[_____] transport chain	release of [_____] to form ATP	ATP and [_____]

Table 9.2

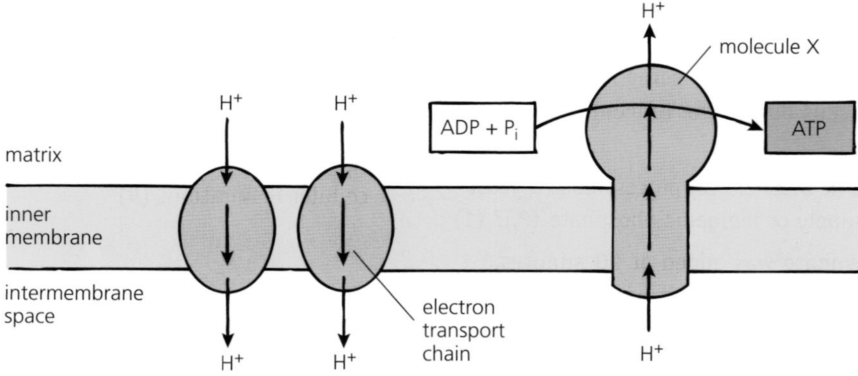

Figure 9.23

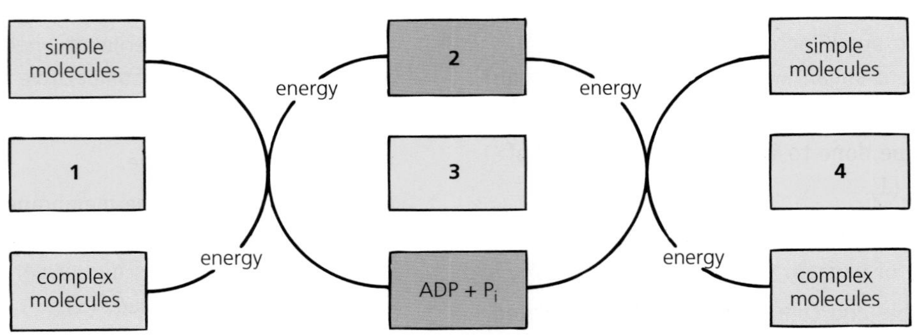

Figure 9.24

a) Copy the diagram and add four arrowheads to show the directions in which the two coupled reactions occur. (2)

b) Complete boxes 1–4 using each of the terms given in purple bold print in the passage. (2)

6 State whether each of the following is an anabolic (A) or a catabolic (C) reaction:

a) destruction of a microbe by enzymes in lysosomes

b) formation of the hormone thyroxin in the thyroid gland

c) conversion of glycogen to glucose in muscle tissue

d) digestion of proteins to amino acids

e) synthesis of nucleic acids. (5)

7 One mole of glucose releases 2880 kJ of energy. During aerobic respiration in living organisms, 44% of this is used to generate ATP. The rest is lost as heat.

a) What percentage of the energy generated during aerobic respiration is lost as heat? (1)

b) Out of a mole of glucose, how many kilojoules are used to generate ATP? (1)

c) Name TWO forms of cellular work that the energy held by ATP could be used to carry out. (2)

8 Figure 9.25 shows the volume of carbon dioxide released by yeast cells during the fermentation of glucose.

a) For which specific biochemical process does a yeast cell require a supply of inorganic phosphate (P_i)? (1)

b) Inorganic phosphate was added at 90 minutes. Name the substance to which it became combined. (1)

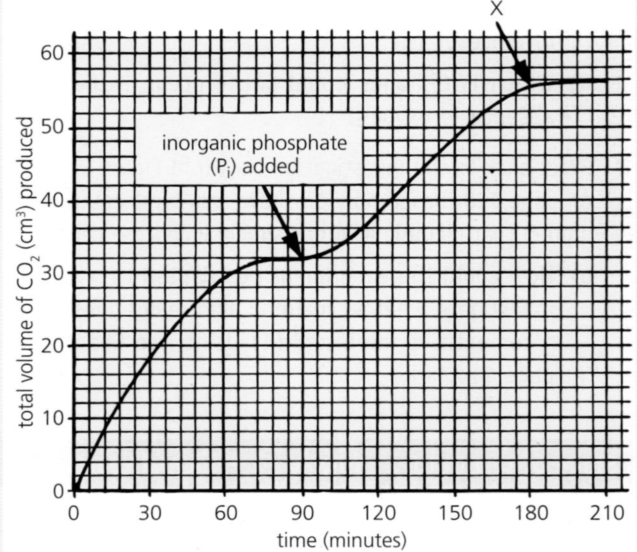

Figure 9.25

c) i) State what happened to the rate of carbon dioxide production following the addition of inorganic phosphate at 90 minutes.

ii) Suggest a reason for this change. (2)

d) Give a possible explanation for the levelling off of carbon dioxide production at point X on the graph. (1)

9 Give an account of the production of NADH and the role played by the electron transport chain during cellular respiration. (9)

What You Should Know

Chapters 8–9

(See Table 9.3 for word bank)

acetyl	enzymes	on
activation	FAD	orientation
ADP	genetic	oxygen
affinity	glycolysis	phosphorylation
anabolism	hydrogen	pores
ATP	induced	product
ATP synthase	inhibit	pumps
break	inhibition	pyruvate
catabolism	investment	regulated
citrate	irreversible	shape
competitive	lactate	structure
complex	lowering	substrate
concentration	metabolism	transferred
electron	NADH	transition
energy	negative	wasted
environment	off	water

Table 9.3 Word bank for chapters 8–9

1 Cell _______ encompasses all the enzyme-catalysed reactions that occur in a cell.

2 _______ consists of biosynthetic metabolic pathways that build up _______ molecules from simpler constituents and need a supply of energy; _______ consists of metabolic pathways that _______ down larger molecules into smaller ones and usually release _______.

3 Some protein molecules in the cell membrane have _______ that allow certain molecules to diffuse through the membrane; others act as _______ and actively transport ions across the membrane against a concentration gradient.

4 For a metabolic reaction to occur, _______ energy is needed to form a _______ state from which end products are produced. _______ catalyse biochemical reactions by _______ the activation energy needed by the reactants to form their transition state.

5 Substrate molecules have an _______ for the active site on an enzyme. The active site's shape determines the _______ of the reactants on it and it binds to them closely with an _______ fit.

6 The enzymes controlling a metabolic pathway usually work as a group. Although some steps are _______, most metabolic reactions are reversible. The direction in which the reaction occurs depends on factors such as concentration of the _______ and removal of a _______ as it becomes converted to another metabolite.

7 Each step in a metabolic pathway is _______ by an enzyme which catalyses a specific reaction. Each enzyme is under _______ control.

8 Some metabolic pathways are required continuously and the genes that code for their enzymes are always switched _______. Other pathways are only needed on certain occasions. To prevent resources being _______, the genes that code for their enzymes are switched on or _______ as required in response to signals from within the cell and from its _______.

9 Molecules of a _______ inhibitor resemble the substrate in _______. They become attached to the active site and slow down the reaction. Their effect is reversed by increasing the _______ of substrate.

10 Some regulatory molecules stimulate enzyme activity or _______ it non-competitively by changing the _______ of the enzyme molecule and its active site(s).

11 Some metabolic pathways are controlled by end-product _______, a form of _______ feedback control.

12 _______ is a high-energy compound able to release and transfer energy when it is required for cellular processes. ATP is regenerated from _______ and P_i by phosphorylation using energy released during cellular respiration. _______ also occurs when P_i and energy are _______ from ATP to a reactant in a pathway.

13 Cellular respiration begins with _______, the breakdown of glucose to _______. This consists of an energy _______ phase and an energy payoff phase with a net gain of two molecules of ATP.

14 In the presence of oxygen, pyruvate is broken down into carbon dioxide and an _______ group. With the help of coenzyme A, the acetyl group enters the citric acid cycle by combining with oxaloacetate to become _______.

15 As one respiratory substrate is converted to another in the citric acid cycle, carbon dioxide is released, ATP is formed and pairs of ______ ions are removed and passed to coenzyme NAD forming NADH.

16 ______ passes electrons to ______ transport chains where the energy released is used to pump hydrogen ions across inner mitochondrial membranes. The return flow of these hydrogen ions makes each ______ molecule catalyse the synthesis of ATP. ______, the final electron acceptor, combines with hydrogen to form ______.

17 In the absence of oxygen, fermentation occurs. The final metabolic products are ______ and carbon dioxide in plant cells and ______ in animal cells and some bacteria.

10 Metabolic rate

Metabolic rate

The quantity of energy consumed by an organism per unit of time is called its **metabolic rate**. Normally this energy is generated by cells respiring aerobically as summarised in the following equation:

glucose + oxygen → carbon dioxide + water + energy

Therefore metabolic rate can be measured as:

- **oxygen** consumption per unit time
- **carbon dioxide** production per unit time
- **energy** production (as heat) per unit time.

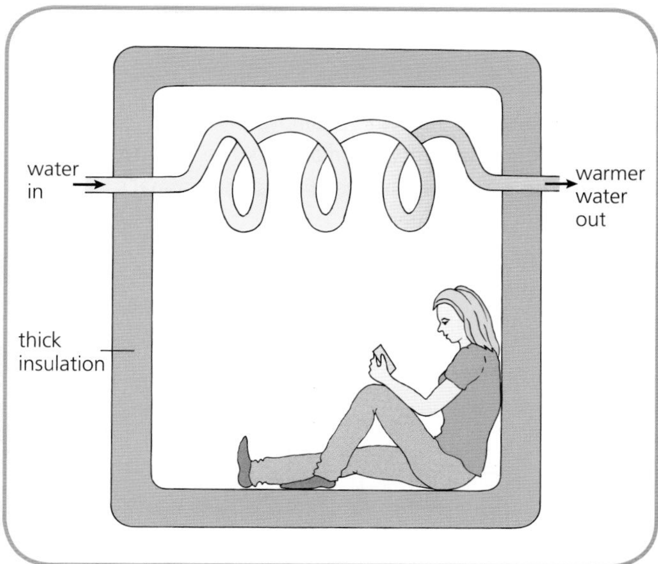

Figure 10.1 Calorimeter

Calorimeter

An organism's metabolic rate can be measured by placing it in a **calorimeter**. This is a well-insulated container containing a pipe through which water flows (see Figure 10.1). Heat generated by the organism causes a rise in temperature of the water in the pipe. By measuring the temperature of the water entering and leaving the calorimeter for a given period of time, the organism's metabolic rate can be calculated from the data collected.

Similarly an organism's metabolic rate can be measured by placing it in a **respirometer**. This is a chamber through which a continuous stream of air is pumped. Differences in oxygen concentration, carbon dioxide concentration and temperature between the air entering and the air leaving the respirometer are detected by **probes** (also see Investigation). Table 10.1 compares metabolic rates (as oxygen consumption) of various animals at rest.

Animal	Volume of oxygen consumed ($mm^3 g^{-1}$ body mass h^{-1})
sea anemone	13
octopus	80
eel	128
frog	150
human	200
mouse	1 500
hummingbird	3 500

Table 10.1 Metabolic rates at rest

Investigation

Measuring metabolic rate using probes (sensors)

The experiment is shown in Figure 10.2. Table 10.2 gives the purpose of each piece of equipment.

Equipment	Purpose
soda lime tube (containing sodium hydroxide)	to absorb all carbon dioxide from incoming air so that its initial concentration is not a variable factor
air pump	to pump a continuous flow of air through the system
flow meter	to maintain the flow of air at a steady rate that is low enough for the carbon dioxide sensor to work
animal chamber	to accommodate the animal whose metabolic rate is to be measured
temperature probe (sensor)	to measure changes of temperature in the animal chamber and send data to computer
condensing bath and drying column	to remove water vapour from passing air since the sensors need air to be dry
oxygen probe (sensor)	to measure percentage oxygen concentration and send data to computer
carbon dioxide probe (sensor) and analyser	to measure carbon dioxide in parts per million and send data to computer

Table 10.2 Purposes of respirometer equipment

The three probes (sensors) are calibrated in advance. The animal is inserted into the chamber and the experiment run for a set length of time (such as 30 minutes). The computer software monitors the data from the three sensors simultaneously and displays the information on the screen. From these data the animal's **metabolic rate** (as volume of oxygen consumed per unit time) can be determined.

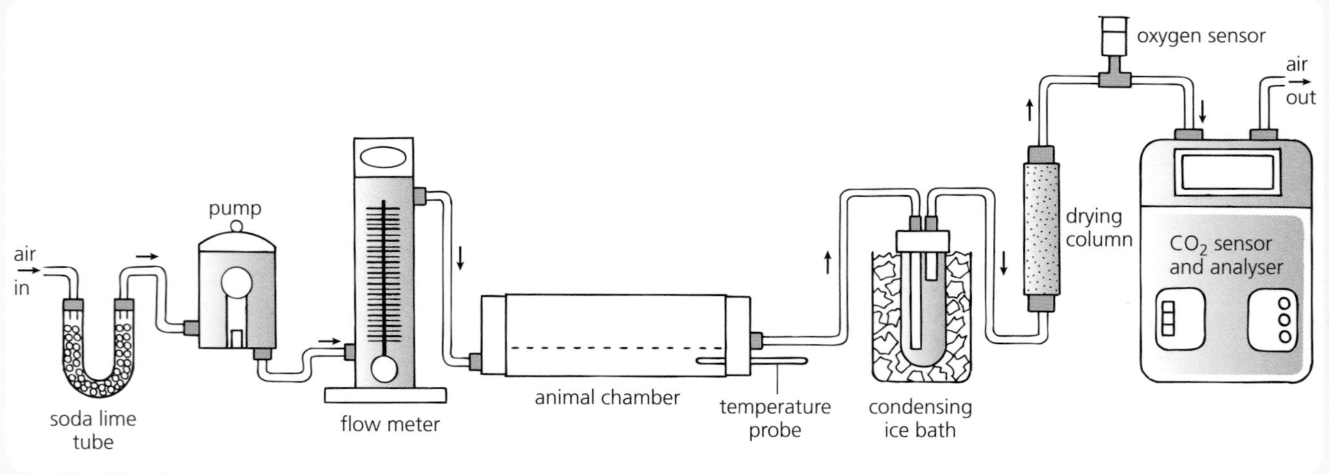

Figure 10.2 Investigating metabolic rate (connections to computer not shown)

Oxygen delivery

As an organism's metabolic rate increases to meet an increasing demand for energy (for example, for rapid movement or maintenance of body temperature), its rate of aerobic respiration and consumption of oxygen increases. Therefore aerobic organisms with high metabolic rates need **efficient transport systems** to deliver large supplies of oxygen to respiring cells.

Circulatory systems in vertebrates

All vertebrates have a **closed** circulatory system where the blood is contained in a continuous circuit of blood vessels and is kept moving by a muscular pump, the heart. In such a cardiovascular system, the heart pumps the blood into large vessels that branch into smaller and smaller vessels. The smallest of these vessels are thin-walled capillaries which allow **oxygen** to pass rapidly from the bloodstream to the fluid which bathes respiring cells and then on into the cells. Carbon dioxide moves in the opposite direction. The arrangement of the heart chambers and the circulatory system vary among the vertebrate groups.

Single circulatory system

The circulatory system in a fish is described as **single** because blood passes through the **two**-chambered heart **once only** for each complete circuit of the body (see Figure 10.3). In any closed circulatory system, a **drop in pressure** occurs when blood passes through a capillary bed because it is a network of narrow tubes which offer resistance to flow of blood.

In a fish, blood flows to the gills at **high** pressure but is delivered next to the capillary beds of the body at **low** pressure. Therefore this is a primitive and relatively inefficient method of circulation compared to more advanced vertebrate groups.

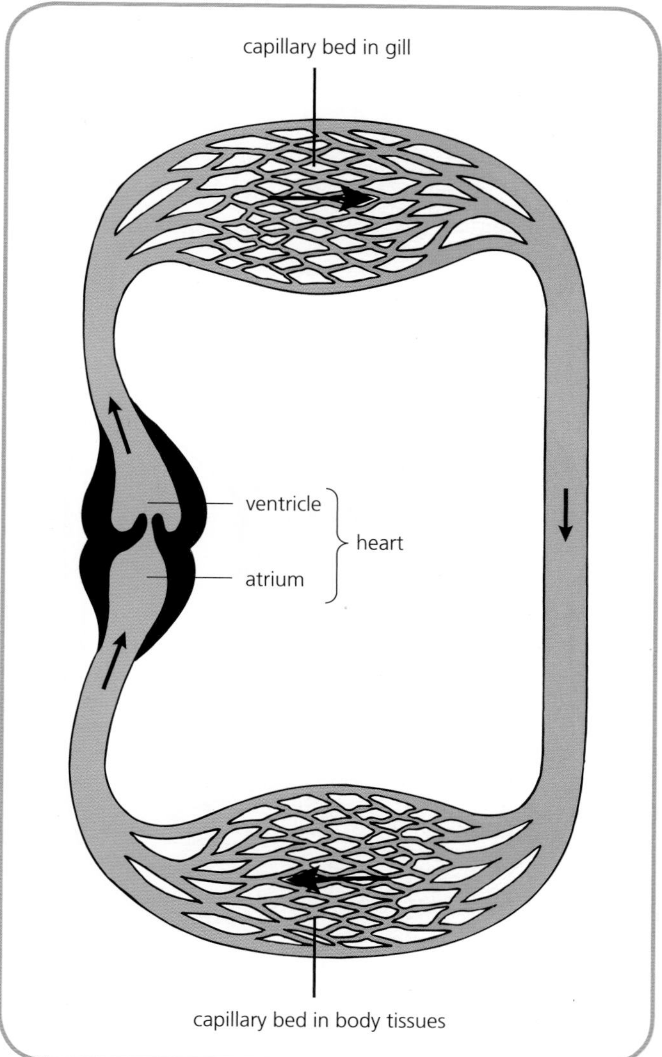

Figure 10.3 Single circulatory system

Double circulatory system

The type of system present in the other vertebrate groups is described as **double** because blood passes through the heart twice for each complete circuit of the body. Blood is pumped to both the lungs and the body's capillary beds at **high** pressure ensuring a vigorous flow to all parts. Therefore a double circulation is more efficient than a single one.

Incomplete

In amphibians and reptiles, the system is described as **incomplete** because there is only **one ventricle** in the heart (see Figure 10.4) and some mixing of oxygenated blood from the lungs and deoxygenated blood from the body occurs.

In amphibians, this mixing is not a major problem because blood returning from the body has been partly oxygenated by gas exchange through the animal's moist skin. In most reptiles, little mixing occurs because the single ventricle is partly divided by a septum.

Complete

In birds and mammals, the system is described as **complete** because the heart has **two ventricles** completely separated by a septum (see Figure 10.5). Therefore no mixing of oxygenated and deoxygenated blood occurs. The complete double circulatory system is the most advanced and efficient circulatory system. It enables an endothermic ('warm-blooded') vertebrate to deliver large quantities of oxygen to respiring tissues which release heat during metabolism and keep its body warm.

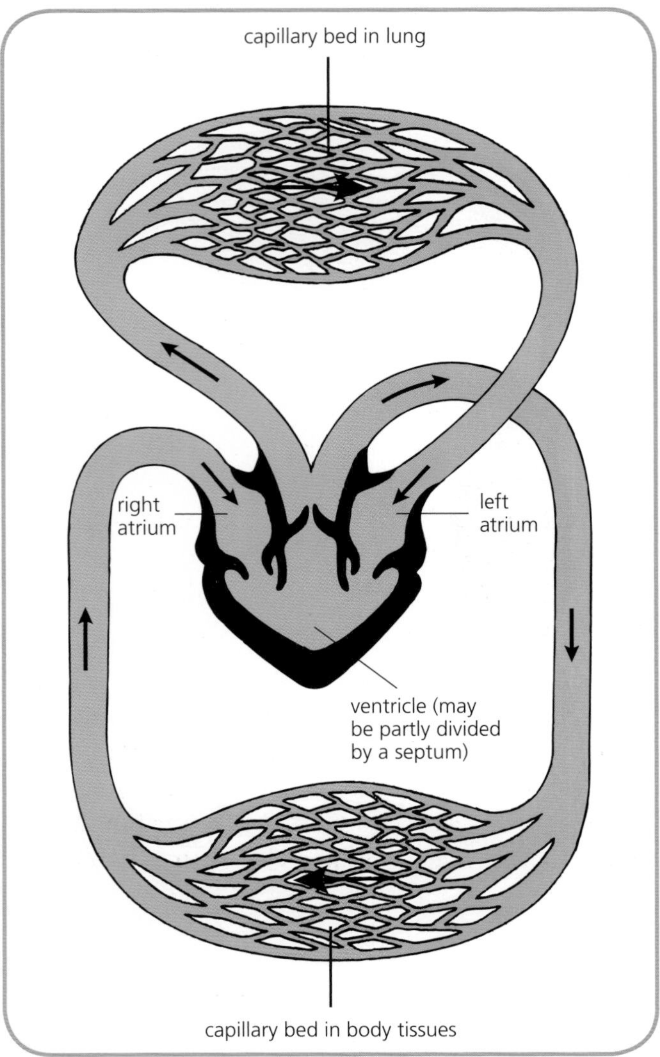

Figure 10.4 Incomplete double circulatory system

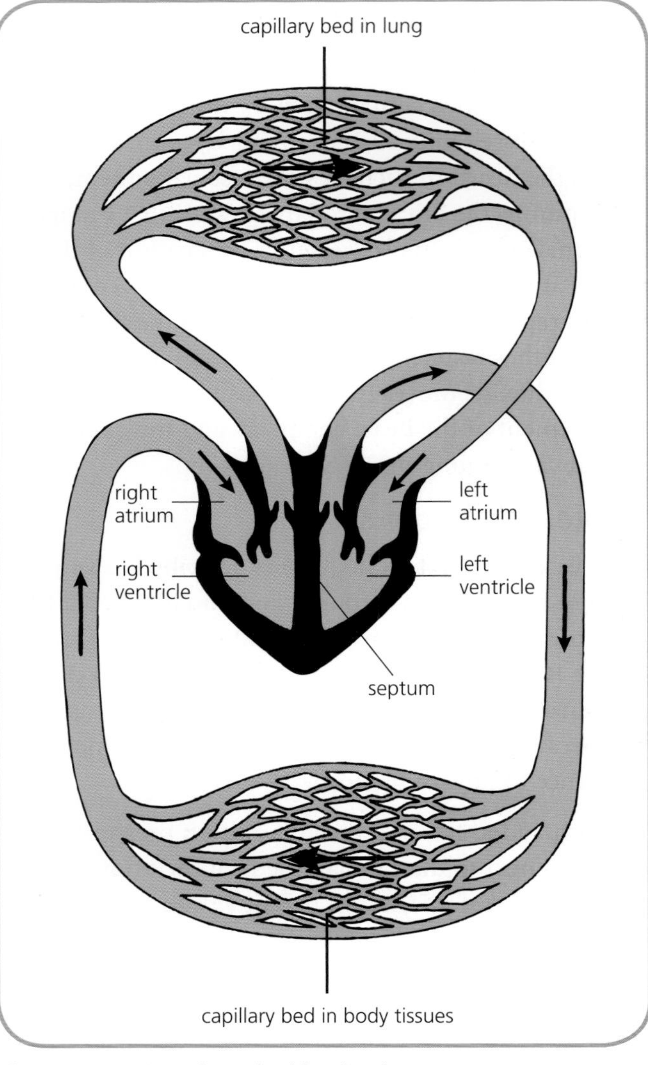

Figure 10.5 Complete double circulatory system

Testing Your Knowledge 1

1 a) Define the term *metabolic rate*. (1)

b) State TWO ways in which metabolic rate can be measured. (2)

2 Rewrite the following sentences, choosing the correct answer at each underlined choice. (3)

The heart of a fish contains <u>two/three</u> chambers. Blood is pumped at <u>high/low</u> pressure to the gills and then on to the body's capillary beds at <u>high/low</u> pressure. The heart of a mammal contains <u>three/four</u> chambers. Blood is pumped to the mammal's lungs at <u>high/low</u> pressure and to the body's capillary beds at <u>high/low</u> pressure.

3 Decide whether each of the following statements is true or false and then indicate your choice using T or F. Where a statement is false, give the word that should have been used in place of the word in bold print. (6)

a) Metabolic rate can be investigated using an oxygen **probe**.

b) For each complete circuit of the body, blood passes through a fish's heart **twice**.

c) Some mixing of oxygenated and deoxygenated blood occurs in the heart of a **bird**.

Metabolism in conformers and regulators

Fluctuations in an external abiotic factor such as salinity or temperature may occur in an organism's environment. When this happens some organisms, called **regulators**, are able to alter their normal metabolic rate and maintain a steady state (see Figure 10.6) by employing physiological mechanisms.

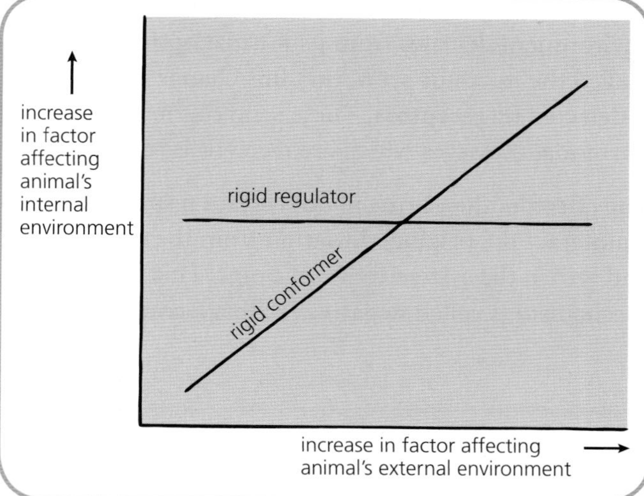

Figure 10.6 Responses to an environmental variable

Other organisms, called **conformers**, are unable to alter their normal metabolic rate by these means.

Conformers

The state of a conformer's internal environment is **directly dependent** upon the abiotic factors that affect its external environment. However, this does not normally pose a problem because conformers live in environments that remain relatively stable (such as the ocean floor).

The advantage of this way of life is that the animal's metabolic costs are **low** since it does not employ energy-consuming physiological mechanisms to maintain its inner steady state. The disadvantage is that the animal is restricted to a **narrow range** of ecological niches and is **less adaptable** to environmental change.

Few organisms are complete conformers. Many employ **behavioural responses** to maintain their optimum metabolic rate. Lizards, for example, are unable to maintain their body temperature by employing physiological mechanisms such as shivering but they can manage it by behavioural means such as basking in sunshine.

Case Study: Response of a conformer to a change in an environmental factor

Thermal biology of *Anolis cristatellus*

Anolis cristatellus is the scientific name of a small lizard (see Figure 10.7) that lives in open lowlands and dense forests in Puerto Rico. Scientists have found that those members of the species living in open, sunny, lowland habitats are not complete conformers. They frequently raise their body temperature by basking in sunshine early and late in the day. For these animals the metabolic cost (energy expenditure) incurred by travelling a short distance to a sunny spot is low compared with the physiological benefits gained from this means of thermoregulation.

On the other hand, those members of the species that live in the shady, forest habitats tend to be almost complete conformers. They passively allow their body temperature to drop to that of their cool surroundings and very rarely bask in the sun. This behaviour is explained by the fact that the **metabolic cost** of travelling a relatively long distance in order to find a rare, sunny spot in the forest **outweighs any benefits** gained by raising their body temperature. It also increases the risk of capture by predators.

Figure 10.7 *Anolis cristatellus*

Regulators

The state of a regulator's internal environment is not directly dependent upon the abiotic factors that affect its external environment. Regulators employ **physiological means** to control their inner steady state. The Atlantic salmon, for example, spends part of its life in fresh water and part of its life in salt water. Yet it manages to maintain the solute concentration of its blood at a steady state by osmoregulation.

This is of advantage because the animal is able to exploit a **wider range** of ecological niches. For example, the Atlantic salmon is able to use the relatively safe, freshwater environment for breeding purposes but migrate to food-rich, marine waters during its growing years. The disadvantage is that the animal has to **expend energy** generated by its metabolism on the physiological mechanisms (such as osmoregulation) needed to maintain its inner steady state.

Physiological homeostasis

Physiological homeostasis is the maintenance of the body's internal environment within certain tolerable limits despite changes in the body's external environment. This regulation is brought about by **negative feedback control** and requires energy.

Principle of negative feedback control

When some factor affecting the body's internal environment deviates from its normal optimum level (called the **norm** or **set point**) this change in the factor is detected by **receptors**. These send out nerve or hormonal messages which are received by **effectors**.

The effectors then bring about certain responses which counteract the original deviation from the norm and return the system to its set point. This corrective mechanism is called **negative feedback control**

Related Topic

Comparison of marine and estuarine invertebrates

The **spider crab** spends its life in sea water. The solute concentration of its blood is equal to that of the surrounding sea water. If it is placed in an environment with a higher or lower solute concentration than sea water, its body is unable to maintain a steady inner solute concentration by osmoregulation (see Figure 10.8). Instead it conforms to the environment by losing water or taking in water until its solute concentration is equal to that of the environment, even if this proves to be fatal.

The **shore crab**, on the other hand, can regulate the solute concentration of its body fluids to some extent. This enables the shore crab to operate in sea water and in river estuaries where the water contains far less salt than sea water. By being a regulator, the shore crab is able to exploit a wider range of environments than the spider crab, a conformer. However, the shore crab must find relatively more food to provide the energy expended during osmoregulation.

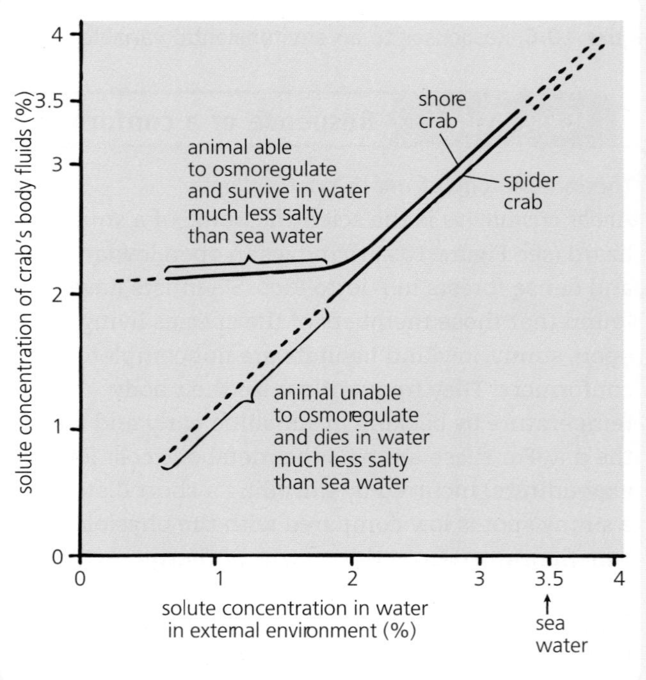

Figure 10.8 Comparison of conformer and regulator

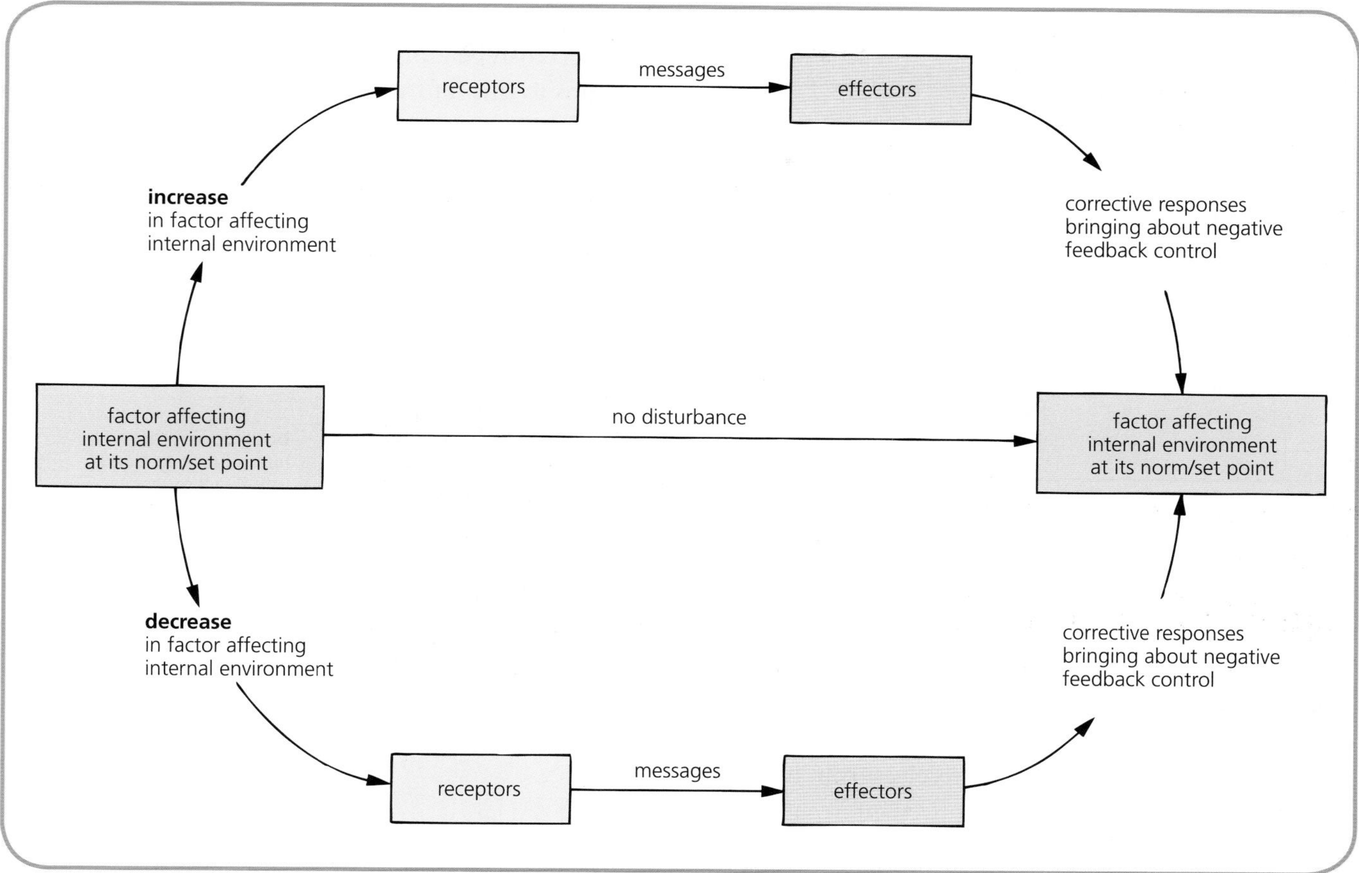

Figure 10.9 Principle of negative feedback control

(see Figure 10.9). It provides the stable environmental conditions needed by the body to function efficiently despite wide fluctuations in the external environment (many of which would be unfavourable).

Thermoregulation

Ectotherm

An **ectotherm** is an animal which is unable to regulate its body temperature by physiological means. Invertebrates, fish, amphibians and reptiles are, almost without exception, ectotherms and their body temperature normally varies directly with that of the external environment. They obtain most of their body heat by absorbing it from the surrounding environment.

Endotherm

An **endotherm** is an animal which is able to maintain its body temperature at a relatively constant level independent of the temperature of the external environment. All birds and mammals are endotherms. They have a **high metabolic rate** which generates most or all of their body's heat energy.

Importance of regulating body temperature

Most **enzymes** work best at 35–40 °C. Animals that can maintain their body temperature within this range possess an efficient and active metabolism. It consists of enzyme-controlled reactions and processes involving molecular diffusion that proceed at **optimal rates** regardless of the external temperature. Therefore terrestrial endotherms are capable of intense physical activity at all times of the day and night. They have an advantage over ectotherms whose metabolic rate slows down when the external temperature decreases.

The graph in Figure 10.10 summarises the effect of an increase in external temperature on body temperature of the two types of animal. Thermoregulation is brought about by **homeostatic control**.

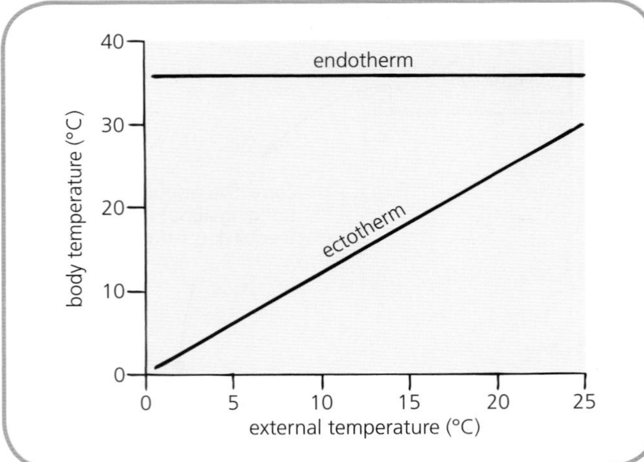

Figure 10.10 Effect of external temperature on body temperature

Role of hypothalamus

In addition to playing many other roles, the **hypothalamus** is the body's temperature-monitoring centre (see Figure 10.11). It acts as a **thermostat** and is sensitive to nerve impulses that it receives from heat and cold receptors in the skin. These convey information to it about the surface temperature of the body.

In addition, the hypothalamus itself possesses central **thermoreceptors**. These are sensitive to changes in temperature of blood which in turn reflect changes in the temperature of the **body core** (see Figure 10.12). The thermo-regulatory centre in the hypothalamus responds to this information by sending appropriate nerve impulses to **effectors**. These trigger corrective feedback mechanisms and return the body temperature to its normal level (set point).

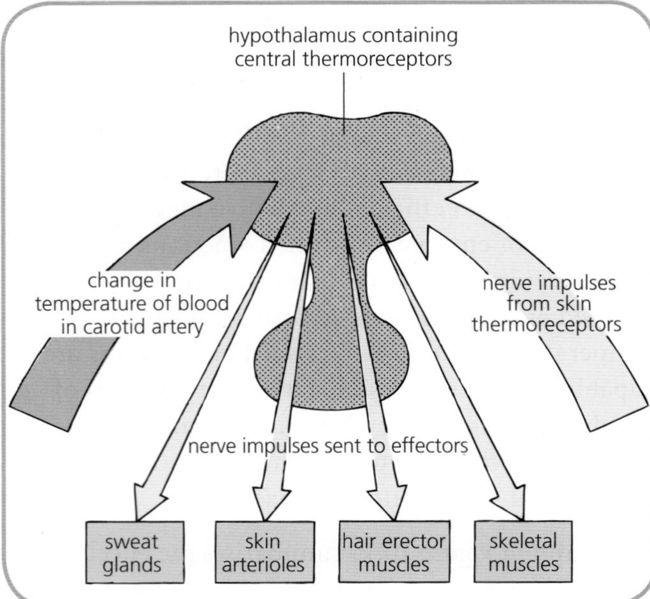

Figure 10.11 Hypothalamus as a temperature-monitoring centre

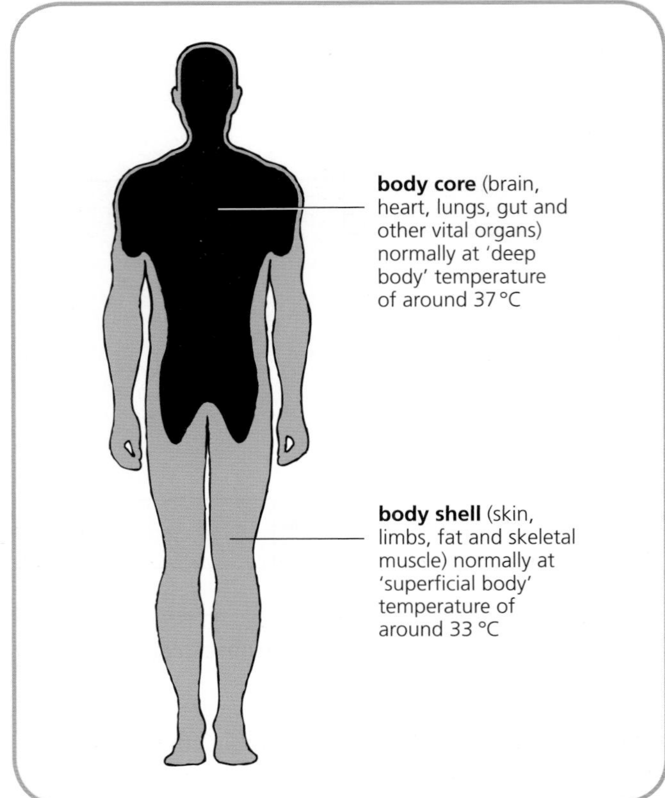

Figure 10.12 Body core and body shell

Role of skin

The **skin** plays a leading role in temperature regulation. In response to nerve impulses from the hypothalamus, the skin acts as an effector.

Correction of overheating

The skin helps to correct overheating of the body by employing the following mechanisms which promote heat loss.

Vasodilation

Arterioles leading to skin become **dilated** (see Figure 10.13). This allows a large volume of blood to flow through the capillaries near the skin surface. From here, the blood is able to lose heat by **radiation**.

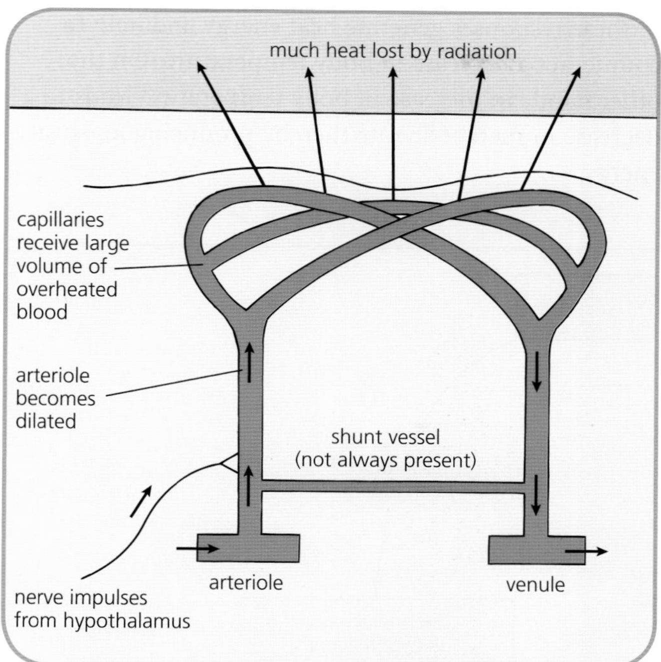

Figure 10.13 Vasodilation in skin

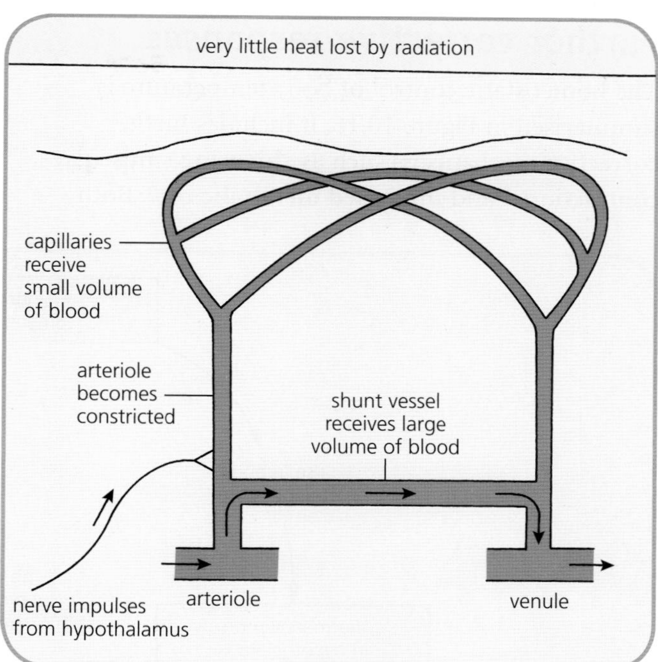

Figure 10.14 Vasoconstriction in skin

Increase in rate of sweating

Heat energy from the body is used to convert the water in sweat to **water vapour** and by this means brings about a lowering of body temperature.

Correction of overcooling

The skin helps to correct overcooling of the body by employing the following mechanisms which reduce heat loss.

Vasoconstriction

Arterioles leading to the skin become **constricted** (see Figure 10.14). This allows only a small volume of blood to flow to the surface capillaries. Little heat is therefore lost by radiation.

Decreased rate of sweating

Since sweating is reduced to a minimum, heat is conserved.

Contraction of erector muscles

This process (see Figure 10.15) is more effective in furry animals than in human beings. It results in hairs being raised from the skin surface. A wide layer of air, which is a poor conductor of heat, is trapped between the animal's body and the external environment. This layer of **insulation** reduces heat loss.

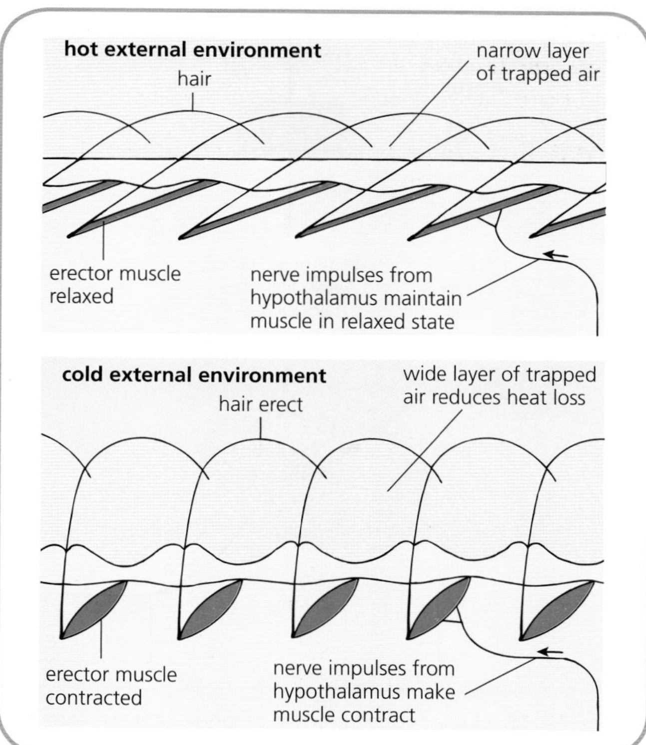

Figure 10.15 Action of hair erector muscles

Further corrective responses

The homeostatic control of body temperature is summarised in Figure 10.16. It includes further corrective mechanisms such as **shivering** (muscular contractions) and **increased metabolic rate**. Both of these responses generate heat energy and help to counteract a decrease in body temperature. On the other hand, an increase in body temperature leads to a decrease in metabolic rate thereby producing less heat energy.

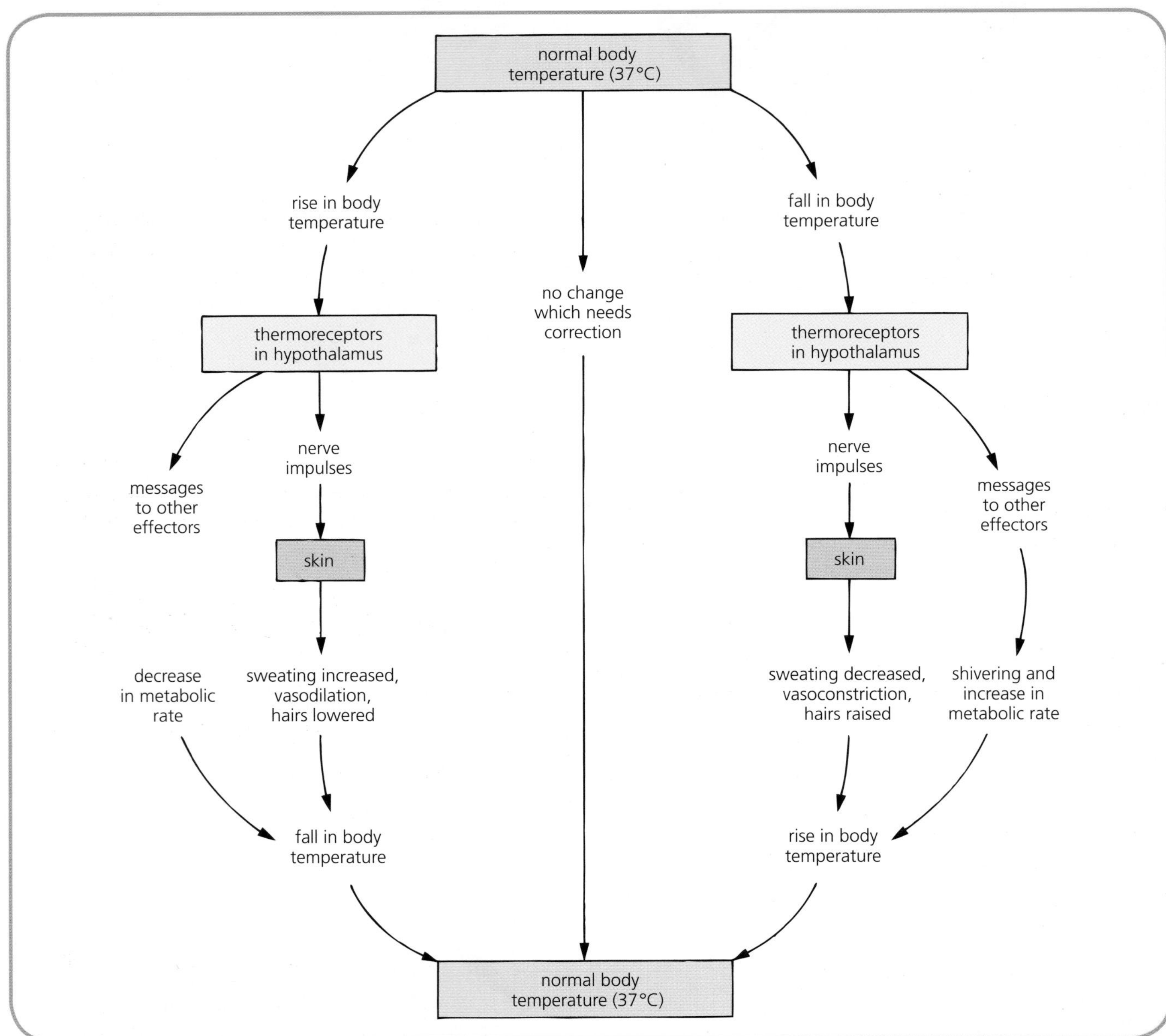

Figure 10.16 Homeostatic control of body temperature

Related Activity

Investigating response to sudden heat loss using a thermistor

A **thermistor** is a device which responds to tiny changes in temperature. In this investigation the thermistor is taped between two fingers of one hand, as shown in Figure 10.17, and the initial temperature of the skin is recorded from the digital meter. (Alternatively the thermistor can be connected up to an interface with a computer.)

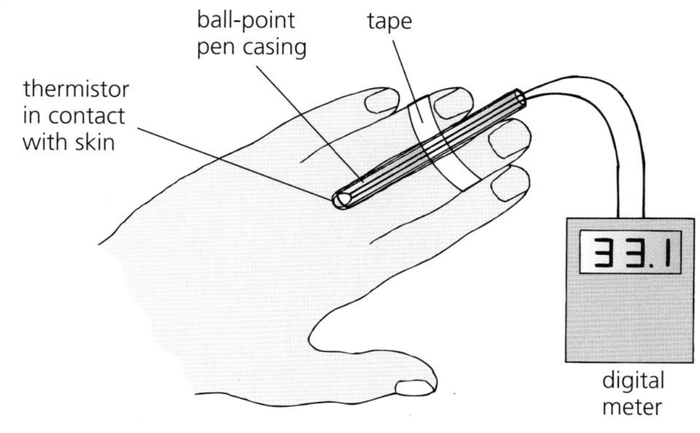

Figure 10.17 Use of a thermistor

The other hand is plunged into a container of icy water to cause a sudden heat loss. Temperature readings are taken every 30 seconds for 5 minutes. The skin temperature of the hand attached to the thermistor is found to drop by around 1 °C. A second thermistor positioned in the armpit during the experiment shows that the temperature of the body core remains constant.

It is therefore concluded that when heat is lost from one extremity (such as a hand in icy water), a **compensatory reduction in temperature** occurs in the other extremity but not in the temperature of the body core. This reduction in temperature is brought about by the following homeostatic mechanism: thermoreceptors in the skin in icy water send nerve impulses to the hypothalamus, which in turn sends impulses to the other hand causing **vasoconstriction** which reduces heat loss.

This response by the body's extremities helps to conserve heat when the body is exposed to extremes of temperature. The temperature of the body's extremities is therefore found to fluctuate more than that of the body core.

Testing Your Knowledge 2

1 Construct a table to compare conformers and regulators with respect to:
   - **a)** ability to control the internal environment by physiological means (1)
   - **b)** relative metabolic costs of lifestyle (1)
   - **c)** extent of range of ecological niches that can be exploited. (1)

2 **a)** What is meant by the term *physiological homeostasis*? (2)
   - **b) i)** Outline the principle of negative feedback control.
   - **ii)** Why is such control of advantage to an organism? (5)

3 **a)** Is a human being an ectotherm or an endotherm? (1)
   - **b)** Explain your answer. (1)

4 **a)** By what means does the hypothalamus in a mammal obtain information about the internal temperature of the body? (2)
   - **b)** Name TWO effectors to which the hypothalamus sends nerve impulses when the body temperature decreases to below a normal level. (2)

Applying Your Knowledge and Skills

1 The graph in Figure 10.18 shows basal metabolic rate for humans aged 1–70 years.

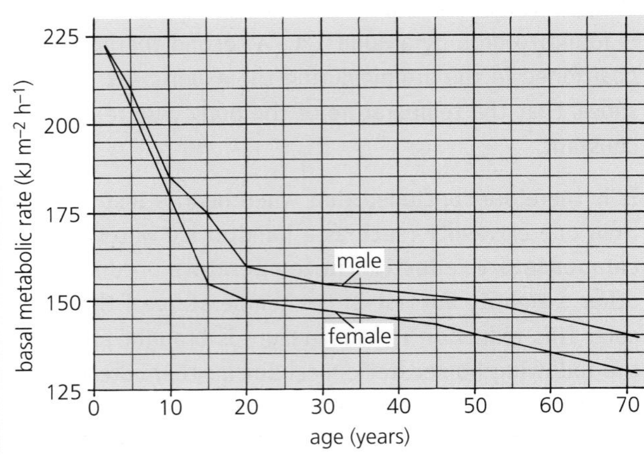

Figure 10.18

a) Draw TWO conclusions from the graph. (2)

b) By how many times is the basal metabolic rate (BMR) of a 5-year-old male greater than that of a 70-year-old male? (1)

c) What percentage decrease in BMR occurs in females between the ages of 10 and 70 years? (1)

2 a) Calculate the total surface area and the total volume of the cube shown in Figure 10.19:
  i) before it was sawn up
  ii) after it was sawn up. (2)

b) Calculate the surface area to volume ratio of
  i) the original large cube
  ii) one of the smaller cubes produced after cutting. (2)

c) Imagine that each size of cube represents a living organism. Which size would tend to lose more heat relative to its body size when the external temperature is low? (1)

d) Figure 10.20 refers to six species of shrew. Which species has:
  i) the smallest body size?
  ii) the largest body size? (1)

e) Which species of shrew has:
  i) the lowest metabolic rate?
  ii) the highest metabolic rate? (1)

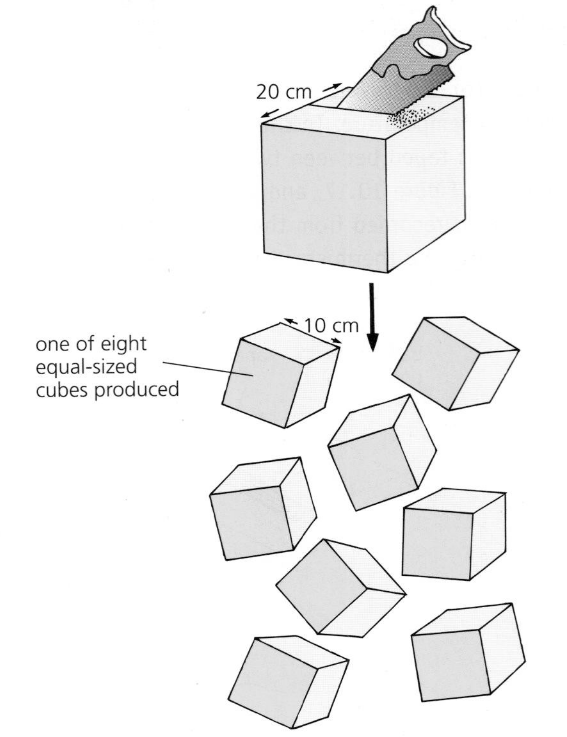

Figure 10.19

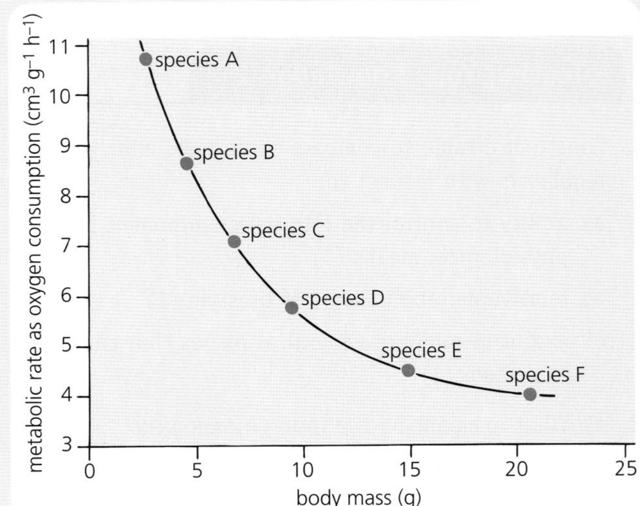

Figure 10.20

f) i) What is the overall relationship between body size and metabolic rate?
  ii) Relate your answer to i) to your answer to part c). (2)

Vertebrate group	Type of circulation	Number of chambers in heart	Pressure of blood arriving at skeletal muscles	Evolutionary level of circulatory system
fish				
	incomplete double			intermediate
mammal				

Table 10.3

3 **a)** Copy and complete Table 10.3. (5)

b) i) Which type of circulatory system is shown in Figure 10.21?

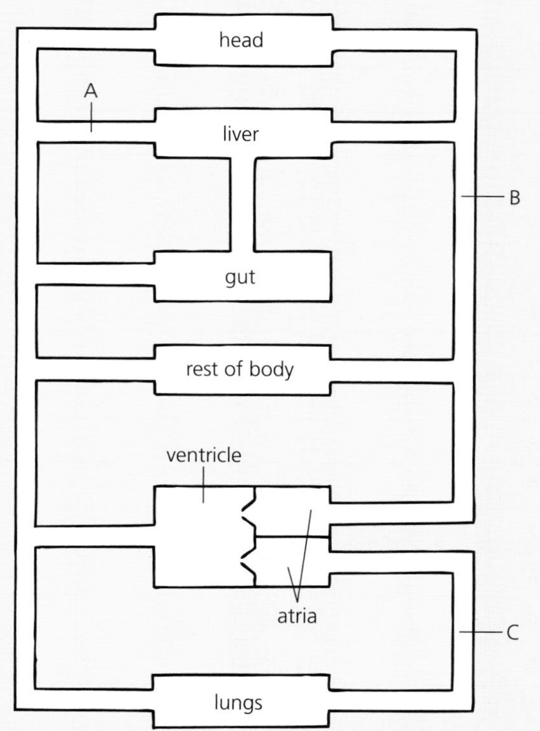

Figure 10.21

ii) Name an organism which possesses such a system.

iii) Will blood pressure at each of points A, B and C be high or low? (5)

4 *Anolis cristatellus* is a lizard that lives in open lowlands and dense forests in Puerto Rico. The graph in Figure 10.22 shows the relationship between mean body temperature and mean external air temperature for two populations. Members of population X often move to nearby sunny spots and bask in sunshine; members of population Y almost never seek out sunny spots (which are rare in the dense forest).

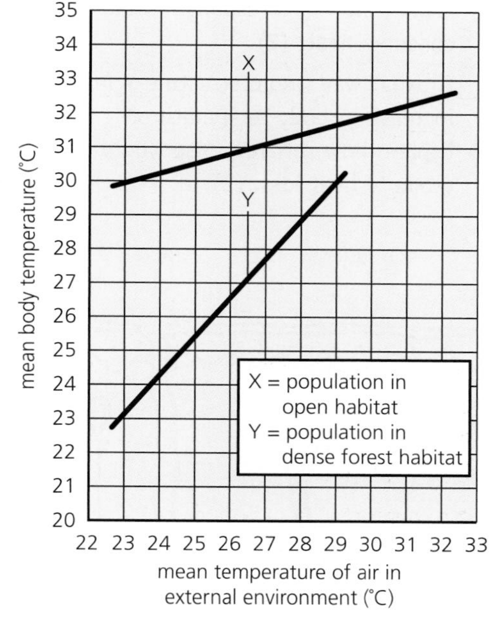

Figure 10.22

a) What is the mean body temperature of a lizard from population X when the external temperature is

i) 23 °C? **ii)** 30 °C? (1)

b) What is the mean body temperature of a lizard from population Y when the external temperature is:

i) 23 °C? **ii)** 30 °C? (1)

c) Which animal's body is:

i) completely dependent on external temperature?

ii) partly independent of external temperature? (1)

d) i) Which animal is not a complete conformer?

ii) Does this animal regulate its body temperature by physiological or behavioural means? (2)

e) Suggest TWO possible reasons why members of population Y rarely move to find a sunny spot when they feel cold. (2)

→

5 Figure 10.23 represents a section through human skin.

a) Name the part of the brain to which heat and cold receptors relay information about the external environment. (1)

b) By what means does the thermoregulatory centre of the brain communicate information to structures X and Y in order to affect control of body temperature? (1)

c) i) In what way should structure X respond following a drop in body temperature?
ii) Explain how this response would help to conserve heat. (2)

d) i) In what way would structure Y respond to an increase in body temperature?
ii) Explain how this response would help to promote heat loss. (2)

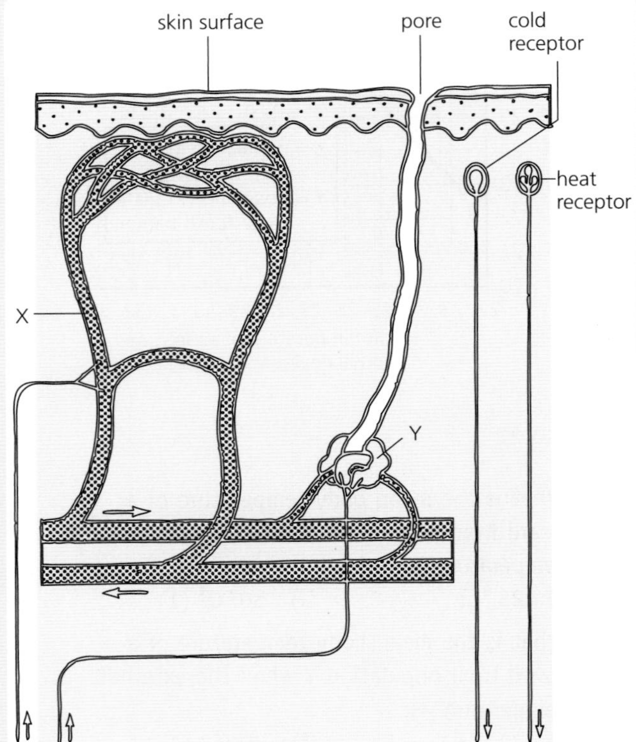

Figure 10.23

6 The data in Table 10.4 refer to the body temperature of a student who exercised vigorously and then, after a rest, plunged into a cold bath. Body temperature was measured every 2 minutes by inserting a sterilised thermistor under the student's tongue.

Time (minutes)	Body temperature (°C)
0	37.00
2	37.00
4	37.00
6	37.05
8	37.10
10	37.10
12	37.15
14	37.20
16	37.30
18	37.40
20	37.50
22	37.60
24	37.60
26	37.50
28	37.45
30	37.40
32	37.40
34	37.35
36	37.35
38	36.90
40	36.65
42	36.70
44	36.80
46	36.85
48	37.00
50	37.00

Table 10.4

a) Plot a line graph of the data. (3)

b) Using FOUR arrows, indicate on your graph that:
i) exercise was begun at minute 2
ii) exercise was stopped at minute 22
iii) immersion in the cold bath occurred at minute 34
iv) exit from the cold bath took place at minute 40. (4)

c) In general what trend in body temperature occurs during:
  i) the period of vigorous exercise?
  ii) the time in the cold bath?
  iii) Why is there a slight delay before each of these trends begins? (3)

d) The student's skin was flushed from minute 18 onwards.
  i) Suggest why.
  ii) What is the benefit to the body of flushed skin? (2)

e) i) At which ONE of the following times in minutes was the student found to be shivering?

  A 2 **B** 22 **C** 32 **D** 42

  ii) What is the survival value of shivering? (2)

7 Write an essay on negative feedback control in the human body with reference to control of body temperature. (9)

11 Metabolism and adverse conditions

In some environments the extreme heat and drought of summer, and in others, the extreme cold and lack of food in winter, create conditions that are **beyond the tolerable limits** of an animal's normal metabolic rate. Homeostatic systems of control would break down when the animal's body could no longer generate enough energy to effect the corrective mechanisms needed to return it to its steady state. However, these cyclic seasonal fluctuations rarely prove to be fatal because animals are either adapted to **survive** them or are able to **avoid** them.

Surviving adverse conditions

A **reduction in metabolic rate** enables an organism to avoid expending excessive quantities of energy trying to stay warm in an extremely cold climate or to stay cool in an extremely hot one. This can be achieved by a period of **dormancy**.

Research Topic Aspects of surviving adverse conditions

As a result of millions of years of evolution and natural selection, living organisms have become adapted to life in their particular ecosystem. If the ecosystem is affected by extreme fluctuations in climate or unpredictable environmental changes, then the organisms possess **adaptations** which enable them to survive the adverse conditions. These adaptations fall into three categories:

- **structural** (involving specialised structures possessed by the organism)
- **physiological** (depending on ways in which the organism's body and metabolism operate)
- **behavioural** (depending on the ways in which the organism responds to stimuli).

The discussion that follows considers these with reference to animals surviving adverse conditions in a cold climate.

Structural

Body size

The body size of birds and mammals tends to be larger in colder climates because a larger body size has a relatively smaller surface area from which heat energy can be lost (see Figure 11.1).

Appendages

Body appendages tend to be smaller in colder regions. For example, an Arctic fox (see Figure 11.2) possesses small ear flaps (pinnae) therefore exposing only a small surface area to potential heat loss. On the other hand, an African bat-eared fox has large pinnae which promote heat loss in a hot climate. The legs and snouts of mammals are frequently shorter and stouter in colder regions to conserve heat energy.

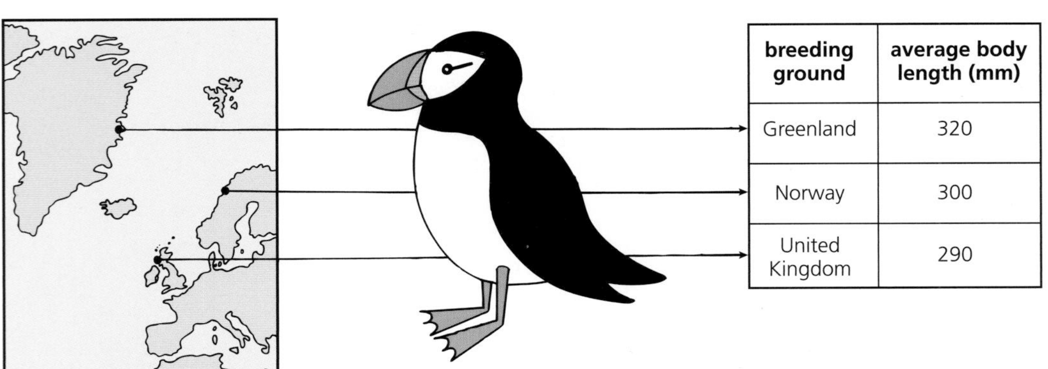

breeding ground	average body length (mm)
Greenland	320
Norway	300
United Kingdom	290

Figure 11.1 Body size in puffins

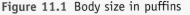

Figure 11.2 Pinna size in Arctic and African foxes

Insulation

A layer of air (a poor conductor of heat) trapped by fur or feathers provides an animal with good insulation by cutting down loss of heat from its body core.

Colour of fur or plumage

Many mammals and birds undergo seasonal changes in their fur or feathers. For example, the hare and the ptarmigan (see Figure 11.3) change from brown to white in winter. Not only does the white colour serve as an effective camouflage against a background of snow, it also reduces heat loss from the animal's body since a lighter-coloured object radiates less heat than a darker one.

Physiological

Many organisms survive adverse cold conditions by spending part of their life cycle in a dormant state. During this time metabolic activity decreases to a minimum thereby conserving energy. In birds and mammals, specialised brown fat, produced and stored during the food-rich seasons of the year, is used as fuel. In addition to **hibernation** (see page 156), some additional forms of dormancy occur among animals as follows.

Figure 11.3 Ptarmigan in snow

Diapause

This is commonly found in insects. It is a period during which growth and development are suspended and metabolism decreased. It is induced by certain stimuli (such as decreasing day lengths) and normally lasts from autumn until spring.

Behavioural

Collective den

Some mammals that are non-colonial during mild weather share a collective den in winter. This helps to reduce loss of heat from their body core by reducing the surface area exposed to the external environment.

Snow roost

Some types of grouse survive periods of extreme cold by resting in groups under the snow in a 'snow roost'.

Migration

Many animals avoid the adverse conditions of winter by migrating to warmer climes (see page 158).

Dormancy

Dormancy occurs as part of an organism's life cycle when its growth and development are temporarily arrested. The organism's **metabolic rate decreases** to the minimum needed to keep its cells alive. Therefore energy is conserved and the plant or animal is able to survive a period of adverse conditions such as winter cold, summer drought or scarcity of food.

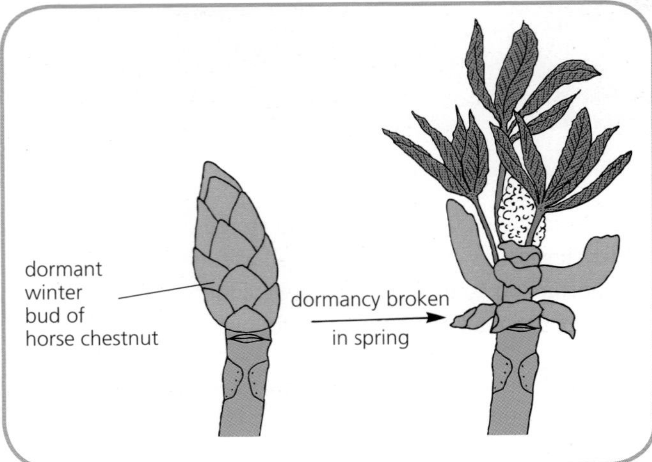

Figure 11.4 Dormant winter bud

Predictive dormancy

When an organism becomes dormant *before* the arrival of the adverse conditions, this is called **predictive dormancy**. For example, many trees respond to decreasing photoperiod (day length) and temperature in autumn by shedding their leaves and entering their dormant phase before the onset of winter. Their **winter buds** remain dormant until the spring (see Figure 11.4). Then the period of arrested growth and development is brought to a halt by the arrival of increasing day lengths and the action of plant growth substances.

Consequential dormancy

When an organism becomes dormant *after* the arrival of adverse conditions, this is called **consequential dormancy**. It is more common in regions where the climate is unpredictable. The advantage of consequential dormancy is that the organism can remain active for longer and continue to exploit available resources. However, a sudden, severe change in environmental conditions may kill off many organisms before they have had time to become dormant.

Dormancy in animals

Two examples of dormancy in animals are hibernation and aestivation.

Hibernation

Hibernation (or 'winter sleep') is a form of dormancy that enables some animals (usually mammals) to survive the adverse conditions of winter. It may last for weeks or even months. Before hibernating (often in a predictive way in response to shortening day length) the endothermic animal consumes extra food which becomes laid down as a **store of fat**.

During hibernation, the animal's **rate of metabolism drops** and this results in a **decrease in body temperature**. These changes are accompanied by a **slower heart rate**, a **slower breathing rate** and a state of general inactivity where the bare minimum of energy is expended to maintain the vital activities of cells. Together, these changes enable the animal (for example, the hedgehog, as shown in Figure 11.5) to survive a prolonged period of low temperature in its surroundings. If, however, the external temperature drops too far, the hibernator will increase its metabolic rate slightly to prevent a fatal drop in body temperature.

Bears do not go into true, deep hibernation. The decrease in their metabolic rate is significant but less than that shown by smaller mammals such as the

Figure 11.5 Hibernating hedgehog

hedgehog. As a result, the bear's body temperature drops from 37 °C to around 31 °C and it can be aroused fairly easily. In contrast, the hedgehog's body temperature drops to around 6 °C and it cannot be wakened easily.

Aestivation

Aestivation (or 'summer sleep') is a form of dormancy employed by some animals to survive periods of excessive heat and drought in summer. For example, during a period of intense heat, a land snail seeks out a safe place (such as a spot high up in vegetation). It then retreats into its shell and the opening becomes sealed with dried mucus except for a tiny hole to allow gas exchange. The snail remains in this state with its **metabolic rate at a minimum level** until favourable conditions return.

Aestivation is also found to occur among vertebrates such as tortoises, crocodiles and lungfish. A lungfish (see Figure 11.6) buries itself in the mud of a dried-up lake and surrounds itself with a cocoon of dried mucus. It exchanges gases through a breathing tube and remains in a dormant state for many months until the arrival of the next rainy season.

Figure 11.6 Lungfish

Daily torpor

Daily torpor is the physiological state in which an animal's rate of metabolism and activity become greatly reduced for part of **every 24-hour cycle.** It is accompanied by a slowing down of heart rate and breathing rate and a decrease in body temperature. It is common among small birds and mammals.

Hummingbirds (see Figure 11.7), for example, feed during the day and exhibit torpor at night; bats and shrews feed at night and become torpid during daylight hours. A small animal has a relatively **large surface area** from which heat is lost rapidly. Therefore, when the animal is an active endotherm, it needs a very high rate of metabolism to maintain its body temperature. A daily period of torpor is of survival value to such an animal because it greatly **decreases the rate of energy consumption** during the time when searching for food would be unsuccessful or would leave the animal open to danger.

Figure 11.7 Hummingbird

Testing Your Knowledge 1

1 Give an example of a set of environmental conditions that would be beyond the tolerable limits for *normal* metabolic activity in a hedgehog. (1)

2 a) Identify
   i) a physiological ii) a structural
   characteristic typical of a deciduous tree during its period of dormancy. (2)

   b) Explain the difference between *predictive* and *consequential* dormancy. (2)

   c) i) Which type of dormancy is shown by deciduous trees?
   ii) Explain your answer. (2)

3 a) *Hibernation* and *aestivation* are forms of dormancy.
   i) Identify a characteristic they have in common.
   ii) Identify a characteristic by which they differ. (2)

   b) With reference to a named animal, explain the meaning of *daily torpor* and its importance to the animal. (2)

Migration

Migration is the regular movement by the members of a species from one place to another over a relatively long distance. By migrating to a favourable environment, an animal is able to avoid the conditions of **metabolic adversity** caused by shortage of food and low temperatures. Long-distance migration is carried out by many vertebrates and a few invertebrates. It normally involves an **annual round trip** between two regions, each of which offers conditions more favourable than the other for part of the year.

Birds

Some migratory birds such as the Arctic skua move all the way from one hemisphere of the world to the other. Figure 11.8 shows the migratory routes which it takes in autumn. It returns by the same route in spring. Other birds such as the yellow wagtail cover less distance by simply migrating from a temperate region to a tropical one within the same hemisphere.

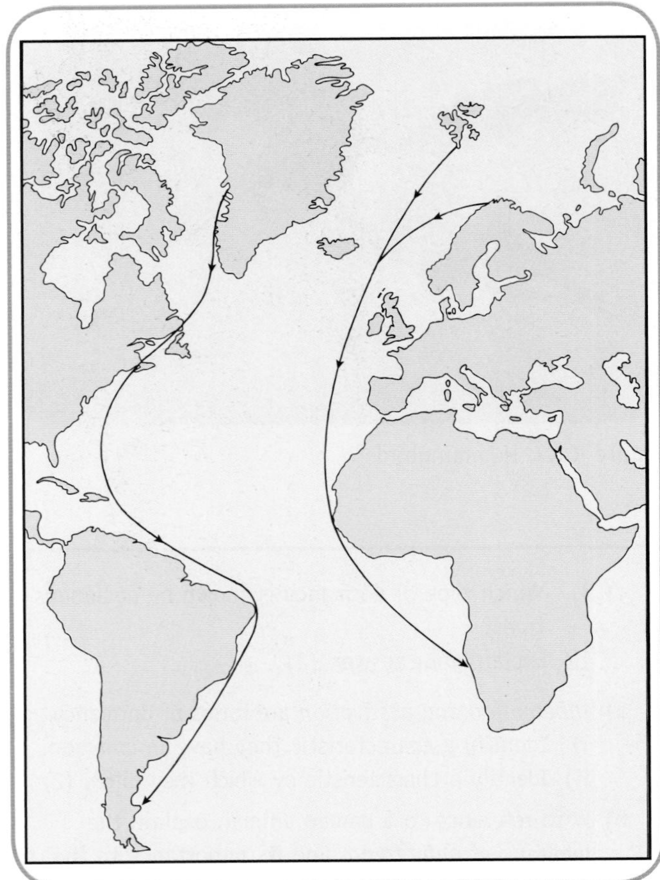

Figure 11.8 Migratory routes of Arctic skua in autumn

Mammals

Many species of whale migrate. For example, humpbacks spend the summer in polar regions where they gorge on plankton and build up a layer of blubber. At the start of winter they migrate up to 7 000 kilometres to subtropical waters where the females give birth and suckle their young. Although food for adults may be scarce in subtropical waters, the warm conditions suit the young whose small bodies would lose too much heat in polar regions.

Invertebrates

Each spring, millions of Monarch butterflies migrate north from their winter territories in Mexico to their breeding areas in the USA and Canada. These areas contain abundant supplies of their exclusive food source – the milkweed plant. In autumn the butterflies migrate back to Mexico to avoid the cold winter conditions of northern regions. A Monarch butterfly only lives for a short time, so the butterflies that are found heading for Mexico are three to five generations removed from the ones that overwintered there the previous year. Somehow the information needed for successful migration has been passed on from generation to generation.

Specialised techniques

When studying long-distance migration, scientists want to find out information such as:

- **when** the animals migrated
- **where** they overwintered
- whether or not they **returned** to their original summer territory
- **how long** they lived for.

Many migratory animals travel thousands of miles which makes it very difficult to follow their route in detail. However, scientists have developed **specialised techniques** to overcome these difficulties.

Individual marking

Ringing with metal bands

This **ringing** technique (also known as **banding**) has been used ever since the beginning of scientific investigation into migration. A metal band (carrying

Figure 11.9 Bird with leg band

the bird's unique number and the investigators' contact details) is attached to the bird's leg (see Figure 11.9). If the bird is recaptured and its information reported, then this record of its movements **contributes to an overall picture** of the migratory behaviour of the species. Details of the migratory flyways used by many species of birds have been built up over the years from information obtained in this way both inside and outside the country of the ring's origin.

Satellite tracking

More recently lightweight **transmitters** have been developed that can be glued to the animal's body or implanted under its skin. The transmitter emits signals that are picked up by receivers on Earth-orbiting **satellites** as they pass overhead. The signals are beamed back to ground stations and the information relayed to scientists (see Figures 11.10 and 11.11). This technique of **tracking** an animal's route from space has yielded the most precise information so far on the exact location of flyways used by birds during their migratory cycle. It has also been used successfully to track several mammals including whales and seals.

Unlike a ringed animal, one bearing a 'high-tech' transmitter does not have to be recaptured to provide scientists with data. However, transmitters are much more expensive than 'low-tech' rings and may have a drag effect on some small birds.

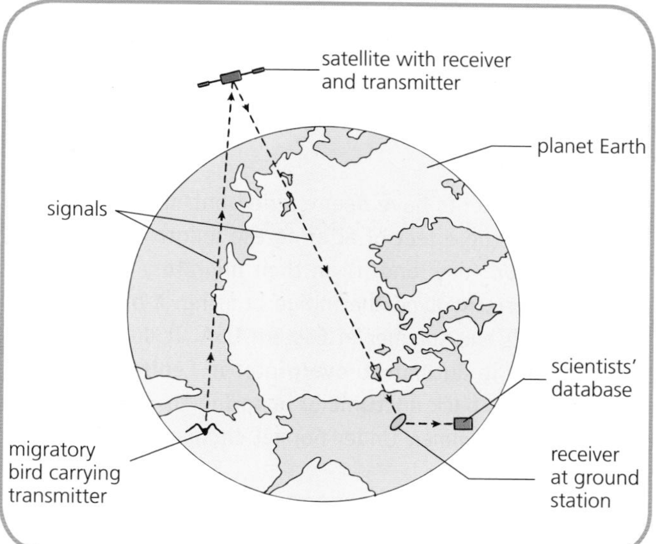

Figure 11.10 Satellite tracking

Figure 11.11 Cat heaven

Related Topic

Migration triggers and adaptations

Photoperiod

Many experiments have been carried out on birds to investigate the effect of altering the length of the **photoperiod** (day length) on their migratory behaviour. One of these involved the indigo bunting, a bird that spends the summer in Eastern USA. It migrates nocturnally in autumn to overwinter in Central America and migrates back nocturnally in spring to Eastern USA for the summer. Under normal circumstances, prior to migration, the bird eats extra food, stores fat and becomes restless at night. Table 11.1 gives a summary of the experiment.

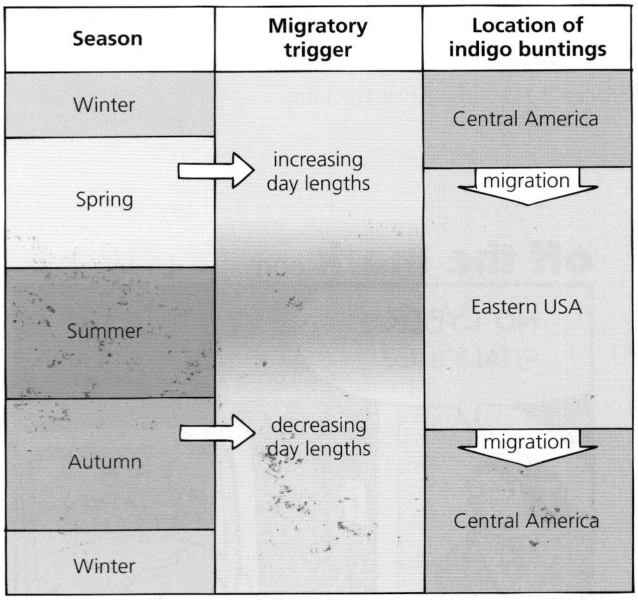

From this experiment it is concluded that in indigo bunting birds, the changes that occur prior to migration are triggered by changes in the photoperiod as summarised in Figure 11.12.

It is now known that **changing photoperiod** is the primary **trigger** for migration in many birds. It causes hormonal changes in the bird's body which result in the behavioural changes (such as night-time restlessness) and physiological changes (including storage of fat) that occur in preparation for migration. Decreasing environmental temperature may, in some cases, play a secondary role but length of photoperiod is a much more **reliable indicator** of time of year.

Season	Migratory trigger	Location of indigo buntings
Winter		Central America
Spring	→ increasing day lengths	migration ↓
Summer		Eastern USA
Autumn	→ decreasing day lengths	migration ↓
Winter		Central America

Figure 11.12 Triggers for migration

Sun

Many experiments have been carried out to investigate whether birds make use of the **Sun** as a **compass** to locate the **correct direction** in which to migrate. In one of these experiments, the importance of sunset as an indicator of direction for autumn migration was investigated using cone-shaped, funnel cages. These are lined with white blotting paper and each has an ink pad at its base, as shown in Figure 11.13. If the bird tries to move in a particular direction, it leaves a tell-tale trail of inky footprints on the paper.

Tests were conducted on sparrows during autumn under the following set of conditions:

Group	Treatment in laboratory	Result
A	Birds exposed to normal day lengths from September to April.	When day length increased in spring, the birds binged on food and laid down a store of fat. In April they showed night-time restlessness.
B	Birds exposed to normal day lengths from September to December and then given artificially longer 'spring' day lengths in December.	The birds binged on food and laid down a store of fat. They then showed night-time restlessness.
C	Birds exposed to normal day lengths from September to March and then given artificially shorter 'autumn' day lengths in March.	The birds binged on food and laid down a store of fat. They then showed night-time restlessness.
D	Birds exposed to photoperiods of constant medium length.	The birds did not binge on food, store fat or become restless at night.

Table 11.1 Investigating effect of photoperiod on migratory behaviour

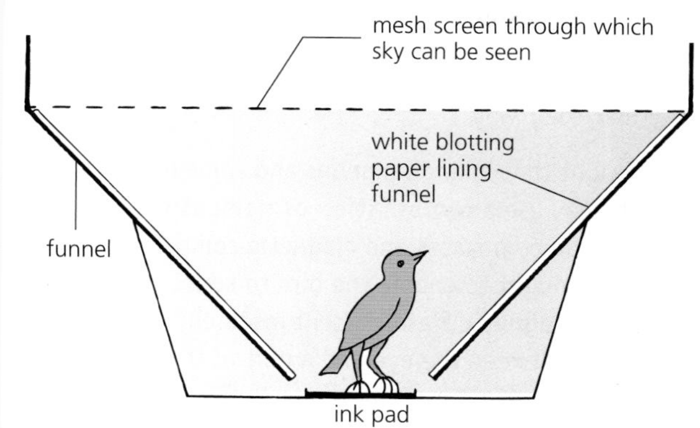

Figure 11.13 Funnel-cage experiment

- birds placed in cages at sunset under a clear sky (group A)
- birds placed in cages at sunset under a cloudy sky (group B)
- birds placed in cages after sunset under a clear sky (group C).

Figure 11.14 shows a typical footprint record for each condition and the resulting vector diagram. (A vector diagram gives a quantified version of the footprints.) The results for group A show a southerly trend in choice of direction. For B and C there is no clear-cut directional choice. Therefore it is concluded that for sparrows, sunset is an important **visual cue** used in their autumnal migration to locate the direction to take. Other similar experiments show that if the perceived position of sunset is altered by using mirrors, then the direction chosen by the birds changes accordingly.

Internal clock

Many migratory animals make use of an **internal clock**. In an experiment where the only light source (acting as the Sun) was held in one position constantly, starlings were found to continue to change their angle of orientation by 15° per hour. This is their normal response in order to compensate for the change in the Sun's position as the Earth rotates.

Stars

The cone-shaped cage shown in Figure 11.13 has been used in many experiments. In one of these, indigo buntings were tested at night time. On clear nights (but not on cloudy nights) during spring, the birds jumped in a north-easterly direction. This is the direction which they would normally take in April when migrating from Central America to Eastern USA. On clear nights (but not on cloudy nights) during autumn, the birds jumped in a south-westerly direction. This is the direction which they would normally take in September when migrating from Eastern USA to Central America.

The experiment was continued in a planetarium where results similar to those obtained for clear nights were obtained using an artificial, starry sky. In addition, it was possible to identify particular **patterns of stars** to

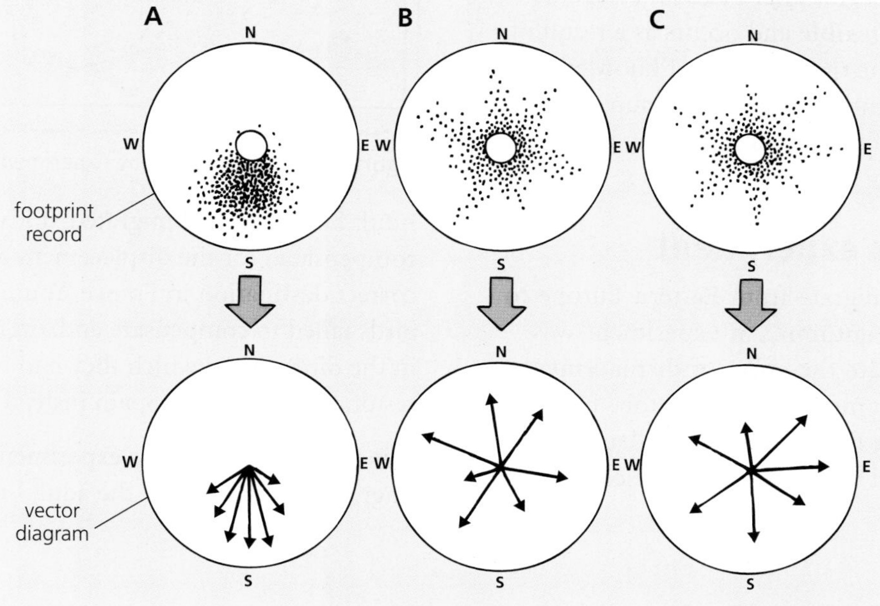

Figure 11.14 Results from an ink pad experiment

which the birds were responding by shutting off various sections of the 'sky'. It was concluded that these birds possess a built-in **genetic mechanism** that makes them head in a particular direction in response to certain star patterns.

Magnetic field

Although many visual clues are used by birds for navigation, experiments with pigeons suggest that their navigational system also contains **non-visual components.** Out of a large group of pigeons released wearing frosted contact lenses, 60% were able to find their way home.

Experiments have also been carried out in which the birds have been subjected to **altered magnetic fields.** In one of these, bar magnets were attached to the backs of a group of pigeons and brass bars to those of the control group. During sunny conditions all the pigeons returned home quickly but on a cloudy day the 'magnet' birds, unlike the 'brass-bar' birds, became disorientated and many lost their way.

It is thought that homing pigeons and some long-distance migrants may use a **combination of navigational devices** including sun compasses and magnetic compasses. The latter are thought to enable the bird to sense changes in the Earth's **magnetic field.** Recent research indicates a connection between the eye and a part of the brain that is active during navigation, and suggests that some birds may actually 'see' the Earth's magnetic field. It is thought that this may appear as areas of light and shade superimposed on the normal images that the bird sees. Monarch butterflies are also thought to use some form of magnetic orientation to find their way during migration.

Innate and learned influences on migratory behaviour

Innate behaviour is **inherited** and **inflexible.** Innate influences are thought to play the primary role in migratory behaviour. This pattern of behaviour is performed in the same way by every member of the species. It occurs in response to an external stimulus such as a change in photoperiod.

Learned behaviour begins after birth and is gained by **experience.** It is **flexible** and occurs as a result of trial and error and the transmission of knowledge and skills among the members of a social group. Learned influences are thought to play a secondary role in migratory behaviour.

Displacement experiment

Starlings normally migrate from Eastern Europe to Northern France in autumn. An experiment was designed to investigate the effect of **displacement** on starlings. A large number of migratory birds were captured using fine netting in Holland and taken to Switzerland where they were released (see Figure 11.15).

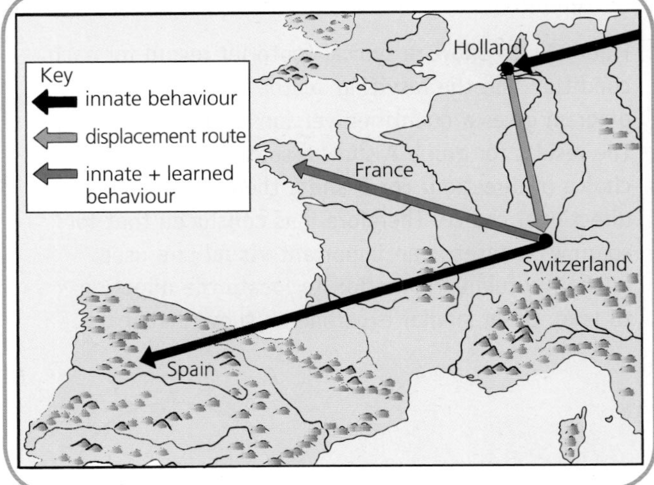

Figure 11.15 Displacement experiment

Adult birds that had migrated at least once before compensated for the displacement and arrived at the correct destination in France. Young, inexperienced birds failed to compensate and, on release, continued in the direction in which they had been travelling. As a result, they arrived in Spain instead of France.

It is concluded from this experiment that the migratory behaviour of the adult birds was based on a combination of both **innate and learned** influences. They made use of knowledge (gained from previous journeys) of familiar geographical features along the route to navigate their way to the correct location. On the other hand, the migratory behaviour of the young birds was **purely innate**. They possessed genetic information about the direction of the destination but had no previously learned information about its actual location that they could put to use.

Investigating directional tendencies

Figure 11.16 shows the **flight paths** taken by two distinct populations (A and B) of blackcap warbler birds during their autumn migration. Members of population A (in part of Germany) always head south-west; members of group B (in part of Austria) always head south-east and then south. An experiment was designed to investigate whether this behaviour is innate or learned.

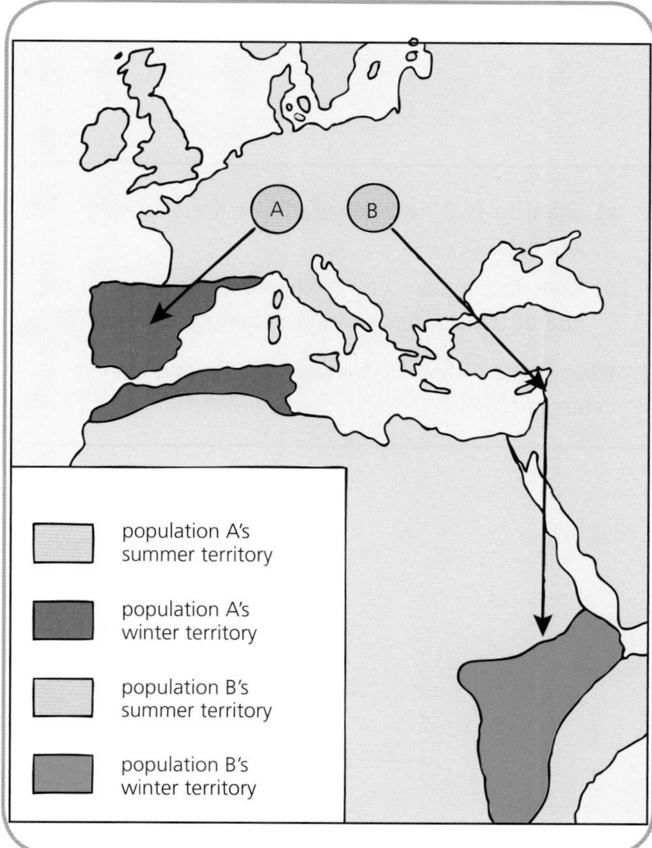

Figure 11.16 Two migratory flight paths

Nestlings from both populations were hand-reared and then their orientation tested using funnel cages (see Figure 11.13). Figure 11.17 shows the results. The directional choices made by the members of each test group (as indicated by the ink marks on the walls of the cages) corresponded to the actual direction of the flight path typical of their population of origin.

These results suggest that the young birds possess genetic information about the direction in which they are to migrate and that heading in this direction is a form of **innate** behaviour.

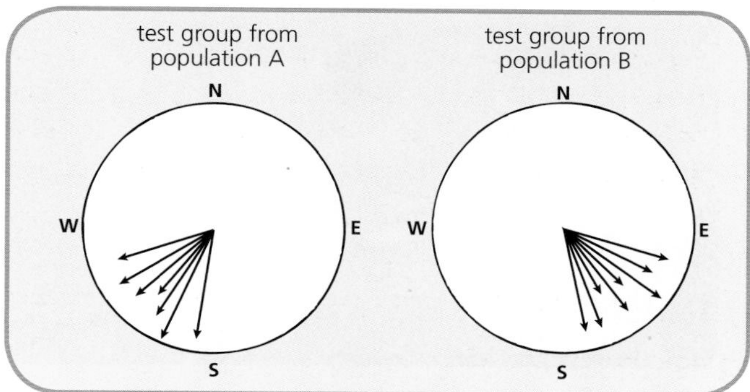

Figure 11.17 Vector diagrams of directional choice

Cross-fostering experiment

Herring gulls are non-migratory; lesser black-backed gulls are migratory. In an experiment, their eggs were switched round and the migratory behaviour of the 'fostered' offspring studied.

The black-backed gulls raised by the non-migratory herring gulls did migrate, supporting the idea of migration being **innate** behaviour. The herring gulls raised by the black-backed gulls did move with their migratory foster parents. They were thought to be simply following their foster parents and to be exhibiting learned behaviour.

Research Topic: Genetic control of migratory behaviour in blackcap warblers

In an investigation, scientists chose six genes from the blackcap, a common European bird (see Figure 11.18), as possible candidates for a '**migration gene**'. These genes were known to influence behavioural traits linked to migration such as increased metabolism and level of night-time restlessness.

Figure 11.18 Blackcap

The scientists sampled the DNA of 14 populations of blackcap ranging from those in the Cape Verde Islands in the Atlantic Ocean that never migrate to those in Russia that travel over 3500 kilometres during migration. They found a link between one of the six genes (called ADCYAP1) and a certain form of behaviour. Different versions of this gene exist. They vary in the number of copies of a two-base repeat that they possess at one end of the genetic sequence. The greatest number of two-base repeats was found to match the highest level of night-time restlessness and the longest distance travelled during migration.

The ADCYAP1 gene codes for a peptide which influences daily rhythms and affects rate of metabolism and usage of fat – changes associated with preparations for migration. Scientists attribute about 3% of migratory behaviour to this gene. Clearly many other genes are also involved, in conjunction with the influence of environmental factors.

Testing Your Knowledge 2

1 a) With reference to a named bird, explain what the term *migration* means. (2)

b) In what way does the bird gain from migration considering that the process requires much energy to be expended? (1)

2 a) Identify TWO methods used to mark migratory animals. (2)

b) Construct a flow chart of the events involved in the use of transmitters to track migratory animals. (3)

3 What is the difference between *innate* and *learned* behaviour? (2)

Applying Your Knowledge and Skills

1 The data in Table 11.2 refer to marmots (a type of European rodent) that hibernate in winter.

Bodily function	State of animal	
	Active	Hibernating
basal metabolic rate ($kJ\ m^{-2}\ day^{-1}$)	1728	113
body temperature (°C)	36	3
breathing rate ($breaths\ min^{-1}$)	26.0	0.2
heart rate ($beats\ min^{-1}$)	80	5

Table 11.2

a) By how many times does heart rate decrease during hibernation? (1)

b) What would be the basal metabolic rate per *minute* in active marmots? (1)

c) By how many times does the breathing rate increase when the animal becomes active in spring? (1)

d) What percentage decrease in body temperature occurs during hibernation? (1)

2 A tomato seed is surrounded by a capsule of juicy 'gel' which can be removed from the seed by careful washing and use of a strainer. In an experiment at a seed bank, seeds from a rare variety of tomato from South America were arranged in a Petri dish, as shown in Figure 11.19, and incubated at the optimum temperature for their germination.

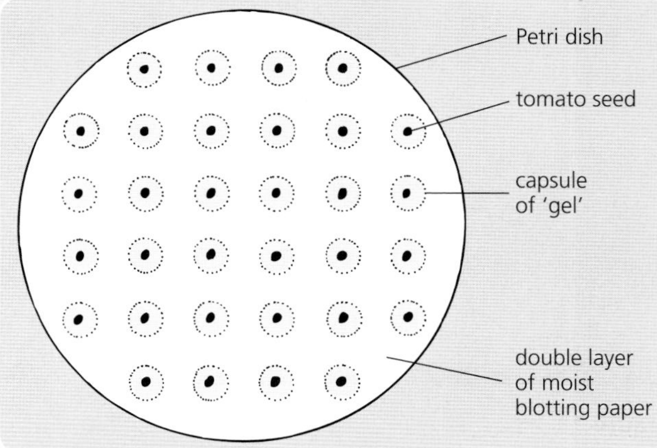

Figure 11.19

a) This experiment was set up to investigate the hypothesis that the juicy gel contains a chemical that keeps the seeds dormant. What result would support this hypothesis? (1)

b) What control should have been set up? (1)

c) What should have been done to increase the reliability of the results? (1)

d) By what means could the experiment be extended to investigate whether the juicy gel present in tomatoes grown in Scotland contains a chemical that keeps the South American seeds dormant? (3)

3 Read the passage and answer the questions that follow it.

The European hedgehog lives in wooded areas; the desert hedgehog lives at the edge of the Sahara desert. The European hedgehog eats earthworms and insects but these become unavailable in very cold weather. It hibernates to survive harsh winter conditions and scarcity of food. Its metabolic rate drops to a minimum, its body temperature decreases to 6 °C, its heart rate drops from 100 to 10 beats min^{-1} and its breathing rate decreases significantly. Like other deep hibernators, it may awaken a few times to feed although some individuals sleep throughout the whole hibernation period.

The desert hedgehog eats scorpions and snakes but these go deep underground when the desert becomes intolerably hot in summer. It is then that the desert hedgehog retreats to a burrow and enters a state of aestivation. Its metabolic rate slows down but not to nearly as low a level as that shown by the European hedgehog during hibernation. If the desert were to get as cold as an extreme European winter, the desert hedgehog would die because it is unable to go into deep hibernation. Even in aestivation, it only sleeps for about a week at a time and may make brief visits outside the burrow before returning to sleep.

a) Construct a table to show FIVE differences between the two types of hedgehog based on the information in the passage. (5)

b) Compared to European hedgehogs, desert hedgehogs have smaller bodies and larger pinnae (ear flaps). Suggest why. (2)

c) Brown fat cells are larger and contain more fat droplets than normal white fat cells.

i) Suggest which type of hedgehog has a layer of brown fat cells under the skin around its neck and shoulders.

ii) Justify your choice of answer. (3)

4 The brambling is a bird which overwinters in Britain and then migrates north to its Scandinavian breeding area in spring. Normally it is active during the day and inactive at night. However, it is found to become active at night just before migration. This behaviour is known to be triggered by changes in photoperiod. The graph in Figure 11.20 shows the results from a series of experiments on the brambling. Table 11.3 gives details of natural day lengths.

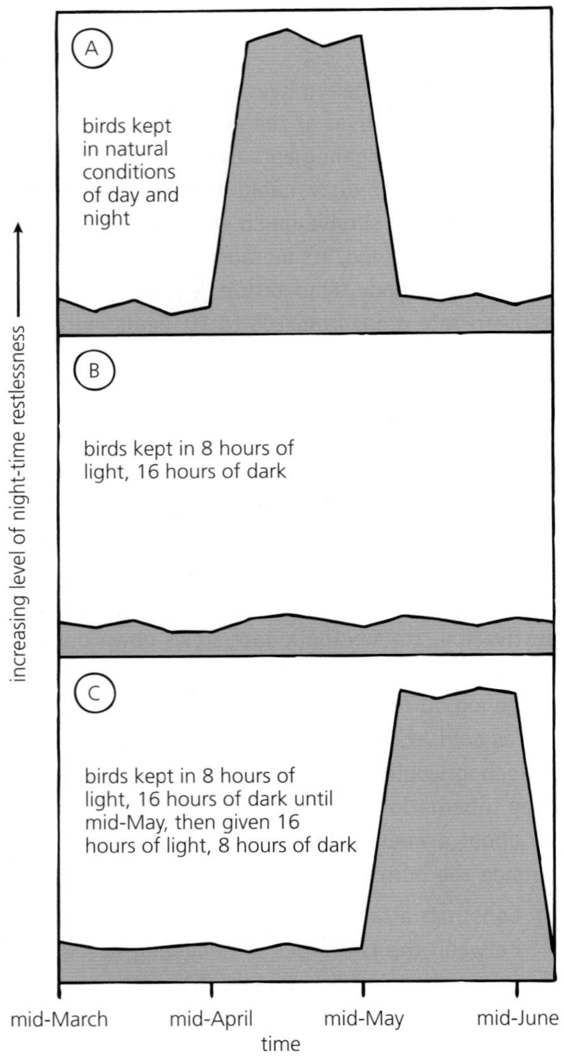

Figure 11.20

a) What happens to day length as the season changes from spring to summer? (1)

b) i) Give precise details of the environmental stimulus that triggers night-time restlessness.
ii) When does this occur under natural conditions? (2)

c) Suggest why group C failed to show night-time restlessness until mid-May. (1)

d) A student studied the data and concluded that the critical environmental stimulus was ten or more hours of darkness. Explain why the data do not support this conclusion. (1)

5 Figure 11.21 shows the results of an experiment carried out on a species of bird that normally migrates in autumn from Sweden to Spain. The birds were captured at point X in Holland and released at point Y in Eastern Europe.

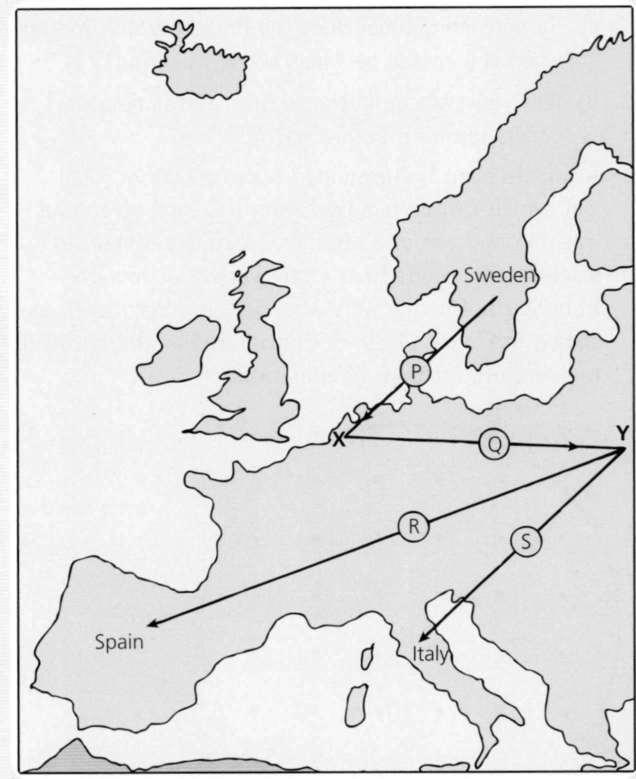

Figure 11.21

	Mid-March	Mid-April	Mid-May	Mid-June
Length of day (h)	12	14	15.5	16.5
Length of night (h)	12	10	8.5	7.5

Table 11.3

a) Which arrow represents the displacement route? (1)

b) i) Is innate behaviour inherited or learned by trial and error?

ii) Which arrows represent migration that could be achieved by innate behaviour alone? (2)

c) Which arrow represents migration by displaced birds based on both innate behaviour and knowledge gained from previous experience? (1)

d) Suggest why some birds failed to reach the correct destination in Spain and arrived in Italy instead. (2)

6 The initial direction that would be taken by a bird on a migratory flight can be investigated in the laboratory using the apparatus shown in Figure 11.22. It consists of a circular cage suspended inside an outer container with transparent windows. Mirrors can be inserted into the windows in such a way that to the birds in the cage the Sun seems to have rotated through 90°. The scientist observes the movements of the birds through the transparent floor. The details in Table 11.4 refer to a series of experiments carried out using a starling that had been hand reared from a nestling. Starlings normally migrate north in spring and south in autumn.

a) Which experiments produced results that are equivalent to the bird's natural migratory behaviour pattern? (1)

b) i) What is the main shortcoming of this investigation?

ii) How could it be overcome? (2)

c) Suggest why this equipment is less useful than that shown in Figure 11.13 on page 161. (1)

d) Which TWO experiments should be compared to find out the effect of the **time of year** on the choice of direction when the windows are transparent and the sky is clear? (1)

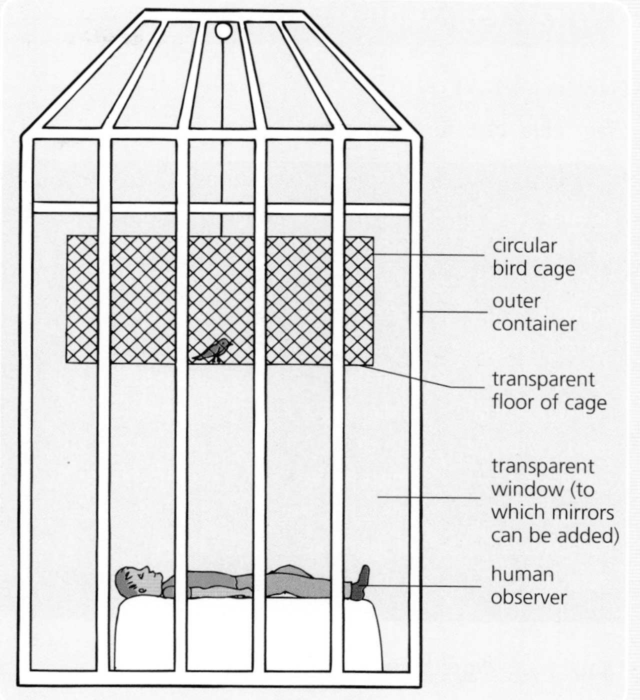

Figure 11.22

e) Which TWO experiments should be compared to find out the effect of the **state of the sky** on choice of direction in autumn when the windows are mirrored? (1)

f) Which TWO experiments should be compared to find out the effect of the **state of the outer container's windows** in spring when the sky is overcast? (1)

g) What TWO further experiments should have been included to improve the design of the investigation? (2)

Experiment	State of outer container's windows	Actual time of year	State of sky	Direction taken by bird	Conclusion
A	transparent	spring	clear	north	normal migration direction chosen
B	transparent	autumn	clear	south	normal migration direction chosen
C	transparent	spring	overcast	random	clouds prevented bird getting a fix on Sun's position
D	mirrored	autumn	clear	west	Sun being at 90° from real position made bird alter direction by 90°
E	mirrored	spring	overcast	random	clouds prevented bird getting a fix on Sun's position
F	mirrored	autumn	overcast	random	clouds prevented bird getting a fix on Sun's position

Table 11.4

What You Should Know

Chapters 10–11

(See Table 11.5 for word bank)

adaptations	feedback	narrow
aestivation	hibernation	niches
avoid	high	optimum
body	homeostasis	oxygen
carbon dioxide	hypothalamus	predictive
conformers	innate	pressure
consequential	internal	probes
deoxygenated	lungs	regulators
dormancy	metabolic	temperature
drought	metabolism	thermoregulation
effectors	migrating	torpor
external	mixing	tracking

Table 11.5 Word bank for chapters 10–11

1 ______ rate is a measure of the quantity of energy consumed by an organism per unit of time. It can be measured as oxygen consumption, _____ production or heat production per unit time by using electronic ______.

2 Aerobic respiration consumes ______. Therefore aerobic organisms with high rates of ______ need systems that can deliver oxygen efficiently to cells.

3 Fish have a two-chambered heart which pumps blood at ______ pressure to the gills. Blood leaving the gills is transported to the ______ at low pressure. Amphibians and reptiles have a three-chambered heart which pumps blood at high pressure to both the ______ and the body but allows some mixing of oxygenated and ______ blood. Birds and mammals have a four-chambered heart which pumps blood at high ______ to both the lungs and the body with no ______ of blood.

4 When fluctuations in an external abiotic factor occur, some organisms called ______ are able to maintain their steady ______ state by using physiological mechanisms. This increases their range of possible ecological ______ but it requires energy. Other organisms, called ______, are unable to maintain their metabolic rate in this way and the state of their internal environment is influenced directly by changes in the ______ environment. Their range of ecological niches is ______ but their metabolic costs are low.

5 The maintenance of the body's internal environment within certain tolerable limits is called ______ and it requires energy. It is brought about by corrective mechanisms called negative ______ control.

6 Birds and mammals are able to maintain their body temperature at the ______ level for enzyme action. Such ______ keeps their metabolism at a high level regardless of low external ______.

7 The body's temperature-monitoring centre is in the ______. It sends nerve impulses to ______ which respond by bringing about changes that warm up or cool down the body as required. In extremely adverse conditions, these responses are inadequate and the organism needs to have further ______ which enable it to survive the conditions or to ______ them. One of these is reduction in metabolic rate during ______.

8 An organism that becomes dormant before the arrival of adverse conditions shows ______ dormancy; one that becomes dormant after their arrival shows ______ dormancy.

9 ______ is a form of dormancy that involves a decrease in metabolism and body temperature that enables an animal to survive a period of extreme cold; ______ is a form of dormancy that enables an animal to survive a period of intense heat or ______. Daily ______ is a period of decreased metabolism for part of each 24-hour cycle.

10 Some animals avoid adverse conditions by ______ to a more suitable environment for part of the year. Patterns of long-distance migration are studied using individual marking and ______ of animals. Migration is basically a form of ______ behaviour but some aspects of it may be learned.

12 Environmental control of metabolism

An enormous variety of micro-organisms exist on Earth. Some are prokaryotes that take the form of **bacteria** (such as *Escherichia coli*) and **archaea**. Others are species of **eukaryotes**. These may be unicellular algae that are able to photosynthesise or unicellular fungi such as yeast (see Figure 12.1) and multicellular filamentous fungi such as *Penicillium* (see Figure 12.2) that need a source of ready-made organic food.

Figure 12.1 Yeast

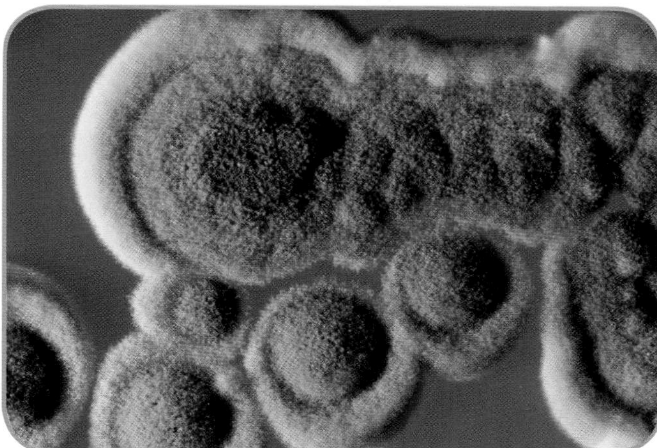

Figure 12.2 *Penicillium* colonies

Adaptability

In general, micro-organisms are **highly adaptable** and able to make use of a wide variety of substrates for their metabolism. Therefore they occur in every environment that supports life.

Useful products

In the course of their metabolic pathways, micro-organisms make a wide range of **metabolic products**, many of which are useful to humans. Employing micro-organisms to convert raw materials into useful substances dates back thousands of years to ancient times when people discovered gradually how to make products such as bread, beer and wine. In recent times, the use of microbes has escalated to an industrial level which provides humans with many products and services. Micro-organisms are ideal for a variety of **research and industrial uses** because:

- they are easy to cultivate (culture)
- they reproduce and grow quickly
- their food substrate is often a cheap substance (or even a waste product from another source)
- they produce many different useful products
- their metabolism can be manipulated relatively easily.

Environmental control of metabolism

Micro-organisms are of particular use in fermentation industries because their metabolism can be **controlled** much more easily than that of larger organisms. Scientists are able to control specific micro-organisms by manipulation of their environmental conditions during culture and, as a result, ensure optimum yield of a useful product.

Growing micro-organisms

Some micro-organisms, such as unicellular algae, are able to derive energy from light by photosynthesis. However, most micro-organisms used in industry are bacteria and fungi which derive their energy from a chemical substrate. It is to these microbes that the following text refers.

Types of growth media

Micro-organisms are normally grown under controlled conditions in a laboratory either in a liquid medium called broth or on a solid medium called agar jelly to which essential nutrients have been added. The **growth medium** is contained within Petri dishes, flasks or bottles (see Figure 12.3) for small-scale laboratory work and within huge, stainless-steel fermenters in large, industrial processes.

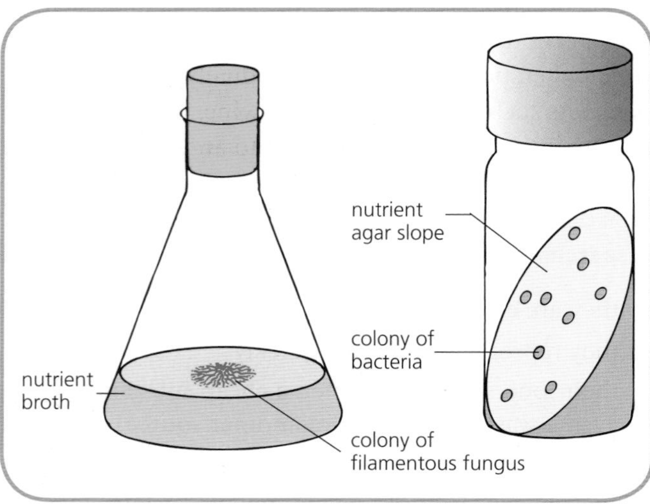

Figure 12.3 Growth media

Culture conditions

The growth of the micro-organism being cultured is affected directly by the chemical composition of its growth medium and by the environmental conditions to which it is exposed.

Chemical composition of growth medium

The microbe requires a supply of an organic compound such as carbohydrate as an **energy source**. It also needs a supply of raw materials to produce cellular building blocks such as amino acids and nucleotides needed in the **biosynthesis** of proteins and nucleic acids for new cells (see Table 12.1). Some micro-organisms can synthesise all of the complex molecules including a full range of amino acids from simple chemicals; others need specific complex compounds such as fatty acids or certain vitamins to be present in their growth medium.

Environmental conditions

Aseptic techniques are employed in the preparation and transfer of growth medium and the inoculation of the medium with the micro-organism. Every effort is made to keep conditions **sterile** to eliminate contaminants that would affect the growth of the microbe. Use of a fermenter allows a variety of environmental conditions to be monitored and controlled. This is done by computers when industrial-sized fermenters are being used (see page 171).

Chemical requirement	Typical source of chemical	Explanation
carbon	organic compound such as carbohydrate	organic compounds provide the microbe with energy and raw materials for biosynthesis
hydrogen	water and organic compounds	essential component of all organic materials
oxygen	water and air	oxygen is a component of many organic materials and the final electron acceptor in aerobic respiration
nitrogen	compound containing ammonium or nitrate group	nitrogen is needed for the synthesis of nucleic acids, amino acids and proteins
phosphorus	compound containing phosphate group	phosphorus is needed for the synthesis of ATP and nucleic acids
sulphur	compound containing sulphate group	sulphur is needed for the synthesis of some amino acids

Table 12.1 Nutritional requirements of micro-organisms

Investigation

Effect of nutrients and temperature on growth of *E. coli*

The experiment is carried out as shown in Figure 12.4. From the results it is concluded that *E. coli* requires the presence of essential nutrients in its growth medium and that it grows better at 30 °C than at 20 °C.

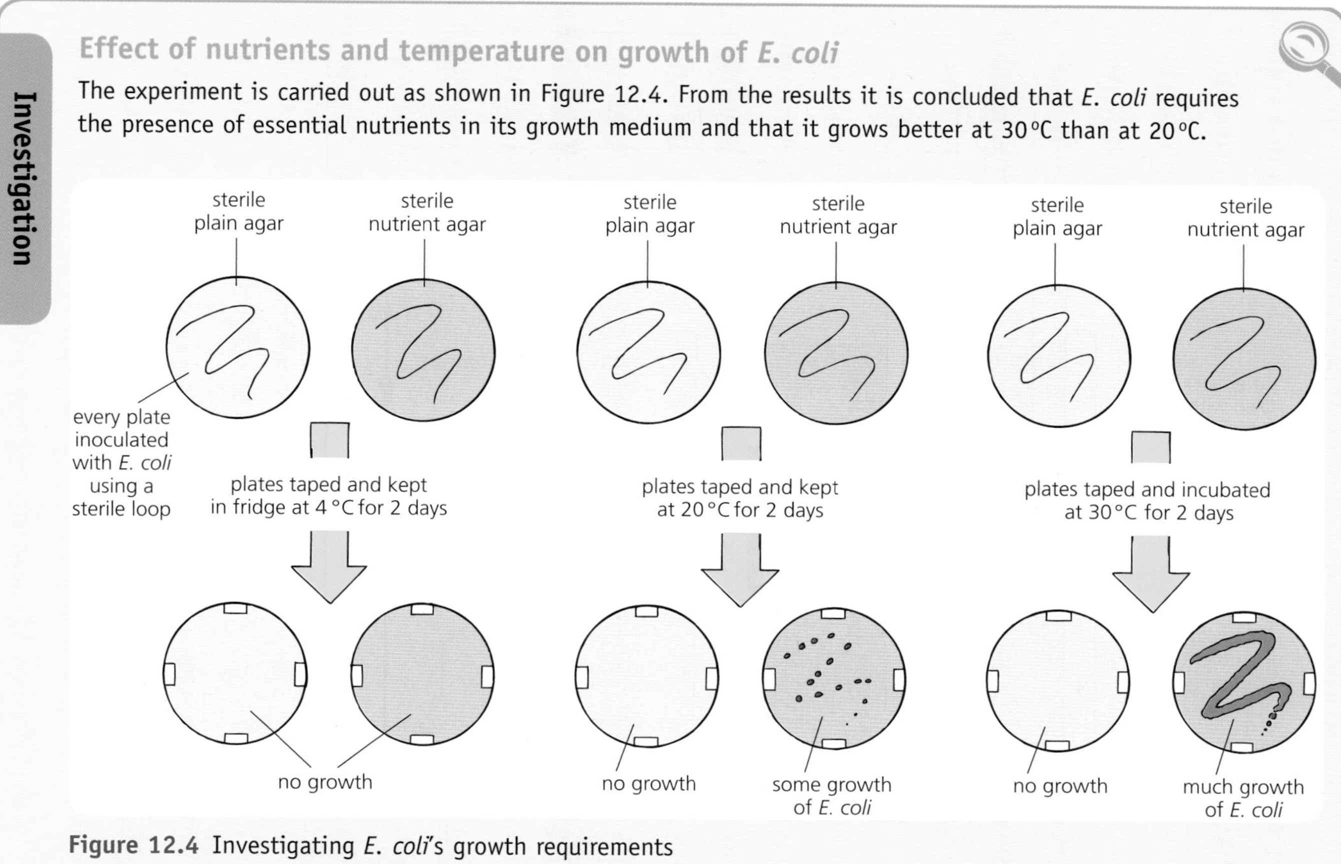

Figure 12.4 Investigating *E. coli*'s growth requirements

Computer-controlled fermenter

In the context of industrial microbiology, the term **fermentation** refers to both aerobic and anaerobic processes (and not specifically to anaerobic reactions such as alcohol production). Fermentation industries grow micro-organisms on a vast scale in order to produce huge quantities of a useful product such as an antibiotic. Such commercial production of antibiotics (and other substances such as enzymes, vaccines and cell metabolites) is made possible by the use of enormous **industrial fermenters** which are often able to hold thousands of litres of nutrient liquid.

Figure 12.5 shows a simplified version of an industrial fermenter being used to grow a fungus. The system is controlled automatically by **computers**. **Sensors** in contact with the nutrient solution monitor the various conditions that are needed for the best growth of the micro-organism. Some examples of factors that affect its growth are:

- temperature
- oxygen concentration
- pH
- glucose concentration.

If any of these factors varies from the **optimum level**, the sensor picks up the change and sends information to the computer. The computer responds by communicating with the source of supply of the essential factor. The supply is adjusted accordingly until the required level is restored.

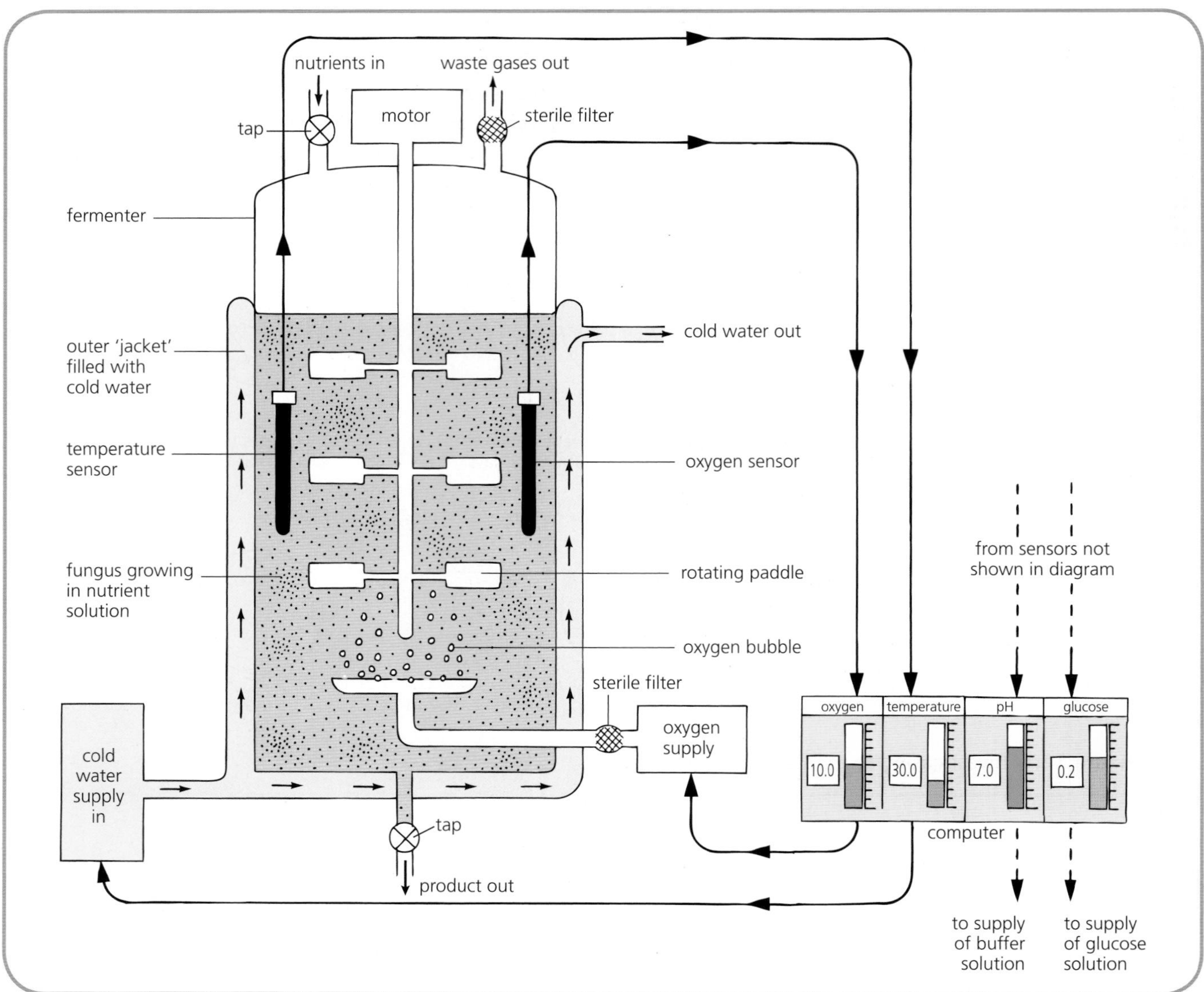

Figure 12.5 Industrial fermenter

Imagine that the fungus being cultured in the fermenter in Figure 12.5 grows best under the following conditions:

- 30 °C
- 10% oxygen
- pH 7
- 0.2 molar glucose solution.

If the temperature rises above 30 °C then the **temperature sensor** picks up this information and sends it to the computer. The computer then sends out a message which causes an increase in the rate of flow of the cold water cooling system. This continues until the temperature drops back down to 30 °C.

If the growing fungus is using up oxygen so rapidly that the concentration has dropped below 10%, this information is picked up by the **oxygen sensor** and sent to the computer. It then sends out a message that brings about an increase in supply of oxygen to the fermenter until the level returns to 10%.

Although not shown in Figure 12.5, further sensors for pH and nutrient levels would be monitoring these factors, allowing them to be adjusted as required.

Related Activity

Investigating the growth of yeast using a simple fermenter

The fermenter is set up as shown in Figure 12.6 and maintained at 25 °C. The magnetic stirrer is set at maximum speed throughout the investigation. Growth is monitored over four to five days by sampling the culture medium and taking optical density readings in order to follow changes in **turbidity** (cloudiness) which gives an indication of cell mass.

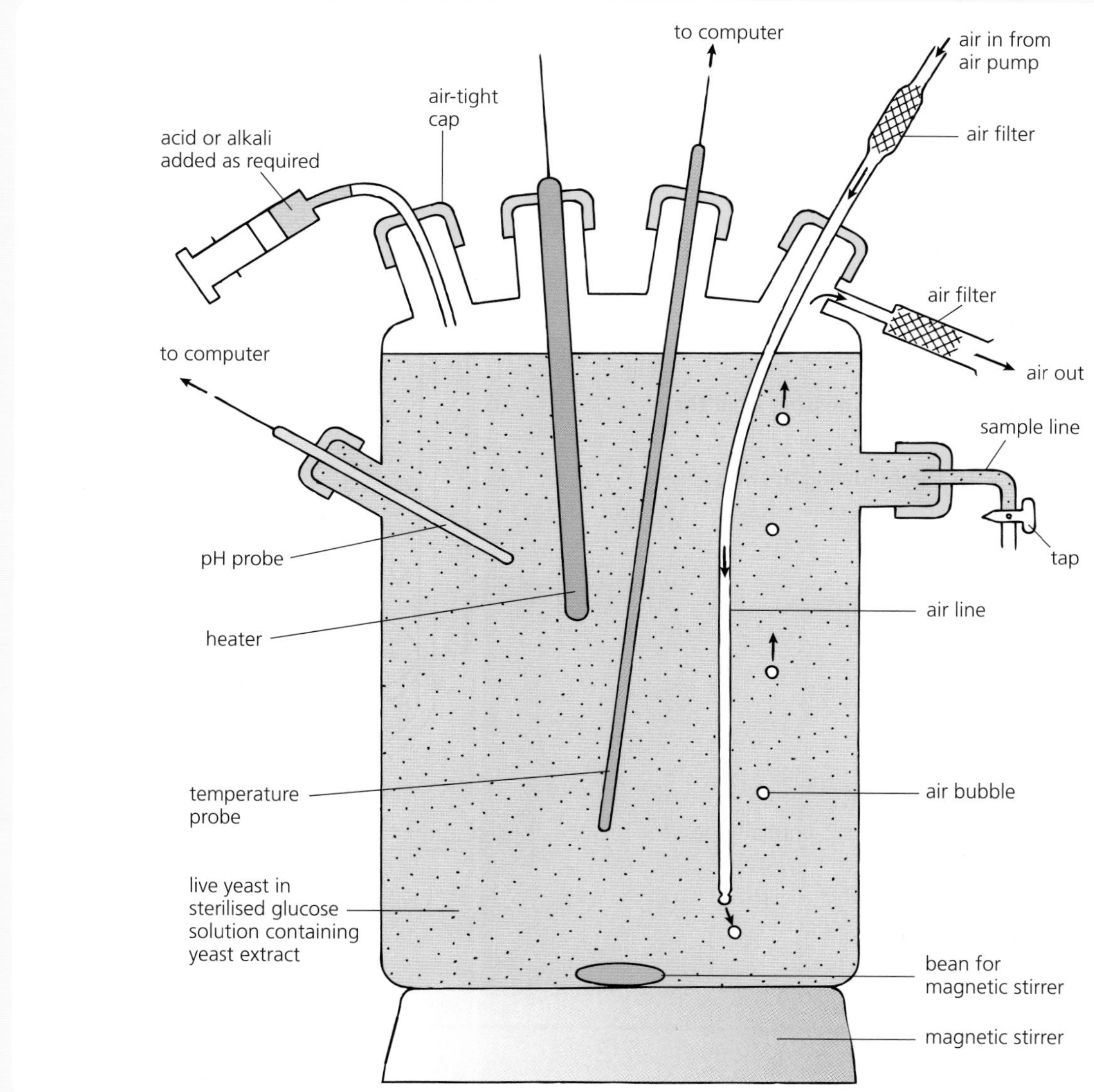

Figure 12.6 Simple fermenter

Monitoring the product

When given constant optimum conditions, the fungus grows rapidly and then produces the product which it normally releases into the surrounding medium. Computer-controlled technology also monitors this release and brings the process to a halt when the required level of antibiotic has been produced. This technology makes the system very efficient because it saves energy and prevents raw materials being wasted.

Testing Your Knowledge 1

1 Identify TWO properties of micro-organisms that make them highly useful for a variety of research and industrial purposes. (2)

2 Nutrient agar and nutrient broth are growth media.

   a) What is the basic difference between the two? (1)

   b) Which would be more likely to be used in an industrial fermenter? (1)

3 a) Explain why a micro-organism needs a supply of:

      i) nitrogen

      ii) phosphorus

      iii) organic compound. (3)

   b) Why must *aseptic conditions* be employed when setting up a culture of micro-organisms? (1)

4 Figure 12.7 shows a simplified version of an industrial fermenter.

   a) Identify THREE types of sensor missing from the diagram. (3)

   b) Match boxes 1–6 with the following answers:

      A products out

      B waste gases out

      C acid in

      D oxygen in

      E cold water in

      F cold water out (6)

   c) i) Identify part X.

      ii) State the function of part Y.

      iii) Of what material would part Z normally be composed? (3)

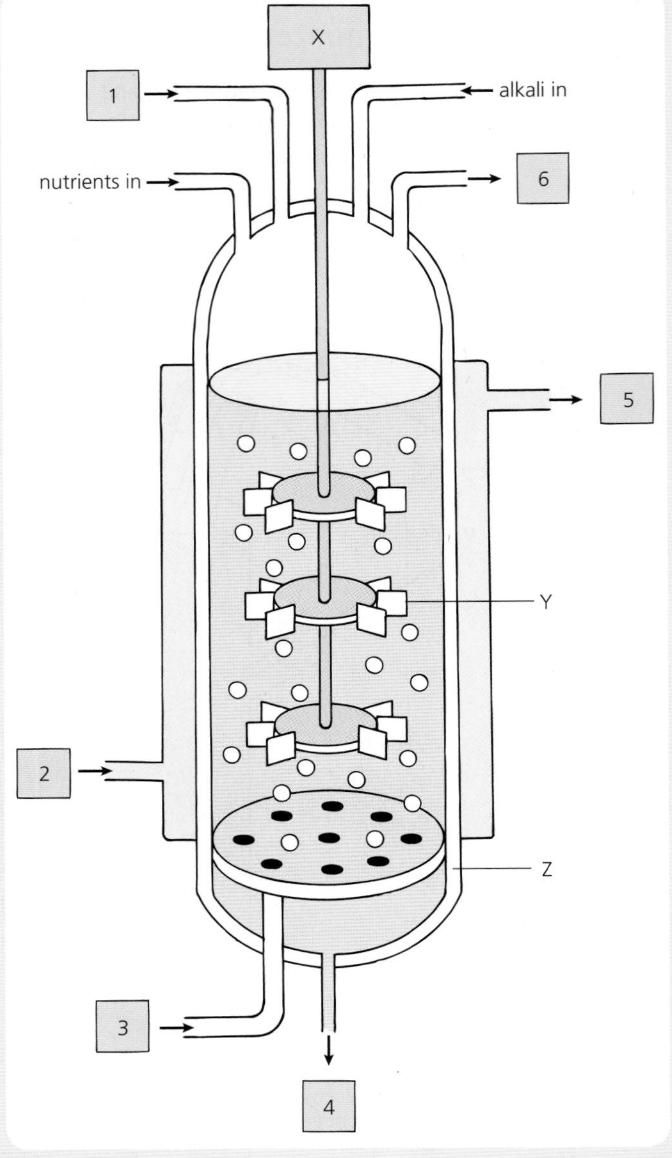

Figure 12.7

Patterns of growth

Dry biomass and fresh biomass

Growth occurs when the rate of synthesis of organic materials by an organism exceeds the rate of their breakdown. Growth involves an **irreversible increase in dry biomass**. Gain in dry biomass is a more reliable indicator of growth than gain in fresh biomass because an organism's fresh biomass varies depending on water availability.

Increase in dry biomass is often used to measure growth of a filamentous fungus. However, this method is less practicable for measuring growth of unicellular organisms. Therefore growth of bacteria and yeast cells is usually investigated by measuring **increase in cell number** over a period of time.

Growth of unicellular organisms

Generation time

The time needed for a population of unicellular organisms to double in number is called the mean **generation time** or **doubling time**. A population of unicellular organisms such as bacteria growing in a liquid medium use up the nutrients provided and secrete metabolites that they have made back into the medium. These changes result in the microbe's **pattern of growth** varying over time. It falls into **four distinct phases**, as shown in Figure 12.8.

Lag phase

During the **lag phase** there is little or no increase in cell number. The cells adjust to the growth medium and show increased metabolic activity. They may need to **induce enzymes** for use in metabolising the new substrate(s). This phase is the flat part at the start of the graph.

Log or exponential phase of growth

During the **exponential phase** the cells grow and multiply at the maximum rate, provided that nutrients are plentiful and no factor is limiting. This phase is represented by the steep incline on the graph. The population doubles its number with each cell division. For example, the cell number of an organism that has a doubling time of 0.5 hours would increase as indicated in Table 12.2.

The cell number resulting from this phase of growth is so large after a few hours that it becomes difficult to plot accurately all the points (especially the lower values) on a sheet of normal graph paper (see Figure 12.9). One way of solving this problem is to create a scale that doubles in number at each interval on the vertical axis. Figure 12.10 shows how the full range of values in Table 12.2 can be plotted accurately on one sheet of graph paper by this means. In addition, this type of graph can depict an equal rate of growth as a line of equal slope.

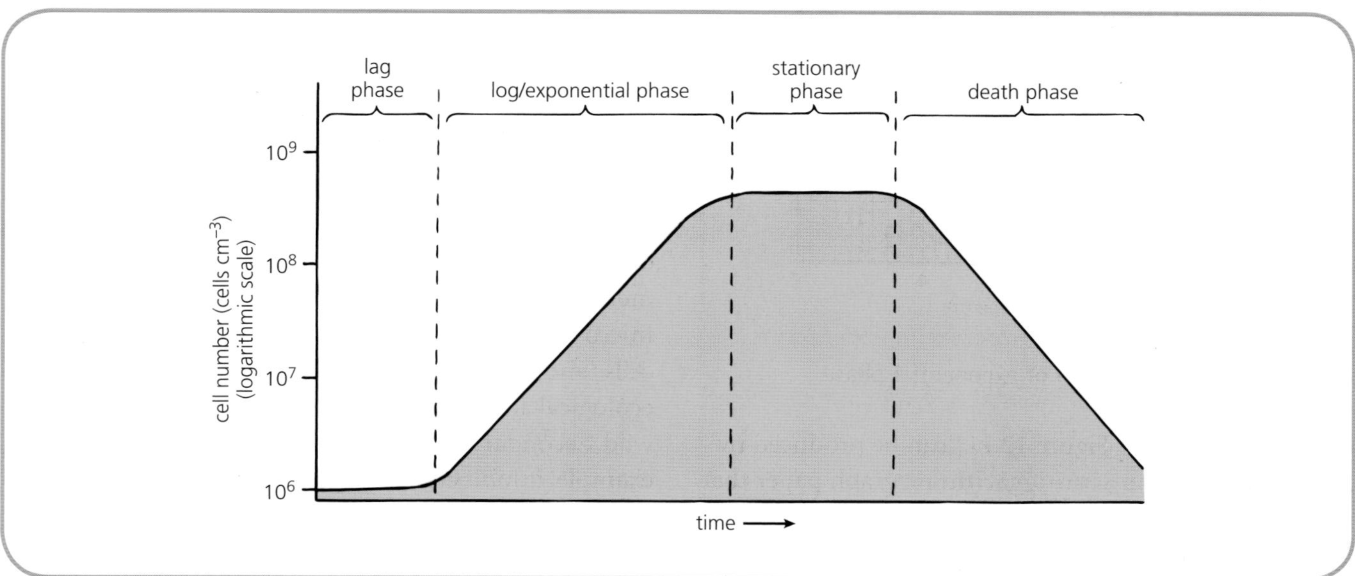

Figure 12.8 Growth pattern of an unicellular culture

Time (h)	Cell number
0.0	1
0.5	2
1.0	4
1.5	8
2.0	16
2.5	32
3.0	64
3.5	128
4.0	256
4.5	512
5.0	1024
5.5	2048
6.0	4096

Table 12.2 Growth of bacteria

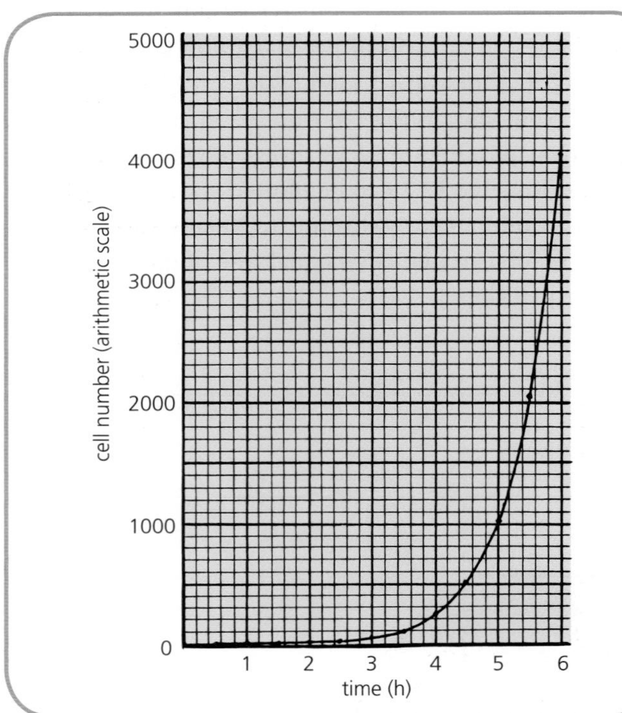

Figure 12.9 'Normal' graph of exponential phase

A similar graph (see Figure 12.11) can be produced by plotting the data on **semi-logarithmic** graph paper that shows all the subdivisions of the y-axis scale. This time the interval on the y axis from 1 to 10 is the same as that from 10 to 100 and so on. In each successive cycle,

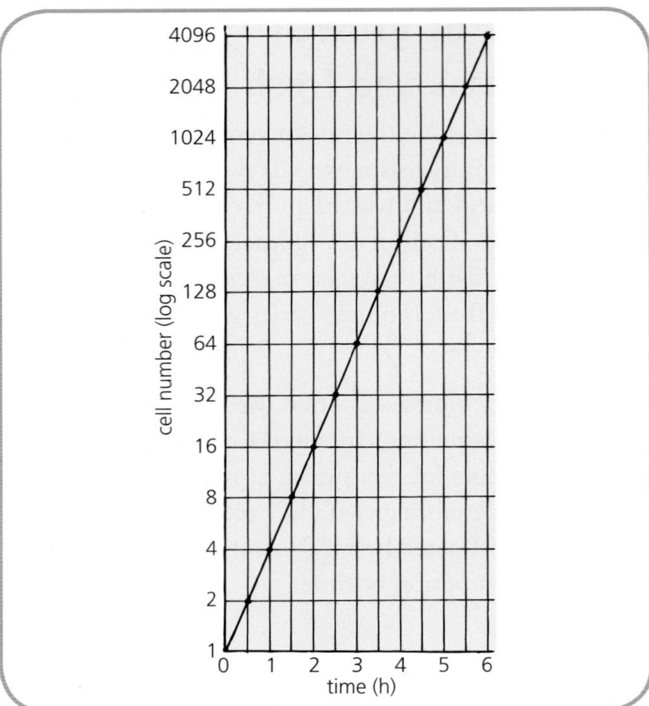

Figure 12.10 Graph with doubling of values on y axis

the values are **ten times** greater than the cycle below. This graph also depicts an equal rate of growth as a line of equal slope.

Stationary phase

The rapid phase of exponential growth does not continue indefinitely (see Figure 12.8). Nutrients begin to run out and/or toxic metabolites produced by the microbe may build up and cause its rate of cell division to decrease. At the point where the rate of production of new cells is equal to the death rate of the old ones, the population has reached the **stationary phase**. The graph's flat shape shows that the population number remains fairly steady during this phase.

Secondary metabolism during the stationary phase of growth results in the production of **secondary metabolites**. Although these are not used by the micro-organism for growth and production of new cells (and may even be toxic), some may confer an ecological advantage on the micro-organism in the wild. Secondary metabolites such as **antibiotics**, for example, inhibit competing micro-organisms (such as bacteria) in the fungus' ecosystem (for example, soil). Other secondary metabolites promote spore formation (see Figure 12.12) which increases the microbe's

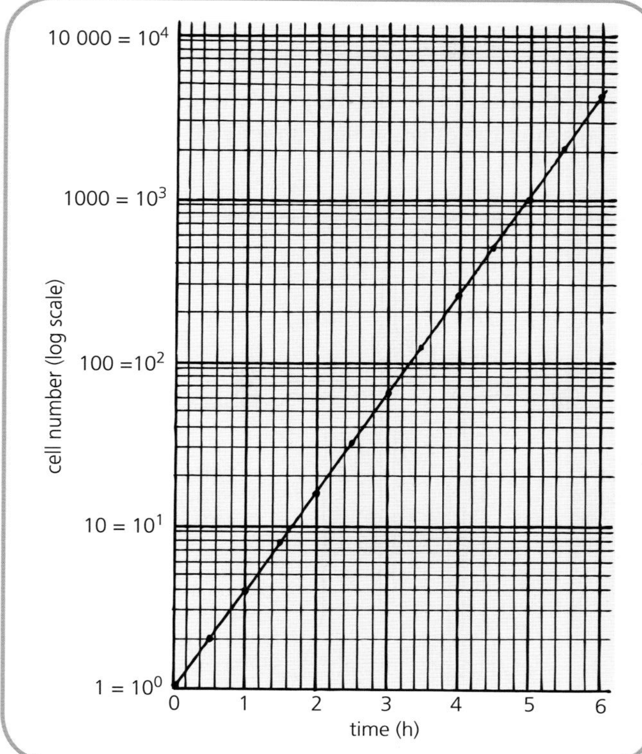

Figure 12.11 Graph using normal semi-log graph paper

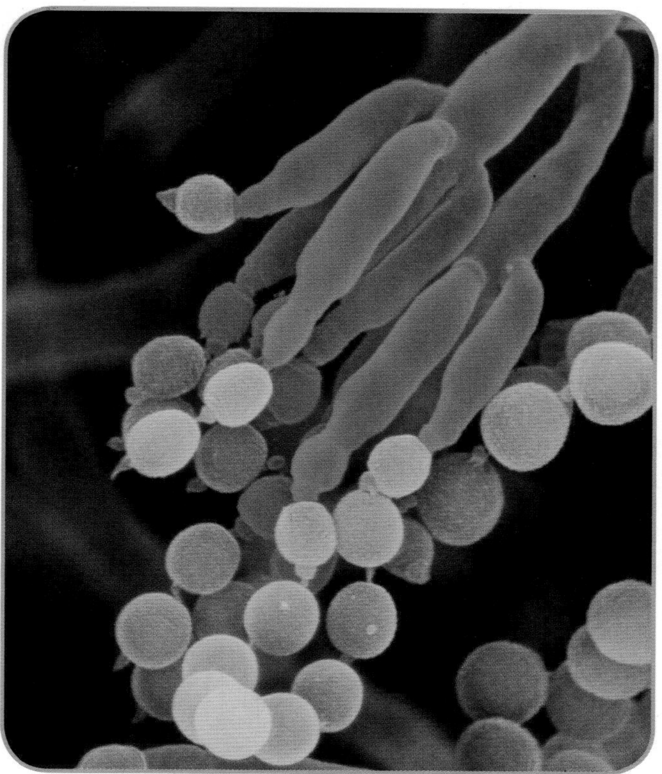

Figure 12.12 Spore formation in filamentous fungus

reproductive potential and chance of survival by expanding its range and outcompeting its rivals.

Death phase

Finally the lack of nutrient substrate and/or the accumulation of a high concentration of **toxic metabolites** lead to the **death phase**. The number of cells dying now exceeds the number of new cells being produced. At this stage the cells are dying at a constant rate and they may undergo lysis (bursting). Figure 12.8 shows how the population enters a period of exponential decline. By the end of this phase, the population may be completely wiped out or it may survive in the form of a few resistant spores. A **viable cell count** gives the number of cells that are alive and capable of reproduction. A **total cell count** refers to all cells, dead or alive. Only a viable cell count shows a death phase.

Testing Your Knowledge 2

1 Define the terms *growth* and *mean generation time*. (2)

2 a) Give ONE difference between the lag phase and the stationary phase of the pattern of growth shown by a population of bacteria. (1)

b) Explain why the graph of the exponential phase of growth takes the form that it does. (1)

c) Give TWO possible causes of death to bacterial cells during the death phase of growth. (2)

3 a) Give an example of a secondary metabolite. (1)

b) Explain how it may confer an ecological advantage on the fungus that produces it. (1)

Applying Your Knowledge and Skills

1 a) Which part of Figure 12.13 shows a fermenter correctly set up and ready for use? (1)

b) With reference to the diagram, give ONE reason why each of the others is not correctly set up. (3)

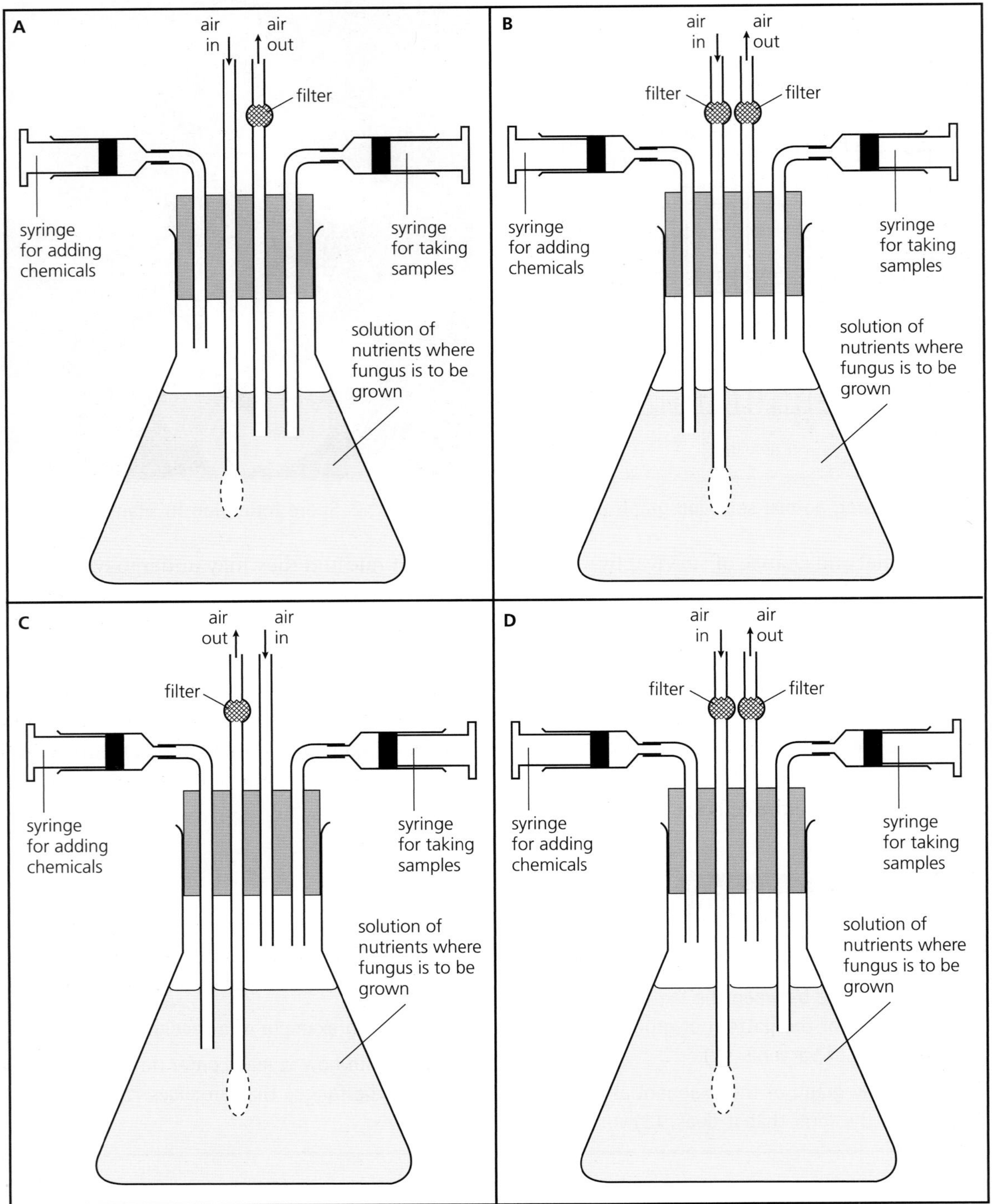

Figure 12.13

→

2 The graph in Figure 12.14 shows the results from closely monitoring the changes that took place during the fermentation of a closed batch of wine.

a) i) Which line represents the number of live yeast cells?
ii) Justify your choice of answer. (2)

b) i) Which line represents the mass of glucose present in the fermenter?
ii) Justify your choice of answer. (2)

c) i) Which line represents the mass of alcohol present in the fermenter?
ii) Justify your choice of answer. (2)

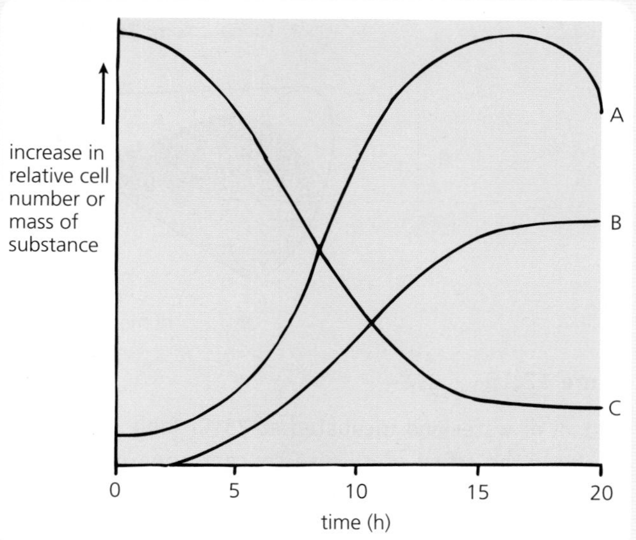

Figure 12.14

3 An inoculum of 3×10^3 bacteria was added to nutrient medium in a fermenter and given optimum conditions for growth. The bacterium's doubling time was 20 minutes.

a) Copy and complete Table 12.3. (4)

Time (in 20-minute intervals)	Cell number ($\times 10^3$)	Cell number (correct to two decimal places)
0	3	3.00×10^3
1	6	
2	12	1.20×10^4
3	24	2.40×10^4
4	48	
5		
6		1.92×10^5
7		
8	768	7.68×10^5
9		
10		
11	6 144	6.14×10^6
12		

Table 12.3

b) How long did it take for the bacterial population to exceed 3 million? (1)

4 The time taken by a generation of micro-organisms to divide and double in number is called the mean generation time.

If **p** = number of bacteria at the start,
q = number of bacteria after **n** generations,
n = number of generations,
g = mean generation time (min)
and **t** = time for **n** generations (min),
then $q = p \times 2^n$ and $g = t/n$.

a) Using the data from question 3 and the above formulae, confirm that after 1 hour:
i) the bacterial population should be 24 000
ii) the generation time is 20 minutes. (Show your working.) (2)

b) Calculate the mean generation time for a colony of bacteria which began as a population of 1000 and had grown to a population of 16×10^3 after 2 hours. (Show your working.) (2)

5 Read the passage and answer the questions that follow it.

Growth and sporulation of a fungus

Figure 12.15 shows the vegetative mycelium of *Trichoderma viride*. It produces antibiotics and therefore vast colonies of this fungus are surface-cultured on liquid media. Although fragments of mycelium can be used as the inoculum, spores have been found to constitute a much more effective method of propagation. It is for this reason that many experiments have been carried out to examine the relationship between nutrition of the fungus and the production of its asexual spores. In such experiments the cultures are grown in triplicate, each flask being inoculated with 0.5×10^6 spores in

→

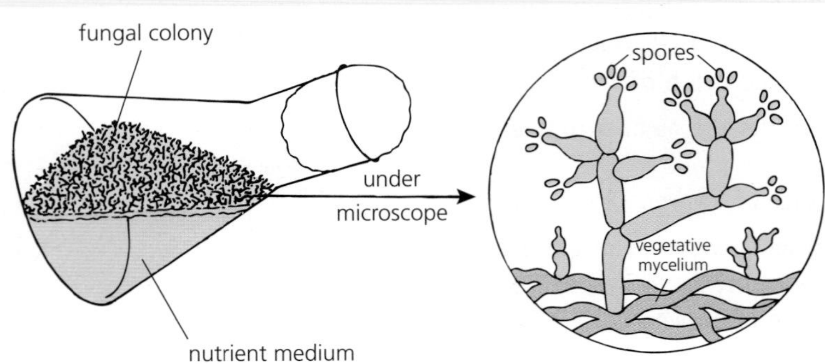

Figure 12.15

1 ml of water and incubated at 25 °C. Table 12.4 shows the effect of glucose concentration on growth and sporulation of *Trichoderma viride*.

a) Why is each culture in triplicate? (1)

b) Explain why culturing the fungus in flasks tilted to one side produces a greater yield of mycelium. (1)

c) Which type of nutrient medium in Table 12.4 brings about most overall growth of the fungus? (1)

Type of nutrient medium	% glucose	Flask	Dry mass of mycelium (mg)	Spore count ($\times 10^6 ml^{-1}$)
A (rich in nitrate)	0.1	1	9	16
		2	11	15
		3	14	16
	1.0	1	111	137
		2	109	131
		3	108	128
	10.0	1	379	65
		2	382	67
		3	378	67
B (rich in sulphate)	0.1	1	8	11
		2	7	12
		3	5	9
	1.0	1	64	99
		2	65	103
		3	68	112
	10.0	1	301	54
		2	301	51
		3	284	47
C (rich in phosphate)	0.1	1	11	19
		2	14	18
		3	17	20
	1.0	1	79	151
		2	81	147
		3	89	143
	10.0	1	357	81
		2	343	76
		3	329	72

Table 12.4

→

d) i) Describe the effect of varying the glucose concentration on dry weight of mycelium.
ii) Explain why this should be so.
iii) What effect does increasing glucose concentration have on sporulation? (3)

e) i) Suggest why on some occasions scientists want to mass produce the vegetative mycelium of *Trichoderma viride*, while on other occasions they wish to obtain maximum sporulation.
ii) From the information given, state the concentration of glucose that would be most suitable in each case.
iii) Briefly describe how an even more accurate estimate of the optimum glucose concentration for sporulation could be obtained. (3)

6 The graph in Figure 12.16 shows the log phase of growth of a species of bacterium.

a) What values should have been entered at points P and Q on the y-axis scale? (2)

b) i) How many cells were present at time 14.15?
ii) Now state this number of cells in words. (2)

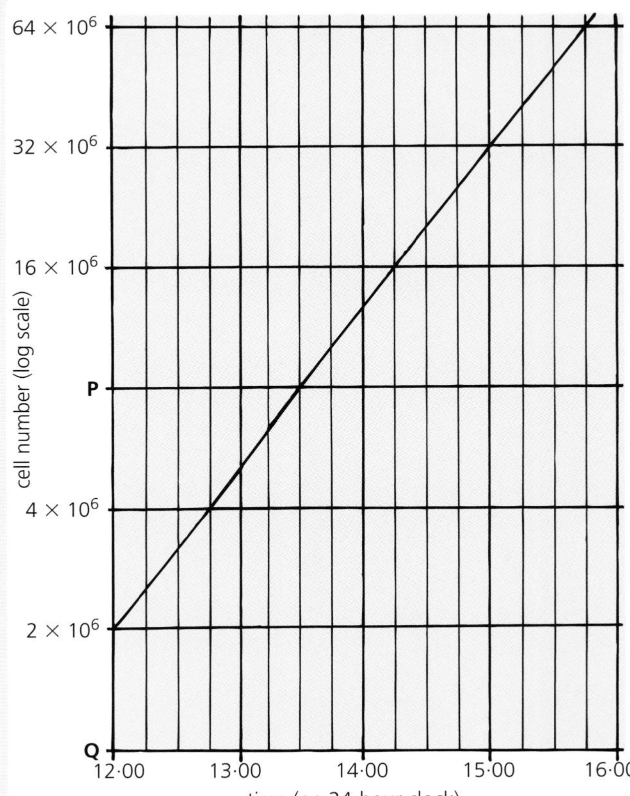

Figure 12.16

c) i) How many more cells were present at 15.00 compared with the number present at 12.45?
ii) By how many times had the population multiplied between 12.00 and 15.45?
iii) How long is the generation time for this micro-organism?
iv) Using the formula $g = t/n$ from question 4, confirm your answer to **iii)**. (Show your working.) (4)

7 Table 12.5 shows the results of growing a batch culture of bacteria.

a) Using a sheet of semi-log graph paper similar to the one shown in Figure 12.17 (page 182), draw a line graph of the results by plotting the points and joining them together with straight lines. (4)

Time on 24-hour clock	Number of viable cells ($\times 10^6$)
08.00	2
10.00	2
12.00	3
14.00	8
18.00	110
22.00	900
08.00	900
12.00	40
14.00	9

Table 12.5

b) On your graph indicate clearly:
i) the exponential phase of growth
ii) the death phase. (2)

c) For how many hours was the experiment run? (1)

d) i) When drawing a conclusion from these data, which part of the graph is least reliable?
ii) Explain your answer.
iii) What could be done in a repeat of the experiment to make this part of the graph more reliable? (3)

8 Give an account of the four phases of growth undergone by a population of bacteria growing in a finite volume of liquid culture medium. (9)

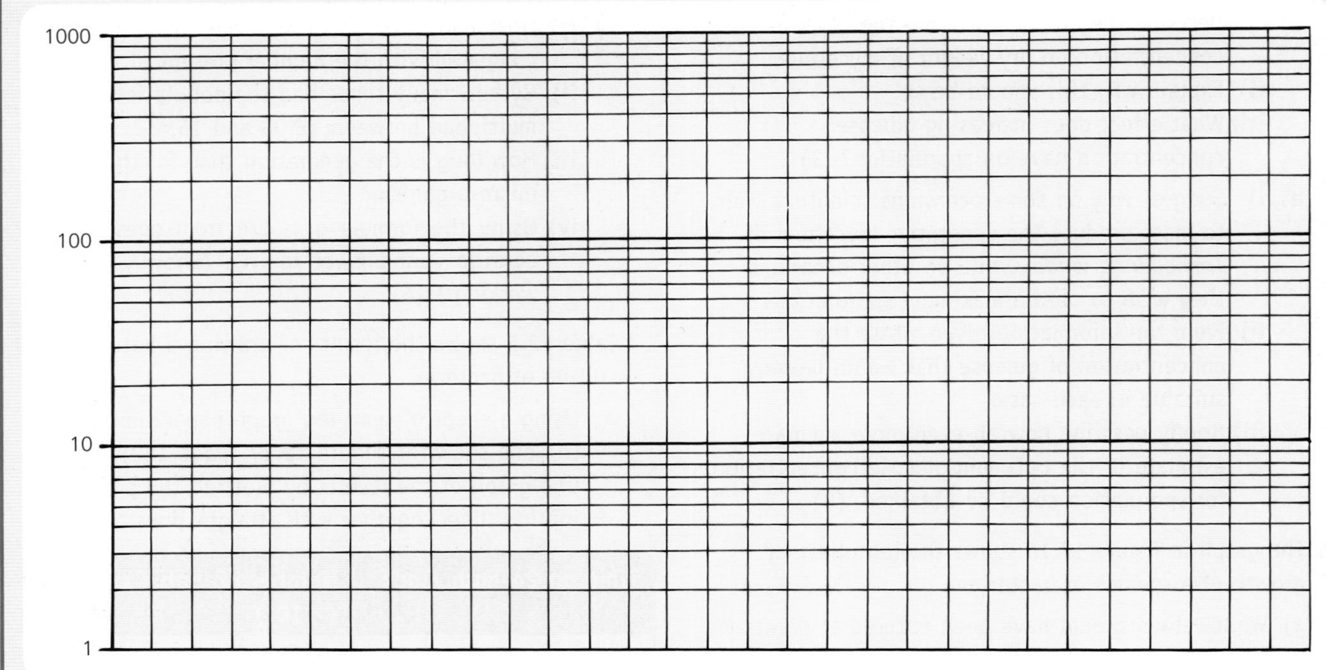

Figure 12.17

13 Genetic control of metabolism

Selection and isolation of micro-organisms

Wild strains of micro-organisms that may be of use in an industrial process are **selected** and cultured in enriched nutrient medium. They are given optimum growing conditions and then pure strains are **isolated** and screened for desired traits.

Strain improvement

Once a pure strain of wild micro-organism exhibiting a desired trait has been isolated, it may still lack many other important features such as:

- genetic stability
- ability to grow on low-cost nutrients
- ability to vastly overproduce the target compound for which it was selected
- ability to allow easy harvesting of the target product following the fermentation process.

Therefore **strain improvement** is employed to try to alter the wild microbe's genome and include the genetic material for these traits. This may be brought about by mutagenesis or recombinant DNA technology.

Mutations and mutagenesis

A **mutation** is a heritable change in an organism's DNA that causes **genetic diversity** (see Chapter 5). Usually the DNA sequence affected is no longer able to function, but on very rare occasions, there arises by mutation a mutant allele that confers an advantage on the organism or endows it with a new property that is useful to humans.

Mutagenesis is the creation of mutations. The rate of mutagenesis can be increased artificially by exposing organisms to **mutagenic agents** such as ultraviolet (UV) light, other forms of radiation or mutagenic chemicals, since all of these alter DNA and induce mutations. The experiment shown in Figure 13.1 shows the production of mutant strains of bacteria following exposure of the original strain to UV light.

Improved strains

Occasionally a mutant strain induced by a mutagenic agent produces an **improved strain** of micro-organism. Many industrial micro-organisms have been improved by subjecting them to repeated sessions of mutagenesis followed by careful selection and screening of the survivors. Normally the improved strain that is selected **lacks an inhibitory control mechanism**. Therefore it no longer expresses some undesirable characteristic or it produces an increased yield of the desired product.

Site-specific mutagenesis

Spontaneous and induced mutations both occur randomly throughout a genome. However, geneticists normally want to study the effect of a mutation on one specific gene. In recent times this has become possible by producing many copies of a gene's DNA *in vitro* (see PCR page 18) and then making a change in the base pair sequence at a specific position in the DNA chain. This process is called **site-specific mutagenesis**. The mutated gene is then introduced back into the cell and the cell cultured to allow any changes in phenotype resulting from the mutation to be studied.

Recombinant DNA technology

This technique enables scientists to transfer gene sequences from one organism to another and even one species to another. By crossing the species barrier, **recombinant DNA technology** makes possible the production of a plant or animal protein by a micro-organism that has been **artificially transformed**.

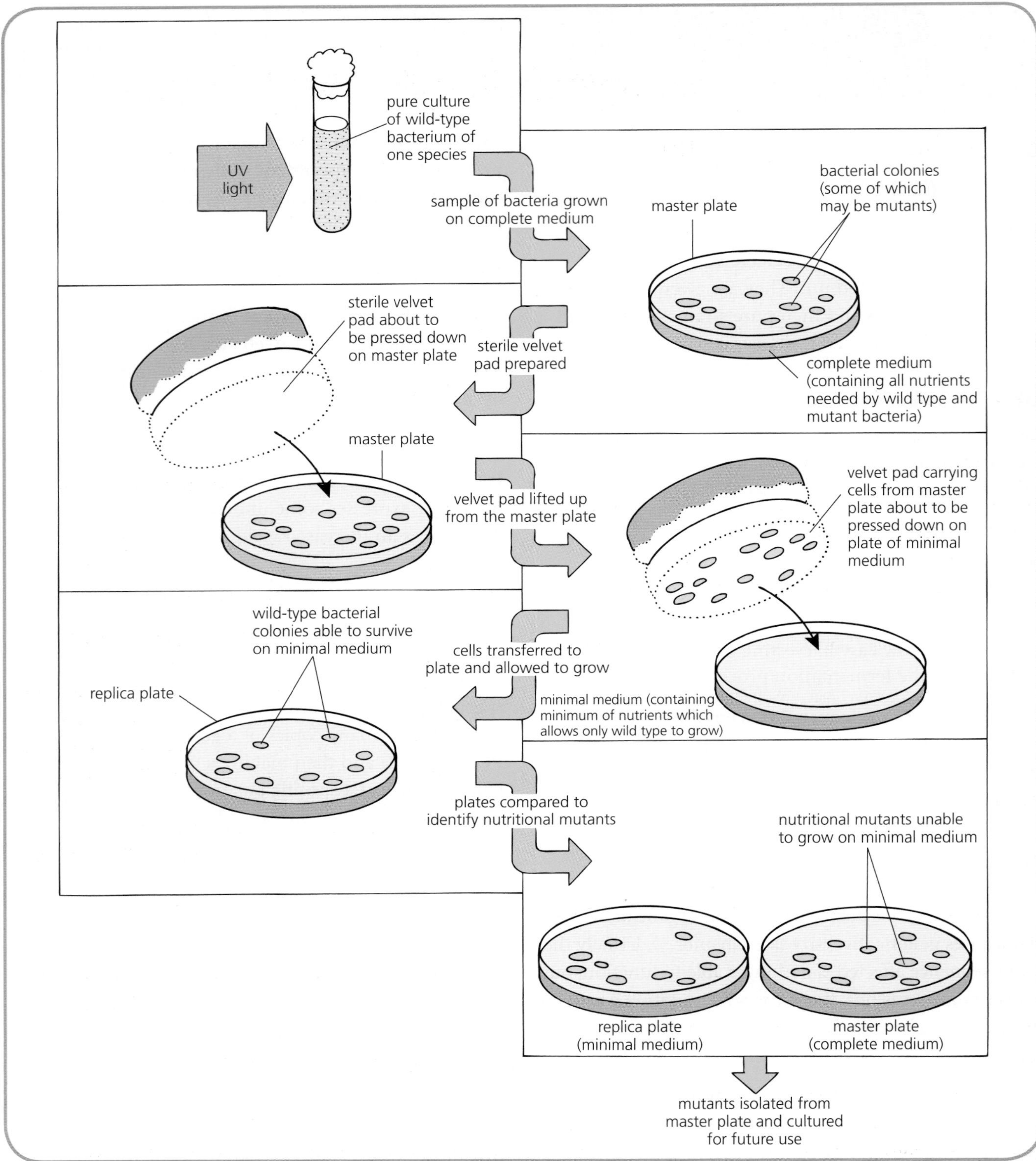

Figure 13.1 Production of mutants following exposure to UV light

Testing Your Knowledge 1

1 Imagine that a micro-organism (Y) is able to make a useful product (y). Name THREE other traits of microbe Y that scientists would regard as desirable when planning to make product y on an industrial scale. (3)

2 a) What is meant by the term *mutagenesis*? (1)

b) Name TWO agents that could be used to increase an organism's rate of mutation. (2)

3 Rewrite the following sentences selecting only the correct answer from the underlined choice. (3)

a) In the absence of outside influences, mutations arise very rarely/frequently.

b) An agent that increases the rate of mutagenesis is called a mutagen/mutant.

c) Mutagenesis can be used to create a wild-type/mutant strain that lacks a particular undesirable characteristic.

Improvement of existing strain

DNA technology can also be used to introduce to an existing micro-organism (such as a bacterium or yeast cell) one or more genes that:

- amplify specific metabolic steps in a pathway thereby **increasing yield** of the target product
- cause the cells to secrete their product into the surrounding medium allowing it to be **easily recovered**
- render the micro-organism unable to survive in an external environment and therefore act as a **safety mechanism**.

Artificial transformation of a bacterium by recombinant DNA technology

Genetic engineers are able to select a particular gene for a desirable characteristic (such as the gene for human insulin), splice its DNA into the DNA of a vector (for example, a plasmid from a bacterial cell) and insert the vector into a host cell (for example, a bacterium such as *E. coli*).

The transformed host cell has its properties and functions altered and when it is cultured, it expresses the 'foreign' gene and produces the product (such as insulin). Since the host cell contains a **combination** of its own DNA and that from another source joined together, it is said to contain **recombinant DNA**. Figure 13.2 illustrates the general principles of recombinant DNA technology involving several 'tools of the trade' as follows.

Restriction endonuclease

A **restriction endonuclease** is an enzyme extracted from bacteria which is used to cut up the DNA (containing the required gene from the donor organism) into fragments and to cleave open the bacterial plasmids that are to receive it.

Each restriction endonuclease recognises a specific short sequence of DNA bases called a **restriction site** (or recognition sequence). This target sequence (four to eight nucleotides in length) is found on both DNA strands but running in opposite directions, as shown in Figure 13.3. The enzyme cuts both DNA strands and may produce blunt ends or sticky ends. If the recognition sequence occurs many times (such as in a long DNA molecule) then the enzyme will make many cuts.

DNA ligase

DNA ligase is an enzyme which seals sticky ends (and blunt ends) together. It is used to seal a DNA fragment into a bacterial plasmid to form a **recombinant plasmid** containing recombinant DNA.

Vector (carrier)

Recombinant plasmids and artificial chromosomes (see below) are used as **vectors** to carry DNA from the genome of one organism (such as a human) into that of another (for example, a bacterial host cell) and

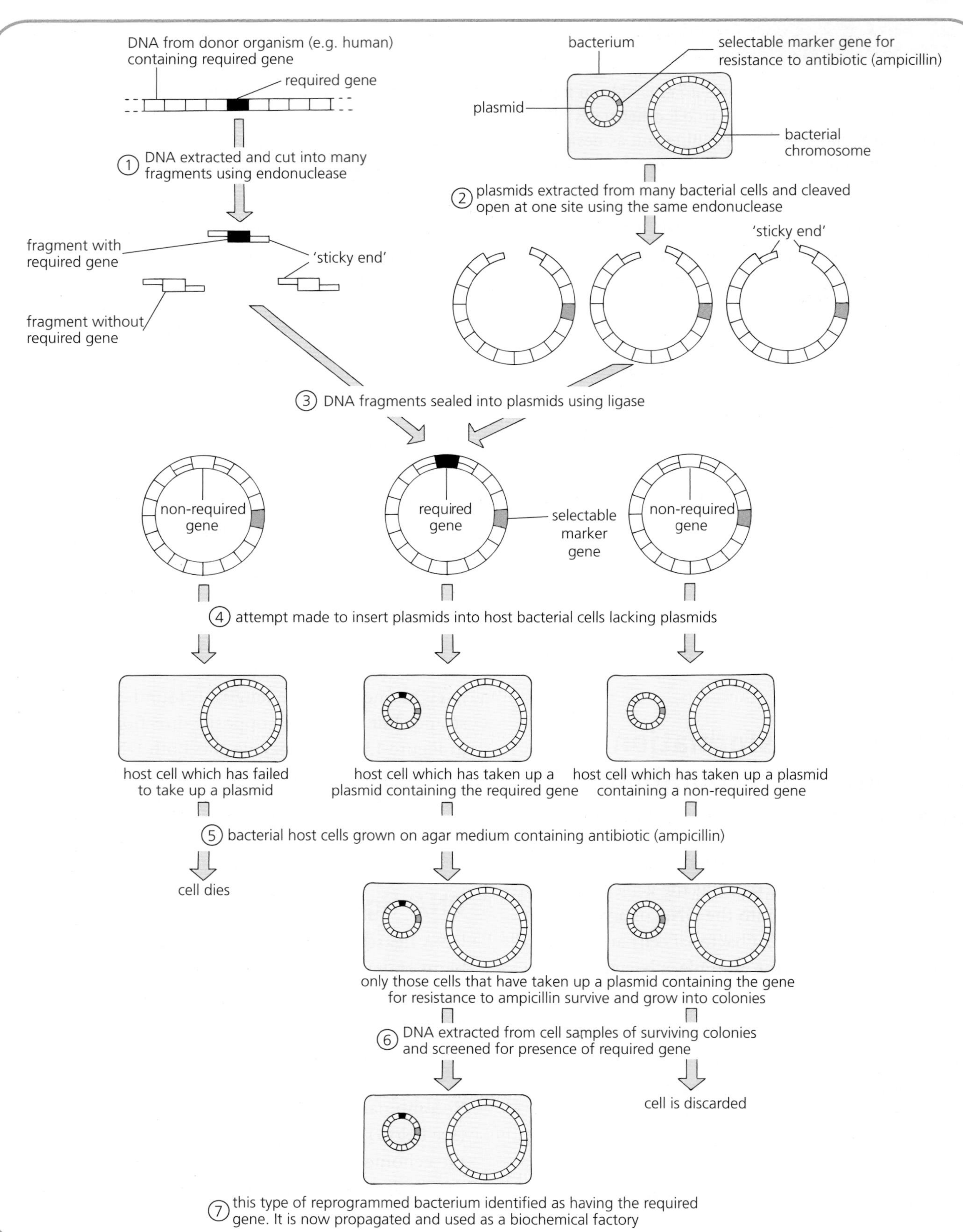

Figure 13.2 Recombinant DNA technology

Enzyme	Recognition sequence on DNA	Position on DNA cut by enzyme	Result
1	**GGCC** **CCGG**	cut AGTCCT**GGCC**ACTGC TCAGGA**CCGG**TGACG cut	AGTCCT**GG** **CC**ACTGC TCAGGA**CC** **GG**TGACG two 'blunt' ends
2	**GATC** **CTAG**	cut GCAT**GATC**TCGT CGTA**CTAG**AGCA cut	GCAT **GATC**TCGT CGTA**CTAG** AGCA two 'sticky' ends
3	**CTGCAG** **GACGTC**	cut AGC**CTGCAG**CCA TCG**GACGTC**GGT cut	AGC**CTGCA** **G**CCA TCG**G** **ACGTC**GGT two 'sticky' ends

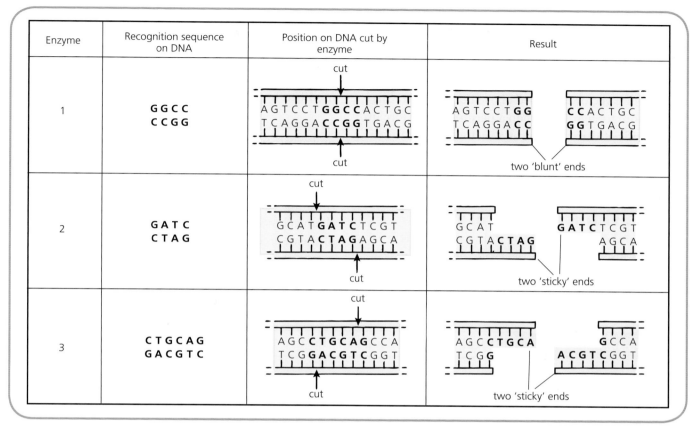

Figure 13.3 Action of restriction endonucleases

bring about **transformation** of the latter. To act as an effective vector, a plasmid must have the features shown in Figure 13.4.

Restriction site

This site must contain the appropriate target sequence of DNA. Then it can be cut open by the same restriction endonuclease used to cut the DNA containing the gene to be transported. If these conditions are met, the sticky ends will be complementary and the gene will become inserted in the plasmid.

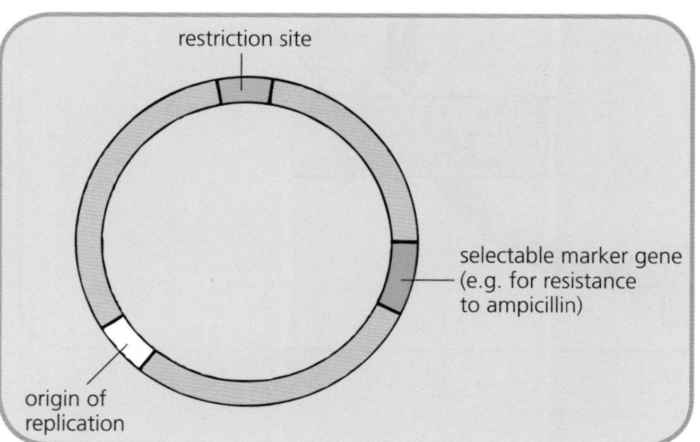

Figure 13.4 Plasmid suitable for use as a vector

Selectable marker

This type of gene protects the micro-organism from a selective agent such as an antibiotic that would prevent it from growing or kill it. The presence of the gene enables the scientist to determine whether a host cell has taken up the plasmid vector or not. For example, if the plasmid contains the **selectable marker** gene for **resistance to ampicillin** then this antibiotic can be used to select for the bacterial DNA. When cultured in a medium containing ampicillin, any host cells that have failed to take up a recombinant plasmid die since they lack the resistance gene (see Figure 13.2).

Origin of replication

This site consists of genes that control **self-replication** of plasmid DNA. It is essential for the generation of many copies of the plasmid (and the required gene) within the transformed bacterial host cell. When many copies of the gene are expressed, more product can be made by fewer cells.

Artificial chromosomes

Scientists have constructed **artificial chromosomes** that can also act as vectors in recombinant DNA technology. Each artificial chromosome possesses all the essential features of a vector described above but is able to carry much more foreign DNA than a plasmid. Therefore use of an artificial chromosome as a vector allows a **much longer sequence** of DNA to be carried from the donor organism to the recipient micro-organism.

Limitations of prokaryotes

Cloning and expressing a gene from a eukaryotic organism inserted into a prokaryotic organism may be problematic. The DNA of eukaryotes contains long stretches of non-coding DNA called **introns** interspersed among the protein-coding regions called **exons** (also see page 27).

The DNA of bacteria has exons but no introns. Therefore transcripts of mRNA in bacteria are not modified by splicing before translation.

As a result, a gene from a eukaryote expressed by a prokaryote may result in the formation of a polypeptide molecule which is **inactive** because it is **incorrectly folded**.

Although many of these problems can be overcome by chemical means, there are situations where production of the desired protein by **genetically transformed eukaryotic cells** (such as **recombinant yeast**) is a preferable option despite their more demanding cultural conditions.

Research Topic — Ethical considerations in the use of micro-organisms

The use of genetically modified micro-organisms (GMMs) and the development of microbiological products by recombinant DNA technology raise issues of ethics (moral values and principles). It also requires careful consideration of potential hazards and the control of risk.

Within the context of biotechnology, a **hazard** is any potential danger to which a person could be exposed as a result of contact with a particular micro-organism or microbiological product. (Hazards are rated on a scale of minor, considerable or severe.) **Risk** is an assessment of the likelihood of such a hazard actually taking place. (Risk is rated on a scale of low, moderate, high or critical.)

It is essential that the level of a potential hazard is correctly identified and that its risk level is reduced to as low a level as possible by adopting appropriate **control measures**. Some examples are given in Table 13.1.

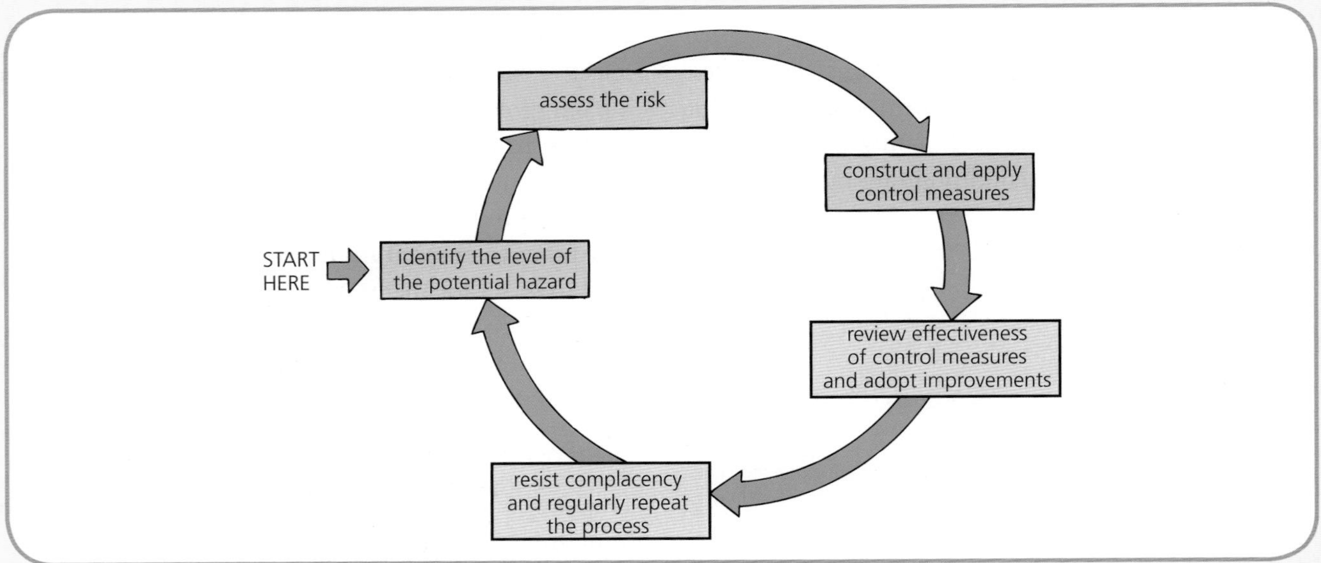

Figure 13.5 Risk assessment approach to safety

→

Organism	Use to which organism is put	Potential hazard	Possible control measures to reduce risk to lowest level
Pathogenic micro-organism	Manufacture of vaccine	The organism could cause the disease if it escaped from the controlled laboratory environment.	All members of staff are highly skilled. Rigorous protocols are adopted by all staff. Strict physical containment of the microbe is ensured by use of negative pressure laboratories containing special features such as safety cabinets and highly efficient air filters.
Genetically modified micro-organism (GMM)	Production of a human protein used to treat a disease	The GMM could mutate and become a pathogen and/or escape into the environment and have an adverse effect on the ecosystem.	Deliberate genetic manipulation of the GMM so that it can only grow under special laboratory conditions and be unable to survive in the wild.
Genetically modified crop plant given a gene from a different plant species making it resistant to a herbicide (also see page 224)	Production of a crop that prospers in the presence of the herbicide and in the absence of weeds	The genetically modified crop plant could exchange genes with its wild relatives during pollination leading to the possible emergence of 'superweeds' resistant to the herbicide.	Deliberate genetic manipulation of the crop plant so that it is unable to hybridise with its wild relatives.

Table 13.1 Potential hazards and control of risk

Risk assessment in the workplace

The safety of the workforce in a manufacturing process is of paramount importance. The assessment of risk (see Figure 13.5) that needs to be applied in a particular biotechnological process depends on the micro-organism being used. The regulations are more complex and the quality of the criteria more stringent for products that are pharmaceutical or foodstuffs or those whose production has involved GMMs. It is important that high standards of risk assessment are maintained and not relaxed in the future.

Testing Your Knowledge 2

1 a) With reference to TWO different species of living organism, describe a transfer of genetic material made possible by recombinant DNA technology. (2)

b) What benefit is gained by humans from the example that you have given? (1)

c) Which part of a bacterium's genetic material is often used as a vector? (1)

2 a) What is the difference between a *restriction endonuclease* and a *restriction site*? (2)

b) Briefly describe the role played by ligase in recombinant DNA technology. (2)

c) What is the function of the selectable marker gene for resistance to the antibiotic ampicillin in the procedure shown in Figure 13.2? (1)

3 a) Why might a gene from a eukaryote and expressed by a bacterium result in the production of an inactive polypeptide? (1)

b) Which unicellular organism could be used to overcome this problem? (1)

Applying Your Knowledge and Skills

1 The graph in Figure 13.6 shows the results and the line of best fit for three versions of the same experiment on bacteria carried out by three different scientific teams A, B and C.

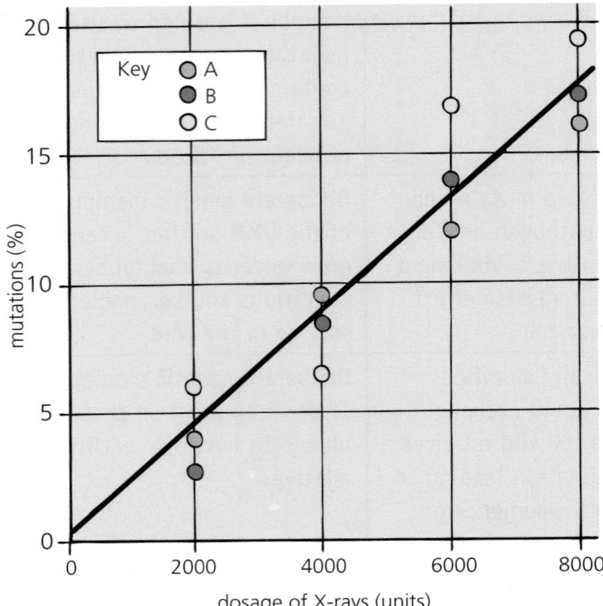

Figure 13.6

a) Identify the
   i) dependent variable
   ii) independent variable. (2)

b) What conclusion can be drawn from the results? (1)

c) Which team's set of data:
   i) deviates to the greatest extent from the line of best fit?
   ii) deviates to the least extent from the line of best fit? (2)

d) Suggest why site-specific mutagenesis is normally a preferable method of strain improvement than exposure of a culture of the organism to a mutagen such as X-rays. (2)

e) The mutation frequency of a bacterium can be expressed as the number of mutations that occur at a genetic site per million cells. In the pneumonia bacterium, it is estimated that the gene for resistance to penicillin arises spontaneously in 1 in 10^7 cells. Express this as a mutation frequency. (1)

2 The following list gives the steps employed during recombinant DNA technology.

- A Host cell allowed to multiply.
- B Required section of DNA cut out of appropriate chromosome.
- C Duplicated plasmids allowed to express a 'foreign' gene.
- D Plasmid extracted from bacterium and opened up.
- E Recombinant plasmid inserted into the bacterial host cell.
- F DNA section sealed into plasmid.

a) Arrange the steps into the correct order. (1)

b) i) During which stages would a restriction endonuclease be employed?
   ii) Why would the same one need to be used for both stages? (3)

c) During which stage(s) would ligase be used? (1)

3 Strains of yeast containing recombinant DNA for insulin, but with one or more codons for certain amino acids altered, have been used to make 'insulin analogues'. A chemical analogue is a substance that mimics another substance. Some insulin analogues act more rapidly than original biosynthetic insulin because they are more readily absorbed from the injection site. The activity of two insulin analogues, X and Y, is shown in Figure 13.7.

a) Identify from the graph:
   i) the prolonged-action analogue
   ii) the rapid-acting analogue. (1)

b) i) For how long does the rapid-acting analogue continue to act?
   ii) When does it have its maximum effect? (1)

c) For how many hours does the prolonged-action analogue continue to act at its maximum level? (1)

d) Suggest why the use of a combination of X and Y is more effective than the use of original biosynthetic insulin on its own. (2)

4 Figure 13.8 shows a plasmid used as a cloning vector. It possesses several restriction sites each of which can be cut open by a specific restriction endonuclease enzyme. Table 13.2 gives information about four of these enzymes.

➜

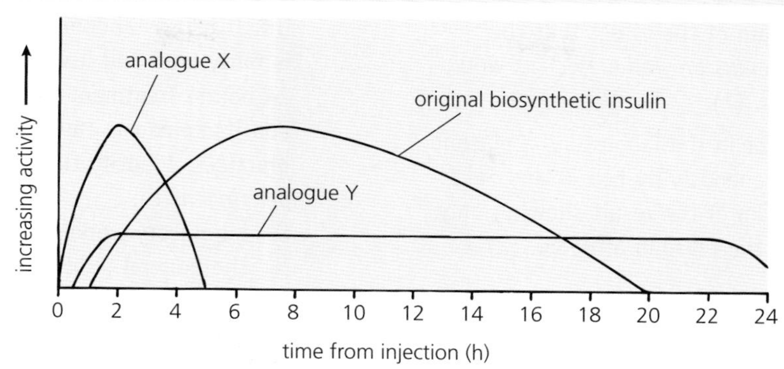

Figure 13.7

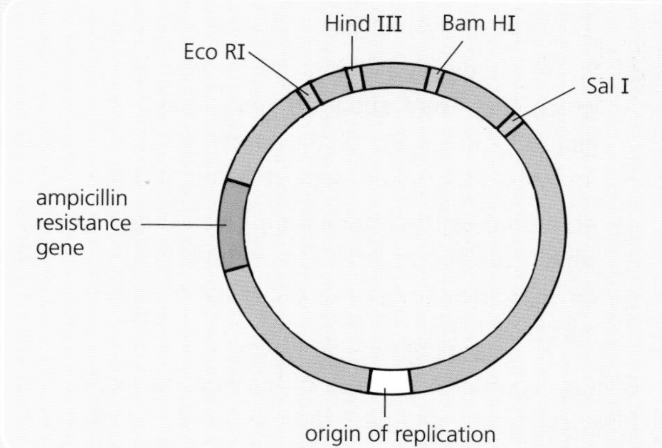

Figure 13.8

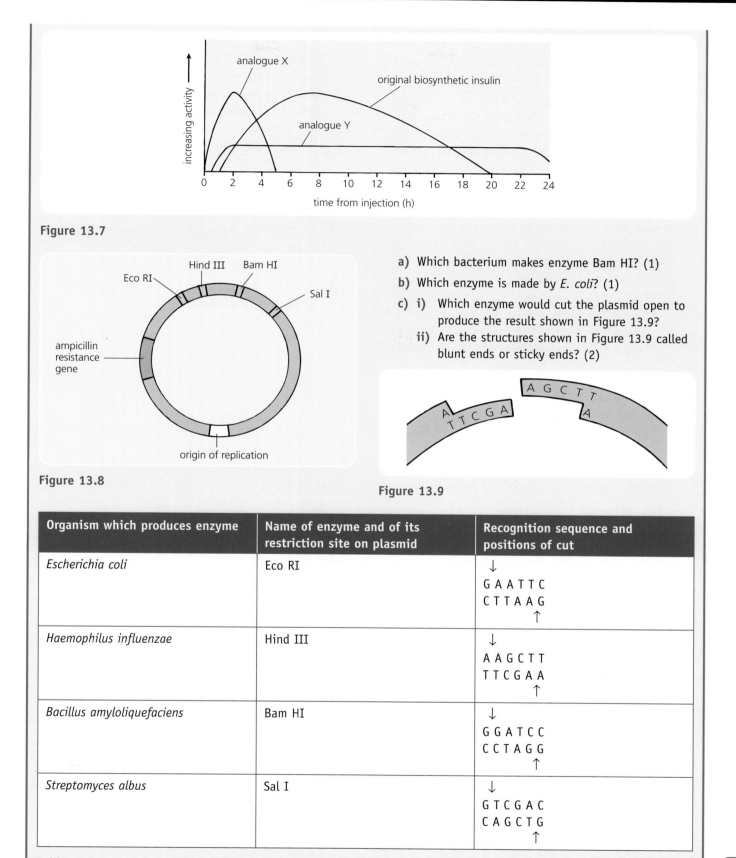

a) Which bacterium makes enzyme Bam HI? (1)

b) Which enzyme is made by *E. coli*? (1)

c) i) Which enzyme would cut the plasmid open to produce the result shown in Figure 13.9?

ii) Are the structures shown in Figure 13.9 called blunt ends or sticky ends? (2)

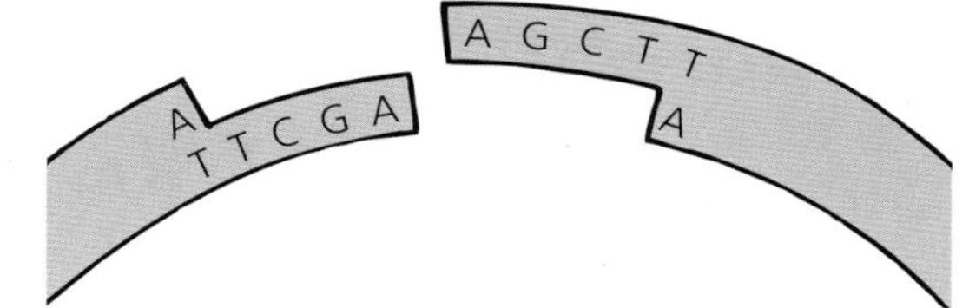

Figure 13.9

Organism which produces enzyme	Name of enzyme and of its restriction site on plasmid	Recognition sequence and positions of cut
Escherichia coli	Eco RI	↓ G A A T T C C T T A A G ↑
Haemophilus influenzae	Hind III	↓ A A G C T T T T C G A A ↑
Bacillus amyloliquefaciens	Bam HI	↓ G G A T C C C C T A G G ↑
Streptomyces albus	Sal I	↓ G T C G A C C A G C T G ↑

Table 13.2

→

d) Draw a simple diagram of the effect that the enzyme made by *Bacillus amyloliquefaciens* would have on the plasmid. (1)

e) As a result of an error in the laboratory, a plasmid was acted on simultaneously by Eco RI and Sal I. Draw a simple diagram of the result showing only the upper region of the plasmid affected. (2)

5 Maintenance of high standards of safety is particularly important in biotechnological industries that make use of micro-organisms. Table 13.3 shows a classification of micro-organisms based on the level of risk and recommended level of containment. Table 13.4 gives a few examples of the safety precautions required at different levels of containment.

a) *Ebola* virus is a pathogen that destroys the inner lining of the blood vessels. It causes the victim to suffer a high fever and chronic internal bleeding. There is no known treatment and the disease is fatal in nine out of ten cases. Identify the risk level of *Ebola* to:
   i) laboratory workers
   ii) the general public.
   iii) What level of containment should be used during research work on this virus? (3)

b) The bacterium *Clostridium perfringens* is the third most common cause of food poisoning in the UK. It causes abdominal pains and diarrhoea but is fairly mild and usually resolves itself within 24 hours. Treatment is available when necessary and its spread is easily prevented by good hygienic practices. Identify the risk level of *Clostridium perfringens* to:
   i) laboratory workers
   ii) the general public.
   iii) What level of containment should be used when working with this bacterium? (3)

c) How many levels of containment require:
   i) filters to be present in the laboratory's air ducts?
   ii) a closed system to be in operation? (2)

d) Scientists decided that a micro-organism with which they were about to work belonged to risk group 3. How many of the safety precautions listed in Table 13.4 would apply to this microbe? (1)

e) According to the information given in the tables, which safety precautions must be taken to contain a risk 4 microbe but are not required for a risk 3 microbe? (2)

6 Give an account of the use of plasmids in recombinant DNA technology and describe the features that a recombinant plasmid must possess in order to carry out its role. (9)

Risk group	Level of risk to laboratory workers	Level of risk to population at large	Description of micro-organism	Level of containment required
1	low	low	It has never been identified as a pathogen and is unlikely to cause disease.	W
2	moderate	low	It can cause disease and may affect laboratory workers but is unlikely to be a serious hazard. Preventative measures and treatment are effective.	X
3	high	low	It presents a severe threat to the health of laboratory workers but not to the community in general. Preventative measures and treatment may be effective.	Y
4	high	high	It is a pathogen that causes severe illness in humans. It constitutes a potential hazard to laboratory workers and the community since it is readily transmitted from person to person. No effective treatment is available.	Z

Table 13.3

→

Safety precaution	Level of containment			
	W	X	Y	Z
written code of practice	+	+	+	+
manual of bio-safety	o	+	+	+
closed system in operation	o	+	+	+
presence of emergency shower facility	o	o	+	+
presence of controlled negative air pressure	o	o	o	+
presence of filters in air ducts	o	o	+	+
presence of air locks and compulsory shower for staff	o	o	o	+

Table 13.4 (+ = required, o = not required)

What You Should Know

Chapters 12–13

(See Table 13.5 for word bank)

adaptable	exponential	origin
amplify	generation	oxygen
bacteria	genetic	polypeptides
biosynthesis	growth	recombinant
death	improved	restriction
depleted	industry	safety
doubles	lag	secondary
endonucleases	ligase	self-replication
energy	marker	stationary
environmental	metabolic	vector
equals	mutagenesis	viable
exceeds	mutations	yeast

Table 13.5 Word bank for chapters 12–13

1 Micro-organisms such as ______, archaea and some eukaryotes are highly ______. They make use of many substrates, produce a wide range of ______ products and occur in any environment that can support life.

2 Much use is made of micro-organisms in research and ______ because they are easy to culture. They grow quickly if given nutrient medium containing a suitable ______ source and a supply of the raw materials needed for the ______ of complex molecules.

3 To grow rapidly, micro-organisms also require ______ conditions such as temperature, ______ concentration and pH to be maintained at optimum levels.

4 A microbe's pattern of ______ falls into four distinct phases. During the ______ phase, the cells adjust to the growth medium, metabolic rate increases and enzymes may become induced.

5 During the log or ______ phase, the cells multiply at maximum speed. The population ______ its number with each cell division. The time needed to do this is called ______ or doubling time.

6 During the ______ phase, the nutrient medium becomes ______, secondary metabolites are produced and the rate of production of new cells ______ the death rate of old ones.

7 During the ______ phase, the number of cells dying greatly ______ the number (if any) of new cells being produced. Only a ______ cell count shows a death phase.

8 Some ______ metabolites may be of advantage ecologically to the microbe that makes them.

9 The processes of ______ and recombinant DNA technology are used to try to improve wild strains of micro-organisms exhibiting useful traits.

→

10 ______ can be induced using ultraviolet light, other forms of radiation or mutagenic chemicals. On very rare occasions the result is an ______ strain.

11 During ______ DNA technology, ______ sequences for useful products are transferred from a plant or animal to a micro-organism using plasmids. Genes that can ______ production of the required product or act as a ______ mechanism may also be introduced.

12 To act as an effective ______, a plasmid must contain a ______ site which will receive the genetic material to be transferred. It also contains a ______ gene to show whether the host cell has taken up the plasmid and an ______ of replication consisting of genes for ______.

13 Enzymes called restriction ______ are used to cut out genetic sequences and cut open plasmids. An enzyme called ______ is used to seal the genetic sequence into a plasmid.

14 Genetically modified prokaryotes make ______ typical of eukaryotes but may be unable to modify them. In these situations, transformed ______ is more useful.

3 Sustainability and Interdependence

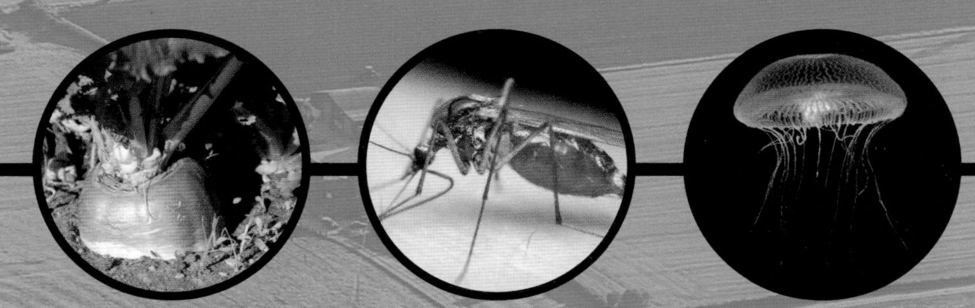

14 Food supply, plant growth and productivity

Food supply and food security

The human population depends on a sufficient and sustainable supply of food for its survival.

Food security may be defined as **access** (both physical and economic) to food of adequate **quantity** and **quality** by human beings (see Figure 14.1). Food security can refer to a small group such as a single household or to a large group such as the population of a country.

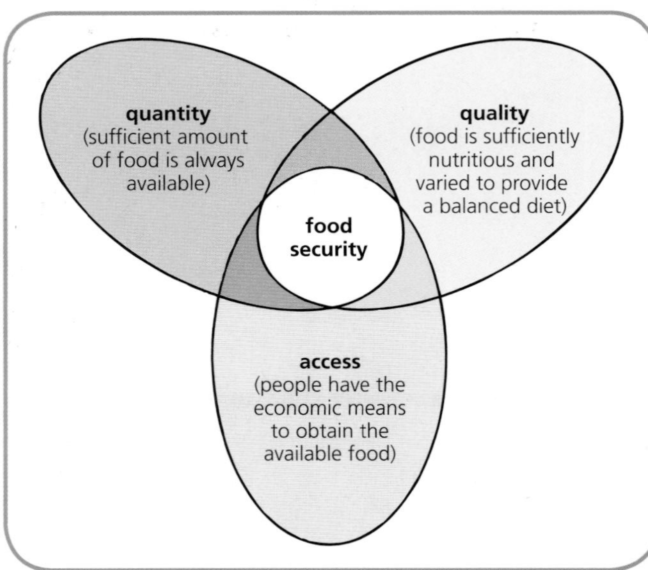

Figure 14.1 Factors affecting food security

Increasing demand for food

As the global human population continues to increase so also does the **demand for food**. The 'green revolution' in the 1960s transformed agricultural practices and raised crop yields but the effect is now levelling off. Current levels of food production will not meet the projected demand and already many people in the world lack food security. Therefore the demand for increased food production continues.

The 'green revolution' depended on the use of vast quantities of fertilisers and pesticides and intensive, multiple-crop farming methods which often led to environmental degradation. Attempts to reverse this unacceptable situation are now underway as society demands methods of increasing food production in a **sustainable way** that does not degrade natural resources such as the land and the water supply.

Increased plant productivity and **manipulation of genetic diversity** (see page 183) are two key factors that will continue to play important roles in the maintenance of sustainable food security.

Agricultural production

Solar radiation drives the biological world. The first organism in every food chain, the green plant producer, obtains its energy from the Sun by photosynthesis. Some of this energy is passed on to animal consumers. Therefore all **food production** (plant or animal) depends on the process of photosynthesis.

Plant crops

Although the Earth possesses at least 75 000 species of **edible plants**, humans depend on only a small number of these to produce 95% of the world's food supply (see Figure 14.2).

These include cereals such as maize and rice, root crops such as cassava and leguminous (pod-bearing) plants such as soya bean.

Limited area

If the area suited to growing crops is **limited**, food production can only be increased by improving efficiency. This may be achieved by adopting one or more of the following practices:

- Identifying any factors that are **limiting** plant growth (for example, shortage of mineral elements and water in the soil) and increasing the supply of these (such as by adding fertiliser and making use of a sprinkler system).
- Replacing the existing strain of crop plant with a **higher-yielding** cultivar.

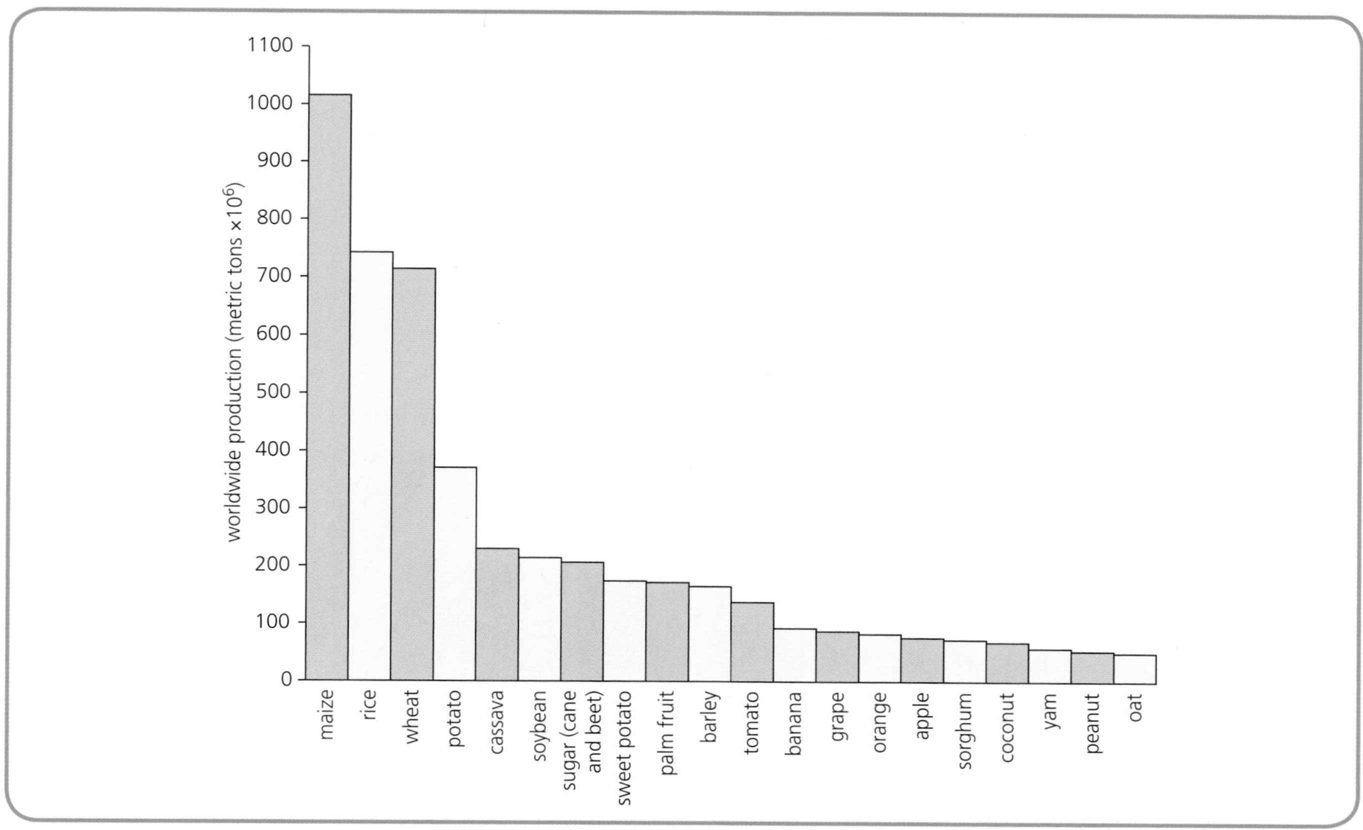

Figure 14.2 Main plant crops

- **Protecting** the crop from pests (such as insects), disease (for example, those caused by fungi) and competition (from weeds) by using minimal doses of biodegradable pesticides, fungicides and herbicides.
- Developing **pest-resistant** crop plants. (Also see page 213.)

Food chains and energy loss

Most of the energy gained by a consumer from its food is used for movement (and in endothermic animals, for maintaining body temperature). Therefore much of the energy taken in by an organism, and released as a result of cellular respiration, is lost and only about **10%**, on average, is incorporated into its body tissues and passed on along the food chain. Figure 14.3 shows a **pyramid of numbers** based on the food chain:

phytoplankton → zooplankton → herring → human

Each layer in the pyramid is called a **trophic level** and energy is lost between one trophic level and the next.

Length of food chain

More efficient use is made of food when humans consume plants rather than animals. In the food chain shown in Figure 14.4, 1000 kg of cereal plant could be used to feed many more people than can be fed by 100 kg of meat from farm animals. As a result of energy loss between trophic levels, livestock production generates far less food per unit area of land than plant production. In general, the **shorter** a food chain, the **greater** the quantity of energy held in the food.

Therefore arable farm land planted with crops produces far more food than the same land planted with grass to feed livestock. Some habitats (such as steep, grassy hillsides) are unsuitable for the cultivation of crop plants but efficient use can be made of them for livestock production (for example, sheep farming).

few large individuals
trophic levels
human
herring
energy loss
zooplankton
many small individuals
phytoplankton

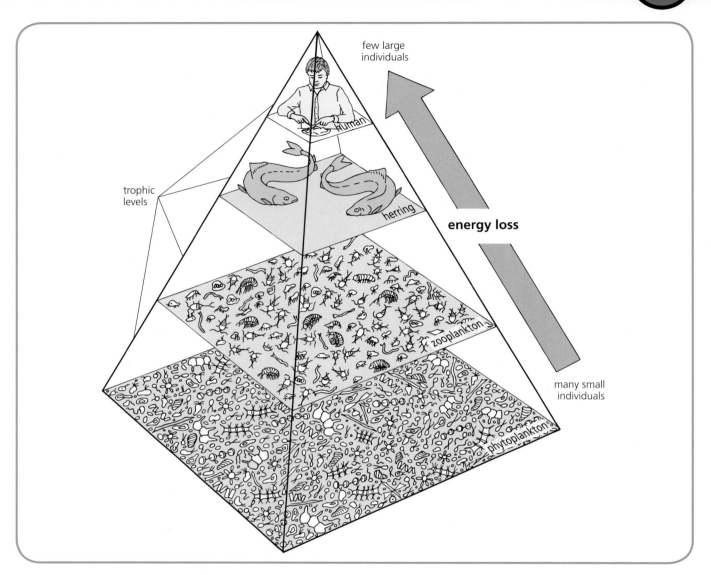

Figure 14.3 Energy loss in a food pyramid (organisms not to scale)

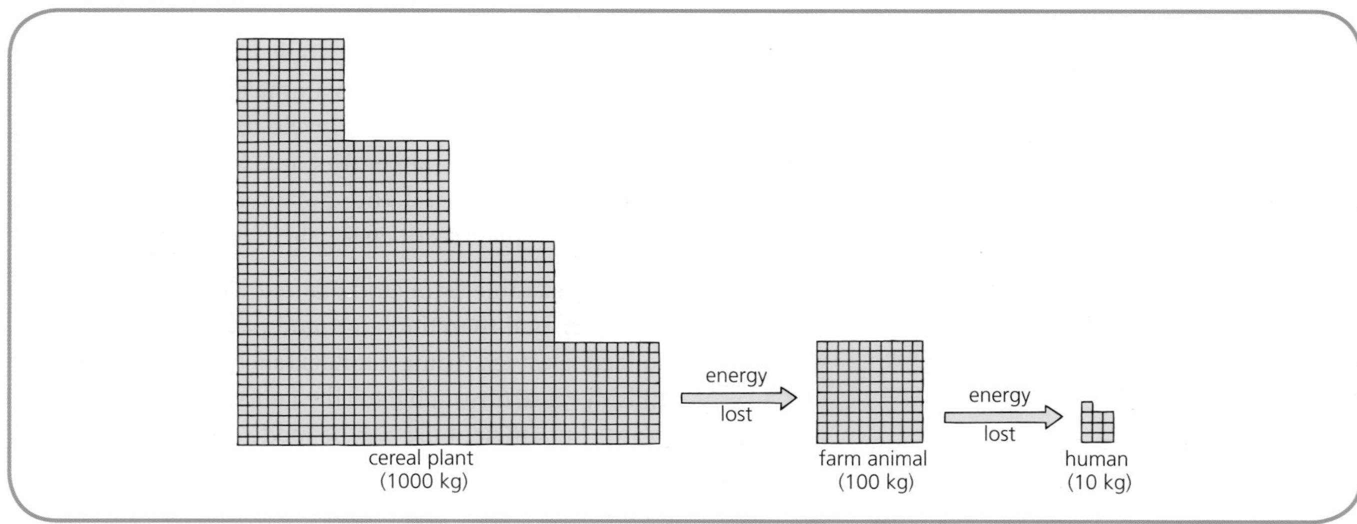

Figure 14.4 Energy loss in a food chain

Testing Your Knowledge 1

1 What is meant by the term *food security*? (2)

2 State TWO practices that can be adopted in order to increase food production in an area of land that is limited in size. (2)

3 Why is more efficient use made of food plants by humans consuming them directly rather than first converting them into animal products? (1)

Photosynthesis

Photosynthesis is the process by which green plants trap light energy and use it to produce carbohydrates.

Light

Light is a form of electromagnetic radiation which travels in waves. **Wavelength** is the distance between two crests on a wave pattern as shown in Figure 14.5. Wavelengths of light are normally measured in **nanometres** (nm) ($1\,\text{nm} = 10^{-9}\,\text{m}$).

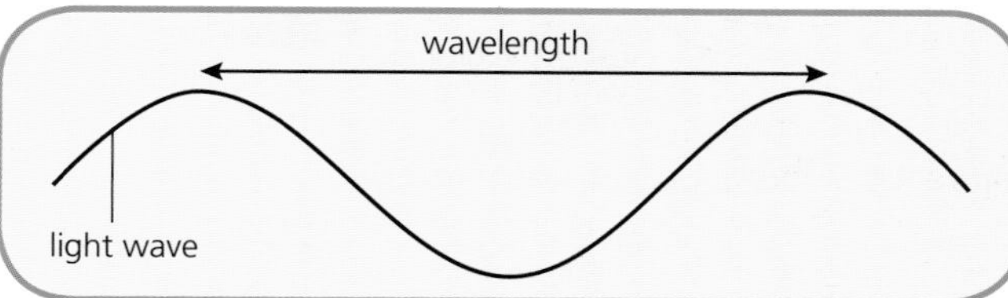

Figure 14.5 Wavelength

Spectrum of visible light

Out of the full range of radiation that falls on the Earth from the Sun and space, the most important, from a biological point of view, is the narrow band from wavelength 380 nm to 750 nm. This is called the **spectrum of visible light** (see Figure 14.6) because it can be seen by the human eye as a variety of colours.

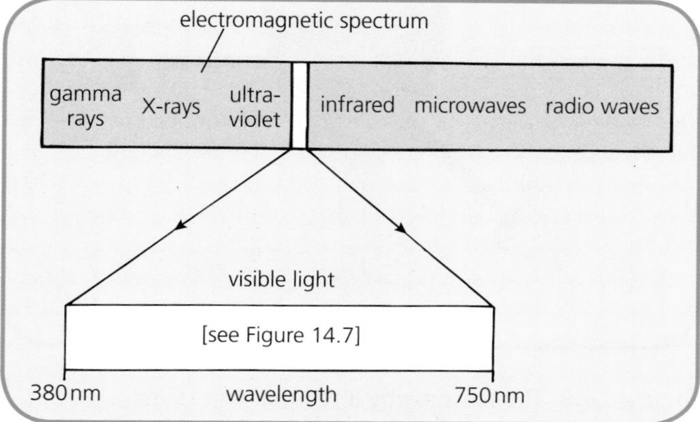

Figure 14.6 Electromagnetic radiation

Absorption, reflection and transmission

When light comes into contact with a substance, it may be **absorbed**, **reflected** or **transmitted**.

Related Activity

Examining the spectrum of visible light

When a beam of white light is passed through a glass **prism** (or **spectroscope**) as shown in Figure 14.7, the **spectrum of visible light** is produced.

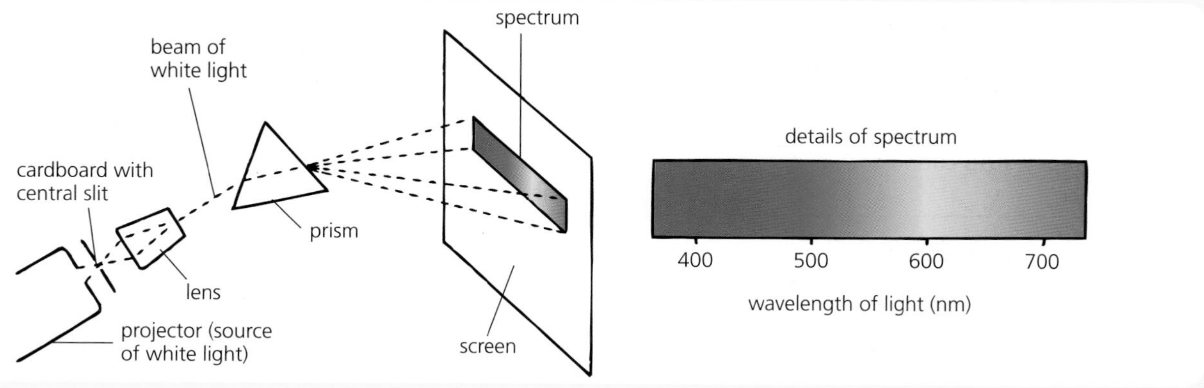

Figure 14.7 Spectrum of visible light

A **pigment** is a substance that absorbs visible light. For example, the wine in Figure 14.8 contains a red pigment which absorbs all the colours in the spectrum of visible light except red which it reflects and transmits. A pigment displays the colour of the light that it does *not* absorb.

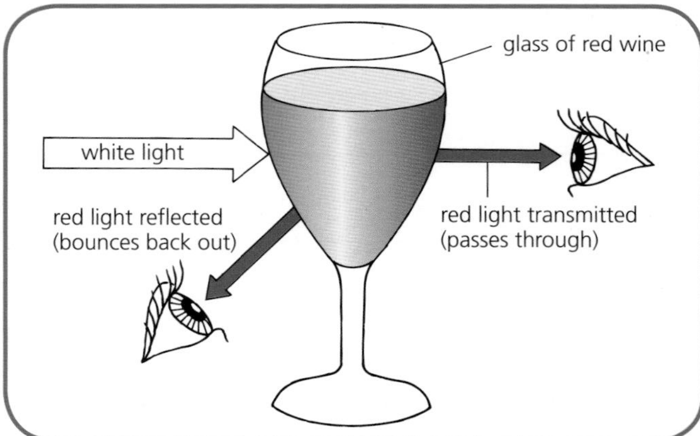

Figure 14.8 The reason why a red pigment is red

Thin-layer chromatography

During **thin-layer chromatography** the mixture to be separated is applied to a thin-layer strip (normally composed of a thin layer of silica gel attached to a backing material). As the solvent passes up through the thin-layer strip, it carries the components of the mixture to different levels depending on the degree of their **solubility** in the solvent (and the extent to which they are absorbed by the silica gel).

Related Activity

Extraction of leaf pigments

Fresh leaves are finely chopped up and then ground in a mortar containing propanone and a little fine sand as shown in Figure 14.9. The extract of soluble pigments is then separated from the cell debris by filtration.

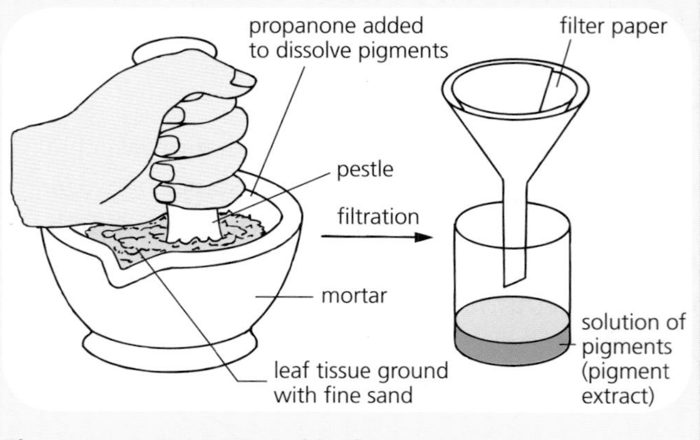

Figure 14.9 Extraction of leaf pigments

Related Activity

Separation of leaf pigments using thin-layer chromatography

A length of thin-layer strip is prepared as shown in Figure 14.10. This diagram illustrates the procedure followed during spotting of the extract and the use of the chromatography solvent.

Spotting and drying of the extract is repeated many times and then the end of the strip is dipped into the solvent. The chromatogram is allowed to run for a few

Design feature or precaution	Reason
plant tissue ground in fine sand	to rupture cells allowing release of contents
chromatography strip cut so that it does not touch sides of tube	to ensure that solvent rises uniformly through the strip rather than more rapidly up its edges
spotting and drying repeated many times	to obtain a concentrated spot of pigments
strip positioned in tube so that pigment spot is above solvent level at the start	to prevent extract dissolving in main bulk of solvent at bottom of tube
naked flames extinguished before starting experiment	to prevent fire risk since solvent chemicals are highly flammable

Table 14.1 Design techniques

→

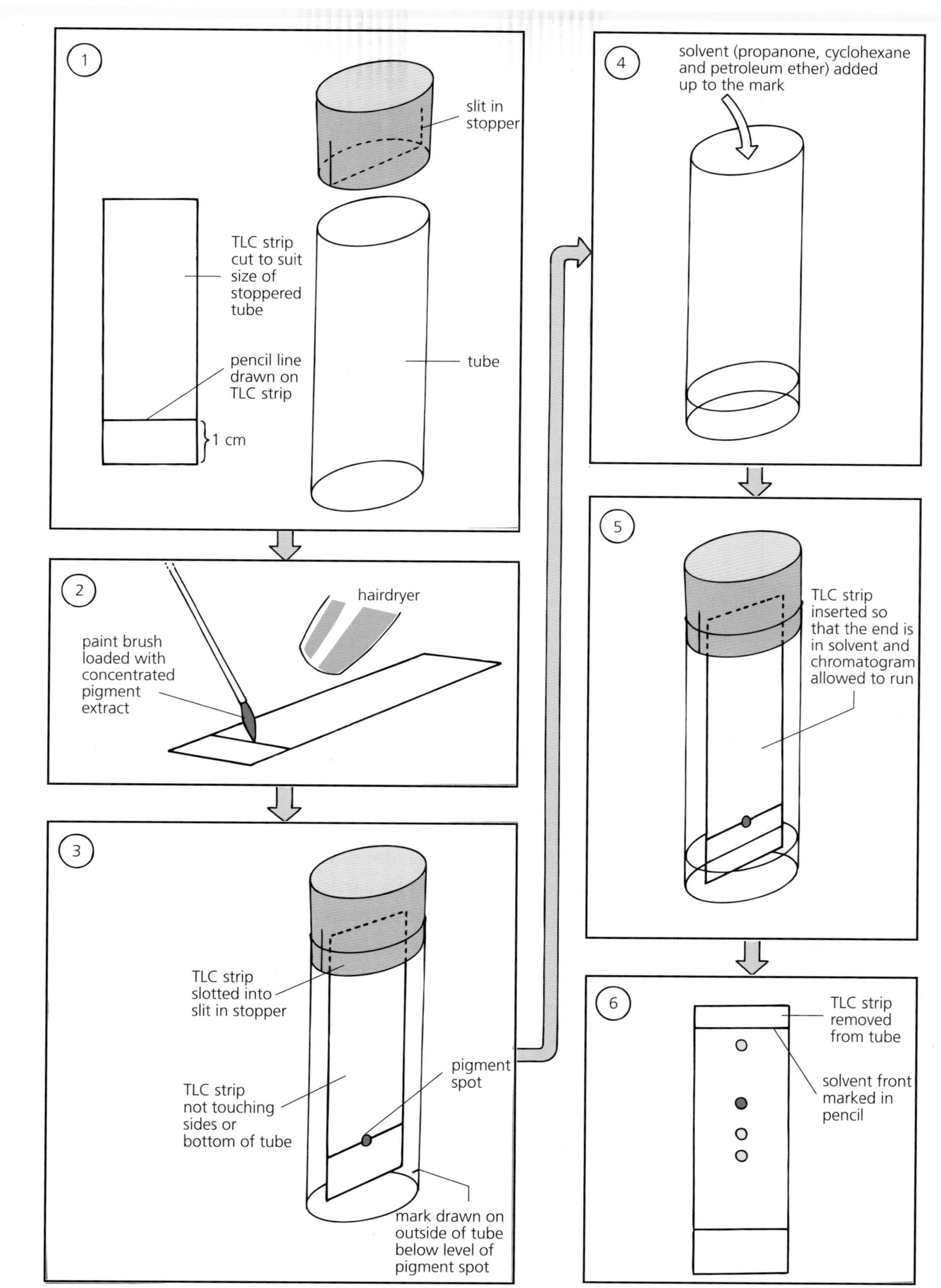

Figure 14.10 Separation of leaf pigments (TLC stands for thin-layer chromatography)

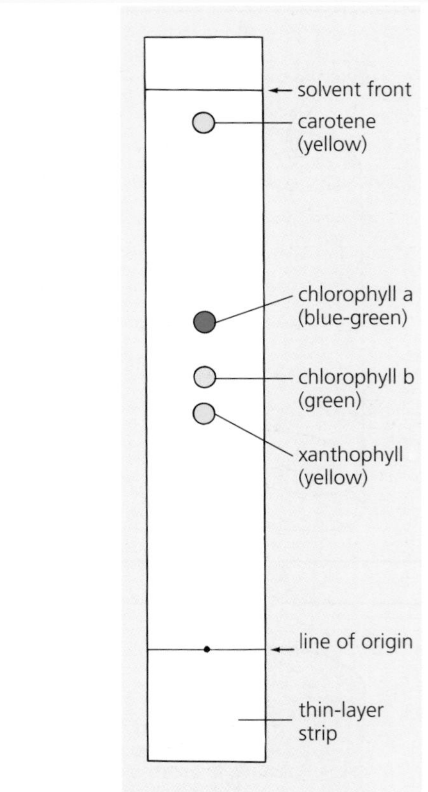

Figure 14.11 Thin-layer chromatogram of leaf pigments

minutes until the solvent has almost reached the top of the strip. The strip is then removed from the tube and a pencil used to mark the position of the solvent front. Table 14.1 gives the reasons for adopting certain techniques and precautions during this experiment.

Chromatogram

Figure 14.11 shows the **chromatogram** formed as a result of thin-layer chromatography with this particular solvent.

The solvent has carried the most soluble pigment (carotene) to the highest position and so on down the strip to xanthophyll, the least soluble. This has been carried the shortest distance.

Rf value

Each pigment separated by chromatography has an **Rf value** (also see Chapter 3, page 38).

Absorption spectrum

When a beam of white light is first passed through a sample of extracted leaf pigments, placed at X in Figure 14.12, and then passed through a glass prism (or spectroscope), an **absorption spectrum** is produced.

Each **black** band is a region of the spectrum where light with a particular wavelength has been absorbed by the leaf pigments, and has therefore failed to pass through the prism and onto the screen. Most of the absorbed light is in the blue and red regions of the spectrum.

Each **coloured** band (such as green) is a region where light of a particular wavelength has *not* been absorbed by the pigment extract. Chlorophyll appears green to the eye because it does not absorb green light but instead reflects and transmits it.

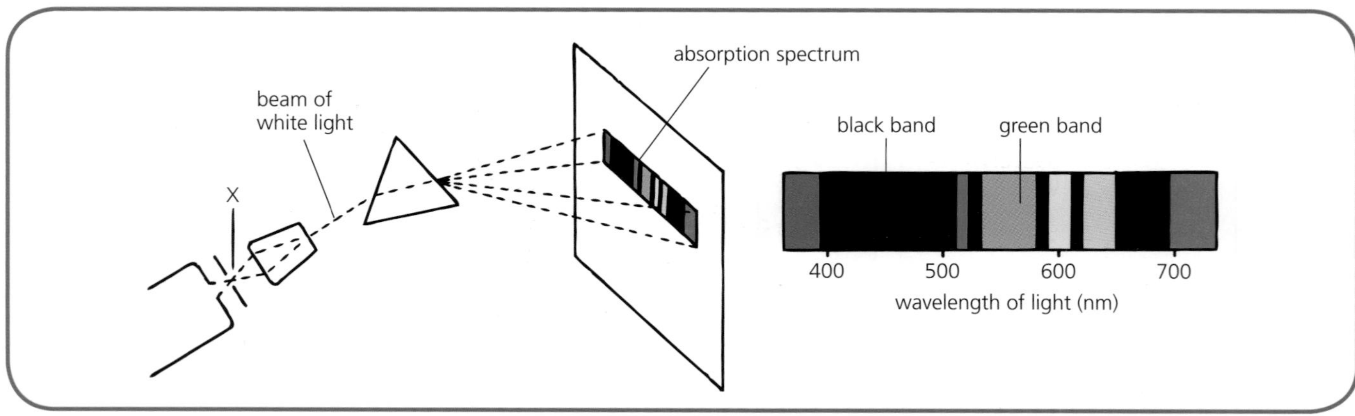

Figure 14.12 Absorption spectrum of leaf pigments

Graph of absorption spectra of leaf pigments

The degree of absorption at each wavelength of visible light by each pigment (chlorophyll a, chlorophyll b and the carotenoids) can be measured using a **spectrometer**. The data obtained allow a detailed graph of each pigment's **absorption spectrum** to be plotted (see Figure 14.13).

Action spectrum

An **action spectrum** charts the effectiveness of different wavelengths of light at bringing about the process of photosynthesis. In the experiment shown in Figure 14.14, a set of coloured filters are used in turn to illuminate *Elodea*, an aquatic plant. Rate of photosynthesis is measured as number of bubbles or volume of oxygen released per minute. Figure 14.15 shows an action spectrum of photosynthesis when the results are graphed.

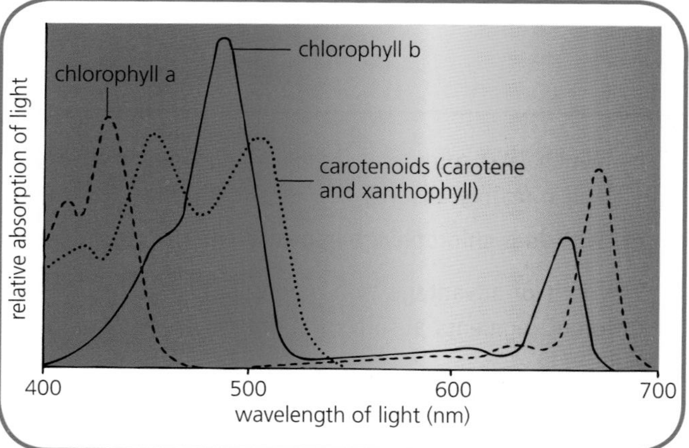

Figure 14.13 Graph of absorption spectra of leaf pigments

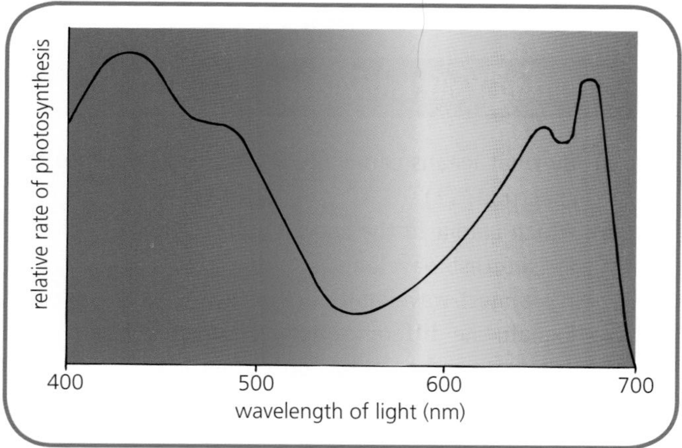

Figure 14.15 Action spectrum

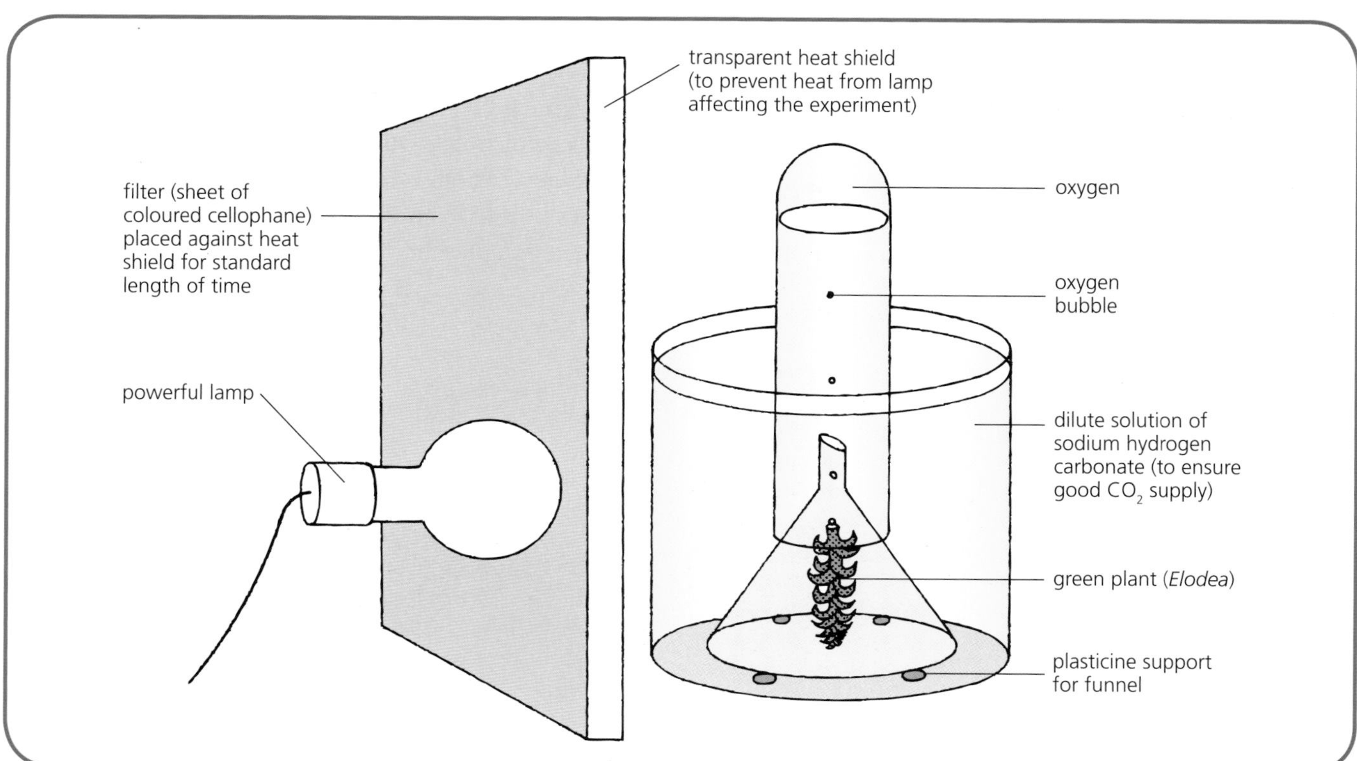

Figure 14.14 Investigating the action spectra of photosynthesis

Comparison of absorption and action spectra

When Figures 14.13 and 14.15 are compared, a **close correlation** is found to exist between the overall absorption spectrum for the leaf pigments and the action spectrum for photosynthesis. It is therefore concluded that the absorption of certain wavelengths of light for use in photosynthesis is the crucial role played by the pigments.

Chlorophyll a and b absorb light energy mainly from the **blue** and **red** regions of the spectrum. The other pigments (such as carotenoids) absorb light energy from other regions (including the blue-green region of the spectrum) and pass the energy on to chlorophyll for photosynthesis.

Advantage to plant

The fact that the different photosynthetic pigments absorb light of different wavelengths from one another is of advantage to the plant because it **extends the range** of wavelengths that the plant can use for photosynthesis.

Testing Your Knowledge 2

1 **a)** By what means can white light be split up into the spectrum? (1)
b) Which colour in the spectrum has light with the
i) shortest wavelength?
ii) longest wavelength? (2)

2 **a)** Explain the difference between an *absorption* spectrum and an *action* spectrum. (2)
b) Which TWO colours of light do chlorophylls a and b mainly absorb? (2)
c) Why does chlorophyll b appear green in colour? (2)

3 Why is it of advantage to a green plant that the different pigments absorb different wavelengths of light? (1)

Capture and transfer of energy during photosynthesis

When light energy is absorbed by a molecule of chlorophyll a, its **electrons** become **excited** and raised to a higher energy state in the pigment molecule. Electrons excited in this way are captured by the primary electron acceptor associated with each chlorophyll a molecule in a chloroplast. This results in the series of events shown in Figure 14.16 taking place.

The transfer of electrons through an electron transport chain releases energy that is used to generate **ATP** by the enzyme **ATP synthase**. Energy is also used for **photolysis** of water. During this process water is split into **oxygen** (which is released) and **hydrogen** which becomes bound to coenzyme **NADP** (acting as a hydrogen acceptor) to form **NADPH**.

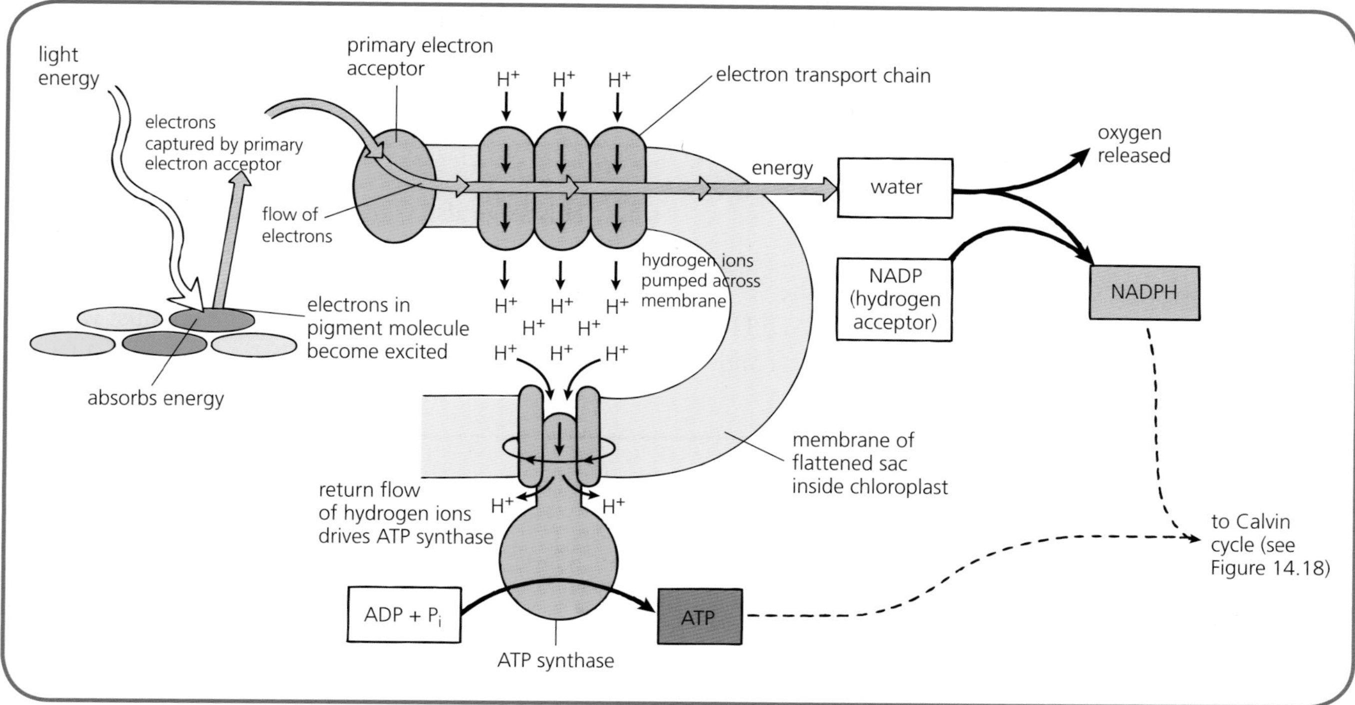

Figure 14.16 Transfer of energy during photosynthesis

Related Activity

Investigating photolysis (the Hill reaction)

DCPIP (dichlorophenol-indophenol) is a chemical that acts as a hydrogen acceptor by undergoing the following chemical reaction:

$$\underset{\text{(dark blue)}}{\text{DCPIP}} + 2H^+ + 2e^- \rightarrow \underset{\text{(colourless)}}{\text{DCPIPH}_2}$$

The experiment is set up as shown in Figure 14.17, with the three test tubes resting against the inner surface of the beaker of crushed ice.

From the results it is concluded that the contents of tube A have lost their dark blue colour because photolysis has occurred. Water in the presence of a hydrogen acceptor has been broken down by light energy into its components as in the equation:

$$2H_2O + 2\text{DCPIP} \xrightarrow[\text{chlorophyll}]{\text{light}} 2\text{DCPIPH}_2 + O_2$$

Tube B shows that light is necessary and tube C shows that chlorophyll is necessary for the reaction to take place.

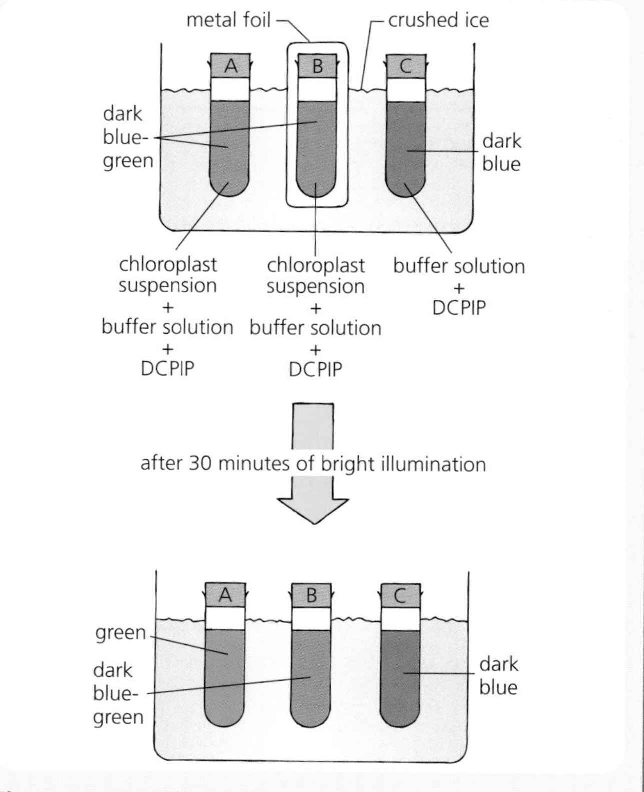

Figure 14.17 Hill reaction

The reaction is also known as the **Hill reaction**, named after the scientist who was the first to carry it out in the 1930s. In addition to showing that hydrogen was released from water, Hill demonstrated that oxygen was released also.

Calvin cycle

At the end of the first stage of photosynthesis (the light-dependent stage), the hydrogen held by NADPH and the energy held by ATP are essential for the second stage. This stage also takes place in the cell's chloroplasts but it is not light dependent. It consists of several enzyme-controlled reactions which take the form of a cycle – called the **Calvin cycle** after the scientist who discovered it. The cycle is summarised in Figure 14.18.

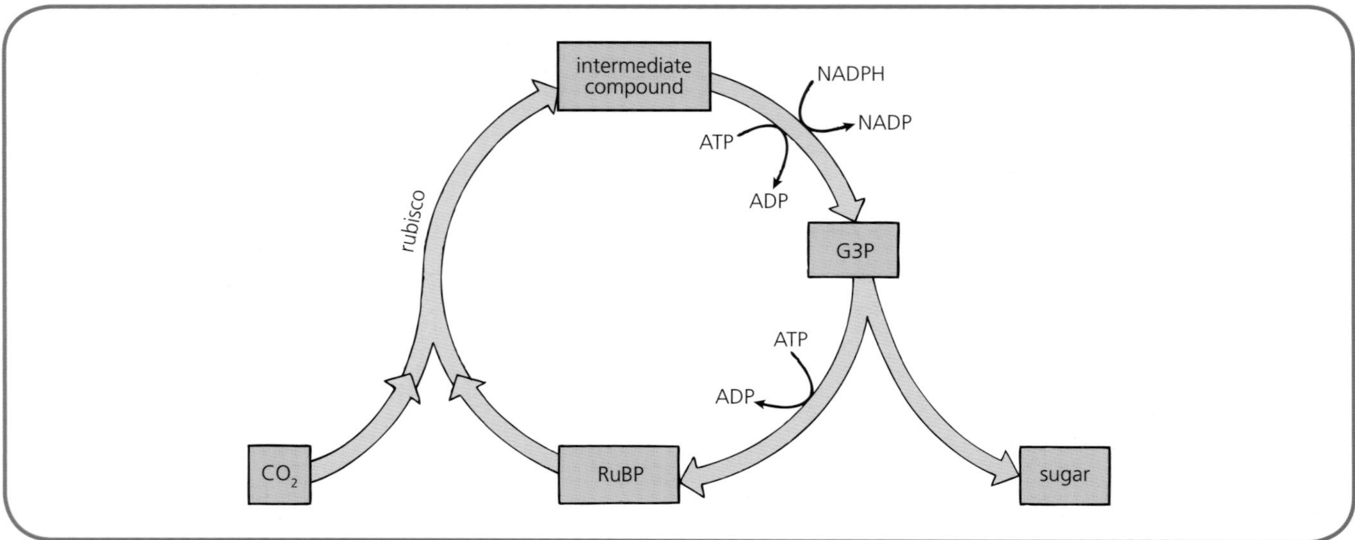

Figure 14.18 Calvin cycle

Carbon dioxide enters the cycle by becoming attached to **RuBP** (ribulose bisphosphate). This chemical reaction is controlled by the enzyme **rubisco** (sometimes written RuBisCO, full name – ribulose bisphosphate carboxylase/oxygenase). The intermediate (3-phosphoglycerate) that is formed becomes combined with hydrogen from NADPH and becomes phosphorylated by receiving an inorganic phosphate (P_i) from ATP which supplies the energy to drive the process. This results in the formation of **G3P** (glyceraldehyde-3-phosphate), some of which is used to regenerate RuBP, the carbon dioxide acceptor. The remaining G3P is used for the synthesis of **sugars** such as **glucose**.

Use of sugar made by photosynthesis

Some of the carbohydrate (for example, glucose) that is formed is used by the plant for cellular respiration to provide the plant with energy for **growth** and **reproduction**. Some of the remaining sugar molecules are synthesised into molecules of complex carbohydrate. For example, they may become built into long chains of **cellulose** and used as a **structural** carbohydrate to build cell walls (see Figure 14.19). Alternatively they may become linked together into long chains and packed into spherical **starch** grains as a **storage** carbohydrate.

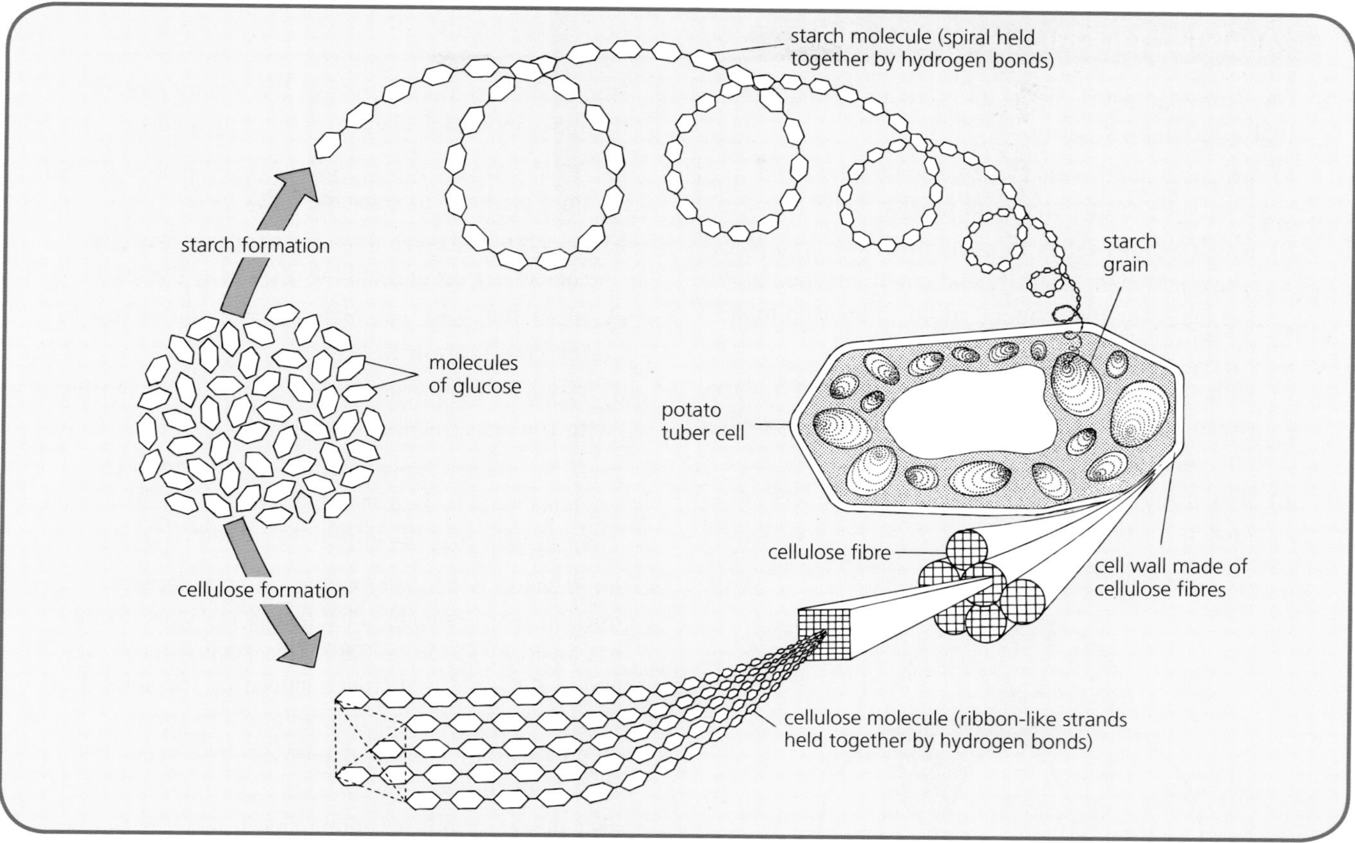

Figure 14.19 Formation of storage and structural carbohydrates

Other molecules of carbohydrate produced by photosynthesis may be passed to other **biosynthetic pathways**. These contribute to the production of a variety of further **metabolites** such as protein, DNA and fat.

Related Activity

Investigating the effect of phosphorylase on glucose-1-phosphate

See pages 124–5.

Testing Your Knowledge 3

1 a) When a molecule of chlorophyll a absorbs light energy, what happens initially to this energy? (1)

b) Electrons captured by a primary electron acceptor are transferred through an electron transport chain releasing energy. Give the equation for the enzyme-controlled process that is then driven by this energy. (2)

c) Some of the energy is also used to split water. What happens to each component of water? (2)

2 a) i) Name the enzyme that attaches CO_2 to ribulose bisphosphate to form an intermediate in the Calvin cycle.

ii) What TWO products from the light-dependent stage of photosynthesis are required to convert this intermediate to G3P? (3)

b) G3P becomes converted into two substances. Name the substance that:

i) remains in the cycle

ii) leaves the cycle. (2)

3 a) Give an example of:

i) a structural carbohydrate

ii) a storage carbohydrate in plants. (2)

b) Name TWO non-carbohydrate metabolites that result from biosynthetic pathways in green plants. (2)

Applying Your Knowledge and Skills

1 The Venn diagram in Figure 14.20 refers to the factors that affect food security.

a) Which number refers to people who have a constantly secure supply of quality food? (1)

b) Which number refers to people who live in a place where sufficient quantity and quality of food are available but they cannot afford to buy the food? (1)

c) Which number refers to people who live in a place where they can afford the food but it lacks quality and variety? (1)

d) How many areas in the diagram indicate lack of food security? (1)

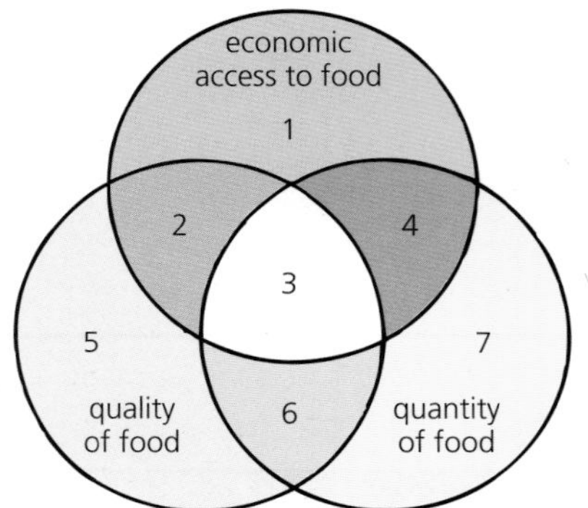

Figure 14.20

2 Figure 14.21 shows the fate of the energy present in the grass in a field when consumed by a cow.

a) Which lettered arrow represents the flow of energy from producer to consumer? (1)

b) Calculate the percentage of energy successfully converted from grass to the cow's body tissues. (1)

c) Name TWO ways in which energy could be lost by the cow at arrow X. (2)

d) Explain why the energy lost at arrow Y is not lost to the ecosystem as a whole. (1)

e) If the energy conversion efficiency of cow to human is 8%, how many of the cow's 120 kJ could have become built into human tissues? (1)

3 An Rf value is normally expressed as a decimal fraction. For example, substance X in Figure 14.22 has an Rf of 48/60 = 0.80. This means that X has moved 80% of the distance moved by the solvent. Figure 14.23 shows an incomplete thin-layer chromatogram of the four photosynthetic pigments from a green leaf.

a) Calculate the Rf of chlorophyll b. (1)

b) Copy or trace the chromatogram in Figure 14.23 and draw a spot to represent xanthophyll which has an Rf of 0.35. (1)

c) Identify pigment Y and calculate its Rf. (2)

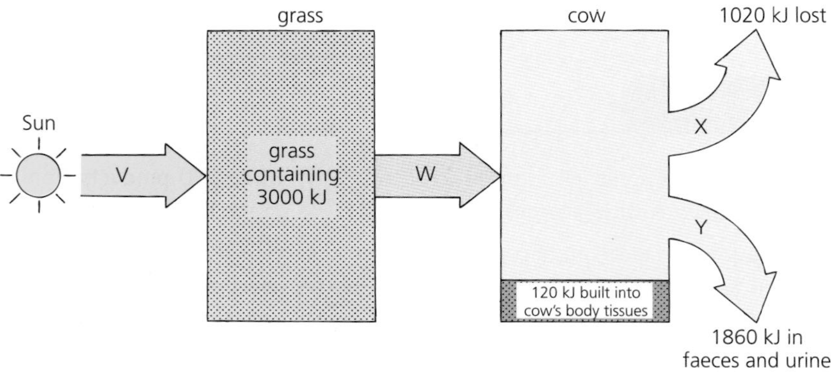

Figure 14.21

→

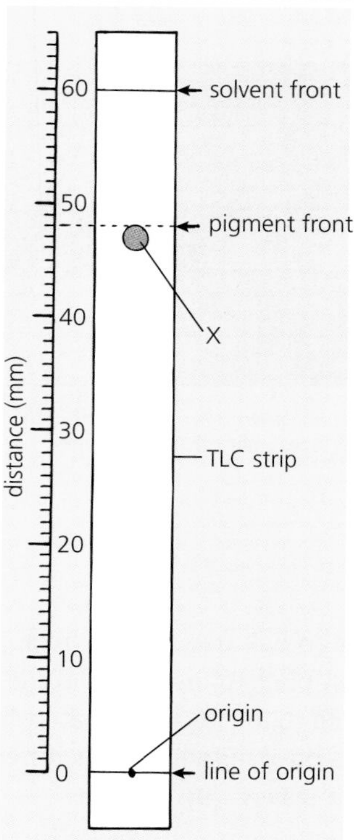

Figure 14.22

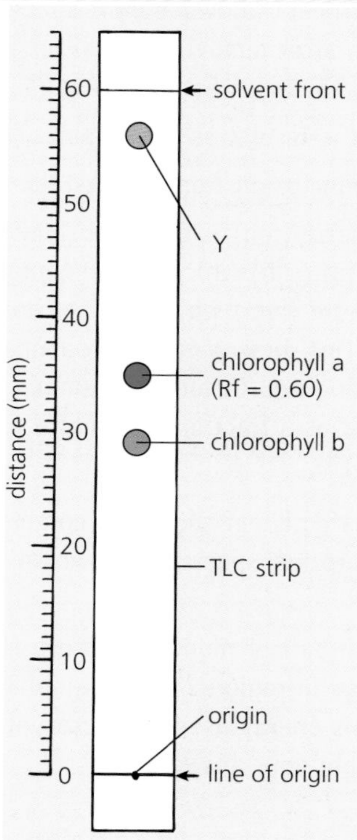

Figure 14.23

d) Why must repeated spotting and drying of pigment extract be carried out when preparing the origin of pigment extract? (1)

e) Why must the origin be kept above the solvent level in the tube when the end of the strip is dipped into the solvent to run the chromatogram? (1)

f) Name a solvent that could be used to remove 'grass' stains from a garment at room temperature. (1)

4 Rate of photosynthesis can be measured by counting the number of oxygen bubbles released per minute by the waterweed *Elodea*. In an experiment, a series of coloured filters were used in turn by inserting each between *Elodea* and the source of white light. Each coloured filter only allows one colour of light to pass through it. The results are shown in Table 14.2.

a) What was the one variable factor investigated in this experiment? (1)

b) If a coloured filter only allows one colour of light to pass through, what happens to the other colours present in white light? (1)

c) Explain why the *mean* number of bubbles was calculated each time. (1)

Colour of light allowed through by filter	Wavelength of light (nm)	Mean number of bubbles of oxygen released per minute
blue	430	15
green	550	1
yellow	600	4
red	640	12

Table 14.2

d) The experiment allows a short space of time to elapse after removing one filter and before inserting the next. Suggest why. (1)

e) Present the results as a bar chart. (2)

f) Draw a conclusion from the results. (1)

g) A nanometre (nm) is one-thousandth of a micrometre (µm) which is one-thousandth of a millimetre (mm) which is one-thousandth of a metre (m). Draw a table to summarise this information and include a column which expresses each unit as a fraction of a metre using negative indices. (3)

5 a) Which pair of leaf pigments combined would give the absorption spectrum shown in Figure 14.24? (2)

b) i) Which pair of pigments absorb most light energy in region Z in the graph?

ii) Is this energy used in photosynthesis?

iii) Explain your answer. (4)

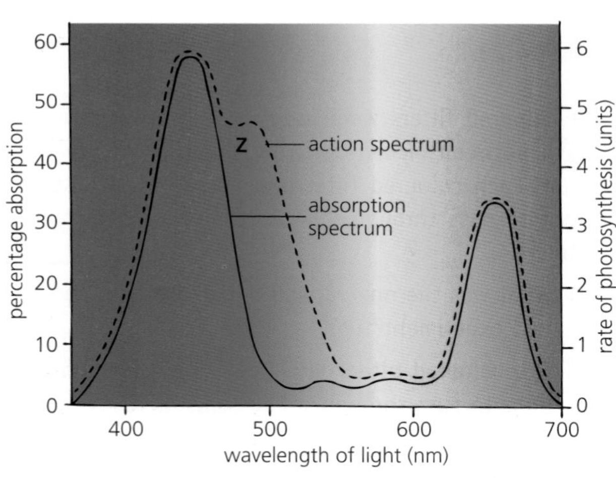

Figure 14.24

6 Figure 14.25 shows the result of placing a strand of alga in a liquid containing motile aerobic bacteria and illuminating the strand with a tiny spectrum of light.

a) In which colours of light did most bacteria congregate? (2)

b) Account for this distribution of the bacteria. (2)

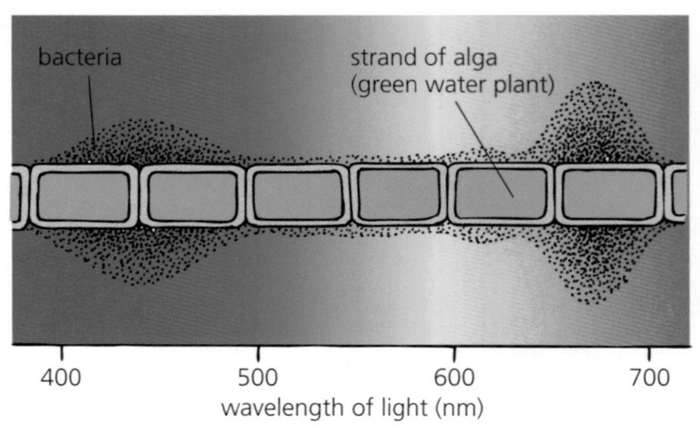

Figure 14.25

7 The leaves of an oak tree change from green to brown in the autumn prior to leaf fall. Outline the procedure that you would follow to investigate whether the pigment content of autumn leaves differs from that of green summer leaves. (9)

8 Redraw Figure 14.26 and construct a diagram of the light-dependent first stage of photosynthesis by completing the seven boxes using the following statements:

a) electrons transferred through electron transport chain releasing energy

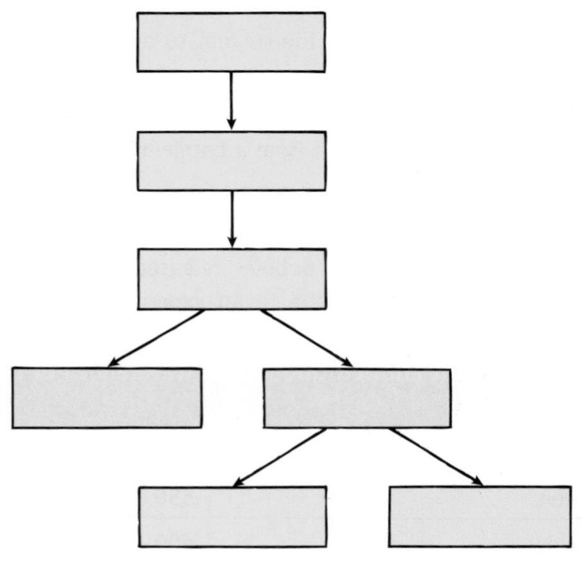

Figure 14.26

→

b) hydrogen transferred to NADP
c) excited electrons captured by primary electron acceptor
d) some energy used to generate ATP
e) light energy absorbed by chlorophyll
f) oxygen released
g) some energy used to split water. (6)

9 Figure 14.27 shows the Calvin cycle.

a) Copy the diagram and complete the blank boxes. (2)

b) i) Which chemical acts as the carbon dioxide acceptor?
ii) Add an arrow and the symbol CO_2 to your diagram to show where CO_2 enters the cycle.
iii) Mark the letter **R** on the arrow which represents the reaction controlled by rubisco. (3)

c) i) Mark **X** on your diagram at TWO points at which ATP is needed for the cycle to turn.
ii) Why is ATP necessary at these points? (2)

d) i) Which substance would accumulate if the plant were deprived of carbon dioxide?
ii) Explain why. (2)

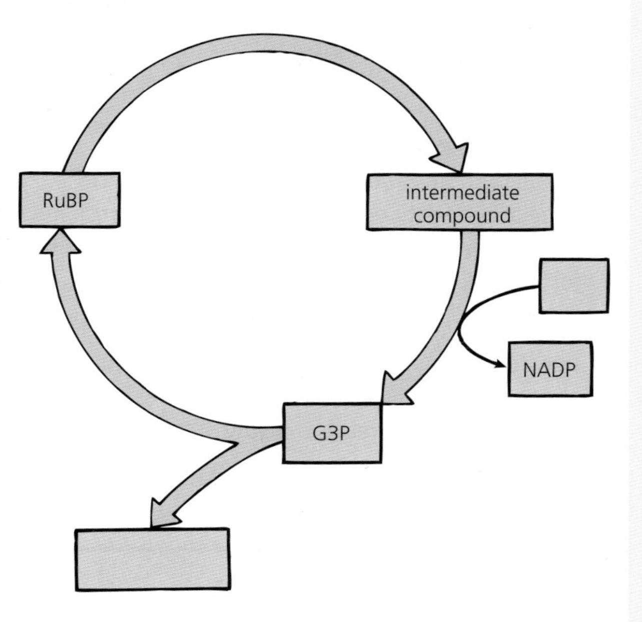

Figure 14.27

e) i) Which substance would accumulate if the plant were placed in darkness?
ii) Explain why. (2)

What You Should Know

Chapter 14

(See Table 14.3 for word bank)

action	electrons	reflected
ATP	G3P	rubisco
ATP synthase	higher-yielding	RuBP
blue	less	security
carbohydrates	NADPH	spectrum
carbon dioxide	phosphorylated	sustainable
Calvin	photolysis	transport
carotenoids	photosynthesis	trophic
cellulose	pigment	weeds

Table 14.3 Word bank for chapter 14

1 The ability to access food of sufficient quantity and quality by humans is called food ______. The continuous increase in the human population is accompanied by a demand for increased food production that is ______ but not damaging to the environment.

2 Food production depends on the process of ______. A small number of crops supply most human food. If agricultural land is limited, increased food production can be achieved by growing ______ cultivars and by protecting crops from pests and competitive ______.

3 As a result of energy loss between ______ levels in a food chain, production of livestock generates ______ food per unit area than crop plants.

4 The process by which light energy is trapped by green plants and used to produce ______ is called photosynthesis.

5 On coming into contact with a leaf, light may be absorbed, ______ or transmitted.

6 Chlorophyll absorbs light primarily in the ______ and red regions of the spectrum of white light. ______ absorb blue-green light and extend the range of wavelength absorbed for photosynthesis. A close correlation exists between the overall absorption ______ for leaf pigments and the ______ spectrum for photosynthesis.

7 Light energy absorbed by a leaf ______ molecule is transferred to ______ which become excited. These electrons are transferred through electron ______ chains where they release energy.

8 Some energy is used to generate ATP under the control of ______; some energy is needed to split water by ______. The oxygen produced is released by the cell; the hydrogen becomes attached to coenzyme NADP to form ______.

9 ______ and NADPH from the light-dependent first part of the process are needed to drive the ______ cycle – the second part of photosynthesis.

10 ______ becomes attached to RuBP under the control of the enzyme ______. The intermediate metabolite that results becomes ______ by ATP. Then it combines with hydrogen from NADPH to form ______.

11 Some G3P is used to regenerate ______. The rest becomes sugar. Some of these sugar molecules are built up into starch, ______ and other metabolites.

15 Plant and animal breeding

Characteristics selected by breeders

Breeders of crop plants and livestock attempt to manipulate a chosen organism's heredity. This is done in order to produce a new and improved **cultivar** (cultivated variety) of plant or **breed** of animal that will provide a source of sustainable food for humans. Examples of the types of characteristics that breeders would select are shown in Table 15.1.

Heritable characteristic	Example
increase in yield	increase in mass of food produced by wheat crop
increase in nutritional value	increase in mass of protein produced by soya bean crop
resistance to pests	resistance of tomato to eelworm
resistance to disease	resistance of potato to late blight
possession of useful physical characteristic	growth of cereal crop to uniform height suited to mechanical harvesting
ability to thrive in a particular environment	ability of maize to grow in cold, damp climate

Table 15.1 Characteristics selected by breeders

Research Topic: Resistance of potato varieties to *Phytophthora infestans*

In recent years agriculture has become overdependent on the use of chemical pesticides to protect crops from pests and diseases. Scientists are constantly seeking new ways of protecting food crops by breeding cultivars that are **resistant** to diseases.

Phytophthora infestans is a fungal pathogen that causes **late blight** in potato plants (see Figure 15.1). It was this devastating disease that led to the Irish potato famine in 1845–46. During the early twentieth century some resistance to *P. infestans* was achieved successfully by crossing existing (domestic) cultivars of potato with wild varieties possessing resistant genes. However, *P. infestans* soon evolved new aggressive races (strains) able to break down the potato plant's resistance.

Scientists in Scotland are investigating the resistance of potato varieties to *P. infestans* using a combination of approaches including the following:

- A variety of potato cultivars from many different parts of the world are being subjected to various races of *P. infestans* to identify resistant strains of potato.

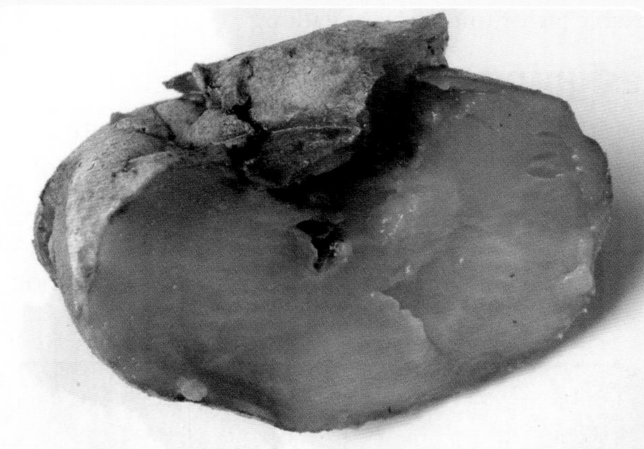

Figure 15.1 Potato infected with late blight

- Marker-assisted selection (MAS) is being used to locate markers (regions of a crop plant's DNA) that are associated with desirable traits, with the aim of transferring these from one variety to another. MAS is an advance on genetic modification involving single genes because it involves a set of interacting genes that contribute to a complex trait such as disease resistance.

→

- Multi-trait pre-breeding programmes are being carried out to combine disease resistance with other quality traits to create new varieties of potato.
- Field trials are being run where new and traditional varieties of potato are grown without fungicide and compared for incidence of late blight and for crop yield.

Field trials

A plant **field trial** takes the form of an investigation. It could be set up, for example, to:

- compare the performance of two different plant cultivars under the same set of environmental conditions
- investigate the effect of different environmental conditions on a new cultivar of a crop plant
- evaluate genetically modified (GM) crops.

Plots and treatments

The area of land to be used for a field trial is divided into equal-sized portions called **plots** (see Figure 15.2). A field trial involves **treatments**. A treatment refers to the way in which one plot is treated compared with other plots. For example, one plot might be given a high concentration of fertiliser and another plot a low concentration. Therefore these plots would represent two different treatments of a variable factor under investigation.

Designing a field trial

Once the plant breeder has established an objective (for example, to investigate the effect of concentration of a nitrogenous fertiliser on a new cultivar of cereal plant), the next stage is to design the field trial. When doing so, the following factors must be taken into consideration.

Figure 15.2 Field trial

Selection of treatments to be used

Each equal-sized plot might, for example, be given a high concentration or a low concentration or no application of fertiliser. If no other factor were varied, then a **fair comparison** could be made between treatments.

Number of replicates to be included

If only one treatment of each concentration of fertiliser were carried out, the results would be unreliable. This is because **uncontrolled variability** exists within the sample. Neither the three plots used nor the methods employed to apply the treatments to them would be exactly identical each time, however hard the scientist tried. Such variability is called **experimental error**.

Several **replicates** (normally a minimum of three) must be set up to take account of the variability and reduce the effect of this experimental error. It also allows valid statistical analysis of the results to be carried out. The more replicates that are set up, the more **reliable** the results.

Randomisation of treatments

If the plots in a field were treated in an orderly sequence then a **bias** could exist in the system. In the field shown in Figure 15.3, for example, soil moisture content decreases along the sequence of plots. Therefore within each replicate, each plot B treatment will have a built-in bias of less soil water compared with its corresponding plot A treatment and more soil water compared with its corresponding plot C treatment. **Randomising** the pattern of replicated treatments, as shown in Figure 15.4, eliminates this bias.

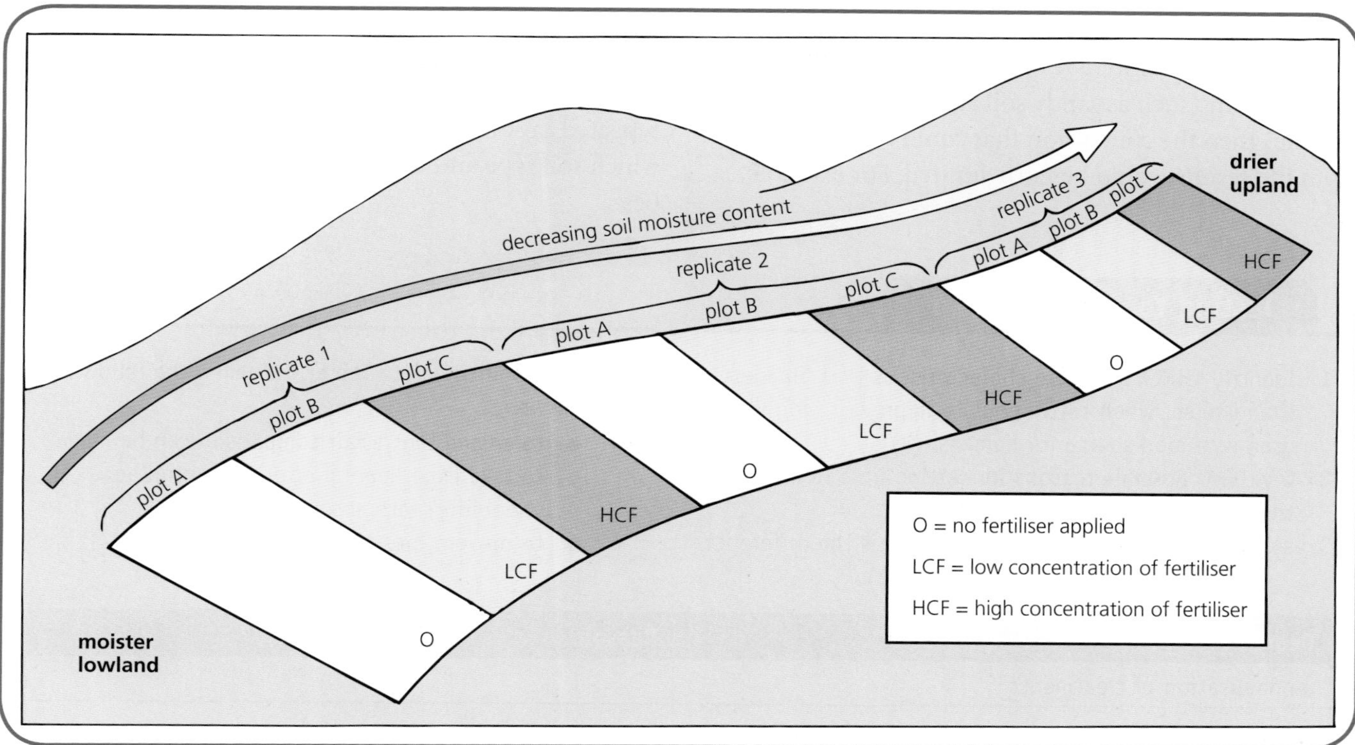

Figure 15.3 Poor design of field trial lacking randomisation

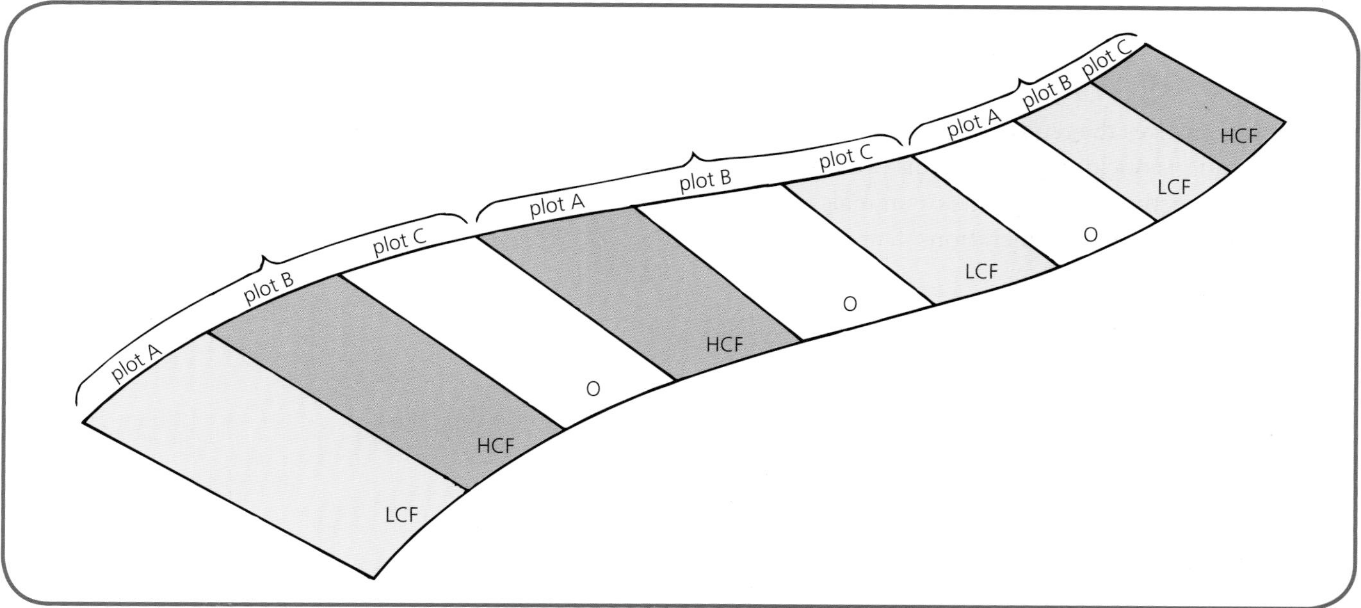

Figure 15.4 Design improved to include randomisation

Repeats in other environments

If a plant field trial were restricted to one environment (such as sandy soil and temperate climate) then the conclusion that could be drawn from the results would be fairly limited. For example, it might be the case that the new cultivar being tested does not grow well in sandy soil regardless of how much fertiliser is added. Therefore a field trial is often repeated in several different environments to find out which soil type and climate conditions suit the plant best.

Testing Your Knowledge 1

1 Identify THREE heritable characteristics that breeders would select when setting out to improve a crop plant used as a food source for humans. (3)

2 Give TWO possible reasons for setting up a field trial. (2)

3 **a)** With reference to field trials, what is the difference between a *plot* and a *treatment*? (2)

b) Copy and complete Table 15.2 using the following answers:

- to ensure that a fair comparison can be made
- to take experimental error (uncontrolled variability) into account
- to prevent bias existing in the system. (2)

Design feature	Reason
randomisation of treatments	
selection of treatments involving one variable factor	
inclusion of several replicates	

Table 15.2

Inbreeding

Inbreeding involves the repeated fusion of gametes from generations of close relatives. This can be achieved by selecting related plants or animals and breeding them for several generations. During this process, heterozygous offspring are eliminated and eventually the population breeds true for the desired characteristics. (Also see Related Information – Selective breeding by inbreeding.)

Related Information

Selective breeding by inbreeding

The flow chart in Figure 15.5 shows a programme of selective inbreeding for desired characteristics. As the programme continues over several generations, alleles for selected traits become concentrated in one line of animal breed.

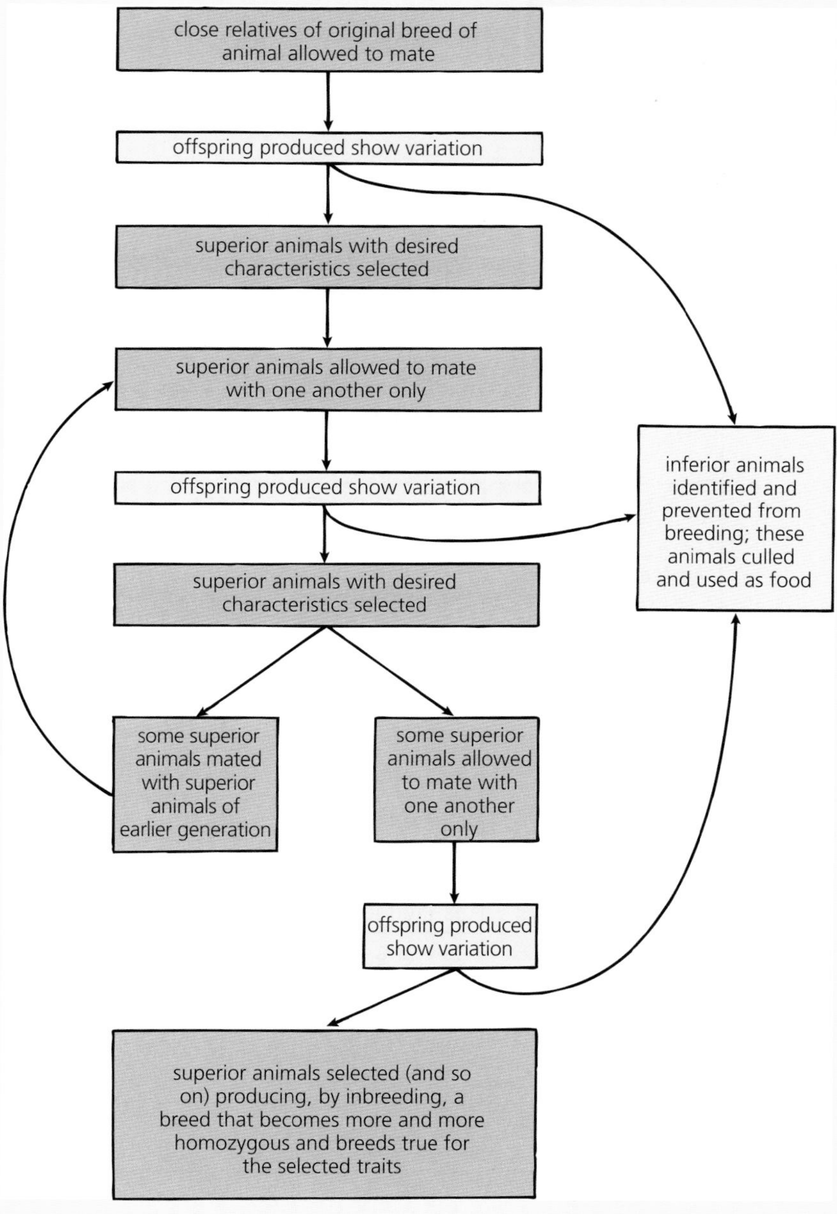

Figure 15.5 Selective breeding by inbreeding

Effects of inbreeding

Loss of heterozygosity

Continuous inbreeding leads to a loss of heterozygosity and an increase in homozygosity (see Related Activity – Analysis of a pattern of inheritance).

Related Activity

Analysis of a pattern of inheritance

Table 15.3 shows the effect of continuous inbreeding. After only three generations of inbreeding, heterozygosity has decreased from 100% to 12.5%. After a few more generations, it will have disappeared almost completely.

Generation	P_1	F_1	F_2	F_3
Genotypes resulting from inbreeding	Aa (selfed)	AA (selfed)	AA (selfed)	4AA
			AA (selfed)	4AA
			AA (selfed)	4AA
			AA (selfed)	4AA
		Aa (selfed)	AA (selfed)	4AA
			Aa (selfed)	AA : 2Aa : aa
			Aa (selfed)	AA : 2Aa : aa
			aa (selfed)	4aa
		Aa (selfed)	AA (selfed)	4AA
			Aa (selfed)	AA : 2Aa : aa
			Aa (selfed)	AA : 2Aa : aa
			aa (selfed)	4aa
		aa (selfed)	aa (selfed)	4aa
			aa (selfed)	4aa
			aa (selfed)	4aa
			aa (selfed)	4aa
Percentage heterozygosity remaining	100%	50%	25%	12.5%

Table 15.3 Effect of continuous inbreeding

Inbreeding depression

Inbreeding can result in an increase in the frequency of individuals who are homozygous for an accumulation of recessive alleles of deleterious (harmful) genes. Expressed phenotypically, these result in a decline in vigour, size, fertility and yield of the plant or animal. This phenomenon is called **inbreeding depression**. It is illustrated in Figure 15.6 for maize. The plant on the left is the original parent forced to inbreed. The other plants represent the next three generations and show a decrease in growth.

Normally inbreeding depression is not a problem suffered by self-pollinating plants. They are natural inbreeders and harmful alleles have been weeded out of their genotypes by natural selection over millions of years.

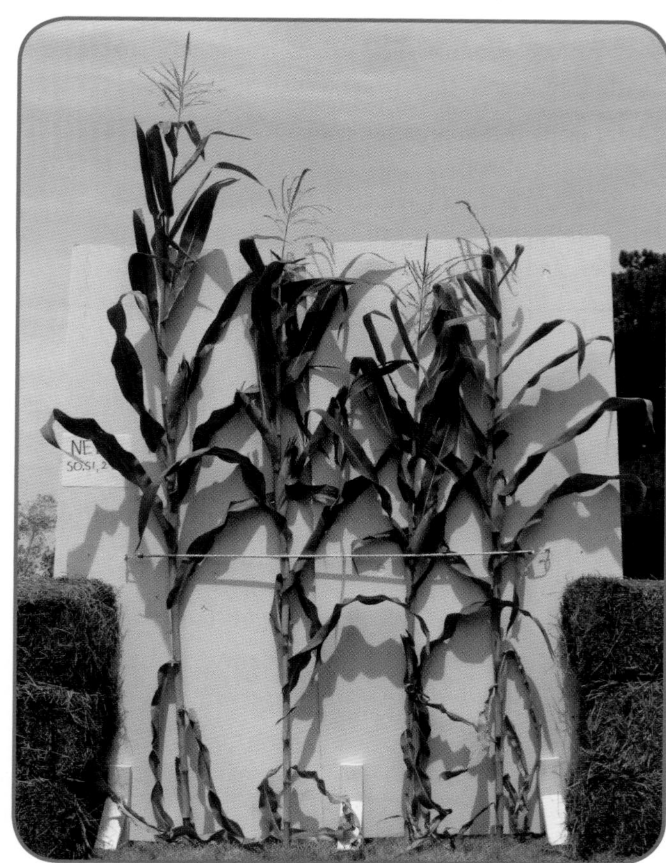

Figure 15.6 Inbreeding depression in maize

Research Topic — Development of crop cultivars

Wheat

Red wheat has been bred over many generations for the excellent baking quality of its flour and its resistance to leaf rust (a fungal disease of cereal plants). In the past it was mainly grown in North America. However, in recent years cultivars of red wheat and other specially bred cultivars have been grown by thousands of British farmers. As a result, above 80% of the wheat used for bread making in the UK is now grown here and only 20% is imported from North America. This trend is expected to continue as a result of advances in wheat breeding and technology.

Swede

Swedes (Swedish turnips) (see Figure 15.7) have been grown in Scotland for more than 200 years. They are an important source of forage for sheep and cattle during the winter in addition to providing a wholesome source of food for people.

Figure 15.7 Cultivar of swede

From the 1930s onwards swede plants were selected and inbred for high dry-matter content. This led to the development of a uniform, high-yielding cultivar. Then in the

→

1990s breeders turned their attention to other characteristics such as resistance to powdery mildew and clubroot.

After many breeding experiments involving both inbreeding of cultivars and crosses between different cultivars, scientists have developed varieties of swede that are high-yielding and resistant to both powdery mildew and the most prevalent strains of clubroot.

These cultivars outperform non-resistant varieties in environments where the diseases occur. However, the susceptible varieties do better than the resistant ones in regions where neither of the diseases is present. When either type of swede is grown under a fleece, it becomes attacked by root flies, so there is still plenty of work for the plant breeder to do to develop even better cultivars of swede.

Crossbreeding

Although continuous inbreeding of a plant cultivar or animal breed may bring about an improvement in a desired trait, eventually the build-up of harmful recessive alleles in natural outbreeders tends to lead to inbreeding depression. It is for this reason that inbreeding is rarely carried out indefinitely. Instead new alleles are introduced to the plant or animal line by **crossbreeding** it with a strain possessing a different but desired genotype.

F_1 hybrids

An **F_1 hybrid** is an individual resulting from a cross between two genetically dissimilar parents of the same species. Breeders often deliberately cross members of one variety of a species which have certain desired features with those of another variety possessing different useful characteristics in an attempt to produce F_1 hybrids which have both.

Research Topic Development of a livestock breed

Ayrshire cattle

Development of the *Ayrshire* breed of cattle (see Figure 15.8) is thought to have begun about 300 years ago. Farmers are known to have crossed native local cattle, which were small and were poor milkers, with several other breeds. Of these, one breed that made a major contribution was the *Teeswater* which in turn contained genetic material from Dutch cattle (later used to develop the *Holstein* breed in Holland). The emerging *Ayrshire* breed was further improved over the years by crossing it with *Highland* cattle and with members of the *Shorthorn* breed.

The cow now known as 'the *Ayrshire*' is the result of breeders crossing, selecting and inbreeding strains of cattle over many generations. Therefore *Ayrshire* cattle are well suited to the climate in Western Scotland and are efficient grazers. In addition, they are excellent milk producers. The milk itself is ideally suited to the production of butter and cheese. Top cows have been known to produce over 9000 litres of milk during their lactation period.

Figure 15.8 *Ayrshire* cow

Hybrid vigour

Plants

Hybridisation between two different homozygous inbred cultivars of plant species produces an F_1 generation whose members are **uniformly heterozygous**. More importantly, they display **increased vigour, yield, fertility** or other characteristics that are improvements on those shown by either parent. This phenomenon is called **hybrid vigour**. Poorer recessive alleles are masked by superior dominant ones, as shown in the example in Figure 15.9.

Figure 15.10 shows hybrid vigour in maize. Since the superior F_1 hybrids are not true-breeding, the hybridisation process using the original parental lines must be repeated every year. The number of F_1 crops produced in this way is limited because of the expense involved. If the F_1 hybrids, which are heterozygous, are allowed to self-pollinate, the F_2 generation produced is genotypically diverse and unsuitable as a crop.

Animals

Similarly in animals, crossbreeding of members of two different breeds may result in the production of a new **crossbreed** with improved characteristics. The crosses in Figure 15.11 show the outcome of hybridisation between different breeds of sheep. Table 15.4 shows some examples of traits enhanced by hybrid vigour in cattle. (The cartoon in Figure 15.12 is wishful thinking!)

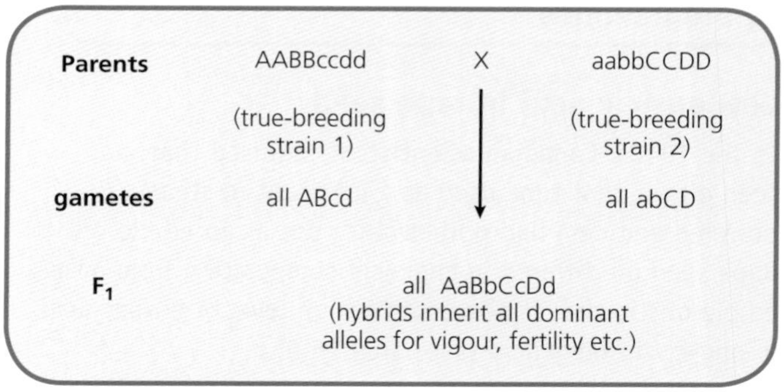

Parents AABBccdd (true-breeding strain 1) X aabbCCDD (true-breeding strain 2)

gametes all ABcd ↓ all abCD

F_1 all AaBbCcDd (hybrids inherit all dominant alleles for vigour, fertility etc.)

Figure 15.9 Hybridisation

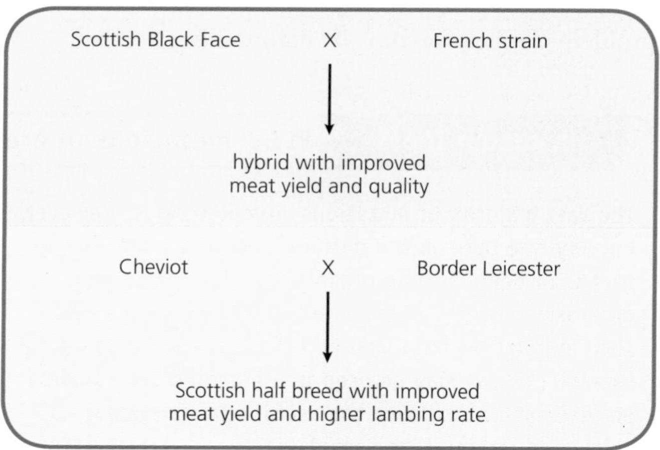

Figure 15.11 Hybrid vigour in sheep

Figure 15.10 Hybrid vigour in maize

Figure 15.12 'Now **that's** what I call hybrid vigour!'

Breed of cattle	Trait	Percentage improvement shown by hybrid
Dairy	birth weight	3–6
	live calves per calving	2
	feed conversion efficiency	3–8
	milk yield	2–10
Beef	birth weight	2–10
	calves weaned per cow mated	9–15
	feed conversion efficiency	1–6

Table 15.4 Hybrid vigour in cattle

If F_1 hybrids are allowed to interbreed, the F_2 generation will contain a wide variety of genotypes with many animals lacking the enhanced characteristics. This problem can be overcome by maintaining the two original parental breeds specifically for the purpose of **crossing them with one another** to produce crossbred animals that express hybrid vigour.

Research Topic Plant mutations in breeding programmes

The vast majority of mutations are deleterious. However, on very rare occasions a mutation arises spontaneously that is of benefit to an organism or that makes the organism useful to humans. A plant found to possess such a mutation for a desired trait (see rape seed example below) may be used in a breeding programme in an attempt to introduce the useful allele(s) into a cultivated strain of crop plant.

Increasing frequency of mutation

In 1928 an American scientist (L.J. Stadler) found that when barley seeds were exposed to X-rays, many more mutants were obtained than occurred naturally. This discovery led to the use of induced mutations in breeding programmes.

In the vast majority of cases, the mutants are inferior to the original wild type of the organism. For this reason deliberate use of radiation as a **mutagenic agent** is not a feasible method of improving breeds of farm animals. However, subjecting many thousands of a useful plant's seeds to radiation poses no such problem. By this means a **few** mutant plants with qualities beneficial to humans have been produced. These include tomatoes with increased yield and a strain of dwarf barley plant with erect ears that is more easily harvested than the original strain which tends to bend and lose its ears before the harvest.

Low euricic acid in rape seed

In the 1940s, Canadian scientists recognised that rape seed oil (at that time used as a lubricant in steam-powered engines) had potential for use as an edible oil. Rape seed oil differed in chemical composition from other edible oils in that it possessed a high level of euricic acid, a long-chain fatty acid (see Table 15.5).

It was known that Asian people had been consuming this type of rape seed oil for centuries with, apparently, no ill effects. However, research carried out on rats at the time indicated that a high intake of the oil rich in euricic acid caused unwelcome side effects such as enlarged adrenal glands.

Because of these nutritional concerns, scientists concentrated on developing a strain of rape plant whose seed oil would contain little or no euricic acid. First they located a low-yielding **mutant** variety of rape from Europe with a much reduced level of euricic acid in its oil. Then they carried out a series of breeding programmes involving careful selection over many generations in order to combine the allele(s) for low level of euricic acid of the mutant strain with the alleles (of the genes) for high yield and vigour found in the original strain. Eventually such a cultivar of rape was developed. Its edible oil, low in euricic acid, is found to be excellent as a cooking oil and as a salad dressing.

➜

Fatty acid	Percentage of fatty acid in vegetable oil		
	Sunflower seed	Soya bean	Rape seed before breeding programme
euricic	0.0	0.0	23.1
oleic	16.0	25.1	33.0
palmitic	7.2	12.0	4.9
stearic	4.1	3.9	1.6

Table 15.5 Fatty acid composition of vegetable oils

It is now known that consumption of a diet rich in **saturated** fatty acids leads to the build up of a fatty sludge on the insides of the artery walls. This restricts blood flow and may lead to circulatory problems and ill health. **Unsaturated** fatty acids have not been linked to this effect. In addition to being low in euricic acid, the developed strain of rape seed oil is rich in unsaturated fatty acids. It therefore enjoys worldwide success as a highly nutritious health food.

Genetic technology

Genome sequencing

Genome sequencing involves the construction of a 'library' of partially overlapping DNA fragments of an organism's genome and then assembling them into sequences of bases with the aid of computer technology (also see page 86). Genome sequencing can be used to identify organisms that possess a particular allele of a gene for a desired characteristic. This organism can then be used in a breeding programme to try to incorporate the useful gene into a new cultivar of crop plant or breed of domestic animal.

Genetically modified crop plants

Several breeding programmes make use of crop plants that have been genetically modified using recombinant DNA technology (see page 183). The following two examples illustrate the potential of this process to make significant improvements to crop plants:

- the Bt toxin gene inserted into crops to make them resistant to insect pests
- the glyphosate-resistance gene inserted into crop plants to make them tolerant to herbicide.

(Also see Related Information – Genetically modified crop plants.)

Related Information

Genetically modified crop plants

Bt toxin gene for pest resistance

Bacillus thuringiensis (Bt for short) is a soil bacterium that makes a protein that is toxic to certain plant-eating insects. Once consumed, the Bt toxin becomes active and binds to receptors in the insect's gut, paralysing it and causing the insect to die of starvation. Many varieties of Bt toxin exist. For example, one type is specific to butterflies whereas another type only works against beetles.

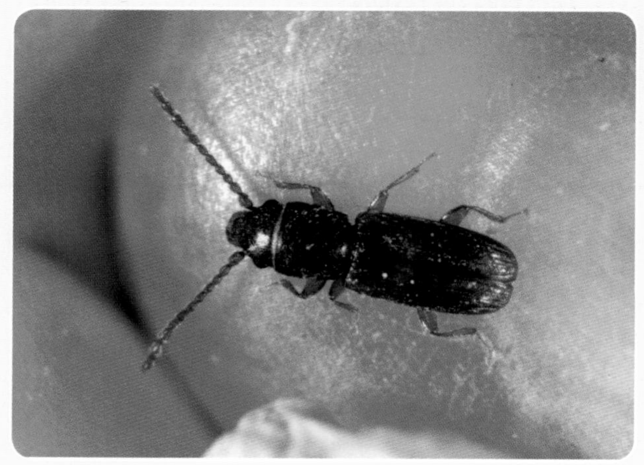

Figure 15.13 Beetle feeding on maize

Genetic engineers have managed to extract the bacterial genes for Bt toxin and to insert them into certain crop plants. These genetically modified plants produce their own Bt toxin making

→

them resistant to certain types of insect. (Consequently, they do not need to be sprayed with insecticide.)

Figure 15.13 shows a beetle feeding on a cob of maize. Some cultivars of Bt maize grown in the USA are resistant to this pest. The toxicity of each type of Bt toxin is limited to one or two insect groups. An insect that lacks the appropriate gut receptors remains unaffected. Bt toxins are non toxic to vertebrates and to most of the beneficial invertebrates.

Glyphosate-resistance gene for herbicide tolerance

Glyphosate is a weed killer (herbicide). It is absorbed through leaves and translocated throughout the plant. It inhibits an enzyme required for the synthesis of several amino acids, resulting in the death of the plant. Glyphosate is broken down into harmless products by micro-organisms within a few days.

Some plants are naturally resistant to glyphosate. Genetic engineers have successfully transferred the gene responsible for this resistance from these plants to crop plants such as soya bean, maize and sorghum.

When the genetically modified crop plants are sprayed with glyphosate, they remain unaffected but any weeds present among them are killed. Unfortunately some varieties of weeds resistant to glyphosate are beginning to emerge.

Testing Your Knowledge 2

1 Distinguish between the terms:
   a) *inbreeding* and *crossbreeding*. (2)
   b) *homozygous* and *heterozygous*. (2)

2 Decide whether each of the following statements is true or false and then use T or F to indicate your choice. Where a statement is false, give the word that should have been used in place of the word in bold print. (4)
   a) **Crossbreeding** individuals of two different animal breeds may produce a new strain.
   b) Continuous inbreeding results in loss of **homozygosity**.
   c) During **inbreeding** selected members of a species are bred for several generations until they breed true.
   d) The accumulation of recessive deleterious alleles in a homozygous **phenotype** results in inbreeding depression.

3 By what means can new alleles be introduced to a plant cultivar that is beginning to show inbreeding depression? (1)

4 a) Why do F_1 hybrids from two homozygous inbred strains often display hybrid vigour? (1)
   b) i) Are such F_1 hybrids *homozygous* or *heterozygous* in genotype?
      ii) Are they *varied* or *uniform* in phenotype?
      iii) When this F_1 generation is selfed, is the F_2 generation varied or uniform? (3)
   c) Why do some farmers continue to maintain the two homozygous parental strains? (1)

5 a) i) What is *genome sequencing*?
      ii) What use can be made of genome sequencing in a breeding programme? (3)
   b) Give an example of an improvement that has been made to some crop plants by modifying them genetically. (1)

Applying Your Knowledge and Skills

1 Weaning weight is the weight of a calf when it changes its food from milk to solids. It is a polygenic characteristic in cattle. The sum of all the additive effects of the genes controlling a polygenic trait is called the breeding value for that trait. Table 15.6 shows the effects of the alleles of four genes on weaning weight in a breed of cattle. Table 15.7 shows the breeding values of weaning weight for five animals. Supply the information missing from boxes **a)**, **b)** and **c)** in Table 15.7. (3)

Gene controlling weaning weight	Alleles of gene	Effect on weaning weight (kg)
1	W^1	+6
	w^1	0
2	W^2	+8
	w^2	0
3	W^3	+3
	w^3	0
4	W^4	+10
	w^4	0

Table 15.6

Animal	Genotype Alleles of gene								Breeding value (for weaning weight)
	1		2		3		4		
P	W^1	W^1	W^2	W^2	W^3	W^3	W^4	W^4	6+6+8+8+3+3+10+10 = 54 kg
Q	W^1	w^1	w^2	w^2	W^3	w^3	W^4	W^4	6+0+0+0+3+0+10+10 = 29 kg
R	w^1	w^1	W^2	w^2	w^3	w^3	W^4	w^4	**a)**
S	W^1	W^1	w^2	w^2	W^3	w^3	W^4	W^4	**b)**
T	**c)**								6+0+8+0+3+0+10+0 = 27 kg

Table 15.7

2 Figure 15.14 shows a field trial of plots set up to investigate the effect of nitrogenous fertiliser and fungicide on a new cultivar of cereal crop. Figure 15.15 shows the results at the end of the growing season as total dry mass (kg) per plot.

replicate 1	replicate 2	replicate 3
35	5	70
70	35	5
5	70	35
70	5	35
35	70	5
5	35	70

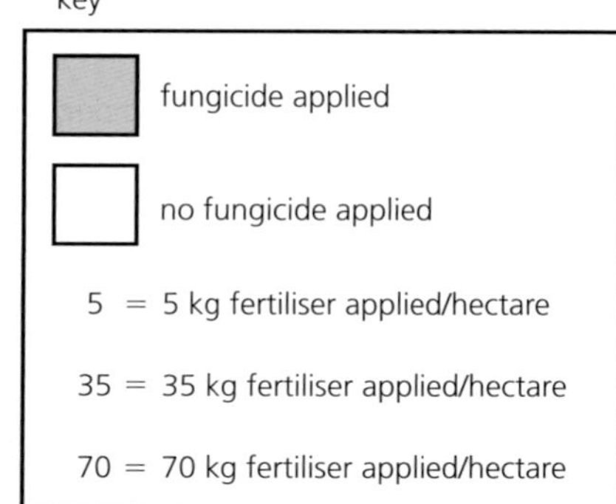

Figure 15.14

→

replicate 1	replicate 2	replicate 3
371	258	329
379	366	207
252	383	305
316	211	364
304	327	249
201	312	386

Figure 15.15

a) How many variable factors were investigated in this crop field trial? (1)

b) How many treatments of
   i) fertiliser were used?
   ii) fungicide were used? (2)

c) State ONE design feature that helped to ensure that a fair comparison of treatments could be made. (1)

d) Identify ONE design feature that took into account variability among samples and reduced the effect of experimental error. (1)

e) What was done to eliminate the chance of the results being biased? (1)

f) i) Table 15.8 has been drawn up to record and view the results (as kg total dry mass per plot) in an organised way. The values of replicate 1's plots have been entered. Copy and complete the table.
   ii) From your completed version of Table 15.8 draw a conclusion about the effect of the fungicide on this new cultivar of cereal crop and explain your answer.
   iii) Why would it be necessary to apply statistical analysis to the results before being able to draw a conclusion about the effect of the two higher concentrations of fertiliser on the crop? (6)

g) What control should have been run in this field trial? (1)

Fungicide or no fungicide	Fertiliser treatment (kg/acre)	Replicate 1	Replicate 2	Replicate 3
no fungicide	5	201		
	35	304		
	70	316		
fungicide	5	252		
	35	371		
	70	379		

Table 15.8

3 The graph in Figure 15.16 shows the effect of continuous inbreeding on heterozygosity.

a) Starting with a genotype heterozygous for 15 genes, what percentage of heterozygosity remains in the F_5 generation? (1)

b) Starting with a genotype heterozygous for 10 genes, what decrease in heterozygosity has occurred by the F_3 generation? (1)

c) Which genotype's heterozygosity has decreased by 75% at the F_2 generation? (1)

d) At which generation was the heterozygosity of all four genotypes first found to be below 10%? (1)

4 The data in Tables 15.9 and 15.10 refer to breeding experiments using maize (*Zea mays*) shown in Figure 15.17.

a) Identify the generation formed by hybridisation. (1)

b) State THREE pieces of information which provide evidence of hybrid vigour. (3)

c) Which of the following was a true-breeding inbred line: P_1, F_1, F_3, F_5? (1)

d) Using an example from the data to illustrate your answer, explain what is meant by the term *inbreeding depression*. (1)

e) Many years ago, American farmers began planting maize grains produced by hybridisation. This practice has continued over the years and has brought about the changes shown in Table 15.10.
   i) State TWO ways in which farmers have benefited by using hybrid maize.
   ii) Explain why farmers must buy expensive hybrid grain every year from supply houses to sow their maize crop instead of simply using grain kept back from the previous year's crop. (3)

→

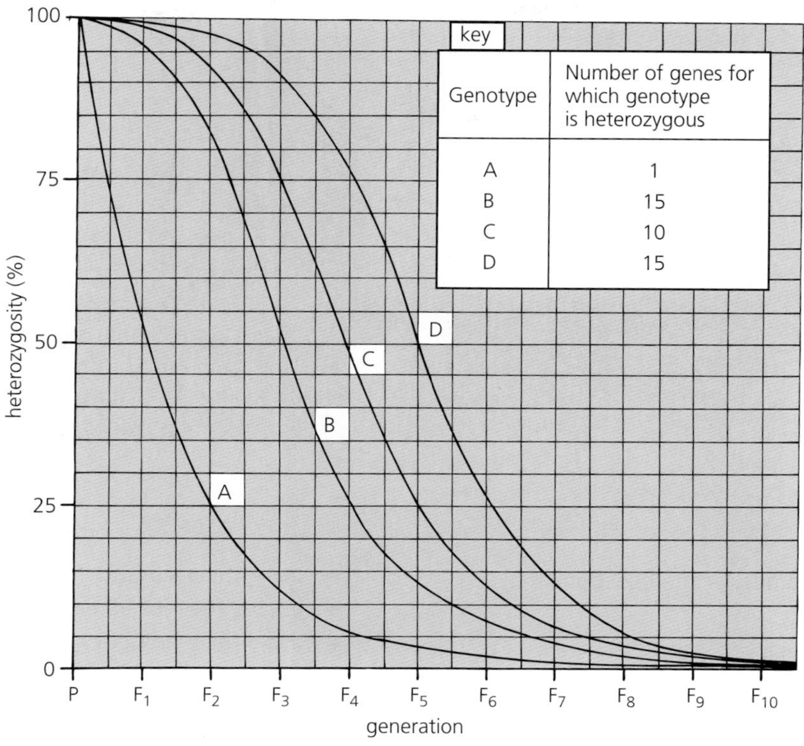

Figure 15.16

		Number of generations selfed	Mean ear length (mm)	Mean yield (metric tonnes ha^{-1})
Parents	P_1	17	84	2.0
	P_2	16	107	2.1
Successive generations (resulting from repeated self-fertilisation)	F_1	0	162	4.3
	F_2	1	141	4.0
	F_3	2	147	3.4
	F_4	3	121	3.1
	F_5	4	94	2.8
	F_6	5	99	2.6
	F_7	6	110	2.5
	F_8	7	107	2.3

Table 15.9

Time	Millions of hectares used for maize planting	Mean yield (metric tonnes ha^{-1})
at start	46	2.4
after 15 years	36	3.4
after 30 years	29	4.2

Table 15.10

Figure 15.17

5 Read the passage and answer the questions that follow it.

New cultivar of maize

Native Americans began growing maize (which originated in Mexico) as a food crop thousands of years ago. This resulted in the development of many strains adapted to a wide range of conditions from tropical to cool, temperate climates. Maize brought to Europe hundreds of years ago was developed by breeding and selection as a Mediterranean crop suited to the warm, dry conditions of Southern Europe. These cultivars were found to be unsuited to conditions such as those that occur in Northern Scotland.

However, production of maize in northern regions is now increasing substantially following the development in France of cultivars developed specially for use in cold, temperate areas. These new varieties are the result of complicated breeding programmes involving crosses between European and North American strains.

The cultivars succeed in the north of Scotland because they contain alleles of genes that:

- allow them to mature early thereby enabling them to make the most of the shorter growing season in the north
- make them suitable for use on marginal sites
- enable them to be easily fertilised
- make them produce good, solid cobs with a high starch content that make excellent feed for ruminant animals.

a) Why were the earlier cultivars, bred directly from maize brought to Europe from America hundreds of years ago, unsuited to conditions in Northern Scotland? (2)

b) By what means has the new cultivar, suited to northern areas, been developed? (2)

c) In what way is this new cultivar able to cope with the shorter growing season in the north of Scotland? (1)

d) **i)** Name TWO ruminants commonly found in Scottish farms.

ii) Why is the new cultivar of maize favoured by farmers of ruminant animals? (3)

e) Figure 15.18 shows one maize plant from cultivar A and one maize plant from cultivar B.

i) Briefly outline the procedure that you would adopt if you wanted to inbreed each cultivar and prevent it from crossbreeding.

ii) Briefly outline the procedure you would adopt to crossbreed the two cultivars and prevent them from inbreeding. (4)

→

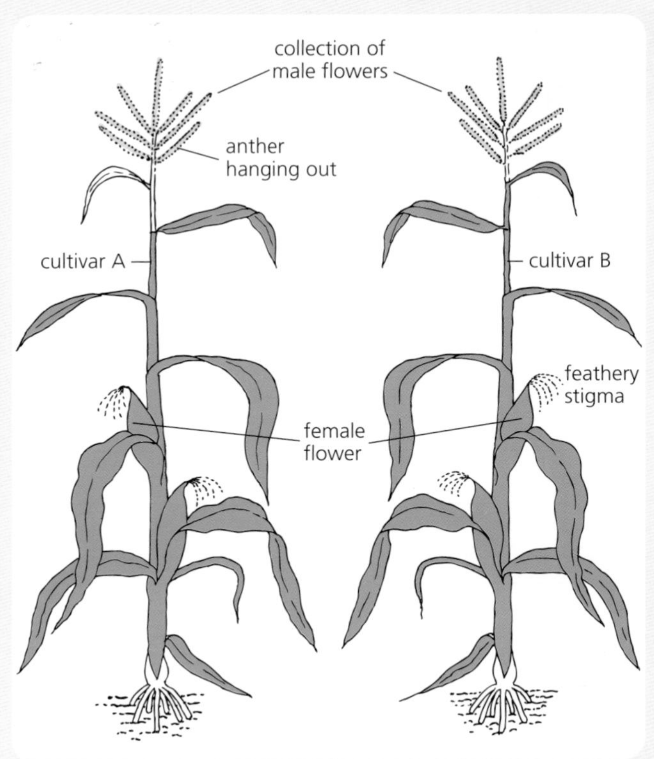

Figure 15.18

6 When a tomato fruit is fully grown, but still green and unripe, it begins to produce ethylene. Ethylene is a growth substance which promotes the ripening of fruit. Genetic engineers have developed a variety of tomato plant which makes very little ethylene by inserting a gene which almost completely blocks ethylene production. Table 15.11 gives the results from one of their experiments. Figure 15.19 shows a typical fruit from each variety of tomato plant.

a) Draw a line graph of the results. (2)

b) Match the terms *control* and *genetically modified variety* with the two lettered varieties in the diagram. (1)

c) State how many units of ethylene were produced on day 6 by:

i) the genetically modified variety

ii) the control. (1)

d) Calculate the percentage decrease in ethylene production at day 6 caused by the blocked gene in the genetically modified variety. (1)

e) Suggest why it is important that ethylene production is not completely blocked in the modified variety. (1)

f) What is the benefit gained from this application of genetic engineering? (1)

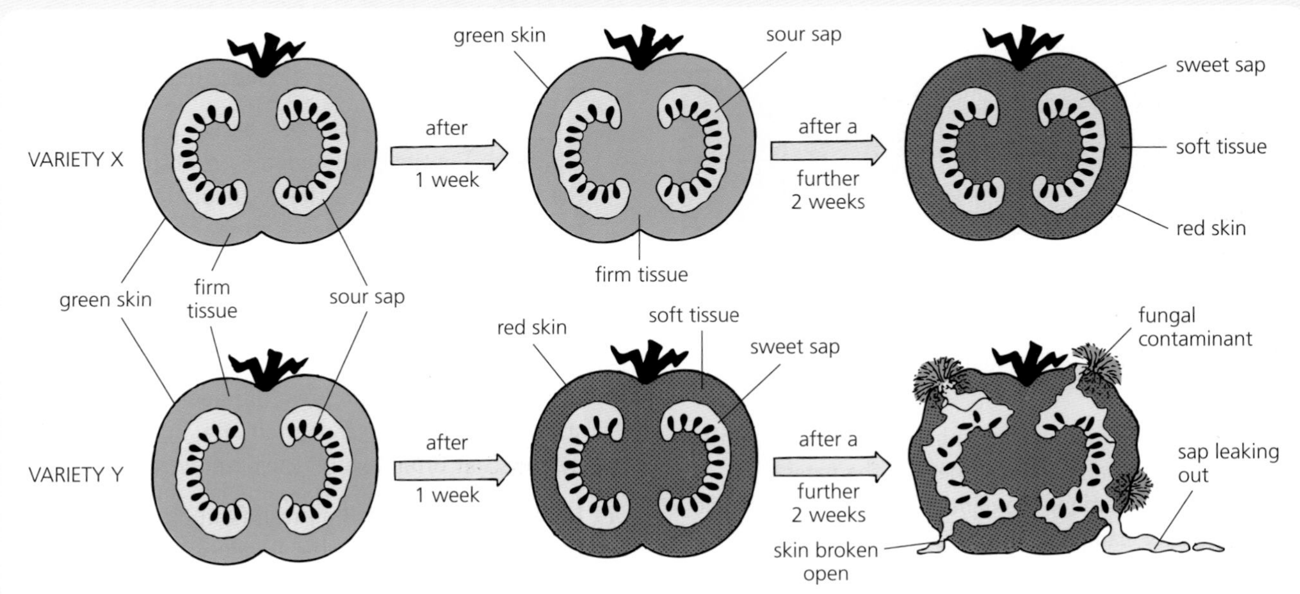

Figure 15.19

Time from green fruit reaching full size (days)	Ethylene production (units)	
	Genetically modified variety	Control
0	1	1
1	4	5
2	7	25
3	5	70
4	6	71
5	8	68
6	7	70
7	6	52
8	7	41
9	7	32
10	8	34
11	6	31
12	7	32
13	7	29
14	8	32
15	7	31
16	6	28
17	8	25
18	7	26
19	7	23
20	6	21
21	6	19

Table 15.11

What You Should Know

Chapter 15

(See Table 15.12 for word bank)

alleles	heterozygosity	replicate
comparison	homozygous	sequencing
crossbred	improved	sustainable
cultivar	inbreeding	treatments
depression	modified	trial
desired	randomisation	uniform
harmful	recombinant	vigour

Table 15.12 Word bank for chapter 15

1 During breeding programmes, breeders select organisms with ______ traits and use them as the parents of the next generation. The genetic material of crops and livestock is manipulated in this way to develop ______ organisms that will provide sources of ______ food.

2 A field ______ is set up to compare the performance of different plant cultivars or the effect of different ______ on one cultivar.

3 The design of a field trial must take into account: the treatments of the factor being investigated to allow for a fair ______, the inclusion of ______ to take uncontrolled variability into account and the ______ of treatments to eliminate bias.

4 ______ involves reproduction between close relatives. Continuous inbreeding eventually results in the loss of ______ and the accumulation of ______ recessive alleles which leads to inbreeding ______.

5 Crossbreeding a plant ______ or animal breed with a member of a different strain that has a desired genotype allows new ______ to be introduced.

6 In plants, crossbreeding two different inbred ______ strains results in the formation of a ______ crop of F_1 hybrids that often exhibit increased vigour and yield.

7 In animals, crossbreeding results in hybrid ______ among the members of the F_1 generation. The two parental breeds may be maintained in order to produce more ______ offspring.

8 Breeding programmes can make use of those organisms that have been shown by genome ______ to possess desired genes and those organisms whose genomes have been genetically ______ by ______ DNA technology.

16 Crop protection

Balanced community

In a natural ecosystem, a **balance** exists between producers and consumers. A **diverse variety** of plant species compete with one another and co-exist with their insect pests and disease-causing micro-organisms. The members of plant, animal and microbial communities tend to live in small mixed populations. If the numbers of a certain species of green plant decrease then the numbers of animals and micro-organisms that depend on the plant fall accordingly. This allows the plant species to recover and soon the balance is restored.

Monoculture

In an agricultural ecosystem the variety of species that make up the community, **the crop**, is greatly reduced and may even take the form of a **monoculture**. A monoculture is a vast population of a single species of crop plant cultivated over a large area for economic efficiency. Often the members of the population are **genetically identical** and are members of a cultivar developed for its productivity.

However, a crop monoculture presents ideal growing conditions to weeds, pests and disease-causing micro-organisms whose activities, in turn, reduce the crop's yield. Insects and micro-organisms, for example, can feed on the plant and reproduce repeatedly without ever running out of food (see Figure 16.1). Therefore farmers employ a variety of control methods to **protect** their crops and avoid economic disaster.

Weeds

Competition

Plants growing side by side in the same habitat compete for light, water, soil nutrients and space. Competition

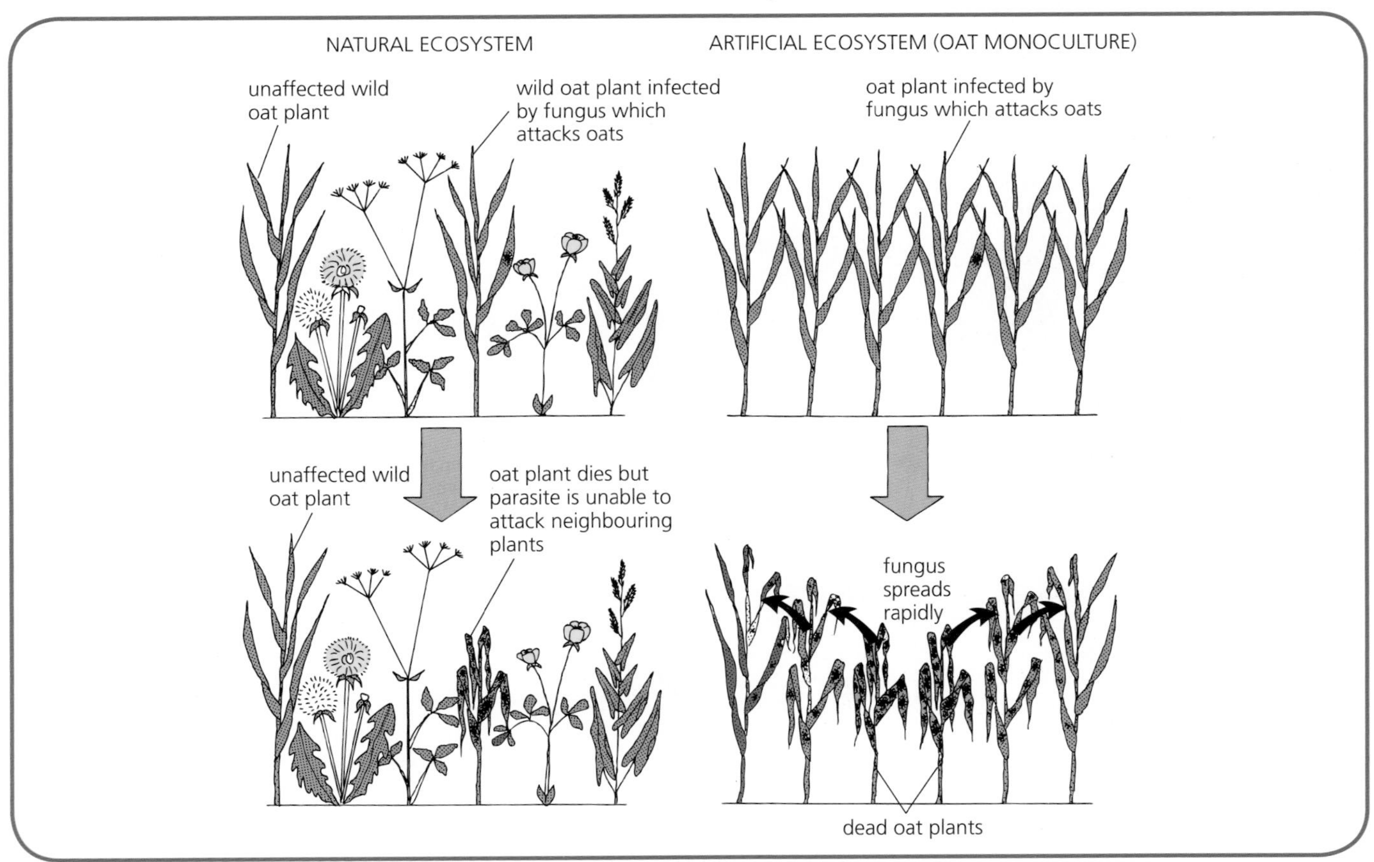

Figure 16.1 Effect of fungal parasite on two ecosystems

among the members of a monoculture is reduced by spacing out the seeds during sowing. However, these spaces tend to become occupied by weeds. A **weed** is any kind of plant that grows where it is not wanted. When weeds overrun a cultivated field, they pose a serious economic problem for the following reasons. They may:

- **cause a significant reduction** in the productivity of the crop due to competition
- **release chemical inhibitors** into the soil which further reduces crop growth
- **contaminate grain crops** with their seeds and reduce the crop's value
- **act as hosts** for crop pests and diseases, for example clubroot (a fungal disease that affects cabbage) is harboured by the weed Shepherd's purse.

Properties of common weeds

Weeds (see Figure 16.2) are perfectly adapted to their life as opportunists.

Annual weeds are able to colonise vacant land or an 'empty' field where the monoculture has yet to establish itself because they:

- grow very **quickly**
- rapidly produce flowers since their **life cycle is short**
- produce **vast numbers** of seeds which are often dispersed effectively by wind
- produce seeds that remain dormant but **viable for long periods** (even years) in the soil.

Perennial weeds are able to compete successfully with the crop plant from the very start of the growing season because the weeds:

- are already **established** in the habitat
- have **storage organs** from the previous year that provide food even if environmental conditions do not favour rapid photosynthesis
- are able to **reproduce vegetatively** using specialised structures such as runners.

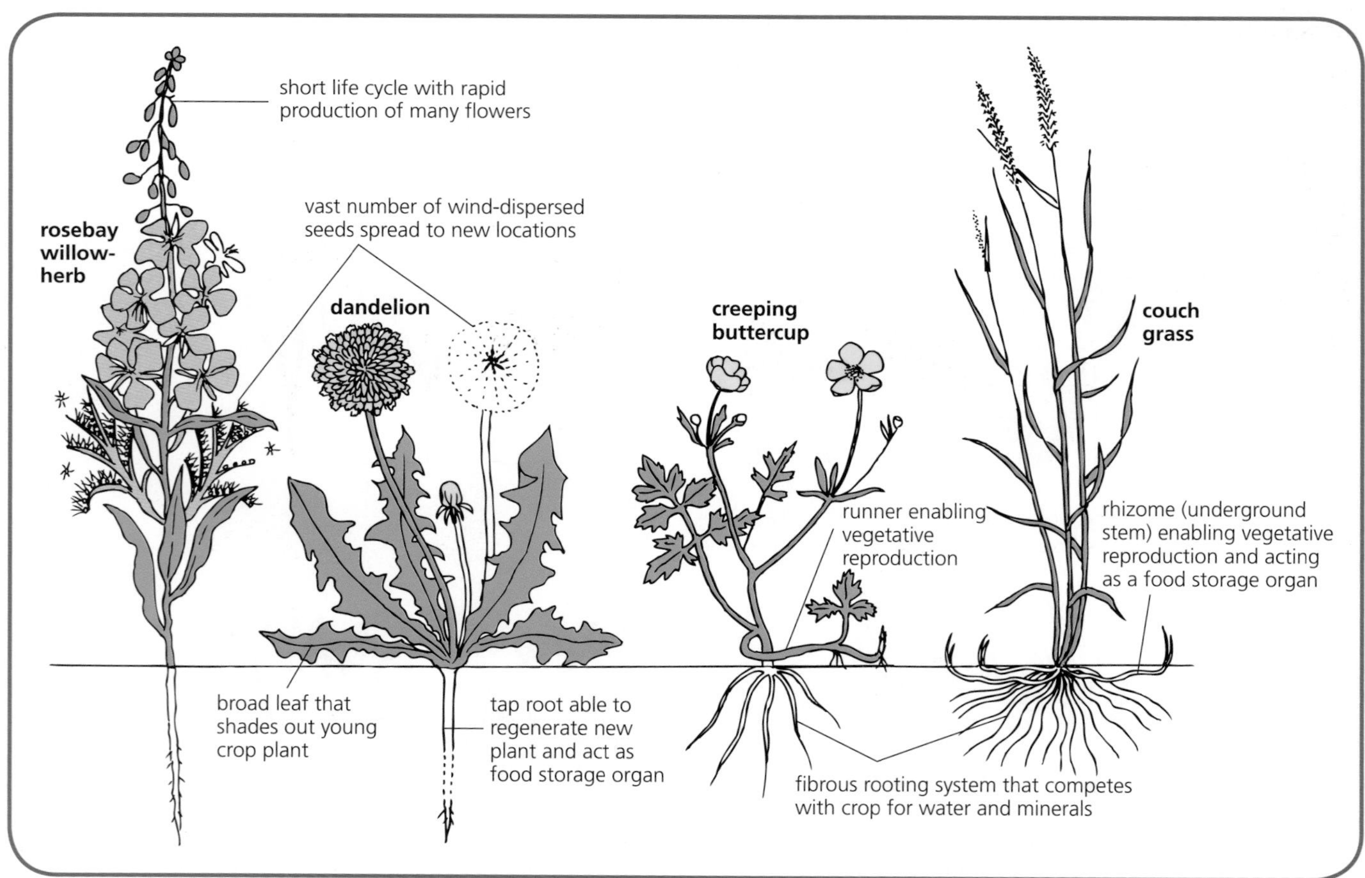

Figure 16.2 Adaptations of weeds (not drawn to scale)

Invertebrate pests

A monoculture of a crop plant presents ideal conditions to herbivorous pests to feed and reproduce extensively. Pests that attack crop plants mainly fall into three groups of **invertebrates**:

- nematode worms
- molluscs
- insects.

Nematodes

Nematode worms (see Figure 16.3) occur in almost all environments and are particularly numerous in soil. Many of these tiny worms attack crops and establish themselves as **parasites** within the host plant's roots. A common example is potato cyst nematode (see page 236) which causes millions of pounds worth of damage annually to the UK potato crop.

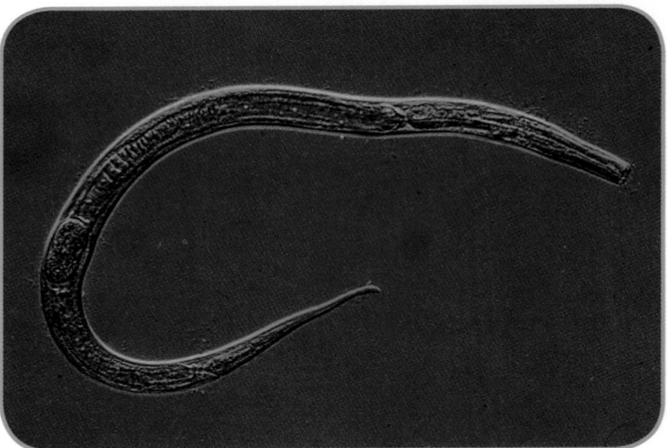

Figure 16.3 Nematode worm

Molluscs

Molluscs such as snails and slugs (see Figure 16.4) can do extensive damage to crops. They are most active at night and possess rasping mouth parts that are ideally suited to dealing with tough green plant parts such as cabbage leaves.

Insects

Herbivorous insects pose the greatest threat to food crops and cause many millions of pounds of damage annually by feeding on leaves, stems, roots and underground storage organs.

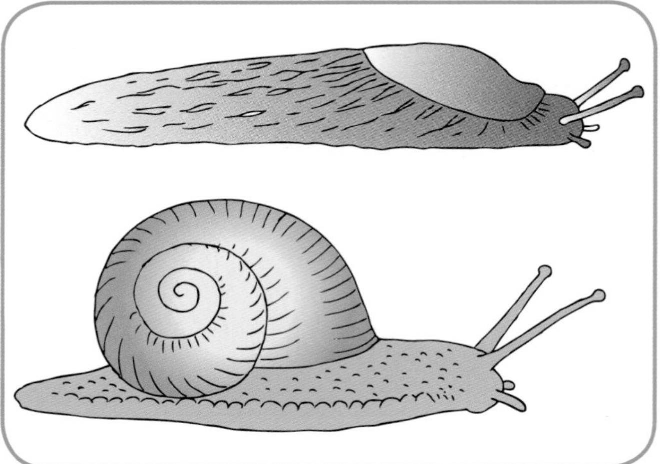

Figure 16.4 Molluscs

An insect often has several distinct stages in its life cycle. In the case of the cabbage white butterfly (see Figure 16.5) it is the **larval** stage that does the damage to the crop's leaves.

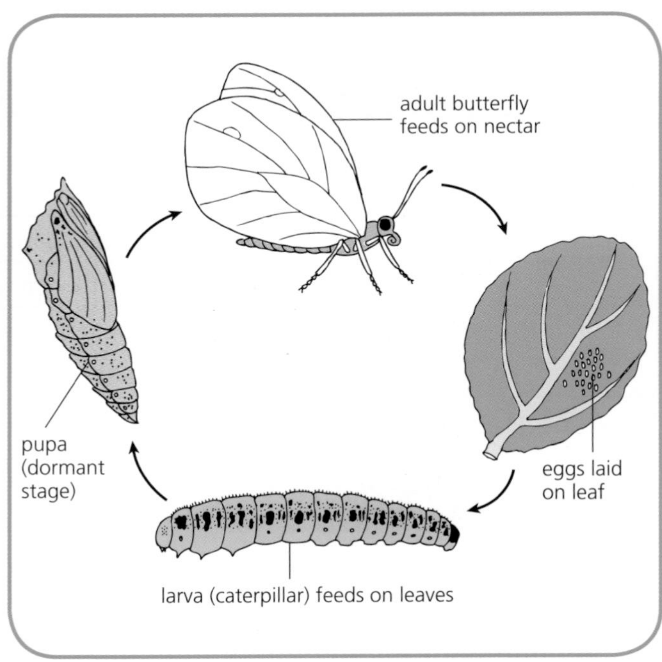

Figure 16.5 Life cycle of cabbage white butterfly

Aphids (greenfly and blackfly) are tiny insect pests that exist as winged and wingless forms (see Figure 16.6). They use their needle-like mouth part to pierce plant tissue in search of the sugary solution present in the phloem tissue.

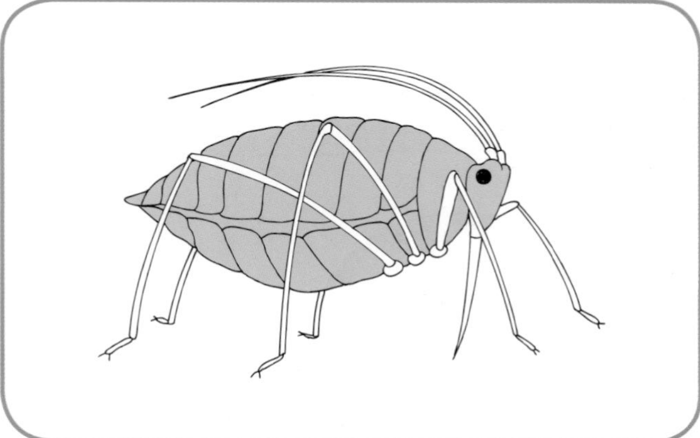
Figure 16.6 Wingless aphid

Decreased yield

Damage to leaves reduces photosynthesis; loss of transported sugar denies growing plant tissues their energy supply. As a result the crop plant's **vigour** and **yield** are adversely affected by insect pests. Some invertebrate pests also act as **vectors of disease**. Aphids, for example, transmit viruses that cause diseases such as potato leaf roll.

Plant diseases

Plant diseases are caused by **pathogens** such as fungi, bacteria and viruses. A few examples are given in Table 16.1. The pathogens may be soil borne, airborne or spread by insect vectors.

Poorer yield

Plant diseases also result in a poorer yield of the crop. In addition, a product that is blemished or infected by a pathogen such as a fungus (see Figure 16.7) is **less**

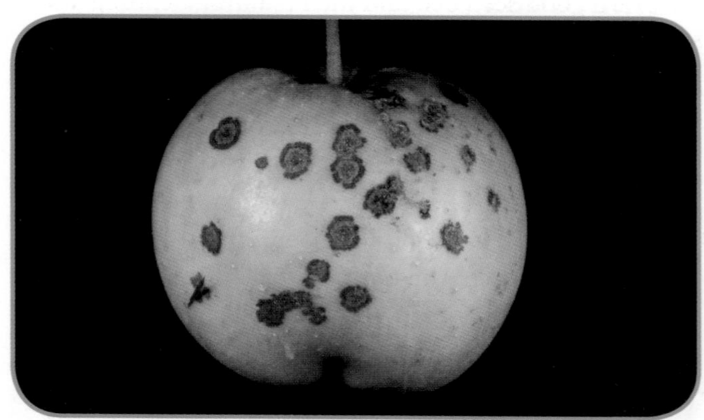
Figure 16.7 Apple scab (fungal disease)

Type of pathogen	Examples of pathogen	Host plant	Disease caused
fungus	*Phytophthora infestans*	potato	late blight
	Puccinia graminis	wheat	black rust
bacterium	*Pseudomonas solanacearum*	tomato	wilt
	Pectobacterium corotovorum	parsley	soft root
virus	tomato bushy stunt virus	tomato	bushy stunt
	potato leaf roll virus	potato	leaf roll

Table 16.1 Plant pathogens

Testing Your Knowledge 1

1 a) Which is normally more complex: the community structure of an *agricultural* ecosystem or that of a *natural* ecosystem? (1)
b) i) Which of these two types of ecosystem offers ideal conditions for the growth of disease-causing micro-organisms?
ii) Briefly explain why. (2)

2 Give TWO examples of adverse effects caused by weeds to a farmer's crop. (2)

3 a) Identify TWO properties of annual weeds that contribute largely to their success. (2)
b) Give TWO reasons why perennial weeds are able to compete successfully with a crop. (2)

4 a) Explain how insects such as aphids cause a decreased yield of a crop plant. (2)
b) Identify TWO other invertebrate groups that damage crops. (2)
c) State TWO means by which pathogenic micro-organisms may be spread from one host plant to another. (2)

marketable than a healthy, unblemished product. The storage life of a crop infected by a pathogen is also reduced because the stored crop is less healthy and more susceptible to attack by a **storage disease**.

Control of weeds, pests and diseases

Cultural means

The process of agriculture involves preparing the soil to provide a good seed bed and then planting, tending and harvesting the crop. **Cultural** means of controlling weeds, pests and diseases have evolved from traditional, non-chemical methods of cultivation by trial and error over time. They tend to be **preventative** rather than **curative** and often require long-term planning. They do not offer the 'quick-fix' solution to the problem associated with chemical sprays. Some examples are given below.

Ploughing

The top 20 cm of soil is turned over by this process (see Figure 16.8) and many of the perennial weeds are **buried** to a depth at which they die and decompose. The crop seed can then be planted and become established before the weeds return.

Early removal of weeds

Some competitive weeds that bring about a decrease in crop yield are known to cause most damage to young crop plants. If these weeds are removed **early** in the life of the crop, often the crop plants become sufficiently sturdy to tolerate competition by weeds when they return at a later stage. Therefore the yield of the crop is largely unaffected.

Figure 16.8 Ploughing (weeds)

Crop rotation

Crop rotation is the practice by which each of a series of very different types of crop plant is grown in turn on the same piece of ground. Once the first crop has been harvested it is followed, the next season, by a dissimilar type of plant that does not act as host to the pest being controlled.

In the plan shown in Figure 16.9, for example, the cabbages follow the pea plants and are followed by the potatoes in the rotation. A pest that can only attack a certain type of host plant (such as brassicas) may be controlled effectively because it is unlikely to survive for four years until its host returns. It is for this reason that crop rotation works best against soil-inhabiting pests that are only able to attack a **narrow range** of host plants. For example, long crop rotations help to control potato cyst nematodes that are specific to one type of crop.

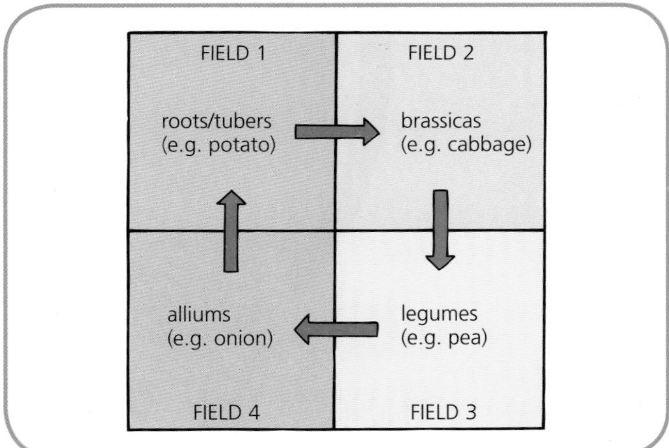

Figure 16.9 Crop rotation

Control of weeds, pests and diseases by chemical means

Unfortunately, traditional cultural control methods alone are often not sufficient to control vast populations of unwanted weeds, pests and pathogenic micro-organisms. Therefore, in addition, the farmer may have to make use of **chemical** agents to ensure high crop yields. Chemicals used to control pests are called **pesticides**.

Herbicides

Pesticides used to kill weeds are called **herbicides** (weed killers).

Investigation

Incidence and viability of potato cyst nematode cysts in two soil types

Potato cyst nematode is a common soil-borne parasite of potato plants. It attacks the roots (but not tubers) of the plant from which it obtains all of its food. This results in a reduced yield of potato tubers. Its life cycle is shown in Figure 16.10.

The investigation is carried out by following the procedure shown (in a simplified way) in Figure 16.11 using soil from an area continuously cropped with potatoes and soil from an area cropped with potatoes as part of a rotation.

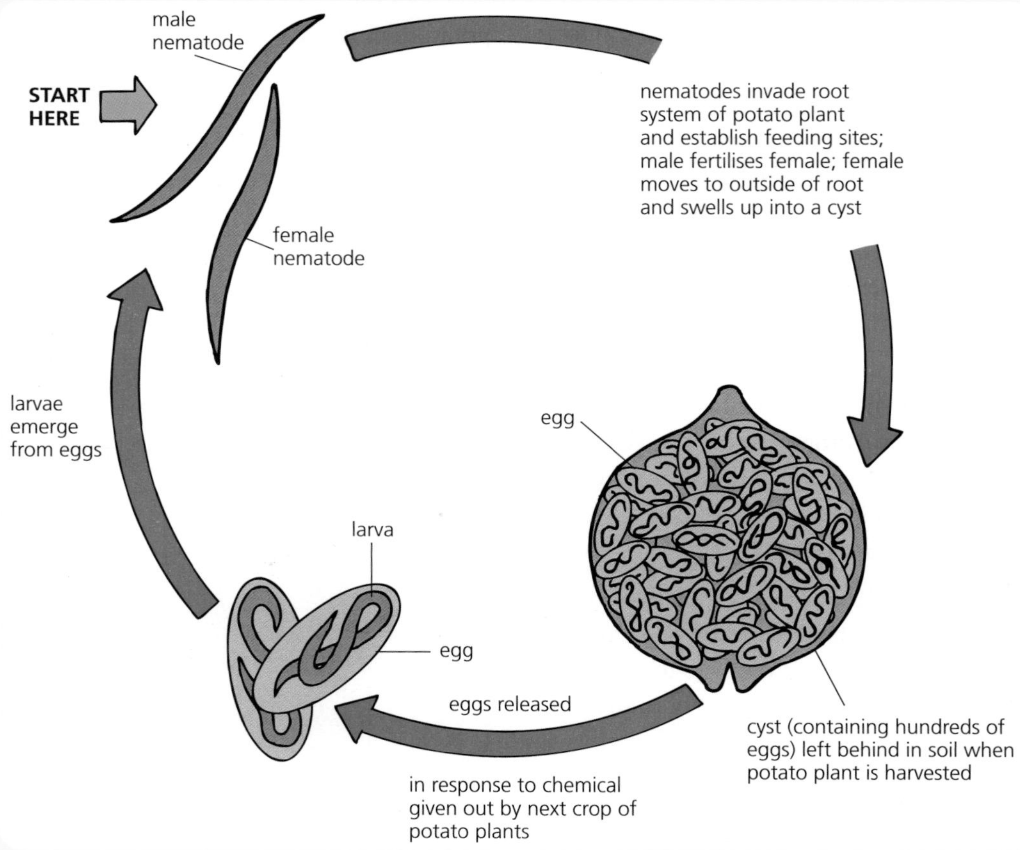

Figure 16.10 Life cycle of potato cyst nematode

Table 16.2 shows a specimen set of results which are presented as a bar chart in Figure 16.12.

	Soil type	
	Continuously cropped with potatoes	**Cropped with potatoes as part of a rotation**
total number of cysts per 100 g of soil	24	10
number of viable cysts out of the random sample of eight cysts	6	2
percentage of viable cysts in the random sample	75	25

Table 16.2

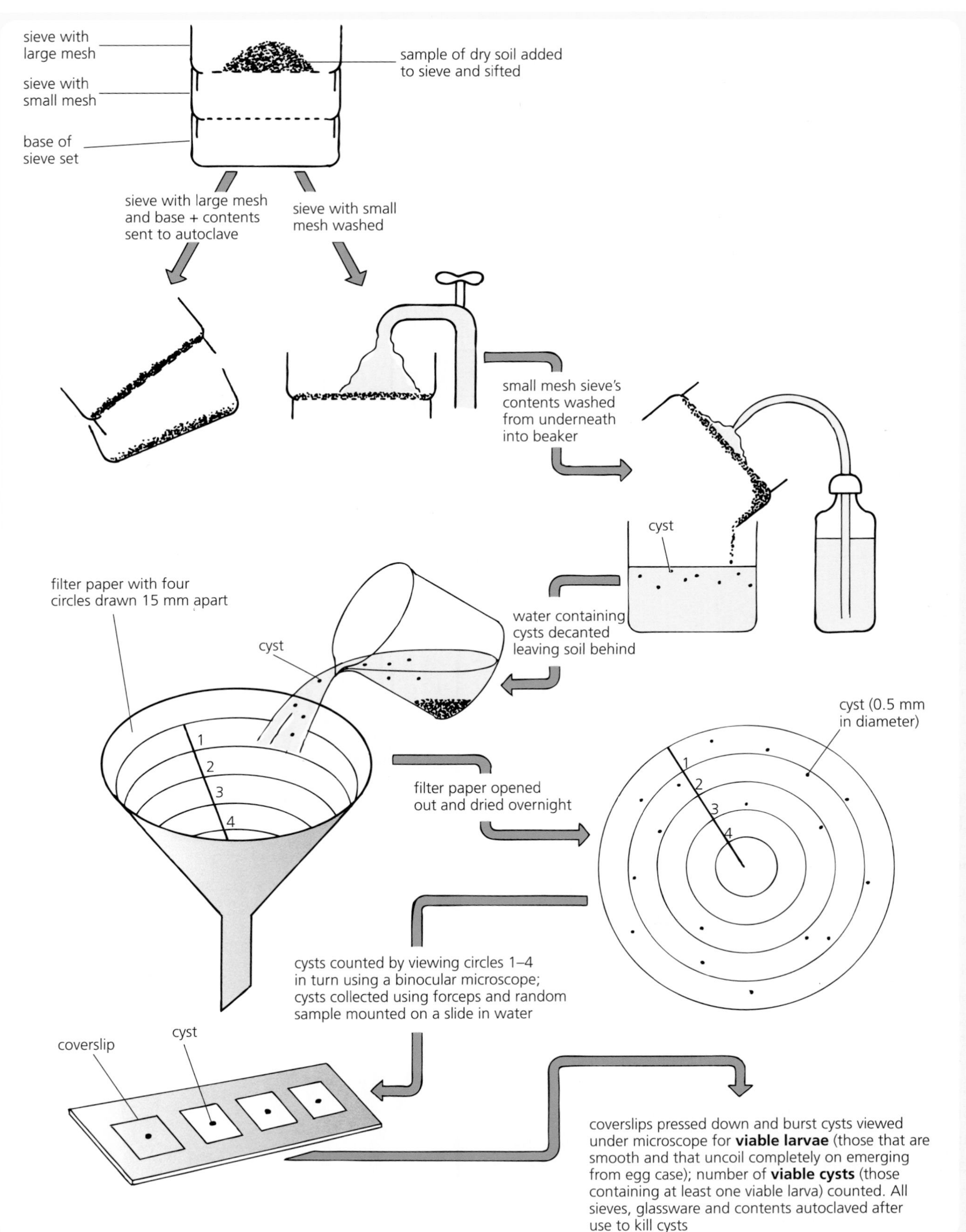

Figure 16.11 Procedure for potato cyst nematode investigation

From the results it is concluded that there is a higher incidence and viability of cysts of the potato cyst nematode in soil that has been continuously cropped with potatoes than soil that has been cropped with potatoes as part of a rotation.

If even **one viable cyst** is found when a field is tested, that land cannot be used to grow seed potatoes. This strict rule is in place because seed potatoes pose a great risk of potato cyst nematode cross-contamination from one location to another.

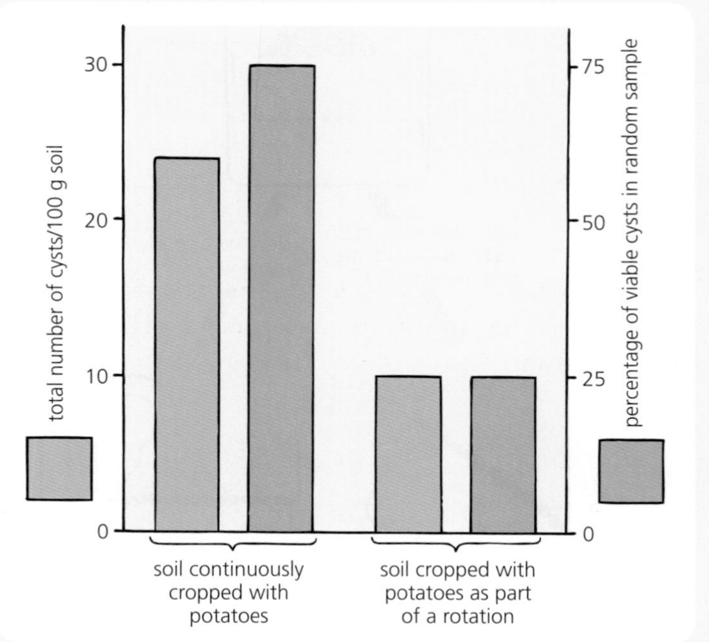

Figure 16.12 Bar chart of nematode cyst investigation results

Selective

Some herbicides act **selectively** by mimicking the action of naturally-occurring plant hormones (growth substances). They stimulate the rate of growth and metabolism of broad-leaved plants (in other words, the weeds) to such an extent that the plants exhaust their food reserves and die (see Figure 16.13). The narrow vertical leaves of the cereal plants absorb little of the chemical and are hardly affected.

Selective herbicides are similar in chemical structure to plant hormones. Therefore they are normally **biodegradable** and broken down by soil bacteria. They do not cause harm to the soil community. However, some may leave residues which enter the food chain (see page 242).

Contact

Contact herbicides act **non-selectively** and destroy all green plant tissue with which they come in contact. Since they are **biodegradable**, their effect is short-lived and they can be used to prepare an area by clearing the ground completely before the crop is sown. However, contact herbicides do not affect underground organs. Therefore well-established perennial weeds with tap

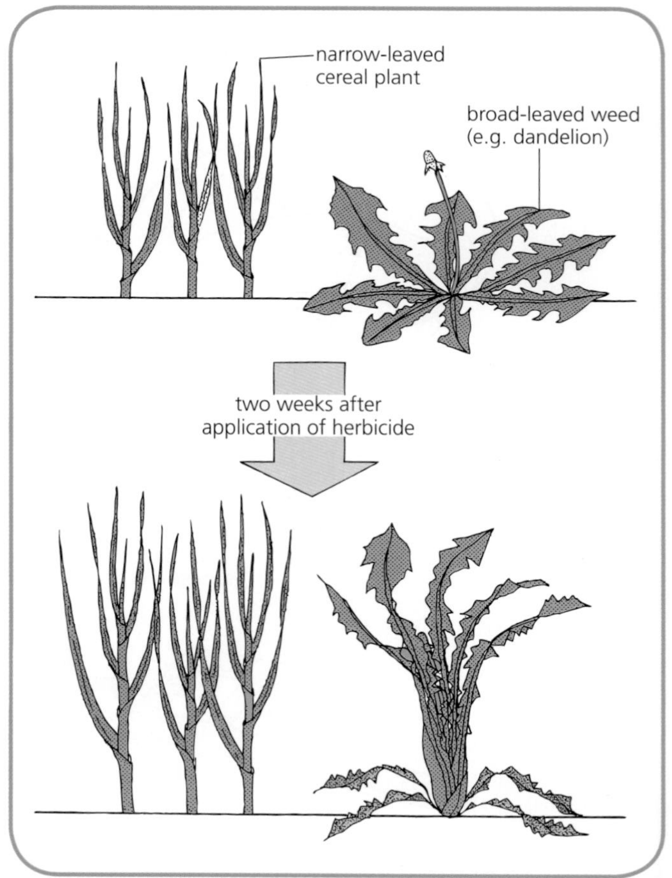

Figure 16.13 Action of selective weedkiller

roots or storage organs soon re-emerge and compete with the new young crop plants.

Systemic

Molecules of **systemic** herbicide are absorbed by an annual or perennial weed and enter its vascular system. The chemical is transported internally to all parts of the plant where it has a **lethal effect**. Although slower to act than contact herbicides, systemic herbicides are more effective because they are able to reach underground storage organs and intricate rooting systems and kill them (see Figure 16.14).

Pesticides

The examples of pesticides given in Table 16.3 refer to those used to kill three groups of common invertebrate pests. Over 60% of crop-growing farms in Scotland apply pesticides to their crops. It is estimated that almost 30% of crops would be lost in the absence of pesticides.

Contact

A **contact** pesticide acts by:

- killing the invertebrate when it comes into contact with the pest (for example, a spray acting directly on aphids)
- remaining as a protective layer of poisonous residue on the plant and taking effect when the pest comes into contact with it.

Systemic

Like a systemic herbicide, a **systemic** pesticide is absorbed by the plant and transported to all of its parts. If a sap-sucking insect such as an aphid pierces the plant's phloem, it will ingest poison along with the plant's sugary sap and die.

Fungicides

Fungicides are pesticides used to kill fungal parasites that cause diseases of crop plants.

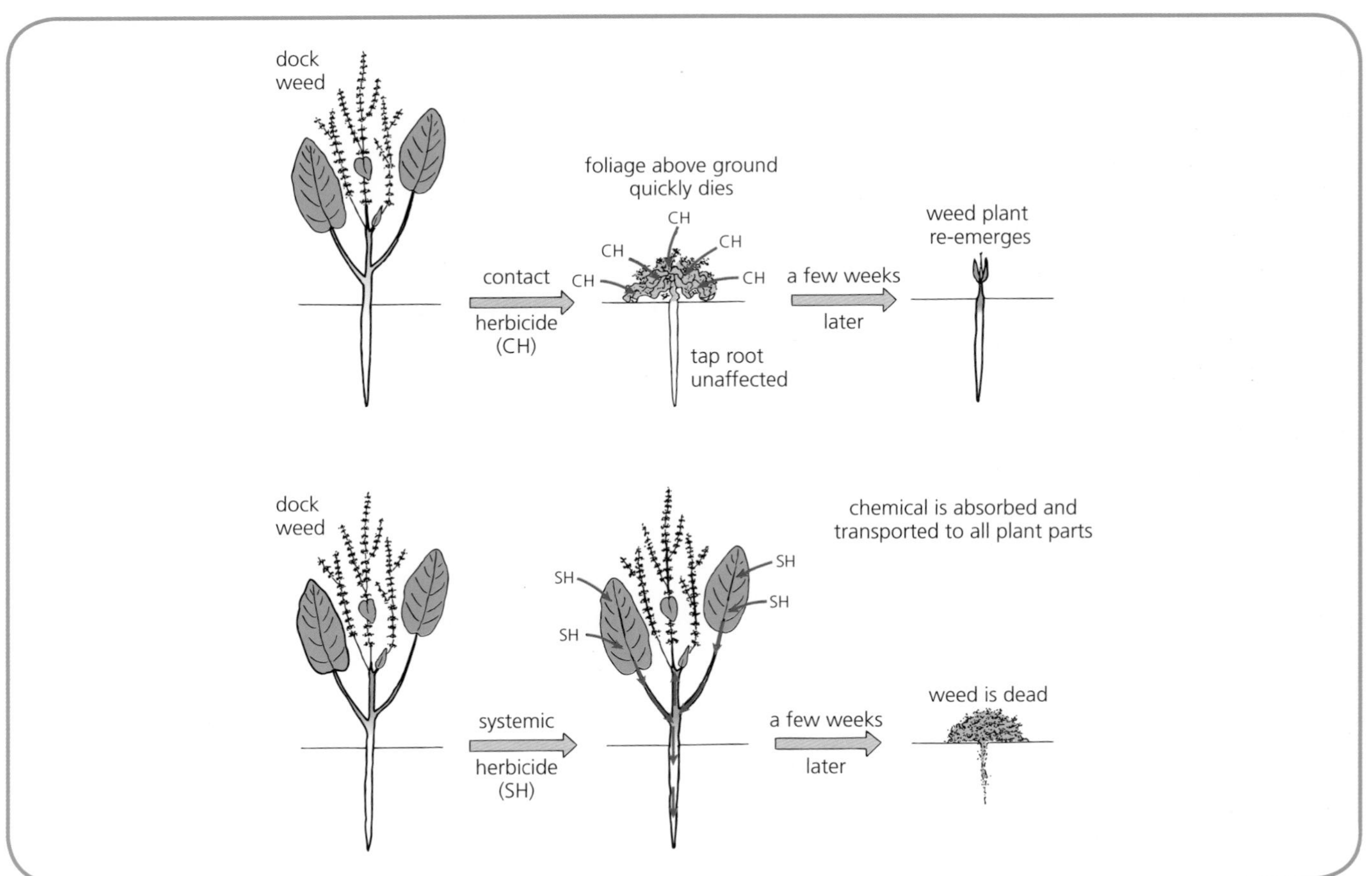

Figure 16.14 Action of contact and systemic herbicides

Pesticide	Pest targeted	Common method of application
insecticide	insects (such as greenfly)	applied as a spray to leaves
molluscicide	molluscs (such as slugs)	mixed with food bait
nematocide	nematodes (such as potato cyst nematodes)	applied as a vapour to fumigate soil

Table 16.3 Pesticides

Contact

Contact fungicide is sprayed onto crop plants prior to fungal attack (see Figure 16.15). When the fungal spores land and begin to germinate they absorb poison and die. However, rain tends to wash the chemical off and new leaves, on emerging, are unprotected and left vulnerable. Therefore repeated applications are required.

Systemic

Systemic fungicides are absorbed by the crop plant and transported throughout its body (in its vascular system). Therefore they tend to give better protection against fungal attack than contact fungicide.

Using weather data to predict disease outbreak

Potato blight is a disease caused by the fungus *Phytophthora infestans* whose spores are airborne. Infection of potato plants by the pathogen depends on the occurrence of a specific set of environmental conditions. This is called a **Smith Period** and it occurs when a minimum air temperature of 10 °C is accompanied by a relative humidity greater than 90% for at least 11 hours on two consecutive days.

It is for this reason that air temperature and humidity in all parts of the UK are monitored every hour by a blight watch service supported by the Potato Council. By this means, Smith Periods can be forecast and a warning system made available to growers. Armed with this information, farmers can spray their potato crops with fungicide **in advance** of infection. Such protective application acting as a **preventative** measure is normally much more effective than waiting until the crop is diseased and then trying to treat it. In addition, it prevents needless applications of fungicide at times when the crop is not under threat.

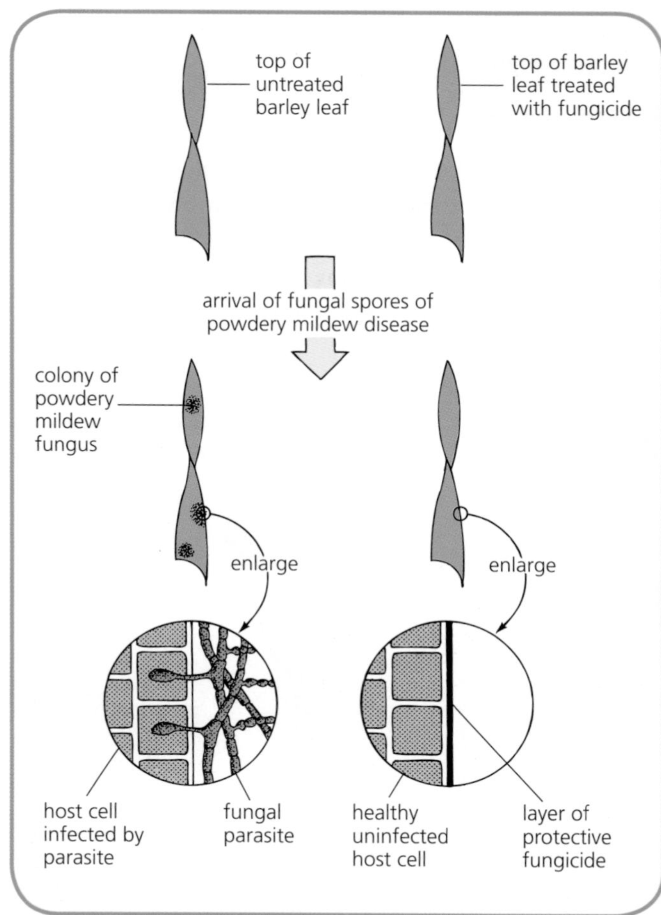

Figure 16.15 Effect of contact fungicide

Research Topic: Control of weeds, pests and diseases of wheat crops

Wheat is widely grown in Britain and needs to be protected from the activities of weeds, invertebrate pests and fungal pathogens.

Weeds

Competition with the crop by perennial weeds is reduced by **ploughing**. **Autumn sowing** of grains gives winter wheat plants an opportunity to establish themselves in a 'weed-free' environment. **Crop rotation** disrupts germination and growth of some weeds.

Invertebrates

If slugs have survived on plant debris left from the previous harvest, they often feed on newly-germinated wheat seedlings. These invertebrate pests can be controlled by cultural means such as **removing the debris** or by chemical means such as the **application of a molluscicide**. Soil-borne pests such as wheat bulb fly are, in part, controlled by crop rotation.

Fungal pathogens

Wheat crops in the UK can become infected by several fungal diseases such as mildew, stem rust (see Figure 16.16), eyespot and leaf blotch via airborne spores. Although fungal diseases rarely kill the host plant, they greatly reduce the yield of wheat grains.

Knowledge of the pathogen's life cycle

Some fungal pathogens (such as rust) spend part of their life cycle on a second host plant (for example, barberry).

Figure 16.16 Stem rust

Removal of this secondary host from agricultural areas helps to control the pathogen by making it difficult for the fungus to complete its life cycle.

Farm hygiene

Some fungal pathogens (such as eye spot and leaf blotch) survive as spores in wheat stubble and debris left on the soil surface after the harvest. They produce airborne spores capable of infecting the next crop of wheat plants especially if the weather is wet. Cultural practices that clean up the field and **reduce wheat residue**, by ploughing it back into the land or by removing it for use as animal feed, help to control the fungal diseases. In the absence of stubble, the fungus is a poor competitor in the soil. A crop rotation giving the field a two-year break from the growth of cereals normally reduces the risk of eyespot, for example, to a very low level.

Timing of nitrogenous fertiliser

A small dose of fertiliser is given in March followed by the main dose in May when growth rate increases. This pattern of application of **nitrogenous fertiliser** has been found to promote the growth of healthy wheat plants and help them to resist infection by fungi.

Cultural versus chemical

On some farms wheat is grown as part of a rotation over several years alternating with crops such as oilseed rape and beans. This wheat is often used to supplement the diet of farm animals. On other farms crop rotation has been abandoned and farmers have chosen to specialise in the **intensive cultivation** of higher-quality wheat for human consumption. However, repeated use of the same field for wheat tends to encourage fungal pathogens. Despite the use of cultivars of wheat resistant to some pathogenic strains, the farmer becomes increasingly reliant on chemical means of control.

Many farmers attempt to make judicious use of fungicide sprays by applying them only when certain risk factors prevail, such as the advance of an epidemic. However, forecasts of epidemics are imprecise and it is tempting to play safe and resort to routine chemical control. Excessive use of chemical sprays leads to short-term gain but long-term problems such as **environmental degradation** and the evolution of **fungicide-resistant** strains of fungal pathogens.

Problems with pesticides

Ideal characteristics of a pesticide

A **pesticide** should be **specific** to the pest, **short-lived** in its action and **safe**. It should not persist when released into the environment but instead break down into simple chemicals that are harmless to the host plant, the environment and the human consumer. However, in reality, some pesticides have proved to be **toxic**, to some extent, to humans and other non-target species.

Persistence of pesticide

Some protective chemicals fail to decompose into harmless substances. Instead they **persist** and affect other species adversely.

DDT

DDT is a pesticide which is now known to be both persistent and extremely toxic. Very dilute concentrations of DDT were widely used as an insecticide on crops in the past. However, the chemical was found to pass easily through food chains and webs and become more and more **concentrated** at each level. This process is called **biomagnification**. Consider the example shown in Figure 16.17. The producers (green leaves of various plant species) become contaminated with a very low concentration of pesticide blown off neighbouring farmland during spraying of crops.

The concentration increases, however, as much plant material is eaten by primary consumers and the chemical **persists** in their cells. Progression on up the pyramid of numbers leads to an ever-increasing concentration of the chemical building up in living cells. This process is called **bioaccumulation**. Finally the few large tertiary consumers (such as the sparrowhawk) at the very top suffer severe poisoning. Once this effect became apparent, use of DDT was banned by developed countries though its use continues in many developing countries.

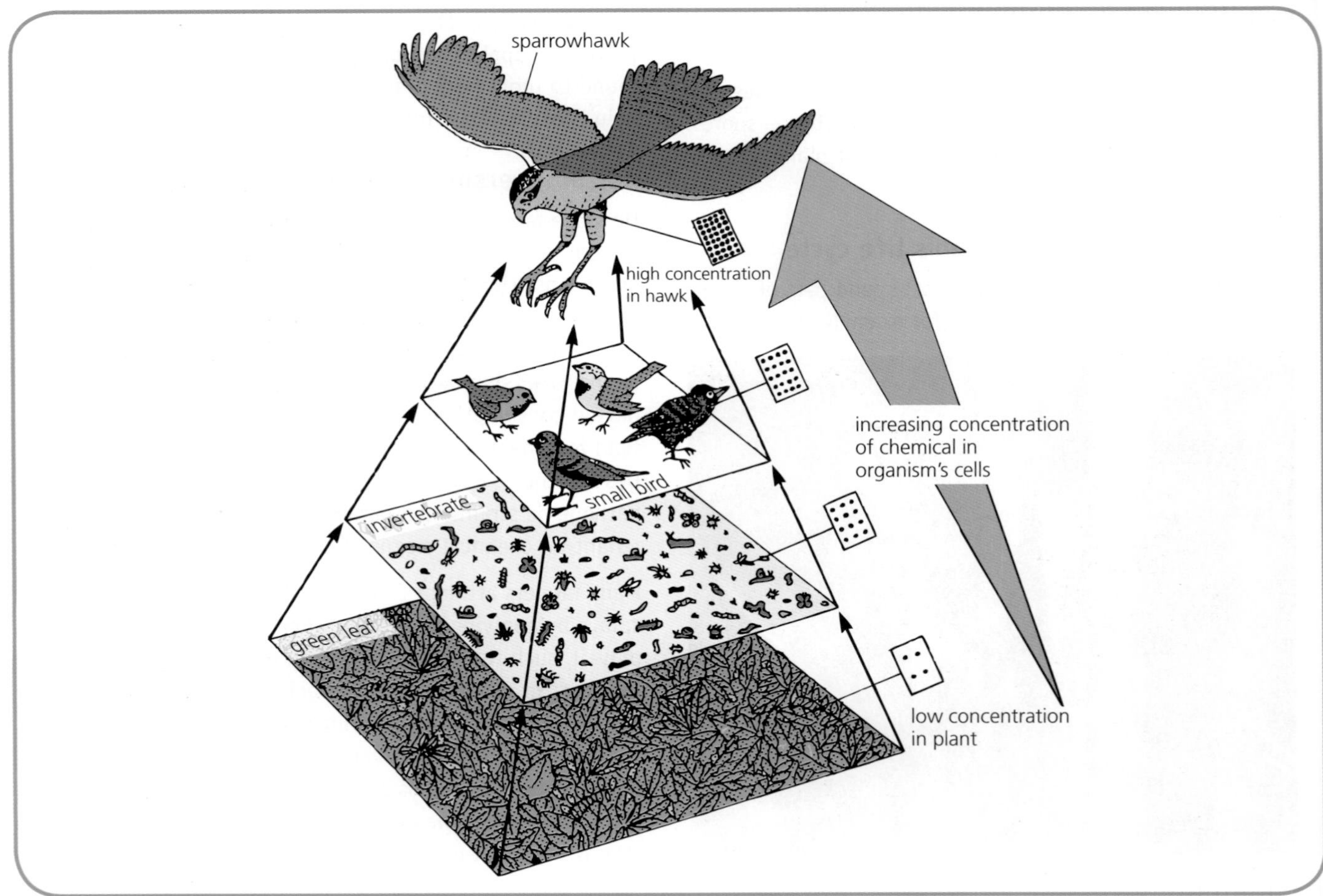

Figure 16.17 Accumulation of chemical in a food pyramid

Resistance to pesticide

When a pesticide is applied to a crop, a few individuals among the pest population may already be **resistant** to the chemical. This is because they just happen to have some unusual feature which is now of advantage to them. This could be, for example, the possession of a thick coat or the ability to produce an enzyme that breaks down the toxin.

These individuals are **naturally selected** and survive to breed the next generation which may also be resistant. As a result, the number of resistant individuals in the population increases generation after generation. Continued use of the pesticide exerts a **selection pressure** and a population of resistant pests is produced.

Biological control

Biological control is the reduction of a pest population by the deliberate introduction of one of its natural enemies. This could be:

- a **predator** (for example the ladybird, an insect that feeds on greenfly – see Figure 16.18)
- a **parasite** (such as *Encarsia*, a wasp that lays its eggs inside a whitefly; the developing wasp larvae feed on the insect host and destroy it – see Research Topic on page 244)
- a **pathogen** (such as *Bacillus thuringiensis*, a bacterium that infects butterfly caterpillars and kills them with a poison called Bt toxin – see Case Study on pages 244–5).

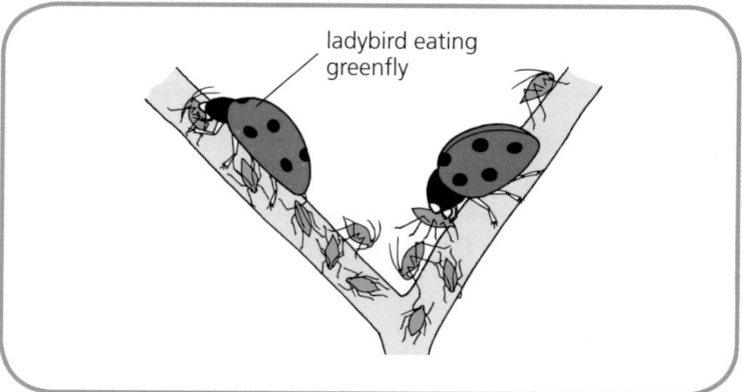

Figure 16.18 Biological control of greenfly by ladybirds

The natural enemy acts as the **control agent** in place of a pesticide. Therefore biological control does not introduce persistent and potentially harmful chemicals to the food chain. In addition, this type of control does not exert a selection pressure that could produce a population of resistant pests.

Timing

In biological control the **timing** of the introduction of the agent of biological control is critical. The predator must be able to find its 'prey' or it will be ineffective. Therefore the agent is normally introduced when:

- infestation of the crop has begun
- environmental conditions are present that favour the establishment of the useful predator, parasite or pathogen.

Glasshouse

A heated **glasshouse** ('greenhouse') provides an environment that can be controlled carefully during the growth of a crop (such as tomatoes or cucumbers). Biological control is very effective in a glasshouse because:

- the system is enclosed so the predator is unable to move away to an alternative environment
- the temperature can be controlled to suit the control agent (for example, control of red spider mites using predatory mite *Phytoseiulus* (see pages 245–6) normally works well when the minimum temperature of 20 °C needed by the predator to reproduce rapidly is maintained).

Risks of biological control

Sometimes biological control of a pest leads to **unintended negative consequences**. The control organism may eventually cause problems for one of the following reasons.

- It may **prey** on a species other than the pest. For example, many years ago the mongoose was introduced to Hawaii to try to control an ever-increasing rat population. Unfortunately the mongooses preferred to prey on the local birds and failed to eat any of the rats.

Research Topic: Control of glasshouse whitefly with the parasite wasp *Encarsia*

Glasshouse whiteflies are tiny insects about 2 mm in length and closely related to greenfly. They establish themselves in greenhouses where tomato or cucumber plants are being cultivated. Whiteflies weaken the crop plant by sucking its sap in the same way aphids do. They also cover the plant surface with excreted 'honeydew' which may promote the growth of moulds. They take in sap greatly in excess of their need for carbohydrate because they also need to obtain amino acids which are much less concentrated in the sap. Each female whitefly lays up to 200 eggs in a circle on the underside of a leaf. The larvae that develop are scale-like and are often referred to as 'scales'.

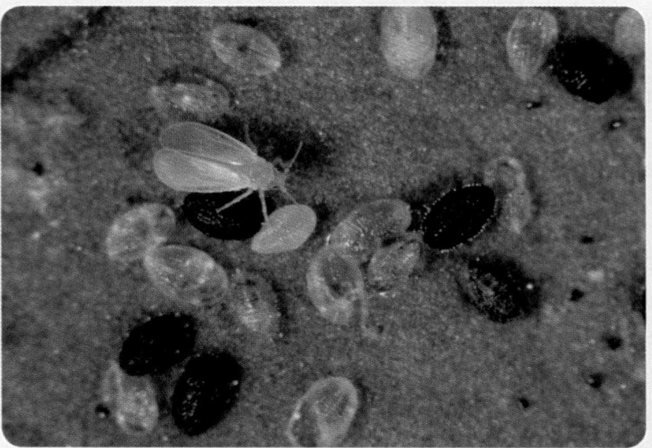

Figure 16.19 Whitefly and *Encarsia* wasp

Control

Glasshouse whitefly can be **controlled biologically** by a minute parasitic wasp called *Encarsia*. The female wasp lays her eggs inside the larvae of the whitefly (see Figure 16.19). Those larvae that are successfully parasitised by the wasp turn black, usefully indicating the effectiveness of the treatment.

To introduce the wasp, special cards bearing whitefly scales already parasitised by wasp pupae are purchased from a supplier and hung up in shady positions among the foliage in the greenhouse. Adult *Encarsia* wasps emerge from the pupae and lay their eggs in any whitefly scales present in the crop plant's leaves. Since whitefly and *Encarsia* are native to tropical regions, the biological control is only effective if the glasshouse temperature is maintained at a minimum of 18 °C during the day and 14 °C at night.

- It may act as a **pathogen** or as a **parasite**. For example, when the parasitic cactus moth was introduced to South Africa to control unwanted populations of prickly pear cactus, the moth larvae destroyed much of a related type of cactus traditionally used as animal fodder.
- It may become an **invasive species** (also see page 296). For example, the cane toad (see Figure 16.20) introduced to Australia failed to control cane beetles on sugar cane plants but outcompeted other native species. As a result it has become so numerous that it is harming the environment.

Figure 16.20 Cane toad

Case Study: Control of butterfly caterpillars with bacterium *Bacillus thuringiensis*

Bacillus thuringiensis is a naturally occurring soil bacterium. Some strains can infest and kill plant-eating insects (see page 223). When *Bacillus thuringiensis* (Bt for short) produces spores, it makes a protein that is toxic to certain insects. The toxin works by paralysing the insect's digestive system. As a result the animal stops feeding and dies of starvation several days later.

Different strains of Bt make different forms of the toxin, each specific to a particular group of insects. For example Bt *kurstaki* only works against

→

leaf- and needle-feeding caterpillars. This strain of Bt is used as an insecticide spray containing a mixture of bacterial spores and crystalline toxin (see Figure 16.21) to protect crops against butterfly caterpillars (see Figure 16.22).

Advantage

This form of Bt insecticide is environmentally friendly in that it is specific to caterpillars and has very little or no effect on beneficial insects and other living organisms.

Disadvantages

Most versions of Bt insecticide are degraded by sunlight and only remain effective for a few days. Constant exposure of a crop to the same toxin creates a selective pressure which favours any pests that are resistant to the toxin.

Figure 16.21 Bt spores and crystals of toxin

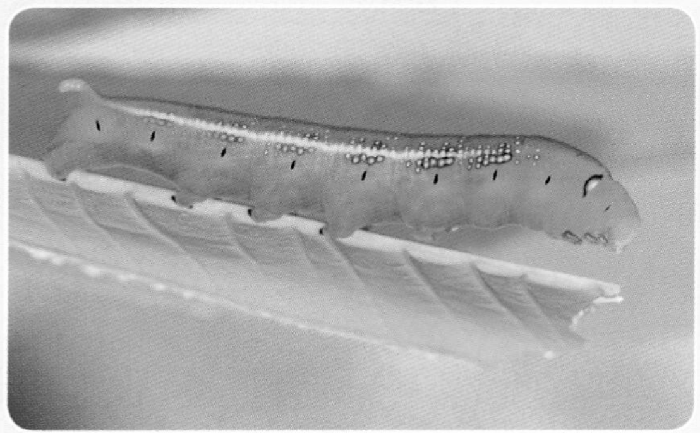

Figure 16.22 Leaf-eating caterpillar

Investigation

Biological and chemical control of red spider mites

Background

- Glasshouse **red spider mites** (see Figure 16.23) are tiny invertebrates that feed on many different species of plant including some that are grown for food.
- Red spider mites pierce cells on the underside of the leaf with their mouthparts to obtain food.
- A female adult lives for several weeks and lays hundreds of eggs.
- A second type of mite, called *Phytoseiulus*, is longer than the red spider mite and has a shiny, orange-red body.
- It is a **predator** and its long legs enable it to move around the leaves quickly, in search of prey.
- A **miticide** is a type of pesticide developed specifically for use against mites.

The investigation is carried out by following the procedure outlined in Figure 16.24 (see page 246).

From the results it is concluded that biological control has been effective. *Phytoseiulus* has predated on the red spider mites in tray A and reduced their number significantly. The results from tray B are not found to differ significantly from those in tray C (the control).

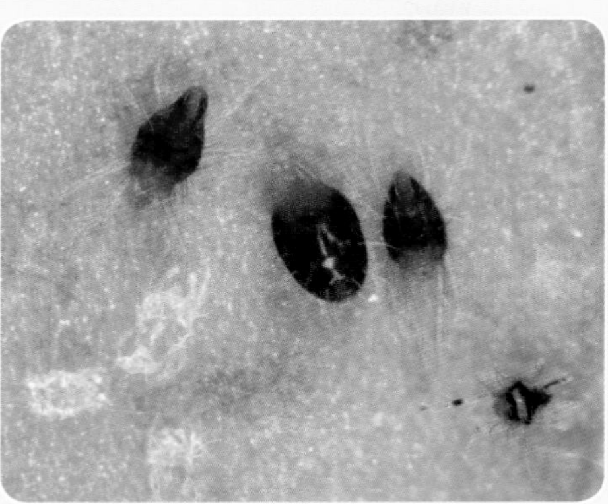

Figure 16.23 Red spider mite

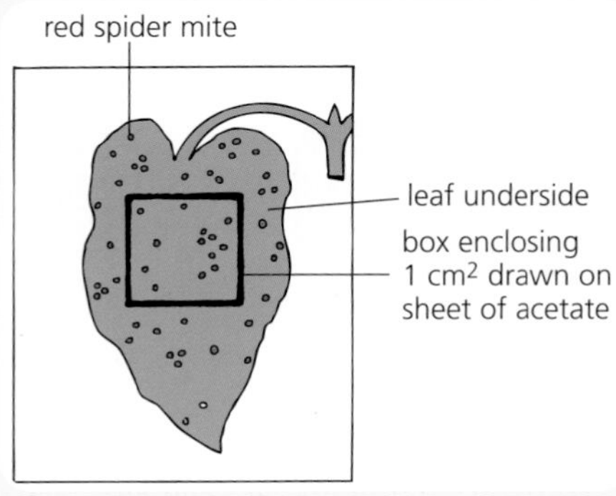

Figure 16.25 Sampling technique

Therefore it is concluded that in this case, chemical control has not been effective, probably because the population of red spider mites was resistant to this type of miticide.

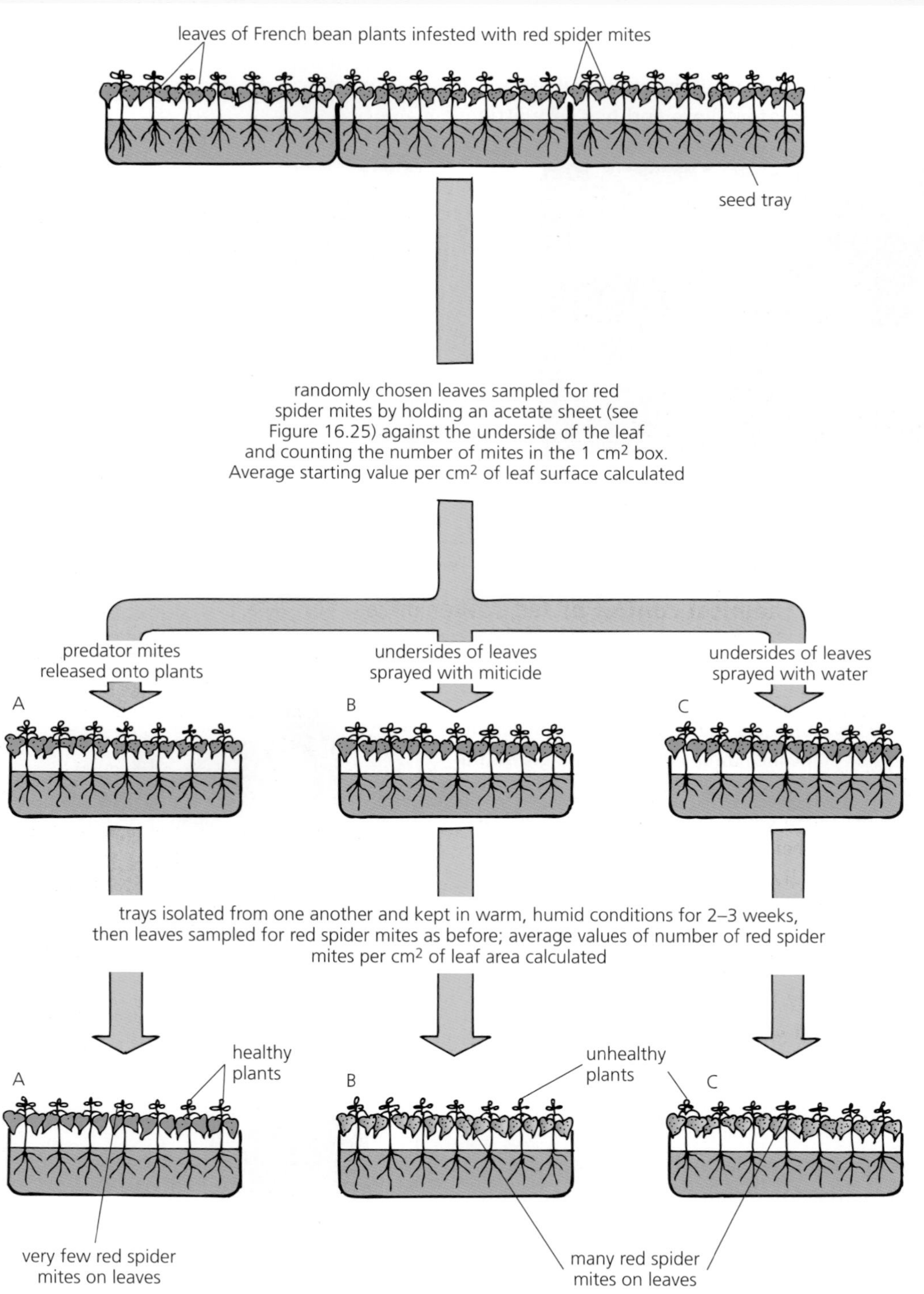

Figure 16.24 Investigating biological and chemical control

Integrated pest management

Integrated pest management (IPM) is a combination of techniques including chemical control, biological control, cultural means and host plant resistance. It sets out to reduce the use of pesticides while bringing pest populations down to a level at which they no longer cause economic damage. The emphasis is on **control** rather than eradication of the pest.

Cautious use of chemicals

Where chemicals need to be used, they are chosen because:

- they are effective even when only used infrequently at specific points in the host's or pest's life cycle
- they show low persistence
- they reduce pest numbers to a level at which methods of biological control can take over
- they do not disrupt biological control since they are selective and kill pests while leaving the useful predators unharmed.

Future

Ideally a situation would be reached where the target species would neither die out nor increase in number to become a pest. It would simply remain under control while at the same time the crop species and the surrounding wildlife would be unaffected.

Testing Your Knowledge 2

1 a) Agricultural pests can be controlled to some extent by *cultural means*. What does this term mean? (1)
b) Crop rotation is a cultural means of control. Briefly explain how it works. (3)
c) Name TWO other cultural means of controlling pests. (2)

2 a) What is the advantage of using a selective weedkiller on a cereal crop? (1)
b) Explain why a *systemic* fungicide may be more effective than a *contact* one on a crop. (2)

3 a) Redraw Figure 16.26 which shows a pyramid of numbers for a food chain and add the following labels: *primary consumer, secondary consumer, tertiary consumer, producer*. (1)

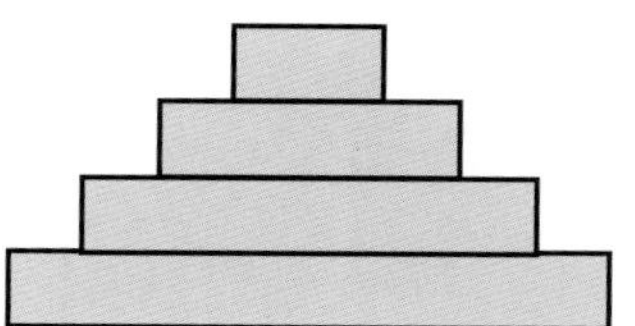

Figure 16.26

b) i) Using dots to represent molecules of persistent pesticide, complete your pyramid.
ii) Explain the distribution of dots you have chosen for your diagram. (3)

4 Copy the following sentences, choosing the correct answer at each underlined choice. (8)
a) Continued use of pesticide exerts a reduction/selection pressure on a population producing a resistant/susceptible population of pests.
b) The reduction of a crop/pest population by the deliberate introduction of one of its natural enemies is called biological/chemical control.
c) The use of a combination of techniques such as chemical and biological control and host plant persistence/resistance is called cultural/integrated pest management.
d) The form of management referred to in c) aims to make maximum/minimum use of chemicals on the farm and to control/eradicate pests.

Applying Your Knowledge and Skills

1 The seven pairs of characteristics listed below refer to a certain **weed species** and to a related **non-weed species.** Construct a table with these two emboldened terms as its headings. Consider each pair of characteristics and then enter each member of the pair in the appropriate column in the table. (Keep in mind that a weed needs to be an opportunist to survive.) (7)

- can grow well on poor soil/needs fertile soil to grow well
- requires short days to flower/able to flower in any day length
- quick to flower and produce many tiny seeds/slow to flower and produce a few large seeds
- self-pollinated/cross-pollinated
- intolerant of drought/tolerant of drought
- tolerant of water-logged soil/intolerant of water-logged soil
- long life cycle/short life cycle

2 'Seed' potatoes are grown and retained to generate future crops. 'Ware' potatoes are grown for consumption and are not retained for further use. Ware potatoes may be grown anywhere but seed potatoes may only be grown on land that has been tested for potato cyst nematodes (PCN) and has been given a clearance certificate. The chart in Figure 16.27 outlines the test in a simple way.

a) Match the following four test outcomes with boxes P, Q, R and S in the chart. (3)

i) Soil may be used for ware potatoes and varieties of seed potato resistant to PCN.

ii) Soil may not be used for seed potatoes and its use for commercial growth of ware potatoes is not advised.

iii) Soil may be used for ware potatoes *and* seed potatoes.

iv) Soil may not be used for seed potatoes but may be used for ware potatoes.

b) Damage to potato crops by PCN is patchy. It tends to be most severe where the soil is very wet or very dry and the growing conditions for the potato plants are stressful. Suggest how PCN damage could be reduced in these areas without using protective chemicals. (2)

c) Hatching of viable cysts can be triggered by a chemical given out by the roots of potato plants in spring. If scientists could find a way of

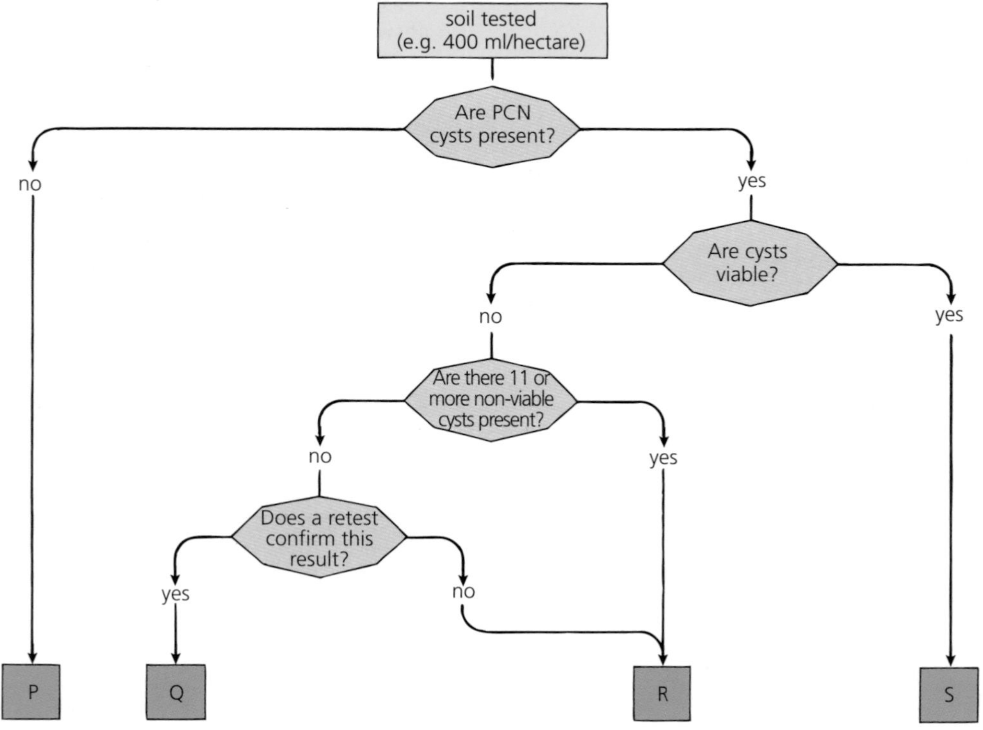

Figure 16.27

synthesising this chemical, suggest how it might be used to control PCN. (2)

d) i) Which would be a more effective means of controlling PCN: a crop rotation of 4 years or of 7 years?

ii) Explain your answer. (2)

3 An investigation was carried out on the effect of concentration of nitrogenous fertiliser and time of weeding on the grain yield of a cereal crop. Table 16.4 shows the results.

a) i) What was the effect of added nitrogen on yield of cereal grain?

ii) Account for this effect. (2)

b) i) What was the effect of time of weeding on yield of grain?

ii) If the farmer could only afford to weed the field once, when would be the best time to do it?

iii) Explain your choice of answer to **ii)** with reference to competition. (4)

c) Calculate the percentage increase in yield that the addition of $5\,kg\,ha^{-1}$ of fertiliser brought about for field 1 compared with the addition of no fertiliser. (1)

d) Suggest TWO factors in addition to nitrogen that the crop plants and the weeds may be competing for. (2)

e) 1 hectare (ha) $= 10\,000\,m^2$. Convert $3400\,kg\,ha^{-1}$ into $g\,m^{-2}$. (1)

4 Figure 16.28 shows a graph of the results from an investigation into the effect of two fungicides on potato plants.

a) How many days did it take for the untreated potato plants to show a level of 75% infection? (1)

b) By how many did the number of infected plants treated with contact fungicide exceed the number of infected plants treated with systemic fungicide 33 days after spraying? (1)

c) By how many times was the number of untreated infected plants at day 21 greater than the number of infected plants treated with systemic fungicide? (1)

d) Among plants treated with contact fungicide, what percentage increase in number of infected plants occurred between day 21 and day 35? (1)

e) i) Which type of fungicide is more effective?

ii) Explain your choice of answer. (2)

5 A wide variety of fungi attack crop plants. A few are listed in Table 16.5.

Such fungal attack can often be prevented by spraying the surface of the potential host plant with fungicide. This reduces the number of fungal spores which germinate. A chemical company set out to investigate the effectiveness of two new fungicides (A and B). Petri dishes of nutrient agar containing fungicide were inoculated with fungal spores. Table 16.6 (page 250) shows the results. Table 16.7 (page 251) gives further information about the two fungicides.

Field number	Time during growing season when field was weeded	Yield of cereal grain ($kg\ ha^{-1}$)		
		Mass of nitrogenous fertiliser added ($kg\ ha^{-1}$)		
		0.0	0.5	5.0
1	all season	2576	2981	3400
2	early season only	2243	2349	3032
3	mid-season only	2065	2197	2842
4	late season only	1472	1774	2087

Table 16.4

Scientific name	Symbol for easy reference	Disease caused	Host plant affected
Erysiphe graminis	W	mildew	wheat and barley
Urcinula necator	X	mildew	grapevine
Puccinia graminis	Y	black rust	wheat
Phytophthora infestans	Z	blight	potato

Table 16.5

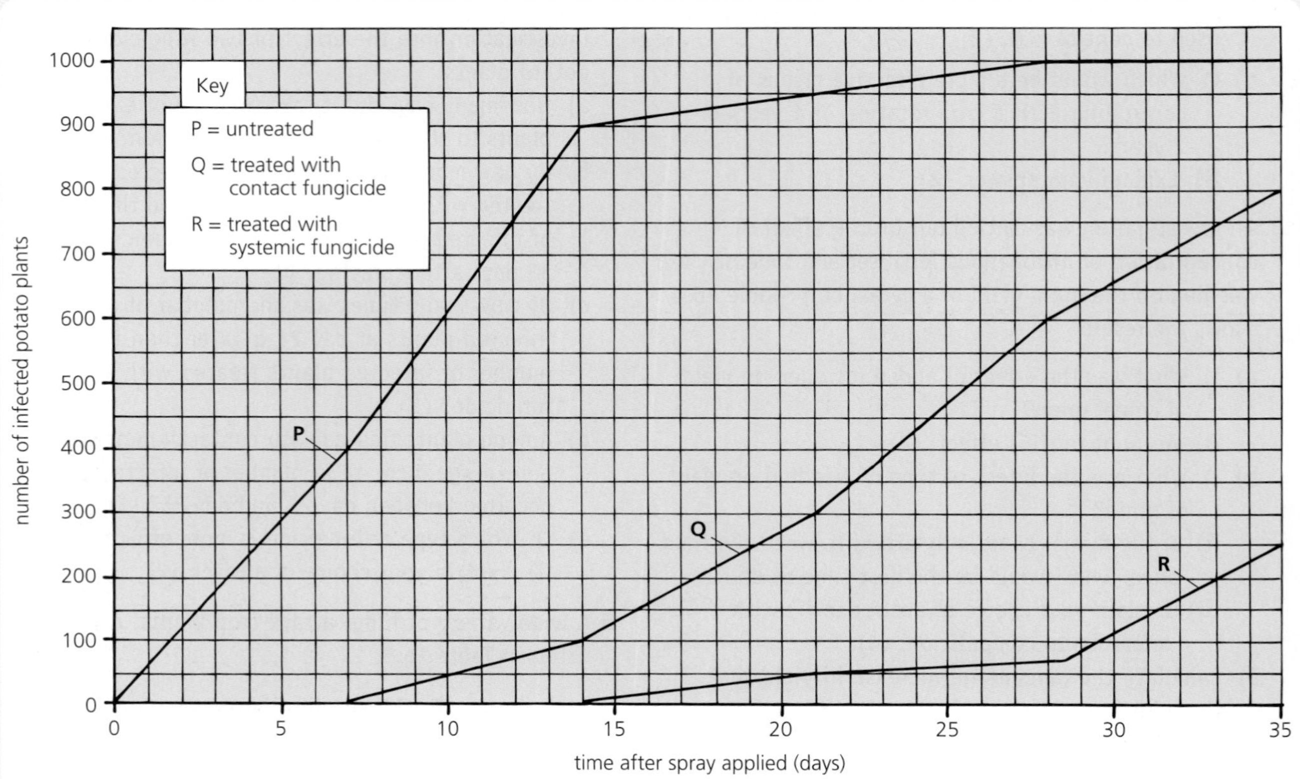

Figure 16.28

	Concentration of fungicide (ppm)	Plate	% germination of fungal spores			
			Fungus W	Fungus X	Fungus Y	Fungus Z
Control	0	1	100	97	77	44
		2	99	100	74	41
		3	100	100	76	43
Fungicide A	5	1	20	5	76	45
		2	23	9	75	41
		3	22	7	76	43
	50	1	0	0	49	44
		2	0	0	48	42
		3	0	0	49	43
Fungicide B	5	1	100	60	76	43
		2	97	62	76	42
		3	100	61	77	44
	50	1	18	5	76	42
		2	17	7	78	43
		3	19	6	74	44

Table 16.6

	Fungicide A		Fungicide B	
Concentration (ppm)	5	50	5	50
Toxic effect on wildlife	none	very slight	very slight	slight
Effect on host plant (% reduction in yield)	0	3	0	1
Projected cost of spray	cheap	fairly cheap	very cheap	cheap
Biodegradable?	yes		yes	

Table 16.7

a) Give the scientific names of TWO fungal pathogens to which wheat can play host. (1)

b) Why was each condition of the experiment set up in *triplicate*? (1)

c) Summarise, in turn, the effect of fungicides A and B on fungus Z. (1)

d) Summarise, in turn, the effect of fungicides A and B on fungus Y. (1)

e) Which fungicide had the greater overall effect on:

i) fungus W?

ii) fungus X?

iii) Which concentration of which fungicide prevented all of fungus W's spores from germinating? (2)

f) i) State the average percentage germination of spores from fungus X under control conditions.

ii) Calculate the reduction in percentage germination of spores caused by applying 5 ppm of fungicide B to fungus X. (2)

g) Experts claim that fungicides A and B tend to show a degree of specificity in their action. Justify this claim with reference to the data. (1)

h) The chemical company running the trials found that they could only afford to continue with one of the fungicides. Suggest TWO reasons why they chose A in preference to B. (2)

6 Dieldrin is a type of pesticide which was used in Britain in the 1950s to dress wheat grain against attack by insects. During each year of its use, thousands of seed-eating birds were found to be poisoned. In the 1960s a severe decline in number of peregrine falcons occurred (see Figure 16.29) and many of the survivors were found to have high concentrations of dieldrin in their bodies. The females produced eggs with thinner shells.

a) Construct a food chain to include the organisms named above. (1)

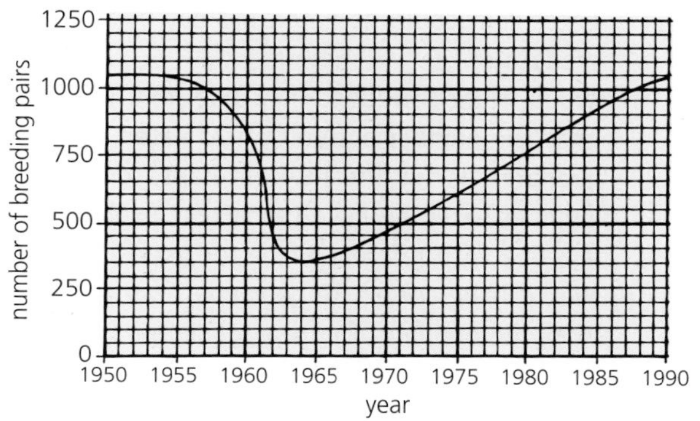

Figure 16.29

b) Low concentrations of dieldrin were used to dress the grain, yet high concentrations were found in the falcons' bodies. Explain fully how this difference arose. (2)

c) During which of the following intervals of time did the number of breeding pairs decrease at the fastest rate? (1)

A 1956–58 **B** 1958–60 **C** 1960–62 **D** 1962–64

d) Calculate the percentage decrease in breeding pairs that occurred between 1957 and 1964. (1)

e) Why does production of thinner egg shells lead to a reduction in number of future breeding pairs? (1)

f) Account for the trend shown in the graph from 1965 onwards. (1)

7 For successful biological control of red spider mites (M) in a glasshouse, the suppliers of *Phytoseiulus* (P), the predator, recommend the use of 200P to deal with a population of 4000M.

a) Express these data as a simple whole number ratio of P : M. (1)

→

b) How many P should be ordered to deal with a population of 32 000M? (1)

c) The bar chart shown in Figure 16.30 includes error bars that each represent the 95% confidence level of the mean. (See Appendix 3 for help.) The chart represents the results of an investigation into the numbers of red mite prey (adults and eggs) eaten by *Phytoseiulus*, the predator.

i) What is the mean number of mite eggs eaten by female predators?

ii) What can be stated (with a 95% level of confidence) as the minimum number of adult mites that could be eaten per day by male predators?

iii) Would it be correct to conclude from the data that the number of adult mites eaten by female predators is significantly different from the number eaten by male predators? Explain your answer, with reference to error bars.

iv) Would it be correct to conclude from the data that the number of mite eggs eaten by female predators is significantly different from the number eaten by males? Explain your answer with reference to error bars. (6)

d) Why is this form of biological control not suitable for use on plants grown outside in the open air? (1)

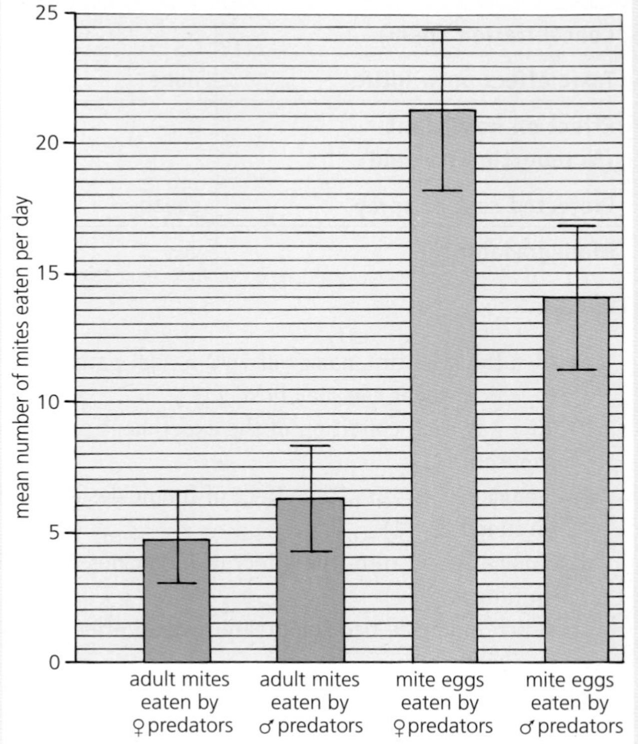

Figure 16.30

8 Give an account of crop protection under the following headings:

a) control by cultural means (3)

b) chemical control (4)

c) biological control. (2)

17 Animal welfare

Wellbeing of domesticated animals

Traditionally the **wellbeing** of domesticated animals was judged solely on their **physical health** as indicated by their ability:

- to grow
- to reproduce and raise offspring successfully
- to resist disease.

However, this narrow view of animal wellbeing has been overtaken in recent times by the work of the **Farm Animal Welfare Council** (see Research Topic – The five freedoms for animal welfare). It is now agreed that an important part of an animal's welfare is the provision of opportunities for the animal to **express its normal, natural behaviour patterns.**

Hens, for example, given adequate space to stretch their wings and wag their tails, take far less time to settle and begin egg-laying than those that are crowded and short of space. An animal's state of wellbeing is therefore regarded as acceptable only if the animal is able to behave in a natural way, live free from disease and grow vigorously.

Costs and benefits

Providing domesticated animals with improved environmental conditions requires an initial financial outlay by the farmer. However, the increased costs of an improved level of welfare result in **long-term benefits.** Contented, unstressed animals (see Figure 17.1) grow better, breed more successfully and generate products (meat, milk, eggs etc.) of higher quality.

The opposite is true of animals that are stressed. Research in pigs, for example, shows that pregnant sows exposed to stress, such as competition for food, produce piglets with a slower rate of growth after being weaned. In addition, these young pigs in turn show poor maternal behaviour towards their own offspring.

The standards of animal welfare maintained in the UK are among the highest in the world but they add to the cost of the produce. It is for this reason that animal products from farms in Britain are normally more expensive than those imported from countries with lower standards of animal welfare.

Figure 17.1 High level of welfare

Ethics

Ethics refer to the moral values and rules that ought to govern human conduct. Clearly it is unethical to subject domesticated animals to a regime of negative experiences such as pain and distress simply to provide humans with supplies of cheap food.

It is hard to believe that chickens raised in the overcrowded conditions of a battery farm (see Figure 17.2), and with their beaks clipped after hatching using a laser to prevent them from injuring one another, are enjoying anything other than a relatively low quality of life.

Figure 17.2 Battery chicken farming

Research Topic: The five freedoms for animal welfare

Farm Animal Welfare Council (FAWC)

The **FAWC** is an independent advisory body set up by the government to review the welfare of farm animals (on farmland, in transit, at the market and at the slaughterhouse). The FAWC advises the government on changes to the law that it considers to be necessary. The Animal Welfare Act 2006 requires that an animal's needs are met. These needs are based on the five freedoms for animal welfare (see Table 17.1) identified by the FAWC. These freedoms place great emphasis on the **avoidance of negative experiences.**

The work of the FAWC continues to press for ever more **humane treatment** of farm animals. It advises farmers to move beyond the minimum standard to the provision of one where the animals have a life worth living during their relatively short life time. This could be achieved by including positive experiences such as opportunities for exercise and play within the overall programme of animal welfare.

Freedom:	Means by which this freedom should be delivered
from hunger and thirst	provision of access to water and to a diet that maintains full health and vigour
from discomfort	provision of an appropriate environment that includes shelter and an area for resting comfortably
from pain, injury and disease	application of preventative measures and, where necessary, rapid diagnosis and treatment of the problem
to express normal behaviour	provision of adequate space, proper facilities and company of other members of the animal's own kind
from fear and distress	provision of conditions and treatment that avoid causing the animal mental suffering

Table 17.1 The five freedoms for animal welfare

Free range versus intensive farming

Compared with free range farming, intensive farming is more cost effective but less ethical owing to the poorer quality of welfare received by the intensively farmed animals. These two different methods of farming are further compared in Table 17.2.

Relative difference	Intensive farming	Free range farming
quantity of land required	less	more
quantity of human labour needed	less	more
quality of animal welfare	often poorer	normally better
quality of animal's life	lower	higher
price at which final goods can be sold	lower	higher
profit generated	higher	lower
ethical value	lower	higher

Table 17.2 Comparison of two farming methods

Behavioural indicators of poor welfare

When animals are kept confined in unnatural or substandard conditions, they often exhibit behaviour patterns that differ from those shown in a natural environment. These unusual forms of behaviour act as **indicators** of poor animal welfare and involve factors such as ill health, stress and a general lack of wellbeing.

Stereotypy

A **stereotypy** is a behaviour pattern that takes the form of **repetitive movement** often lacking variation. At first sight it may appear to the onlooker to lack purpose. A stereotypy is often displayed by an animal housed in bare and/or confined quarters. Pigs in small pens, for example, often make continuous chewing movements without having food in their mouths, many cattle tethered in stalls are found to roll their tongues continuously (see Figure 17.3) and some animals in small enclosures pace the same monotonous path endlessly.

It could be argued that such behaviour does have a purpose in that it enables the animal to express frustration and unconsciously communicate its need for sensory stimulation. In addition, pacing back and forth in a confined space exercises muscles, generates heat energy and releases endorphin hormones that block pain. Some experts claim that rather than being abnormal, this form of behaviour indicates a natural response to confinement in an unnatural environment.

The incidence of stereotypies is greatly reduced when the animal's environment is **enriched** by including features present in its natural habitat and by increasing the size of its environment to supply space for normal, natural exercise.

Figure 17.3 Stereotypy

Misdirected behaviour

When a normal pattern of behaviour is directed inappropriately towards the animal itself, another animal or its surroundings, this is described as **misdirected behaviour**. It is a common occurrence among animals that are confined or kept in isolation. Such an animal may **mutilate** itself by excessively licking, plucking or chewing its own feathers, hair or even limbs. Many birds in confinement overgroom themselves to a damaging level; some performing monkeys have even been known to chew off their own tail.

Misdirected behaviour of one animal towards another may occur in response to boredom or stress. Chicks raised in cages without access to suitable flooring material for foraging will peck one another's feathers and skin (see Figure 17.4). Since fowl are attracted to damaged feathers, this can lead to a rapid spread of injurious pecking throughout a flock of chickens. This activity is thought to be misdirected foraging behaviour.

Figure 17.4 Chicken's skin exposed by inappropriate pecking

Behaviour may be misdirected towards the animal's surroundings. This may take the form of excessive sucking of inanimate objects (especially among mammals separated very early from their mother), chewing cage bars (see Figure 17.5) and gnawing any solid object available.

Misdirected behaviour can be reduced by enriching the animal's environment. One way of doing this is to provide the animal with companions of its own type

Figure 17.5 Misdirected behaviour

in a large, stimulating enclosure that includes objects, sounds and scents found in the animal's natural environment.

Failure in sexual or parental behaviour

Poor welfare and prolonged isolation can be responsible for the failure of animals to reproduce successfully. On the other hand they may produce young but then **reject** them and fail to act as effective parents. This affects the farmer whose livelihood depends on the production of generation after generation of livestock.

During their early development young mammals and birds need **social contact** with members of their own kind in a spacious environment in order to develop into normal healthy adults capable of successful reproduction. When they are given these conditions, reproductive failure is normally overcome.

Level of activity

If an animal displays a very high level of activity such as **hyper-aggression** (see Figure 17.6) or a very low level of activity such as **excessive sleeping** this indicates that it may be suffering as a result of poor welfare. Very high levels of activity are also referred to as **hysteria**; very low levels of activity as **apathy**.

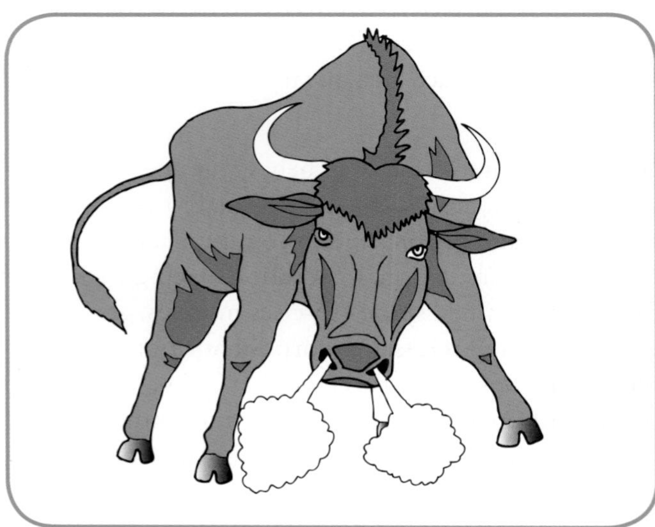

Figure 17.6 Hyper-aggression

Testing Your Knowledge

1 a) Why does it pay in the long run to invest money in facilities and accommodation that will improve the welfare of farm animals? (1)

b) Some people claim that it is unethical to eat meat and eggs from chickens raised in battery farms. Explain this point of view. (1)

2 a) Which of the forms of behaviour listed below is an example of:

i) stereotypy

ii) misdirected behaviour

iii) failure in parental behaviour

iv) altered level of activity? (4)

A a sow in a small enclosure smothering her newborn litter

B a mare repeatedly biting the door-frame of her small stall

C a bull alone in a small field charging aggressively at a fence post

D caged game birds pecking one another's feathers and causing injury

b) In general, what could be done in an attempt to eliminate these forms of undesirable behaviour from future generations of domesticated animals? (2)

Applying Your Knowledge and Skills

1 In the UK a small percentage of dairy cattle are housed all year round with little or no access to grazing. This makes the management of the cows easier for the farmer but it begs the question: *'Do these cows enjoy a satisfactory standard of welfare?'*

Here are some comments made about this practice:

1) The cows are protected from adverse weather.

2) Insufficient space is available for the cows to move around freely.

3) The risk of injury and lameness is increased by hard, concrete flooring.

4) There is a lower risk of exposure to airborne diseases and TB from badgers.

5) The composition of feed can be carefully controlled to suit the cows' requirements.

6) The cows are denied the opportunity to walk on soft, non-slip pasture.

7) The risk of infestation by parasites is greatly reduced.

8) The cows cannot pursue their natural urge to forage.

a) Classify statements 1)–8) into:

i) those in support of and

ii) those against the practice of all-year-round housing for dairy cattle. (2)

b) i) In your opinion, which TWO of the five freedoms for animal welfare are cows that are housed all year round being denied?

ii) Justify your choice of answer. (2)

c) Suggest a compromise that might be acceptable to both sides in this argument. (1)

2 Table 17.3 refers to the results of a survey carried out on risk factors associated with three types of henhouse.

a) i) Risk factors 3 and 5 show the same pattern of risk level. Suggest why.

ii) Which other TWO risk factors show the same pattern? (2)

b) Which risk factor appears to be unaffected by the type of accommodation that the chickens receive? (1)

c) i) Which type of accommodation runs a high risk of the birds being affected by internal parasites?

ii) Suggest why this should be the case. (2)

d) i) Which birds are least likely to suffer low bone strength?

ii) Suggest why. (2)

e) i) What effect does leaving beaks untrimmed have on feather-pecking in caged birds?

ii) Explain why. (2)

f) i) Using the system high risk = 2 points, medium risk = 1 point, low risk = 0 points, work out the overall level of risk for each type of accommodation.

ii) Suggest how risk of bumble foot disease could be reduced.

iii) If bumble foot were eradicated completely, in what way would this shift the balance of overall level of risk for the three types of accommodation? (4)

3 Read the passage and answer the questions that follow it.

'Chemical welfare'

Many diseases caused by bacteria, viruses and protozoa (single-celled animals) are carried by insect vectors and affect livestock. Where intensive farming methods are practised, it is not unusual for several thousand poultry or pigs to be reared on a farm at the same time. This situation creates the perfect conditions for an epidemic to occur. Therefore, in addition to maintaining high standards of hygiene, farmers often make use of chemicals to prevent an outbreak of disease. These treatments include pesticides, vaccines and, in some countries, antibiotics in the animal feed.

Free-grazing animals such as sheep and cattle are also vulnerable to various ailments. Some of these are caused by pests such as scab mites and ticks that pierce the animal's skin. These external parasites can be controlled by dipping the farm animal in pesticide which works by paralysing the pest's nervous system.

However, some parasites live inside the host. Liver flukes, for example, spend one part of their life cycle in the mammalian host and another part in a type of snail that thrives in water-logged pasture. This problem can be tackled by using drugs to kill the parasite within the sheep or cow and/or draining the pasture and spraying it with molluscicide.

→

Risk factor that could affect animal's welfare	Level of risk		
	Type of accommodation		
	A conventional cage with minimal facilities in henhouse	B furnished cage with nesting boxes and perches in henhouse	C uncaged nesting boxes and perches in henhouse with free access to outside area
1 feather-pecking in beak-trimmed flock			
2 feather-pecking in non-beak-trimmed flock			
3 restriction of wing-flapping movements			
4 prevention of foraging			
5 restriction of tail-wagging movements			
6 bumble foot disease caused by poorly-designed perches			
7 low bone strength			
8 predation			
9 internal parasites			
10 external parasites			

Key:

low risk	
medium risk	
high risk	

Table 17.3

a) To which of the five freedoms for animal welfare do the chemicals mentioned in the passage make a contribution? (1)

b) i) Why are antibiotics added to the animal feed in some countries, when the animals are perfectly healthy?

ii) If some bacteria for a disease just happened to be resistant to the antibiotic, what problem could arise? (3)

c) i) Identify a precaution that must be taken by the workforce operating a sheep dip.

ii) Explain your answer to i). (2)

d) The concentration of active ingredient in a brand of sheep dip chemical is 250 mg l^{-1}. Express this as

i) g l^{-1} ii) μg l^{-1}. (2)

e) i) With reference to the control of liver flukes, which treatment is *preventative* and which is *curative*?

ii) Why is prevention better than cure? (3)

4 Read the passage and answer the questions that follow it.

Birds such as fowl are often housed in conditions of excessively high stocking density and bright lighting even when they are known to prefer dim light. In these crowded conditions the birds lack the opportunity to forage. This is because they are denied access to flooring material such as straw that would enable them to peck in a natural way. In addition, they have no access to alternative pecking substrates such as bunches of string and pecking blocks.

→

Birds in an environment lacking basic facilities may express their discomfort in several ways. One of these is inappropriate pecking behaviour which may take the form of pecking at or plucking out their own feathers or those of nearby birds.

Fowl are strongly attracted to damaged feathers and the presence of a few among a flock can be enough to set off a chain reaction of injurious pecking and cannibalism. This rarely happens if the strain of bird has been selectively bred for a reduced level of inappropriate pecking. If the inappropriate pecking continues then chicken farmers using intensive methods and very basic standards of housing for their chickens may have to resort to the use of anti-pecking sprays if things get out of hand.

a) Why do some chickens peck one another in an inappropriate way? (1)

b) Copy and complete Figure 17.7 to show FIVE ways in which attempts could be made to reduce inappropriate pecking behaviour. (Base your answers on information in the text.) (5)

c) i) Which of your answers to **b)** is a short-term method suitable for use only in a crisis?

ii) Identify such a crisis from the passage. (2)

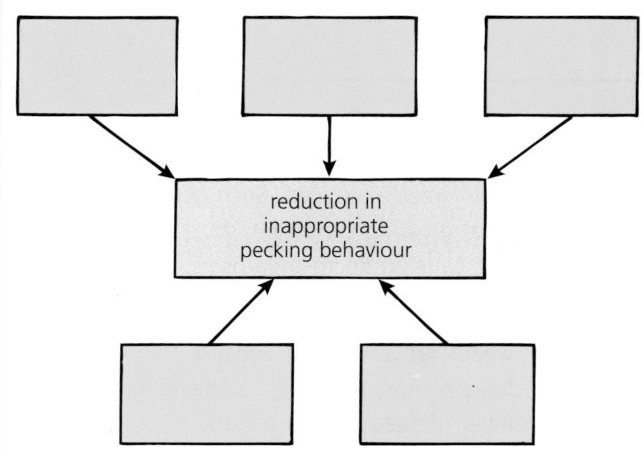

Figure 17.7

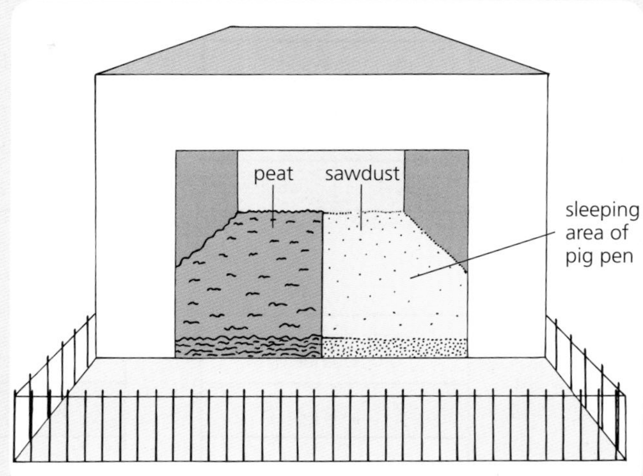

Figure 17.8

5 An investigation was carried out to discover which flooring material (substrate) was preferred by pigs in their sleeping area. The pens were designed to offer the pigs a choice between two substrates at a time (see Figure 17.8). The eight pigs in each group were left for one week to become habituated to their pen. During the second week, the time spent by the pigs in each substrate was recorded. Three replicas were used for each comparison. The six tables in Figure 17.9 show the results.

a) Why were three replicates of each comparison set up? (1)

b) Why were as many as eight pigs used in each group? (1)

c) i) What words in the passage mean that the pigs were given time to become acclimatised to their surroundings?

ii) Suggest why this is important. (2)

d) Identify TWO design features of the pig pen in this investigation that must be kept constant. (2)

e) Put the flooring materials (substrates) into descending order of preference according to the results. (1)

6 Give an account of the ways in which the behaviour of domesticated animals indicates that their level of welfare is poor. (9)

→

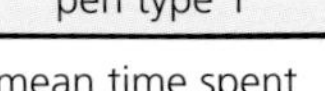

pen type 1

mean time spent on material (h)	
P	S
99(+)	69(–)

pen type 2

mean time spent on material (h)	
P	M
82	86

pen type 3

mean time spent on material (h)	
P	W
141(+)	27(–)

pen type 4

mean time spent on material (h)	
M	S
102(+)	66(–)

pen type 5

mean time spent on material (h)	
M	W
138(+)	30(–)

pen type 6

mean time spent on material (h)	
S	W
114(+)	54(–)

key
P = peat
M = mushroom compost
S = sawdust
W = woodbark

\+ = result significantly higher than would be expected by chance alone
– = result significantly lower than would be expected by chance alone

Figure 17.9

What You Should Know

Chapters 16–17
(See Table 17.4 for word bank)

agricultural	forecast	resist
annual	fungi	resistant
behaviour	indicators	rotation
biological	integrated	selective
biomagnify	natural	stereotypies
crop	nematodes	systemic
cultural	parental	toxic
cycle	perennial	vegetatively
diseased	persist	viable
domesticated	pest	weeds
ethically	pressure	welfare

Table 17.4 Word bank for chapters 16–17

1 Compared to a natural ecosystem, the variety of species that make up the community in an ______ ecosystem is greatly reduced. Such conditions present ideal growing conditions to ______, pests and disease-causing micro-organisms. Their activities reduce the productivity of the ______.

2 ______ weeds are successful because they grow rapidly, have a short life ______ and produce a huge number of seeds which remain ______ for a long time. ______ weeds owe their success to the possession of storage organs, the ability to reproduce ______ and the fact that they are already established and therefore at a competitive advantage.

3 Animal pests that attack crops are mainly invertebrates such as ______, molluscs and insects. Diseases that affect crops are caused by bacteria, ______ and viruses.

→

4 Weeds, pests and diseases of crops can be controlled to some extent by ______ means such as crop ______ and by the use of pesticides. Use of a ______ herbicide is advantageous because it kills broad-leaved weeds but not the narrow-leaved crop. Use of ______ fungicide is advantageous because it protects all parts of the crop from within.

5 Applying a fungicide in response to a disease ______ is more effective than treating the crop once it has become ______.

6 Some protective chemicals are ______ to other living organisms in addition to the pest. They may ______ and ______ in food chains. They may exert a selection ______ on the pest population and produce a ______ population.

7 The use of a natural enemy of the ______ or parasite as the agent of control is called ______ control. A combination of several control techniques is called ______ pest management.

8 In the past, the ability of a ______ animal to grow, to reproduce and to ______ disease successfully were regarded as adequate indications of its wellbeing. However, in recent times, the animal being able to behave in a ______ way has been added to the list.

9 The provision of a high standard of animal ______ is costly but worth the financial outlay both economically and ______.

10 Animals kept in conditions of low level welfare express their discomfort as behavioural ______. These take the form of ______, misdirected ______, altered levels of activity and failure in sexual or ______ behaviour.

18 Symbiosis

Symbiosis

Symbiosis is an ecological relationship between organisms of two different species that live in direct contact with one another. Symbiotic relationships are intimate relationships that have evolved over millions of years. In some cases the two species involved have become adapted in such a way that their metabolisms are complementary to some extent. This coevolution of adaptations is essential because a change in one partner is likely to affect the survival of the other.

Two categories of symbiosis are:

- **parasitism** – one organism, the **parasite**, benefits at the expense of the other organism, the **host**, which is often damaged
- **mutualism** – both organisms benefit from the relationship.

Parasitism provides illustrations of **dependence** since the parasite is always dependent on the host whereas mutualism illustrates **interdependence** since the two partners are mutually dependent upon one another.

Parasitism

In parasitism one organism, the parasite, derives its nutrition from another organism, the host, which it exploits. The host is harmed or at least loses some energy and/or materials to the parasite.

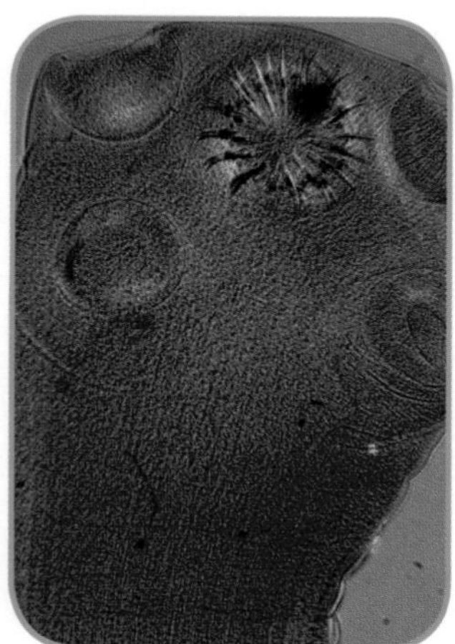

Figure 18.1 Head of tapeworm

In many cases of parasitism, a relatively stable relationship has evolved and a **balance** exists between defence mechanisms used by the host and damage inflicted by the parasite. Parasites often possess a **limited metabolism**. For example, a tapeworm (see Figure 18.1) lacks a digestive system since its host has already digested the food. However, this means that many parasites such as an adult tapeworm cannot survive if they lose contact with their host. It is for this reason that the most effective parasite is one that does not cause its host to die (or at least not until completion of the parasite's life cycle is ensured).

Some parasites such as malarial *Plasmodium*, liver fluke and tapeworm live and feed *inside* their host. Some parasites such as aphid, flea and mosquito (see Figure 18.2) live and feed *outside* their host.

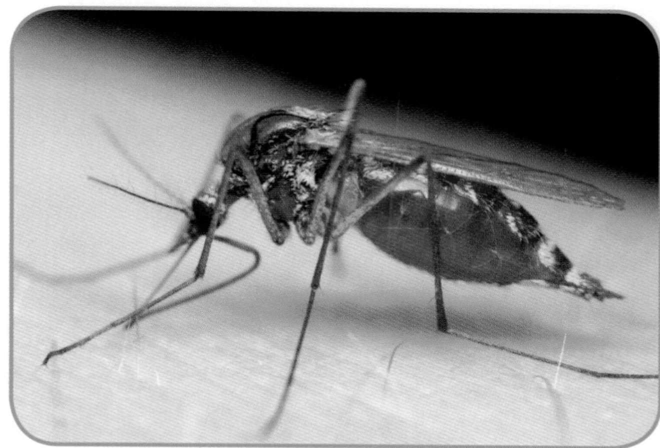

Figure 18.2 Mosquito

Transmission

The transmission of parasites to new hosts involves a variety of mechanisms such as:

- **direct contact** (for example, head and body lice are passed from person to person during physical contact)
- **release of resistant stages** able to survive adverse environmental conditions for long periods until they come into contact with a new host (for example, the resistant larvae and pupae of the cat flea shown in Figure 18.3 transmit the parasite to new hosts)

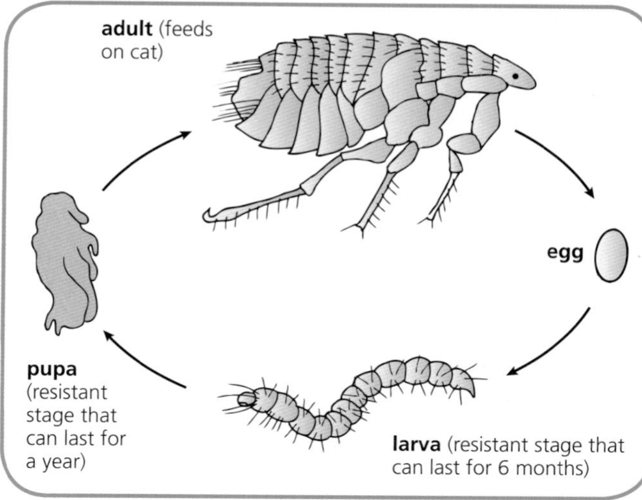

Figure 18.3 Life cycle of cat flea

- **use of a vector** (for example, mosquitoes carry *Plasmodium*, the unicellular organism that causes malaria, from human to human).

Evolution of parasitic life cycles involving a secondary host

There are two types of parasitic life cycle. In a **direct** life cycle, eggs are shed and pass to a new member of the host species. This cycle involves one species of host only, and is common among parasites such as flea and louse that live and feed outside of their host.

More highly evolved is the **indirect** life cycle. In addition to using a **primary** host species as the site of the sexual stage of its reproduction, the parasite has

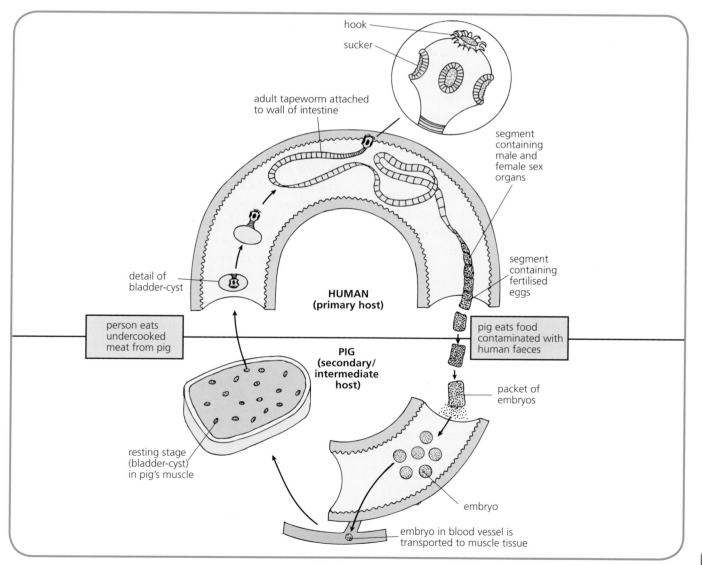

Figure 18.4 Life cycle of tapeworm

become adapted by evolution to employ a **secondary (intermediate)** host species in its life cycle.

A new primary host becomes infected when it is invaded by or consumes the infected **secondary** host (or the parasite that has been released from it). This type of life cycle is commonly found among parasites such as tapeworm (see Figure 18.4) that live and feed inside their host. The involvement of a second (or even third) host in the cycle would seem to complicate matters needlessly for the parasite yet this type of life cycle has evolved by natural selection over millions of years. Therefore the parasite must be gaining some advantage.

Advantage

It is often a difficult, risky business for a parasite that lives inside its host to move from one primary host to another. An adult tapeworm, for example, has no means of locomotion and its eggs are unlikely to be consumed by potential human hosts unless their food or drink is contaminated with raw sewage.

However, the parasite can complete its life cycle with ease once it is established in the secondary host which can be eaten by humans, the primary hosts. Completion of the cycle depends only on the infected meat being **inadequately cooked** and eaten by humans.

In addition to reproducing sexually within its primary host, the parasite may further exploit its secondary host by using it as a site of **asexual reproduction**. By this means it increases enormously its reproductive capacity and chance of survival.

Mutualism

Cellulose digestion

Many herbivorous mammals lack the genes required for the synthesis of cellulose-digesting enzymes despite the fact that cellulose, in the cell walls of plants, is a major constituent of their diet. These herbivores are found to possess special gut chambers containing **cellulose-digesting micro-organisms** (bacteria, archaea and protozoa) that form a **mutualistic** relationship with the herbivore. They are found, for example, in the first two chambers of a cow's complex, four-chambered 'stomach' (see Figure 18.5).

The micro-organisms produce enzymes which digest cellulose to simple sugars used by the herbivore as its energy source. The microbes also convert some of the

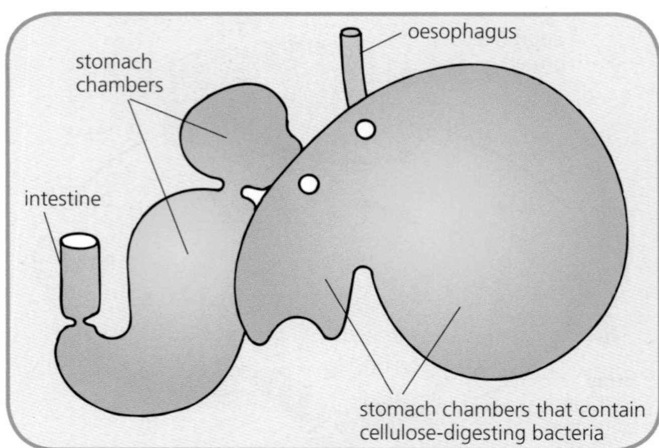

Figure 18.5 'Stomach' of cow

sugar to other metabolites essential to the animal (and make by-products such as methane). In exchange the microbes receive protection, warmth and a constant supply of food.

Humans depend on **ruminants** such as cattle and sheep for supplies of food and other products. Therefore the symbiosis between these animals and cellulose-digesting micro-organisms is of great economic importance.

Coral polyps

A **coral polyp** (see Figure 18.6) is an animal that resembles a sea anemone. However, it is unable to move from place to place. It secretes a hard skeleton which attaches it to the skeletons of earlier polyps forming a **coral reef**. It feeds on microscopic organisms and tiny bits of organic debris in the sea water.

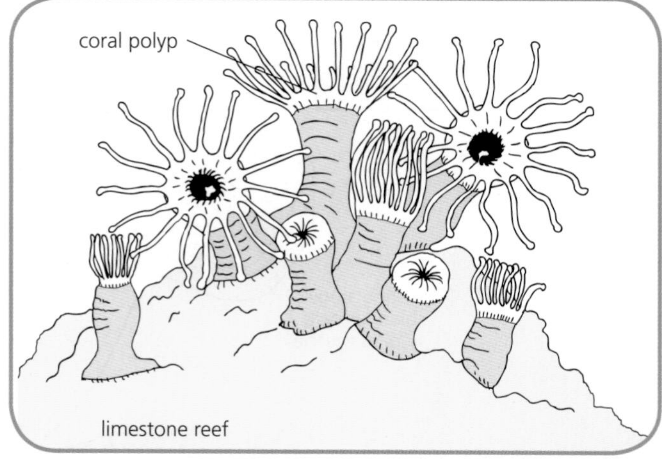

Figure 18.6 Coral polyps

It has a **mutualistic** relationship with a type of unicellular alga called **zooxanthella** which lives within and between its cells (see Figure 18.7). The polyp may use as much as 80% of the carbohydrate made by the algae for energy. In return the algae are provided with a secure habitat and a supply of the polyp's nitrogenous waste which they convert to protein.

The fact that coral reefs are known to have been in existence for over 200 million years and that they occupy more than 280 000 km^2 of tropical and sub-tropical waters demonstrates the success of this symbiotic relationship.

Figure 18.7 Coral polyps containing zooxanthella

Testing Your Knowledge

1 a) What is meant by the term *symbiosis*? (2)

b) Copy and complete Table 18.1, adding type of symbiosis and using the symbols: + = benefit and − = harm. (2)

Type of symbiosis	Species 1	Species 2
		+
parasitism		

Table 18.1

2 Briefly describe THREE methods by which parasites may be transmitted from one host to another. Include named examples in your answer. (6)

3 Explain why:

a) a parasite such as an adult tapeworm is unable to survive out of contact with its host (1)

b) coral polyps are restricted to waters less than 100 metres deep (1)

c) it is of advantage to a parasite to involve a secondary (intermediate) host in its life cycle. (1)

Applying Your Knowledge and Skills

1 Classify the following examples of symbiosis into TWO named categories: (4)

a) Certain bacteria in the human colon feed on unwanted food, releasing vitamin B absorbed by humans.

b) Female sandflies suck mammalian blood to nourish their eggs.

c) The life cycle of the Chinese liver fluke involves three different hosts (human, snail and fish).

d) Egyptian plover birds clean leeches from between the teeth of crocodiles.

e) Dodder is a flowering plant which grows attached to stinging nettle plants from which it obtains all of its food.

f) Certain fungi on the roots of pine trees aid water absorption by the tree and receive carbohydrate from it.

g) Wood-eating termites possess populations of cellulose-digesting microbes in their gut.

h) Rust fungi infect cereal crops and replace many of the ears with fungal spores.

2 An experiment was carried out to investigate the effect of varying the percentage of protein in the diet of young rats infected with a type of tiny parasitic roundworm. The results are shown in graphs 1 and 2 in Figure 18.8.

a) i) What effect does an increase in percentage of dietary protein have on a young rat's growth rate?

ii) When did the scientists infect the rats with the parasitic roundworms?

iii) What effect did the parasitic infection have on a rat's growth rate?

iv) Give a possible explanation for your answer to **iii)**. (4)

➜

graph 1

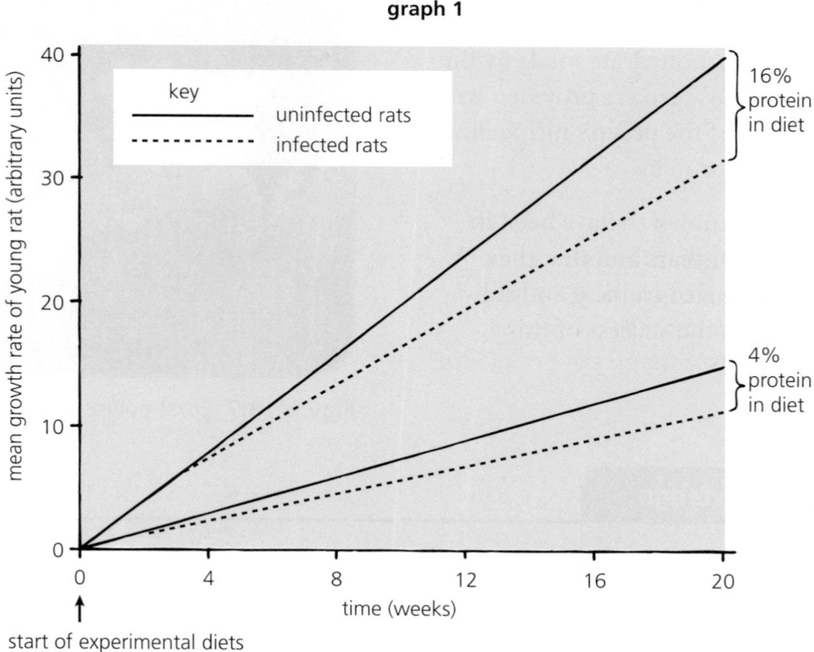

graph 2

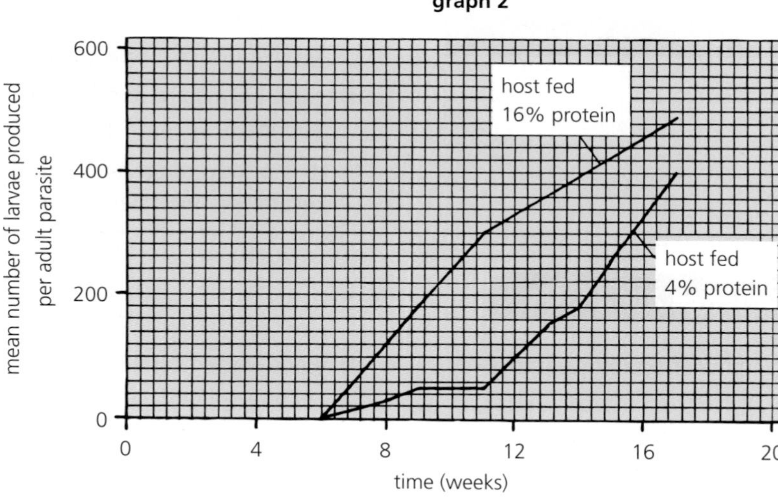

Figure 18.8

b) Suggest why no parasitic larvae appeared until week 6. (1)

c) By how many times was the mean number of larvae produced by an adult parasite in a rat fed 16% protein greater than that in a rat fed 4% protein in week 11? (1)

d) What percentage increase in mean number of larvae produced by an adult parasite occurred in a rat fed 4% protein between weeks 13 and 17? (Round your answer to two decimal places.) (1)

e) If the trends in graph 2 continue, on which week will the mean number of larvae be equal for both concentrations of dietary protein fed to their hosts? (1)

3 Malaria is a debilitating and often fatal disease caused by the unicellular organism *Plasmodium*, a blood parasite. Its life cycle comprises a sexual stage in a mosquito and an asexual stage in a human as shown in Figure 18.9 (see page 267).

a) Why is it wrong to say that 'mosquitoes cause malaria'? (1)

b) In what way is the proboscis of a mosquito very different from the needle of a hypodermic syringe? (1)

c) Suggest why the release of many parasitic spores (merozoites) into the human blood stream is:
   i) accompanied by an attack of fever
   ii) followed by anaemia. (2)

d) The organism within which the fusion of a parasite's gametes occurs is normally regarded as its primary host. Identify the primary and secondary/intermediate hosts in the life cycle of *Plasmodium*. (1)

e) Endoparasites live and feed inside their host; ectoparasites live and feed on the outside of their host. Which organism in Figure 18.9 is
   i) an endoparasite?
   ii) an ectoparasite?
   iii) Identify the ectoparasite's host. (3)

4 In an investigation to find out if a species of hard coral was able to acquire tolerance to an increase in environmental temperature, samples of the coral were exposed for 15 days to four different temperatures. In the first experiment the coral samples all contained type A zooxanthellae and in the second experiment they all contained type B zooxanthellae. Figure 18.10 summarises the results.

a) Which of the four temperatures is the normal mean temperature of the water in which this species of hard coral lives? (1)

b) Describe the condition of the coral in each experiment after 15 days at:
   i) 27 °C
   ii) 32 °C. (2)

c) What is the mean number of type B zooxanthellae:
   i) at 30 °C?
   ii) at 32 °C? (2)

d) The error bars are based on a 95% level of confidence.
   i) At 27 °C the mean number of type A zooxanthellae can be said, with 95% confidence, to vary between which two extremes?

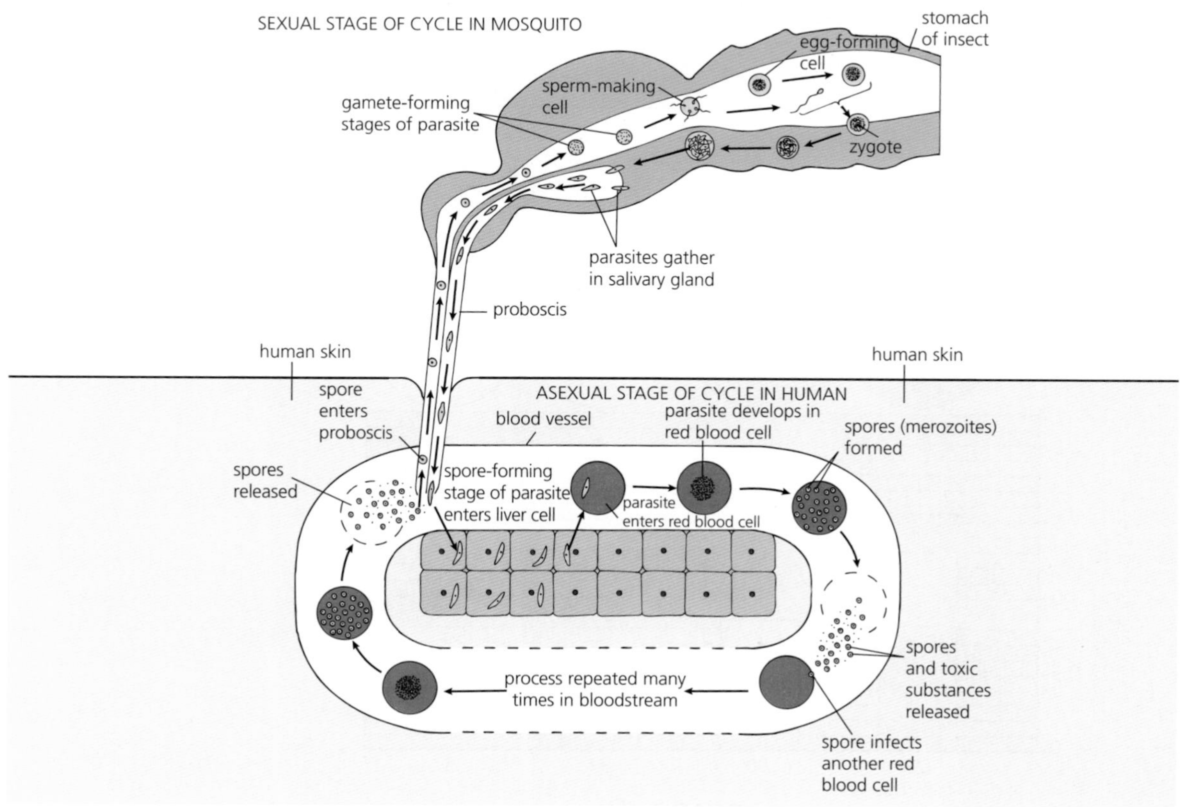

Figure 18.9

ii) Is this mean significantly different from that for type A zooxanthellae at 30 °C?

iii) Explain your answer to ii). (3)

e) i) At 27 °C the mean number of type B zooxanthellae can be said, with 95% confidence, to vary between which two extremes?

ii) Is this mean significantly different from that for type B zooxanthellae at 30 °C?

iii) Explain your answer to ii). (3)

f) What is the effect of increasing temperature on the mean number of type A zooxanthellae? (1)

g) What is the effect of increasing temperature on the mean number of type B zooxanthellae? (1)

h) What conclusion can be drawn about this species of hard coral's ability to acquire tolerance to increasing environmental temperature? (1)

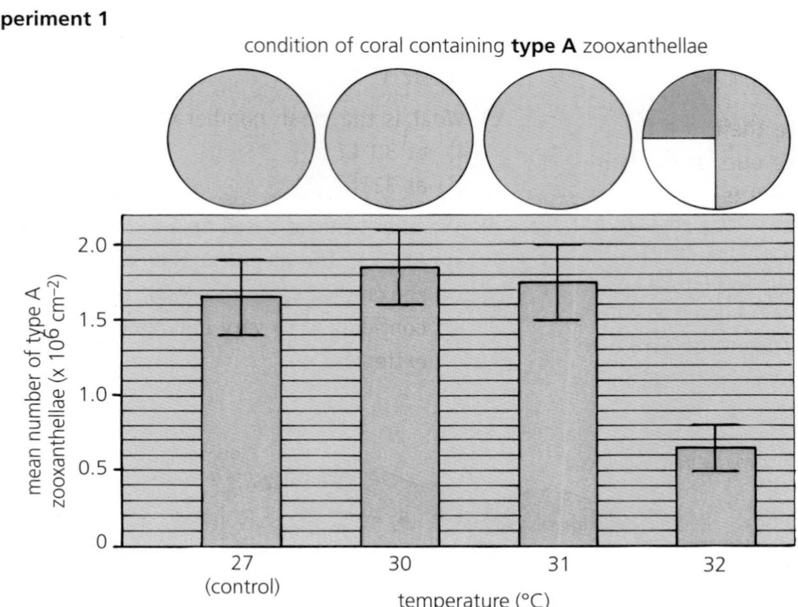

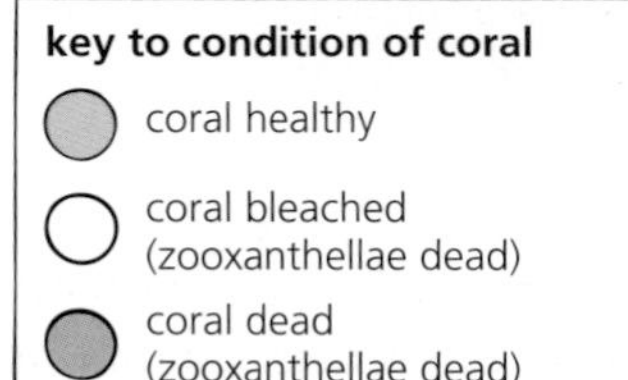

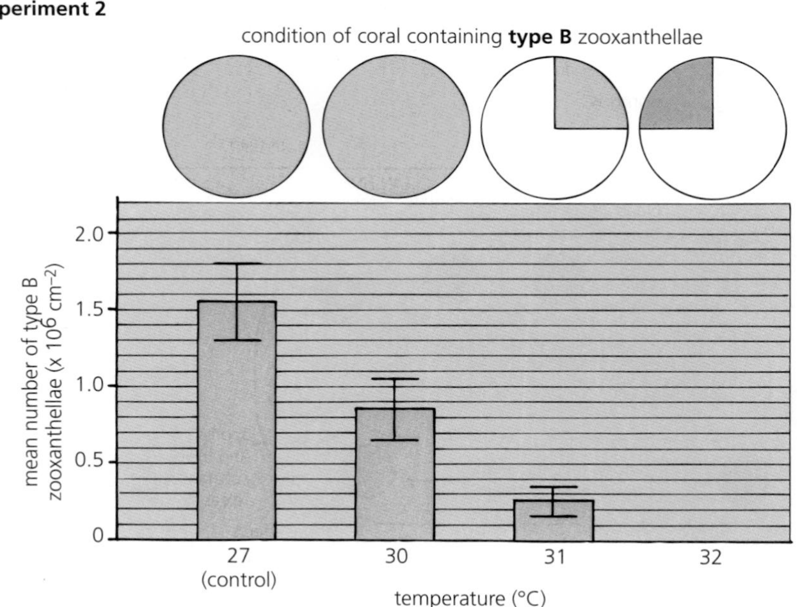

Figure 18.10

19 Social behaviour

Many animals live in **social groups**. A social group may consist of as few as two members or as many as several thousand. The members of a group react to **social signals** given by other members of the same species. The successful cohesion of the group requires its members to exhibit certain **behavioural adaptations** as in the following examples of social behaviour.

Social hierarchy

This is a system where the members of a social group are organised into a graded order of **rank** resulting from aggressive behaviour between members of the group. An individual of higher rank **dominates** and exerts control over other **subordinate** individuals of a lower rank by performing **ritualistic threat displays**. The subordinate animals respond by carrying out **appeasement** behaviour. This demonstrates their acceptance of the individual's dominant status and reduces conflict among the members of the social group. Some animals may increase their social status within the group by forming **alliances** (also see page 279).

Birds

If newly-hatched birds such as pigeons are kept together, one will soon emerge as the **dominant** member of the group. This bird is able to peck and intimidate all other members of the group without being attacked in return. It therefore gets first choice of any available food. Below this dominant bird there is a second one which can peck all others except the first and so on down the line. This linear form of social organisation is called a **pecking order**.

Table 19.1 summarises the results from observing a group of hens over a period of time. Bird A dominates all of the others, bird B dominates all others except A and so on down the line to bird J which is dominated by all the other birds. A pecking order is an example of a **social hierarchy**.

Mammals

Although not so clear cut, a similar system of social organisation exists among some mammals such as wolves. Because of his rank, the dominant male has

		Hen receiving pecks									
		A	B	C	D	E	F	G	H	I	J
Hen giving pecks	A		✓	✓	✓	✓	✓	✓	✓	✓	✓
	B			✓	✓	✓	✓	✓	✓	✓	✓
	C				✓	✓	✓	✓	✓	✓	✓
	D					✓	✓	✓	✓	✓	✓
	E						✓	✓	✓	✓	✓
	F							✓	✓	✓	✓
	G								✓	✓	✓
	H									✓	✓
	I										✓
	J										

Table 19.1 Pecking order for a group of ten hens

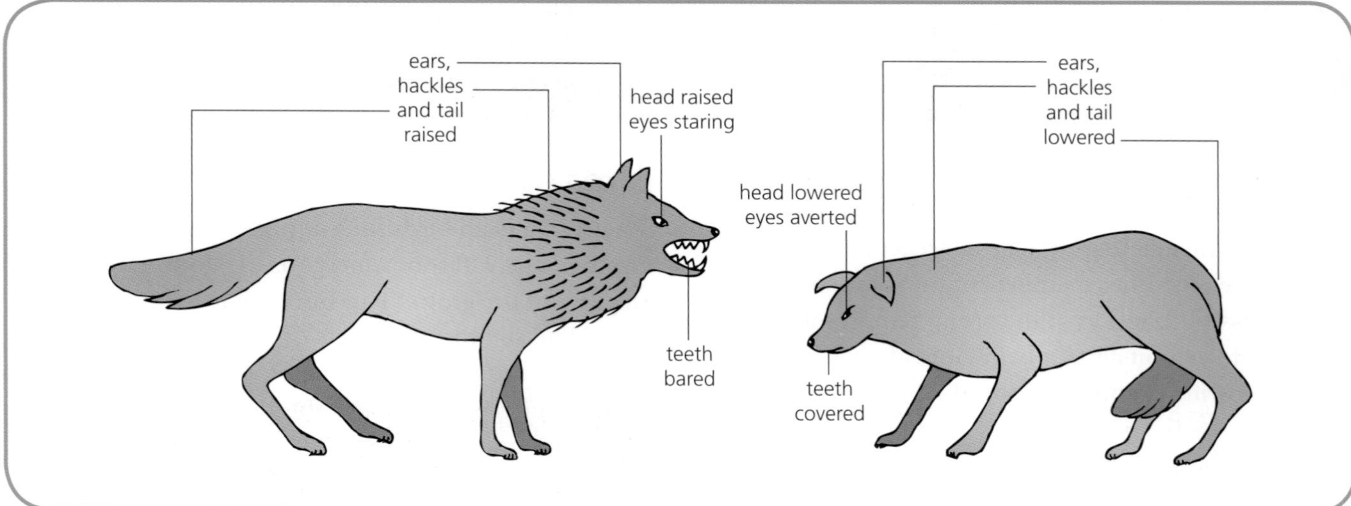

Figure 19.1 Ritualistic displays and submissive responses

certain rights such as first choice of food, preferred sleeping places and available mates. This dominant individual asserts his rank by employing social signals, as shown in Figure 19.1.

Advantages of social hierarchy

A system of social hierarchy increases a species' chance of survival because:

- aggression between its members becomes ritualistic
- real fighting is kept to a minimum
- serious injury is normally avoided
- energy is conserved
- experienced leadership is guaranteed
- the most powerful animals are most likely to pass their genes on to the next generation.

Co-operative hunting

Some predatory mammals such as killer whales, lions, wolves and wild dogs rely on **co-operation** between members of the social group to hunt their prey. Killer whales hunt in packs called pods (see Figure 19.2) and employ various strategies. In a river mouth, a pod will sweep along in a line to catch migrating salmon. In coastal waters the same pod will encircle a shoal of herring and concentrate them into a seething mass. The whales then thrash the herring with their tails to stun them and gorge themselves on a catch of food that would be unavailable to a solitary predator.

Figure 19.2 Pod of killer whales

The **ambush strategy** employed by lions involves some predators driving prey towards others that are hidden in cover and ready to pounce. Dogs and wolves, on the other hand, take turns at **running down** a solitary prey animal to the point of exhaustion and then attacking it (see Figure 19.3). In the case of lions, wolves and dogs, the group of predators tends to concentrate its efforts on a prey animal that has become separated from the rest of the herd. This is often a young and inexperienced or old and infirm animal, making it an easy target.

Figure 19.3 Co-operative hunting

Advantages of co-operative hunting

When a kill is achieved, all members of the predator group obtain food (some of which may be disgorged later by females to feed the young). Thus **co-operative hunting** benefits the subordinate animals as well as the dominant leader of the group. By working together, the animals expend less energy per individual and are able to tackle large prey animals. As a result, all members of the social group gain more food than they would by hunting alone. Provided that the food reward gained by co-operative hunting exceeds that from hunting individually, the social group will continue to share food regardless of the fact that the dominant member(s) receives a much larger share than the subordinate ones.

Social mechanisms for defence

Animals employ various social defence strategies which increase their chance of survival. Within a large group, some members can be foraging while others are keeping watch for predators. Many types of animal rely on this principle of '**safety in numbers**' as a means of defence. Among a flock of birds, for example, there are many eyes constantly on the lookout for enemies. Following an **alarm** call, the bunching and swirling tactics adopted by the flock confuse the predator who finds it much more difficult to capture a member of a large, unpredictable, milling crowd than a solitary individual.

Defence is strengthened further by the members of a social group adopting a **specialised formation** as in the following examples.

Musk ox

Musk oxen (Figure 19.4) are native to Arctic regions of Canada and Greenland. Their natural environment is completely open and offers no scrub or woodland to use for concealment. Their natural enemy (apart from human beings) is the wolf. When threatened, a herd of musk oxen form a **protective group** with cows and calves in the centre and mature males at the outside with their huge horns directed outwards. Individual wolves are gored and packs are driven off by a combined charge. This form of social defence is called **mobbing**.

Figure 19.4 Social defence (musk oxen)

Quail

Bobwhite quails roost in **circles** with their heads to the outside as shown in Figure 19.5. If disturbed, the circle acts as a defensive formation by 'exploding' in the predator's face. By the time their enemy has recovered from the confusion, the birds have flown away to safety.

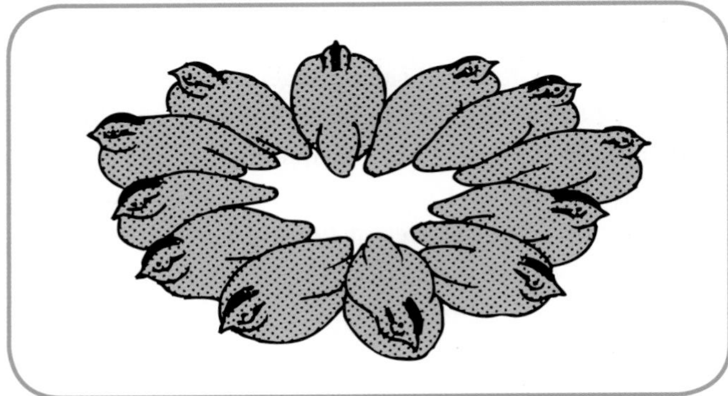

Figure 19.5 Social defence (quail)

Baboon

A strict social hierarchy is observed by the members of a troop of baboons. When the baboons are on the march (see Figure 19.6), the dominant males stay in the centre close to the females with infants. Lower-ranking adult

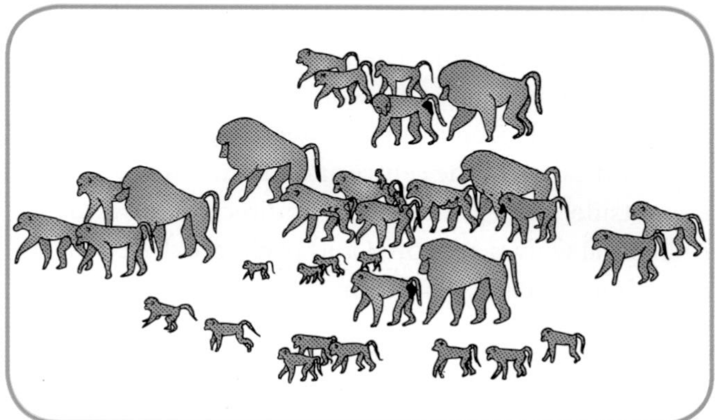

Figure 19.6 Social defence (baboons)

males and juveniles keep to the edge of the troop and raise the alarm if the group is threatened (for example, by a leopard).

Altruism

Members of the same species compete with one another because they need the same resources. However, variation exists among the members of a species. Those that are able to compete most successfully would be expected to selfishly exploit to excess the available resources and produce the maximum number of offspring at the expense of the poorer competitors who would lose out. As a result of this process, the fittest members would pass many copies of their genes on to the next generation.

However, on some occasions, an animal is found to behave in a way that is disadvantageous (harmful) to itself, the **donor**, and beneficial (helpful) to another animal, the **recipient**. This unselfish behaviour is called **altruism** (see Figure 19.7) and at first glance it would seem to contradict the idea of the survival of the fittest. However, on closer inspection, this is not found to be the case.

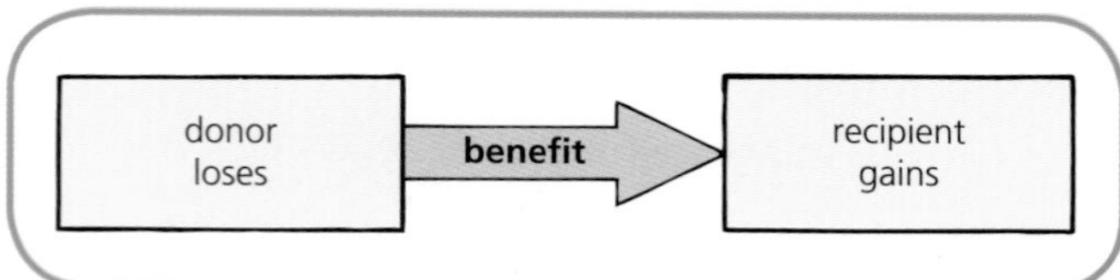

Figure 19.7 Altruism

Reciprocal altruism

Reciprocal altruism involves one individual, at some cost to itself, giving help to another provided that there is a very real prospect of the **favour being returned** at a later date when the roles of donor and recipient become reversed. It requires members of the social group to be sufficiently intelligent to be able to remember who is indebted to whom and how to recognise and expel cheats.

Vampire bats

After a night of hunting, vampire bats that have been successful may return to the roost bloated with blood while other unsuccessful members of the same social group may return hungry. The hungry individuals then beg blood from the bloated ones by licking their face (see Figure 19.8 part a). The latter are often willing to regurgitate some blood (see Figure 19.8 part b) but tend to give it only to those who had been altruistic to them in the past.

In reciprocal altruism, the benefit gained by the recipient **exceeds the cost** to the donor. This is illustrated by vampire bats because a hungry bat (the potential recipient) is likely to die if it fails to find a blood meal on two consecutive nights.

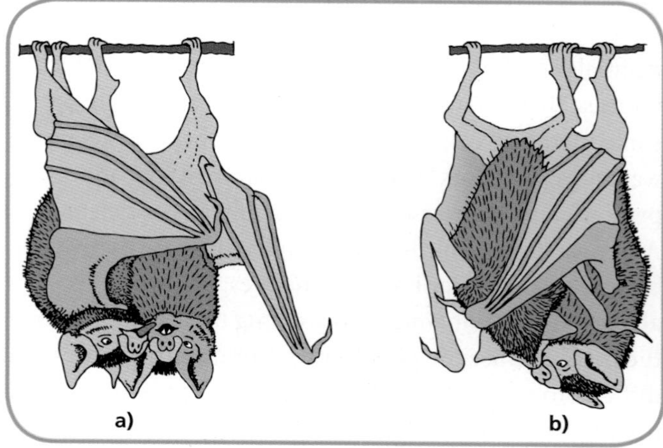

Figure 19.8 Reciprocal altruism in vampire bats

Research Topic The prisoner's dilemma

Imagine two suspects (X and Y) jointly charged with a crime which they *did commit*. The police do not have sufficient evidence to be sure of a conviction for the crime but they do have enough to be sure of a conviction for a minor offence. The police interview the prisoners separately and offer each the following deal:

- If you testify for the prosecution against your accomplice and your accomplice remains silent, you will go free and your accomplice will receive the full 8-year jail sentence.
- If you testify for the prosecution against your accomplice and your accomplice testifies against you, you will each receive a 4-year jail sentence.
- If you both remain silent, you will each be sentenced to 1 year in jail for the minor offence.

Each prisoner must choose whether to betray their accomplice or remain silent. The situation is summarised in Table 19.2.

If the two members of the group both make a self-interested decision and betray the other in an attempt to gain their own **best personal outcome** (complete freedom regardless of the 8-year jail sentence for their accomplice) then this results in the **worst possible mutual outcome** for both (a 4-year jail sentence each).

On the other hand if they trust one another, they can choose to co-operate and remain silent. This course of action results in the **best possible mutual outcome** (a 1-year jail sentence each) though not the best possible personal outcome. It is a form of **reciprocal altruism**.

	X remains silent	X betrays Y
Y remains silent	X and Y each serve 1 year	X goes free; Y serves 8 years
Y betrays X	Y goes free; X serves 8 years	X and Y each serve 4 years

Table 19.2 Four outcomes of the prisoner's dilemma

Investigation

Reciprocal altruism using the prisoner's dilemma

The **prisoner's dilemma** can be investigated by playing it as a game. The jail sentences in Table 19.2 are converted into a scoring system of points as shown in Table 19.3.

	X remains silent	X betrays Y
Y remains silent	X and Y each receive 3 points	X receives 5 points; Y receives 0 points
Y betrays X	Y receives 5 points; X receives 0 points	X and Y each receive 1 point

Table 19.3 Points for prisoner's dilemma game

Within each group there are two players and a banker. Each player holds two cards, one bearing the words '*remain silent*', the other bearing the words '*betray accomplice*'. Each player puts one card face down in front of the banker who then turns them over and awards points accordingly. The game is played repeatedly to represent a series of interrogations. The score from each round is recorded and running totals kept. The object is to gain the maximum number of points before reaching game over. Game over is determined by the banker choosing a random number between 20 and 30 before the game begins which remains unknown to the players. The banker brings the game to a halt at the round indicated by the random number.

Since many rounds are played, each player has several opportunities to punish the other player for previous non-co-operative play. Eventually the incentive to make purely self-interested decisions may be overcome by the threat of punishment.

The final scores are compared with the scores that would have been obtained if X and Y had co-operated with one another and remained silent throughout. Table 19.4 shows a typical set of results. From the results it is concluded that the prisoner's dilemma is a simple model of **reciprocal altruism**. By co-operating from the start, X and Y would have both been better off at the end than they were by operating independently. Table 19.5 shows precautions that are 'adopted' during this investigation and the reasons for doing so.

Round	X				Y			
	Actual score	Running total	Score if silent and co-operating throughout	Running total	Actual score	Running total	Score if silent and co-operating throughout	Running total
1	1	1	3	3	1	1	3	3
2	5	6	3	6	0	1	3	6
3	3	9	3	9	3	4	3	9
4	5	14	3	12	0	4	3	12
5	3	17	3	15	3	7	3	15
6	0	17	3	18	5	12	3	18
7	1	18	3	21	1	13	3	21
8	1	19	3	24	1	14	3	24
9	5	24	3	27	0	14	3	27
10	5	29	3	30	0	14	3	30
11	0	29	3	33	5	19	3	33
12	0	29	3	36	5	24	3	36
13	1	30	3	39	1	25	3	39
14	1	31	3	42	1	26	3	42
15	1	32	3	45	1	27	3	45
16	3	35	3	48	3	30	3	48
17	3	38	3	51	3	33	3	51
18	5	43	3	54	0	33	3	54
19	0	43	3	57	5	38	3	57
20	1	44	3	60	1	39	3	60
21	0	44	3	63	5	44	3	63
22	5	49	3	66	0	44	3	66
23	1	50	3	69	1	45	3	69
24	3	53	3	72	3	48	3	72
25	3	56	3	75	3	51	3	75

Table 19.4 Results of prisoner's dilemma game

Precaution adopted	Reason
the decision of one player remains unknown to the other player until they have both made their decision	to prevent the second player's decision being influenced by that made by the first player
the exact number of rounds is not known in advance to the players	to prevent the players choosing to betray one another repeatedly in a last ditch attempt to enhance their score

Table 19.5 Precautions adopted during investigation

Kin selection

When one member of a flock of birds nesting in long grass in a marsh gives a warning call on spotting the approach of a predator, the other birds respond by remaining motionless, which keeps them safe. However, the bird that made the call has put itself in harm's way and may be caught by the predator. Therefore if an '**altruistic' gene** for making the warning call exists, it would be reasonable to expect that the animals possessing it (and the gene itself) would become **eliminated** from the population. However, if the useful gene is also possessed by the bird's offspring and close relatives, the loss of one individual is easily offset by the survival of many.

Apparent altruism

When a parent sacrifices itself to ensure the survival of its offspring, this **apparent altruism** is not a sacrifice in terms of evolution because the parent's genes are successfully passed on to the next generation. But what about an individual that is harmed as a result of being helpful to other close relatives? A useful gene would only spread through the population if the cost to the individual that sacrifices its chance to reproduce and pass on the gene were more than made up for by the additional number of close relatives also possessing the gene and passing it on to future generations.

Coefficient of relatedness

The **coefficient of relatedness** (see Table 19.6) refers to the proportion of genes on average that are identical in two individuals because of shared ancestors. For example, for siblings it is 0.5 because they have, on average, 50% of their genes in common.

Natural selection favours an altruistic individual helping its siblings or helping its parents (or close relatives) who have the potential to produce more offspring. The donor benefits indirectly provided that many copies of the shared genes reach the next generation, ensuring their 'immortality'. This process, which favours acts of apparent altruism carried out to help close relatives, is called **kin selection** (see Figure 19.9).

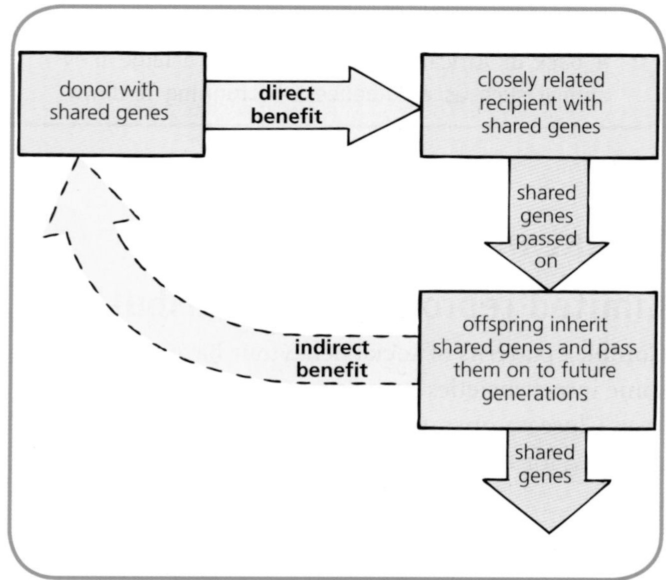

Figure 19.9 Kin selection

Relationship	Coefficient of relatedness
parent – son/daughter	0.5
brother/sister – brother/sister	0.5
half brother/half sister – half brother/half sister	0.25
uncle/aunt – nephew/niece	0.25
grandparent – grandchild	0.25
cousin – cousin	0.125

Table 19.6 Coefficients of relatedness

Testing Your Knowledge 1

1 Instead of fighting, wolves perform a ritual.

a) Describe FOUR features of a dominant wolf's ritualistic display that are on show when it is asserting its authority. (2)

b) Describe the corresponding responses displayed by a subordinate animal. (2)

c) i) What name is given to the type of social organisation that results from this behaviour pattern?

ii) State TWO ways in which it is of advantage to the animals concerned. (3)

2 A pack of African wild dogs catches a large prey animal such as a wildebeest by running it down to the point of exhaustion. Give TWO advantages gained by the dogs from this form of *co-operative* hunting. (2)

3 Musk oxen live in a completely open environment. How do they defend themselves against wolves? (2)

4 a) Using the terms *harmful, recipient, behaviour, helpful* and *donor*, explain the meaning of altruism. (3)

b) What is *reciprocal* altruism? (1)

c) What is the survival value of behaviour that may require an individual to die in order to save its close relatives? (2)

Social insects

Limited reproductive contribution

Complex patterns of social behaviour have evolved in some insect societies such as termites, ants, wasps and bees. Close co-operation occurs between the individuals in caring for the young. A **division of labour** exists as follows.

- Food gathering and defence are carried out by numerous sterile members of the society.
- Reproduction is the responsibility of a few fertile individuals.

Honey bee

Among honey bees, three castes exist – **queens**, **workers** and **drones**. Within a colony the single fertile queen and the many thousands of workers are females and the few hundred drones present are males. Only the queen produces eggs. When these are fertilised by drones they develop into workers.

All sister workers are sterile. They share the same mother (the queen) and therefore have very many of their genes in common. The most efficient method of ensuring that these shared genes are passed on to future generations is by the workers raising their relatives in a communal hive rather than each being fertile and attempting to produce and look after her own brood.

The drones play a purely reproductive role but the workers display a complex series of behavioural acts within their short adult life (4–6 weeks). These acts are all aimed directly or indirectly at **ensuring the survival of the offspring**. They include:

- feeding and grooming larvae (see Figure 19.10)
- building new cells
- collecting and storing pollen
- defending the hive from enemies
- carrying out waggle dances to indicate the direction of food
- foraging for food.

Similarly among colonies of ants, termites and wasps, only a few individuals contribute reproductively to the society. Most members are workers who co-operate with one another to raise the young – not their own offspring but those of their close relatives. This is a further example of **kin selection**.

Figure 19.10 Worker bees feeding larvae

Primate behaviour

The **primates** are animals that belong to an order of placental mammals. They normally possess dextrous hands and feet with opposable first digits, stereoscopic vision and, in the higher apes, a highly developed brain. Examples include lemurs, monkeys, apes and humans.

Parental care

Unlike less highly evolved animals that produce an enormous number of young on the basis that a few will survive, primates produce a small number of young and then take great care of them. Primate offspring are almost helpless initially although they do possess the strong hand grip needed to grasp on tightly to their mother. During the long period of **parental care** (see Figure 19.11), the parents feed their young, keep them clean, protect them from extremes of temperature, transport them from place to place and defend them against enemies.

During this time many opportunities arise for the young primates to **learn** complex social behaviours essential for their survival. At quiet times, very young, playful, tree-living primates, for example, are allowed to explore under the watchful eye of their mother and crawl short distances along the branches by themselves.

Figure 19.11 Parental care

As the young primates grow older, they watch and learn while the adults perform important behaviour patterns such as **foraging**, **hunting** and **recognising danger**.

Language

As the young primates experiment and imitate others, they learn how to communicate with the members of their social group using **language**. Language is a system used to express thoughts and feelings and normally consists of a mixture of sounds and gestures.

Play

Eventually there comes a time when the young primates are well enough developed to leave their mother for long periods each day and **play** with other juveniles (see Figure 19.12).

Figure 19.12 Young primates at play

During social play, the youngsters **practise** the rudiments of adult social behaviour. They chase or flee from companions, play-fight with one another and test the physical limits of their own bodies. The behaviour is exaggerated so that it is recognised by other members of the social group as harmless and 'not-for-real'. By practising the rudiments of adult social behaviour during play, the youngsters learn skills such as **communication**, **co-operation** and **sharing** that they will need to survive as adults.

Reducing conflict

If rival animals engage in a real fight, the loser may be killed but, at the same time, the winner may be seriously injured. This means that the winner may be unable to find food or escape from enemies and therefore, like the loser, fail to survive. It is for this reason that many higher animals have evolved behaviours that make them go to great lengths to **reduce conflict** and avoid engaging in serious fighting.

Ritualistic display

When two social primates find themselves competing for a resource such as a mate, they are likely to exhibit a **threat display** to one another. Such a display normally makes them look larger and fiercer and involves the adoption of certain postures (see Table 19.7). Eventually one of the rivals succeeds in making itself more intimidating and the other responds by conceding defeat, abandoning its threat display and adopting **appeasement behaviour**.

Primate	Features of threat display between rival males of near-equal rank
chimpanzee	bipedal swaggering, shoulders hunched, arms held out, hair bristling, mouth open, teeth covered by lips
gorilla	chest-beating, roaring, strutting walk, hair bristling, eyes staring
marmoset monkey	back arched, tail raised, fur erect, eyes staring
vervet monkey	head bobbing, mouth open, eyes staring, tail arched over body

Table 19.7 Threat display features

Appeasement

Appeasement behaviour consists of a **submissive** display that is the reverse of a threat ritual. The animal's body is made to look smaller, flatter, motionless and as **unthreatening** as possible. A vulnerable part of the body may even be exposed to further defuse the hostile situation.

Once a male has established himself at the top of the **social hierarchy** within a social group of primates, females and subordinate males display their acceptance of the dominant male by employing appeasement behaviours (see Figure 19.13).

Figure 19.13 'He's still to learn that appeasement means grovelling not growling.'

Grooming

Chimpanzees and other primates employ **grooming** (see Figure 19.14) as an effective way of reducing tension within the group. One animal picks plant material, fleas and scabs from the fur of another. This often takes the form of reciprocal altruism. In addition to maintaining hygiene, it **cements friendships** between grooming partners who are more likely to assist one another in a crisis than non-grooming partners. Social grooming may even be used to bring about reconciliation after a fight.

Figure 19.14 Grooming

Facial expression

If a primate closes its eyes during an encounter with a rival, it is indicating that it accepts that it has been dominated and is giving up the struggle. It wants to be regarded as subordinate and submissive. Within a social group of monkeys, if an individual opens and closes its lips rapidly making a gentle smacking noise, this is recognised by others as a friendly, submissive greeting. It is derived from the act of grooming and acts as a form of appeasement helping to keep hostilities at bay. A grinning, open mouth exposing the teeth is employed by a chimp to appease a more dominant individual that it fears.

Body posture

Among chimpanzees, subordinate males greet a dominant male in a **servile** manner. They emit soft, grunting noises, make a series of quick bows and prostrate themselves low enough to be able to look up respectfully at the dominant male.

Sexual presentation

Female chimpanzees, on the other hand, employ a **sexual** approach to appease a dominant male by offering their rumps for sniffing and possible mounting. By doing this they are arousing a response that is an **alternative** to aggression.

Survival value

Ritualistic displays and appeasement behaviours are of survival value because they **reduce conflict** to a minimum within a closely-knit social group. They make it possible for weaker members to live in close proximity to stronger members without threats of needless hostility constantly erupting.

Social status and alliances

Within a primate society, a social hierarchy exists and each member accepts its place (in other words, its rank or status) on the social ladder. The individual tends to refrain from challenging those members of the group that are of higher rank. However, in reality such a hierarchy is both **complex** and in a state of **flux**. An individual's status does not necessarily remain fixed at a certain level.

Change of status

In a society of vervet monkeys, the males are forced to leave the group on reaching puberty. The females remain in the group and form a social hierarchy where a daughter inherits her mother's rank. Individuals of **high status** enjoy priority access to resources such as food. Female vervets of all ranks form long, close relationships with their offspring. These relationships are maintained by frequent grooming and may be extended to include non-relatives which results in the formation of **alliances**.

Females of high status attract more non-kin grooming partners into their alliances than do females of low status. In Figure 19.15, for example, vervet monkey D is a high-ranking female who attracts grooming from five other females, whereas P is a low-ranking female who attracts grooming from only two other females.

However, early in the breeding season when there are very few babies present in the group, a newborn baby, regardless of the mother's rank, acts as a magnet to all ranks of females. Therefore a mother of low rank who gives birth at this time can **increase her status** because many other females are keen to become her grooming partners (see Figure 19.16). If a low-ranking female continues to give birth, season after season, to more daughters than her rivals, she may continue to improve her status within the hierarchy and retain it.

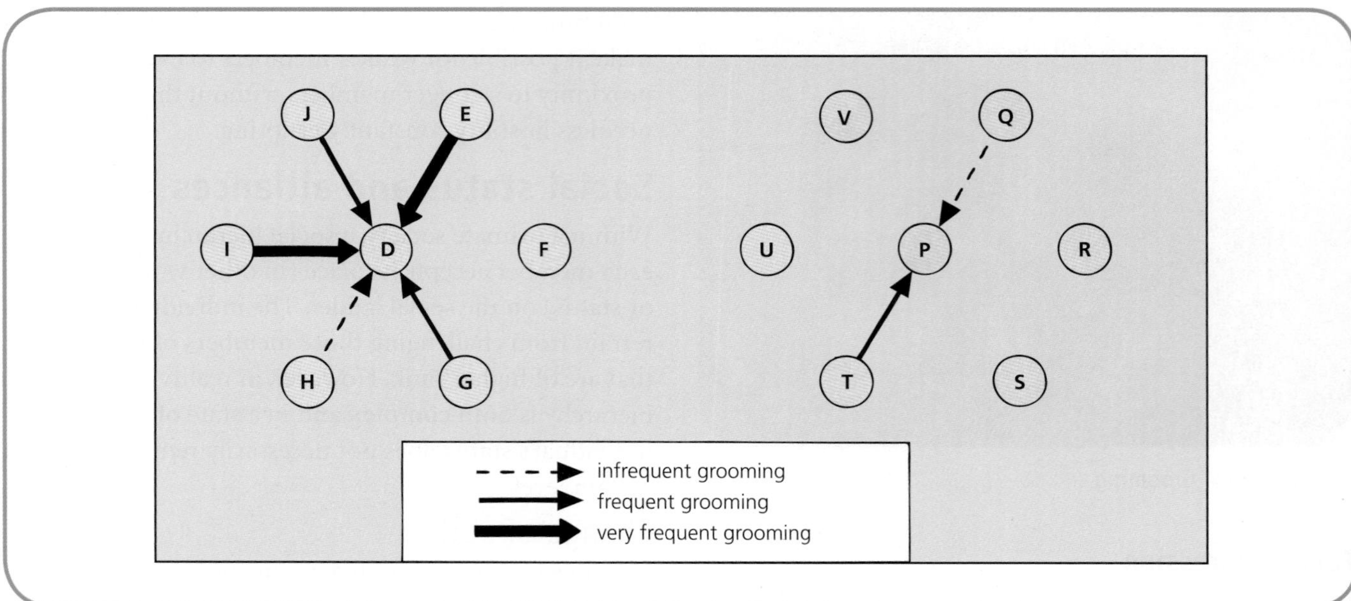

Figure 19.15 Grooming received by two vervet monkeys of different rank

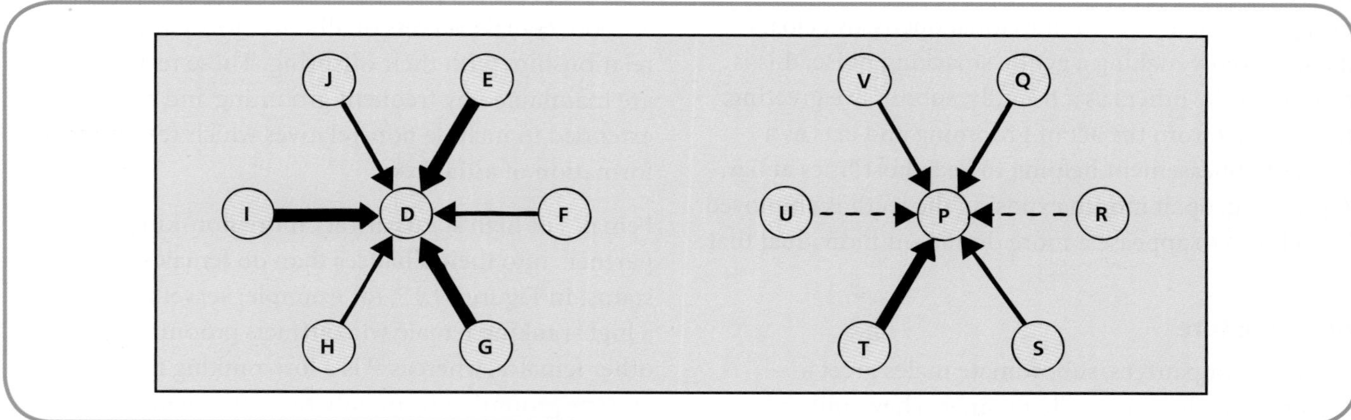

Figure 19.16 Grooming received by two vervet monkeys after giving birth

Testing Your Knowledge 2

1 The housefly lives a solitary life during which it finds food and searches for mates. In what TWO main ways does the behaviour of the members of a society of social insects differ from that of a housefly? (2)

2 **a)** Give TWO examples of complex social behaviours learned by young primates during the period of parental care. (2)

b) What is the survival value of appeasement behaviour among the members of a group of primates? (1)

3 Decide whether each of the following statements is true or false and then use T or F to indicate your choice. Where a statement is false, give the word that should have been used in place of the word in bold print. (7)

a) When mounting a threat display, a primate makes its body look as **small** as possible.

b) Grooming reinforces **rivalry** between social primates.

c) Social primates use appeasement behaviour to reduce **conflict** to a minimum.

d) A primate that closes its eyes during an encounter with a rival is indicating **aggression**.

e) Some female primates use a sexual approach to **appease** a dominant male.

f) Female vervet monkeys use **subservience** to improve their rank within their social group.

Applying Your Knowledge and Skills

1 Five male zebra finches, P, Q, R, S and T, were kept together and observed over a period of several days. During this time, a record was kept of the results from 20 confrontations between each pair of birds. The bird which successfully dominated its rival in each contest was given a score of one point. The results are shown in Table 19.8.

a) Copy and complete the two right-hand columns in the table. (The first example has been done for you.) (1)

b) i) Which bird has the lowest status and is at the bottom of the pecking order?

ii) Explain your choice. (2)

c) Give the complete pecking order for the five birds in descending order. (1)

Contest	Score out of 20 points	Winner	Net number of contests won
T v Q	T 17, Q 3	T	14
T v R	T 3, R 17		
P v Q	P 18, Q 2		
Q v R	Q 0, R 20		
Q v S	Q 8, S 12		
R v P	R 13, P 7		
P v T	P 14, T 6		
S v T	S 5, T 15		
R v S	R 19, S 1		
S v P	S 4, P 16		

Table 19.8

2 Vampire bats need regular feeds of blood. In the absence of food they reach starvation point at about 60 hours from their previous meal as shown in the graph in Figure 19.17.

a) Bat Q has just fed on the blood of a horse by biting its neck and is now at 120% of its pre-feeding body mass. How many hours does it have until it reaches starvation point? (1)

b) It is 45 hours since bat P last fed on blood. How many hours does it have before reaching starvation point? (1)

c) If bat Q regurgitates a donation of blood to bat P, P's pre-feeding body mass increases by 6%. How much time has P gained? (1)

d) By donating this meal to P, bat Q's pre-feeding body mass has decreased by 11%. How much time has it lost? (1)

e) Which is greater, the recipient's gain or the donor's loss? (1)

f) Two days later the roles are reversed and P donates blood to Q. What name is given to this form of social behaviour where favours are returned at a later date? (1)

3 Table 19.9 shows the results of students A and B playing the prisoner's dilemma game using the same system of scoring points as shown in Table 19.10.

a) During which round(s) did:

i) A betray B who remained silent?

ii) B betray A who remained silent?

iii) A and B both betray one another?

iv) A and B both co-operate and remain silent? (4)

b) Identify THREE rounds which could be interpreted as:

i) A punishing B for a betrayal

ii) B punishing A for a betrayal. (2)

c) How many points in total would each student have scored if they had co-operated with one another during every round? (1)

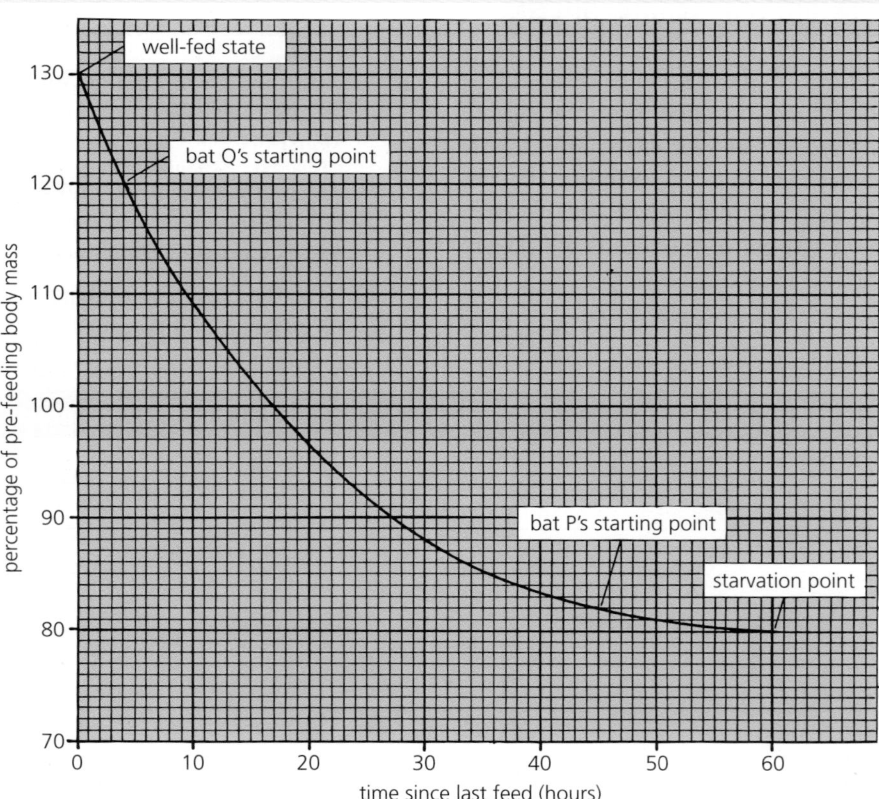

Figure 19.17

Round	Student A's actual score	Student B's actual score
1	3	3
2	5	0
3	0	5
4	1	1
5	1	1
6	3	3
7	3	3
8	0	5
9	5	0
10	0	5
11	1	1
12	1	1
13	3	3
14	5	0
15	3	3
16	0	5
17	0	5
18	1	1
19	3	3
20	3	3

Table 19.9

d) i) Draw a bar chart which allows the actual total scores achieved by A and B to be compared with one another and with the total scores that they would have achieved had they co-operated throughout.

ii) Draw TWO conclusions from your bar chart. (5)

e) Copy the following paragraph and complete the blanks using the answers that follow it.

Even knowing that ______ results in the best outcome for the group as a whole, many players of the prisoner's ______ find it difficult to reciprocate acts of ______ by choosing to ______ the sentence of their accomplice at the cost of staying a little ______ in jail themselves. In the hope of achieving their personal optimal outcome, they tend to make ______ decisions and continue to ______ their accomplice which results in both players ending up ______ off than they would have been by acting co-operatively.

altruism, betray, co-operation, dilemma, longer, reduce, self-interested, worse (4)

4 The stick graph in Figure 19.18 shows a record of five of the tasks (A–E) performed by a worker bee during the first 30 days of her adult life and the time she spent resting.

a) Put the five tasks into the sequence in which she carried them out during these 30 days. (1)

b) Identify the days when she was employed on two tasks and name the tasks. (2)

c) Calculate the total number of hours that she spent cleaning cells in the hive. (1)

d) During the days that she built combs, what was her percentage mean time per day devoted to this task? (1)

e) i) Compare days 23 and 24 with reference to percentage time spent foraging and resting.

ii) Give a possible explanation for the differences you gave as your answer to **i)**. (4)

5 Read the passage and answer the questions that follow it.

A typical group of gorillas consists of one mature, silver-back male, three adult females and several youngsters. Normally the adults have all originated from different social groups because male and female gorillas, on reaching puberty, emigrate from their troop of birth.

The first female to be made part of his group by a lone silver-back male holds the highest rank. The last female to join the group holds the lowest rank. The highest-ranking female and her offspring are allowed to remain closest to the silver-back male. When a

	X remains silent	X betrays Y
Y remains silent	X and Y each receive 3 points	X receives 5 points; Y receives 0 points
Y betrays X	Y receives 5 points; X receives 0 points	X and Y each receive 1 point

Table 19.10

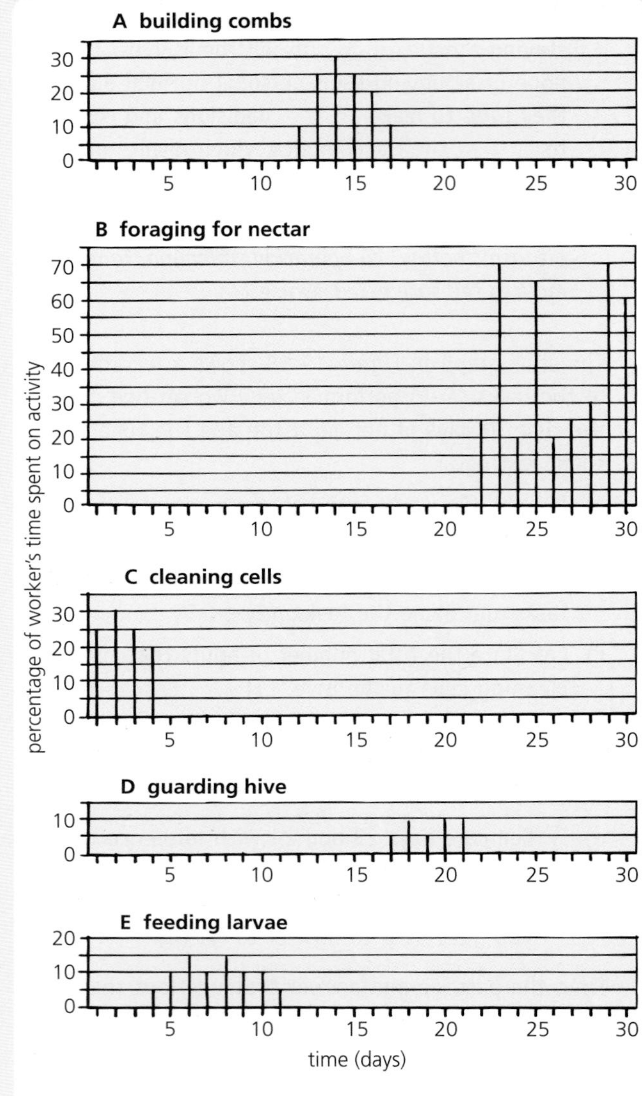

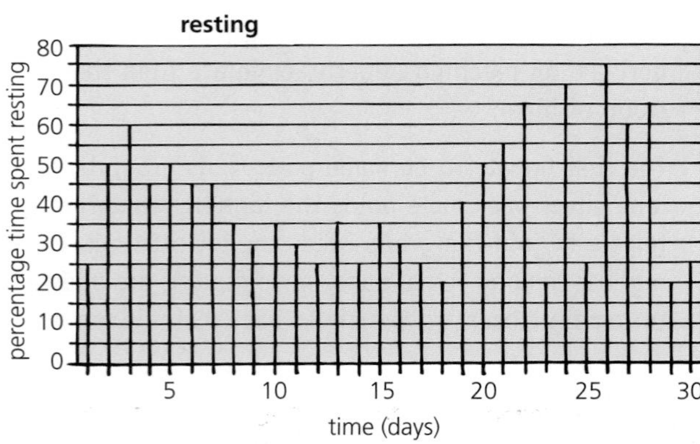

Figure 19.18

silver-back is old and ready to die, one of his sons may inherit the group on reaching sexual maturity.

a) i) Describe the social hierarchy that exists among the females in a group of gorillas.
ii) Which gorilla is least likely to benefit from this arrangement when the group is under attack?
iii) Suggest why the silver-back does not offer equal protection to all the females and their offspring in such a crisis. (3)

b) i) By what means is inbreeding among the members of a social group of gorillas prevented?
ii) When might this mechanism break down and allow inbreeding to occur? (2)

6 In an investigation into the ability of apes to read emotional expressions on a human face, a scientist sat on one side of a transparent screen with two boxes on the table in front of her. On the other side of the screen sat an ape, unable to see inside the boxes. The scientist looked inside one box which, unknown to the ape, contained a piece of banana and smiled with pleasure. Next she looked inside the other box which, unknown to the ape, contained a dead spider and expressed disgust. The ape was then allowed to reach through a hole in the screen and choose a box. The procedure was repeated ten times for each ape and many apes were used. The results are shown in Table 19.11.

a) i) What conclusion can be drawn about the effect of age on the ability of apes to read emotions on a human face?
ii) Explain your answer to i) with reference to the data.

Age and gender of ape	Mean number of times box was chosen	
	Banana	Dead spider
mature males	7.9 (+)	2.1 (−)
juvenile males	5.2	4.8
mature females	7.4 (+)	2.6 (−)
juvenile females	5.3	4.7

(+/− = value significantly higher/lower than expected by chance alone)

Table 19.11

iii) Construct an hypothesis to account for the conclusion you arrived at in **i**). (3)

b) i) What conclusion may be drawn about the effect of gender on the ability of apes to read emotions on a human face?

ii) Explain your answer to **i**). (2)

c) In repeats of the procedure, why did the scientist vary the positions of the two boxes and the order in which she looked into them before expressing an emotion? (2)

7 Give an account of ritualistic display and appeasement behaviours as exhibited by social primates. (9)

What You Should Know

Chapters 18–19
(See Table 19.12 for word bank)

alliance	kin	ritualistic
altruism	metabolism	secondary
co-operative	mutualism	shared
donor	parasite	status
genes	parental	subordinate
grooming	related	symbiosis
hierarchy	reproduction	vectors
host	resistant	workers

Table 19.12 Word bank for chapters 18–19

1 An intimate, coevolved relationship between members of two different species is called ______.

2 In parasitism, one organism, the ______, depends on another organism, the ______, for its food and harms the host. Parasites often have a limited ______. They are transmitted from host to host by direct contact, by ______ and by ______ stages in the parasite's life cycle which often involves a ______ (intermediate) host.

3 In ______ the two organisms help and depend on one another and they both benefit.

4 Many animals are adapted to life in social groups by exhibiting behaviours such as social ______ and ______ hunting. This form of hunting allows food to be ______ and benefits both dominant and ______ animals.

5 A form of behaviour that benefits the recipient but harms the donor is called ______. Behaviour that seems to be altruistic may be of benefit to both the ______ and the recipient if they are closely ______. Since they have many genes in common, the behaviour increases the chance of the donor's ______ surviving in the recipient's offspring. This phenomenon is called ______ selection.

6 Within societies of certain social insects, only a few members of a colony contribute to the ______ of the group. Most members are sterile ______.

7 Young primates learn complex social behaviours during the extensive period of ______ care that they receive. Unnecessary conflict among social primates is reduced by ______ displays and appeasement behaviours which may involve ______, facial expressions and sexual presentation.

8 An ape or monkey's social ______ within a group may increase by it forming an ______ with other non-related members of the group.

20 Components of and threats to biodiversity

Biodiversity

Biodiversity (a contraction of **biological diversity**) can be defined as the **total variation** that exists among all living things on Earth. It refers to the total number, the complexity of structure and the underlying genetic variation that exists among all living organisms. It also takes into account the variability of the ecosystems to which they belong. It has taken about 3 billion years for the vast variety of life forms that exist on Earth to evolve.

The conservation of biodiversity is critical, not only to preserve the wonders of the planet but also to ensure that as many species as possible are able to avoid extinction and adapt to changing environmental conditions. This will maintain a rich variety of resources which humans may be able to find uses for in the future.

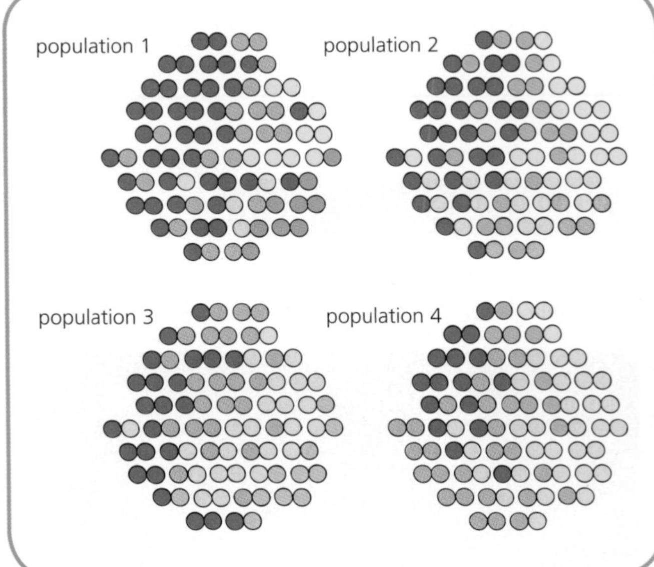

Figure 20.1 Genetic variation in four populations

Measurable components of biodiversity

Most biologists agree that the **measurable components** of biodiversity are threefold:

- genetic diversity
- species diversity
- ecosystem diversity.

Genetic diversity

The **genetic diversity** of a population results from the genetic variation shown by the **number** and **frequency** of all the alleles of the genes possessed by its members.

Figure 20.1 represents four populations of a species. Each symbol represents a diploid individual which possesses two alleles of one gene. Six alleles of this gene exist among the members of the population in different frequencies. The red allele, for example, occurs most frequently in population 1 and least frequently in population 4. The green, orange and purple alleles are rare and each is only present in one population.

If any one of populations 1, 2 or 3 dies out then an allele vanishes. If the lost allele is no longer present in any other population, then the species has lost some of its genetic diversity as illustrated by population 4. Loss of a useful allele from a population may limit the species' ability to adapt to changing environmental conditions in the future. This problem is particularly serious when a species is reduced in number to a few scattered populations.

Species diversity

When quantifying the **species diversity** of an ecosystem, the two factors that are taken into account are:

- the **richness** of species – the number of different species present in the ecosystem
- the relative **abundance** of each species – the proportion of each species in the ecosystem.

Species	Relative abundance (%)	
	Community P	Community Q
A	50	20
B	10	15
C	10	15
D	10	20
E	10	15
F	10	15

Table 20.1 Differing relative abundances

Table 20.1 compares two communities from two different ecosystems. They share the same species richness in that each possesses species A, B, C, D, E and F. However, they differ in relative abundance. Community P is dominated by species A and has a lower species diversity than community Q which is not dominated by one particular species.

Ecosystem diversity

Ecosystem diversity refers to the number of distinct ecosystems present in a defined area.

Figure 20.2 shows equal-sized samples of two city parks. When these areas are compared, Park B is found to contain a higher number of different ecosystems than Park A and is therefore likely to possess a higher level of biodiversity.

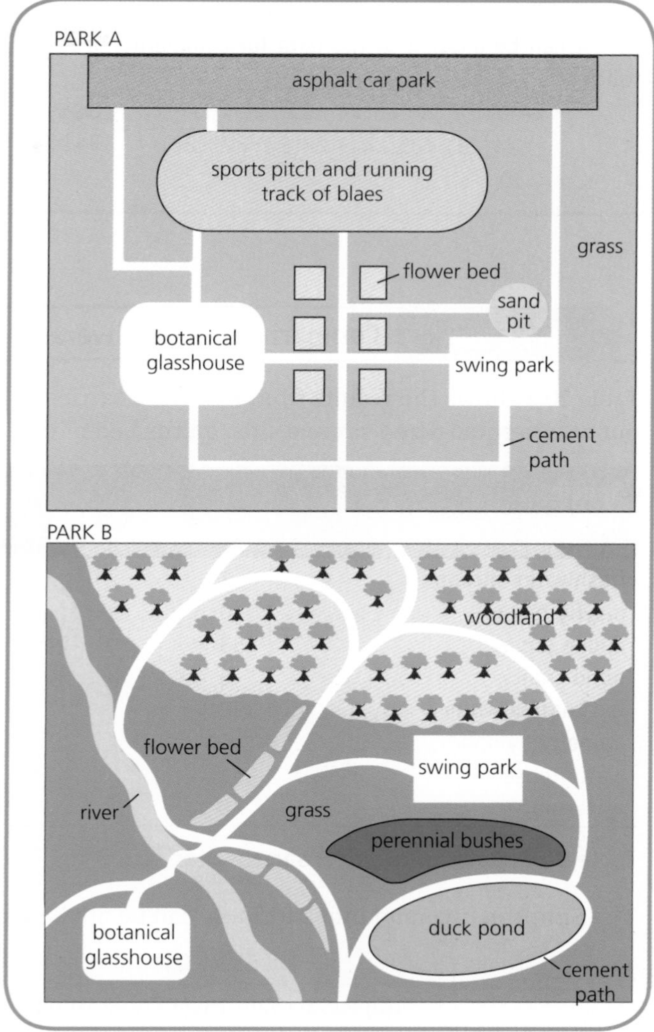

Figure 20.2 Ecosystem diversity

Related Activity

Simpson diversity index

The **Simpson index** is a measure of biodiversity. It uses number of individuals of each species as a measure of abundance. It is obtained by using the following formula:

$$D = \frac{N\,(N-1)}{\Sigma n\,(n-1)}$$

where D = biodiversity index

N = total number of individuals of all species

n = number of individuals per species

Σ = sum of

This formula can be applied to the data in Table 20.2 for community X as follows:

$$D = \frac{90+3+4+3(90+3+4+3-1)}{90(89)+3(2)+4(3)+3(2)} = \frac{9900}{8034}$$

$$= 1.23$$

Species	Number of individuals	
	Community X	Community Y
J	90	24
K	3	25
L	4	28
M	3	23

Table 20.2 Data for Simpson index

And it can be applied to the data for community Y as follows:

$$D = \frac{24+25+28+23(24+25+28+23-1)}{24(23)+25(24)+28(27)+23(22)} = \frac{9900}{2414}$$
$$= 4.10$$

These calculations show that community X with dominant species J has a lower biodiversity index than community Y with the same species richness but no dominant species.

Case Study Comparison of biodiversity indices

Table 20.3 shows the results from fieldwork carried out at two equal-sized sample sites on the bed of the same river. One of the sites was polluted with organic waste, the other was non-polluted.

Species	Number of organisms	
	Polluted site	Non-polluted site
freshwater shrimp	1	27
mayfly nymph	0	53
midge larva	153	0
pond snail	0	9
red tubifex worm	138	0
stonefly nymph	0	18
water louse	241	10
water mite	0	25

Table 20.3 Data for two river sample sites

The Simpson diversity index formula can be applied to the data in Table 20.3 as follows.

$$D \text{ (polluted site)} = \frac{553(532)}{1(0)+153(152)+138(137)+241(240)}$$
$$= \frac{283\,556}{100\,003}$$
$$= 2.84$$

$$D \text{ (non-polluted site)} = \frac{142(141)}{27(26)+53(52)+9(8)+18(17)+10(9)+25(24)}$$
$$= \frac{20\,022}{4526}$$
$$= 4.42$$

From these results it is concluded that the polluted site has a lower species diversity than the non-polluted site. (However, many more sites would need to be sampled to give results that were reliable.)

Research Topic Need for central database

The total number of different species on Earth is unknown. Most large organisms have been studied, classified and named. However, only a rough guess can be made about the number of the many types of smaller organism (such as protozoa and bacteria) and about the number of organisms of all sizes that live in ecosystems yet to be fully explored by humans (including the deep ocean bed and remote tropical rainforest).

It is estimated that there are about 2 million **known** species on Earth. Of the 50% that are animals, most are species of insect. Among the vertebrates, the most numerous are fish. The remaining 50% of known species is made up of bacteria, fungi and green plants. Of these about 250 000 are species of flowering plant.

However, there remains an enormous lack of accurate information particularly about groups of smaller organisms. Therefore estimates of the total number of species on Earth vary. Many experts believe that the true figure lies somewhere between 5 and 20 million.

Storehouse

The world's biodiversity comprises a **natural storehouse** of genetic variation. So far, humans have barely scratched the surface of this vital resource. The vast majority of living species have not been tested for their potential use (for example, as producers of food, medicine, useful chemicals etc.) yet many of them are in danger of becoming extinct and the potential riches stored in their genes being lost.

There is no way of knowing in advance which species are most likely to be of value to us. Unexpected ones can turn out to be of great use. For example, scientists have developed a fluorescent chemical made by a species of jellyfish (see Figure 20.3) which acts as a marker for tracing cancer cells.

Database

Clearly a detailed profile of every known species held in a **central database** is needed. Many databases of species already exist. The United Nations World Conservation Monitoring Centre's database contains details of over 75 000 animal species and about 90 000 plant species of conservation interest. The European Environment Agency runs a database containing details of European species. The European Register of Marine Species records all the species known to be present in European seas. The International Species Information System is a database of wild species held in captivity.

Figure 20.3 Fluorescent jellyfish

However, the ideal database would be one that is centralised and that holds comprehensive information on **every known species** on Earth. To construct such a database would require vast international co-operation and funding. In addition it would be difficult to ensure that every item of information entered was completely accurate because an observer may fail to identify a species correctly. Previously undiscovered species might be wrongly classified due to inadequate identification guides. Therefore the information from all contributors would need to be checked thoroughly by experts to ensure that the database maintained high standards of quality control.

Related Activity

Analysis of data on island biogeography

Island biogeography is the study of the factors that affect the distribution and diversity of species on islands.

Colonisation of new islands

Figure 20.4 shows a new uninhabited volcanic island formed at some distance from the mainland that is populated by a diverse community. It is likely that many of these mainland species will migrate to the island and try to colonise it. However, once on the island, many of them will fail and become extinct. At the point where the rate of **immigration** is equal to the rate of **extinction**, the **equilibrium species number** that the island can support is reached (see Figure 20.5).

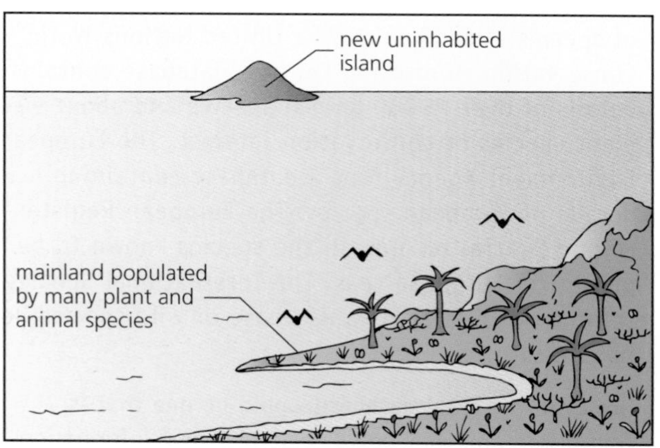

Figure 20.4 Potential colonisation of new island

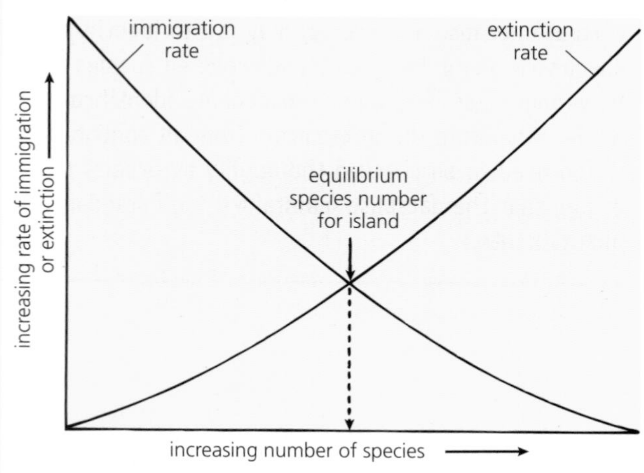

Figure 20.5 Equilibrium species number

Effect of area on species diversity

The equilibrium species number for two islands equidistant from the mainland is also affected by the **size** of the island. An island with a large surface area has a higher immigration rate because potential colonisers are more likely to discover a larger island than a smaller one. A large island is also likely to have a lower extinction rate than a smaller island. This is because the larger one will tend to possess more resources and a wider range of habitats for use by the immigrants.

As a result a smaller island has a lower equilibrium species number (in other words, a lower level of species diversity) than a large island (see Figure 20.6). This prediction is supported by evidence from several studies. For example, Figure 20.7 shows a graph of the results from an investigation into species diversity of amphibians and reptiles on islands in the West Indies.

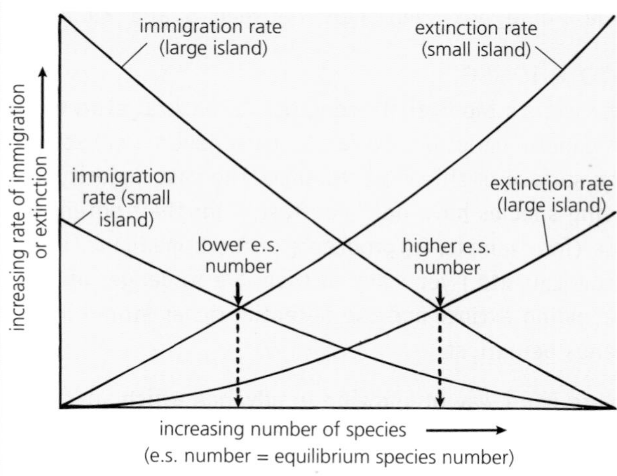

Figure 20.6 Effect of size (area)

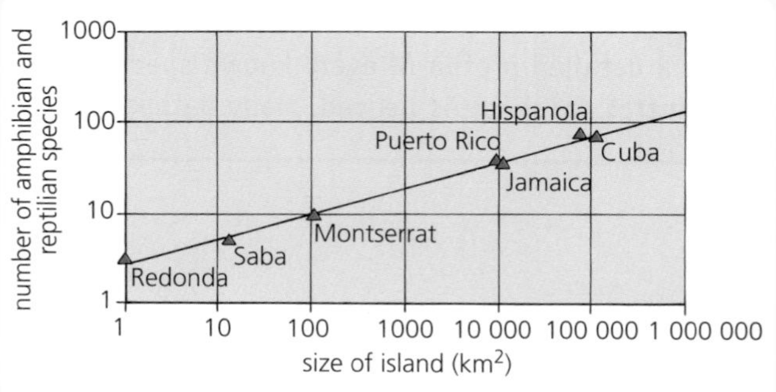

Figure 20.7 Effect of island size on biodiversity

→

Table 20.4 shows the results from a study of species diversity among long-horned beetles on islands off the coast of Florida, USA.

In both of the above examples, the larger the island's surface area, the higher the level of its species diversity.

Island	Area (km^2)	Number of species of long-horned beetle present
A	0.9	3
B	2.9	15
C	3.1	16
D	4.3	16
E	17.1	24
F	55.1	44

Table 20.4 Effect of island size on biodiversity

Testing Your Knowledge 1

1 a) Identify TWO components of biodiversity that are measurable. (2)

b) Why would the loss of population 4 in Figure 20.1 (page 286) be less serious to the species' chance of survival than the loss of population 1, 2 or 3? (1)

2 a) If two communities are equal in species richness but only one of them possesses a dominant species, which community would have the higher species diversity? (1)

b) Return to Figure 20.2 on page 287 and identify THREE types of ecosystem present in Park B but absent from Park A. (3)

3 Decide whether each of the following statements is true or false and then use T or F to indicate your choice. Where a statement is false, give the word that should have been used in place of the word in bold print. (4)

a) The total variation that exists among the living things on Earth is called **biodiversity.**

b) The number of different species in an ecosystem is called the species **abundance.**

c) The proportion of each species in an ecosystem is called the relative **richness.**

d) Genetic variation in a population is represented by the number and frequency of all the **alleles** possessed by its members.

Threats to biodiversity

Overexploitation

In biology, the expression 'to **exploit** a natural resource' means to make the best use of it. 'To **overexploit** a species' means to remove and use up individuals at a rate that exceeds the species' maximum rate of reproduction.

A common example of overexploitation of species is **overharvesting** (also called **overfishing** in the case of fish, whales and marine invertebrates). This process depletes some species to such a low level that their continued exploitation is no longer sustainable. If the overexploitation is stopped in time the population may be able to recover; if not they may become extinct.

In the past the seas were fished freely without thought for the ability of stocks to replenish themselves. The situation reached a peak with the development of sophisticated sonar techniques to locate shoals and the use of enormous nets (some larger than the size of an Olympic sports stadium) to catch them in. Eventually it was agreed that this overexploitation of fish stocks could not continue unchecked.

Recovery

It is estimated that at present around 70% of the world's marine fishing grounds (fisheries) are fully exploited

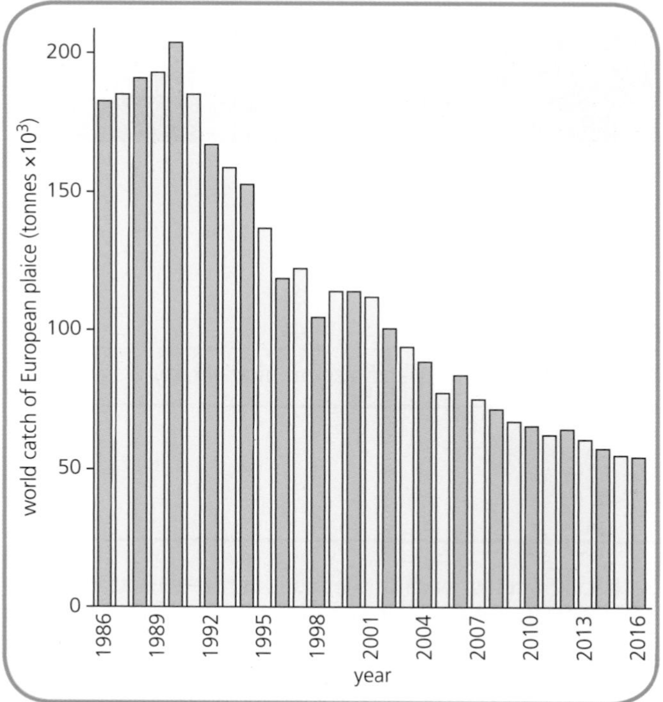

Figure 20.8 Effect of overfishing on plaice

and that many fish populations have been reduced by 60–95% resulting in lower catches (see Figure 20.8). In recent years attempts have been made to promote the recovery of depleted fish stocks. Strict regulations insist that catches must stay within certain **fixed quotas**. Further measures limit fishing by controlling the number of boats, the length of time they spend at sea and the areas where they are allowed to fish.

When these practices are fully supported by the authorities, estimates of some fish populations are found to show **signs of recovery**. In recent years estimated stocks of cod in the North Sea have shown a promising rise and have reversed the previous downward trend (see Figure 20.9).

Whales

Although estimates of whale populations tend to be subject to a large margin of error, it is known that their numbers have declined dramatically since the start of commercial whaling.

In 1986 the International Whaling Commission imposed a **moratorium** (suspension of activity) on commercial whaling. However, several countries have ignored the ban

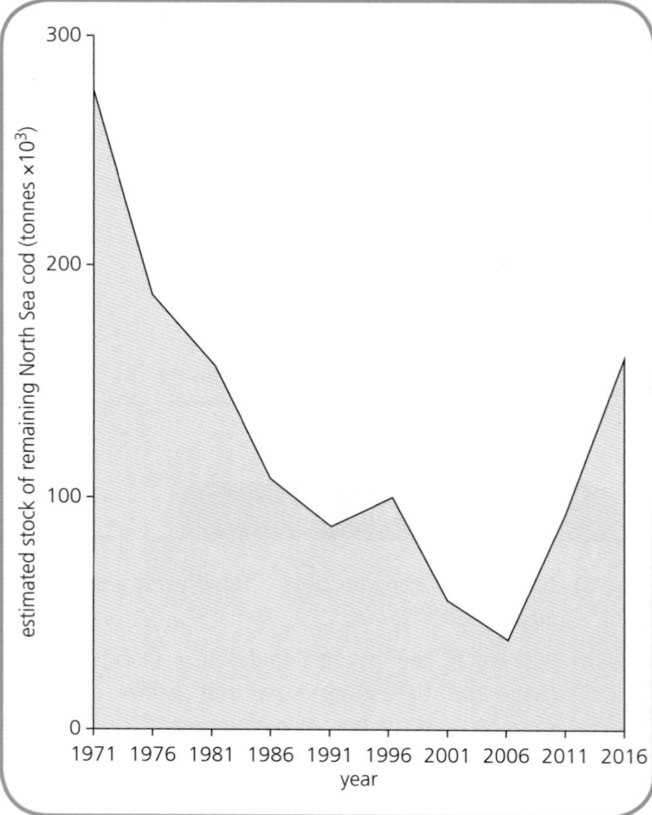

Figure 20.9 Recovery of cod stocks

or found ways around it (see Figure 20.10) and many thousands of whales have been killed since 1986. The downward spiral in whale numbers will only be halted if the moratorium is retained and all countries of the world respect it.

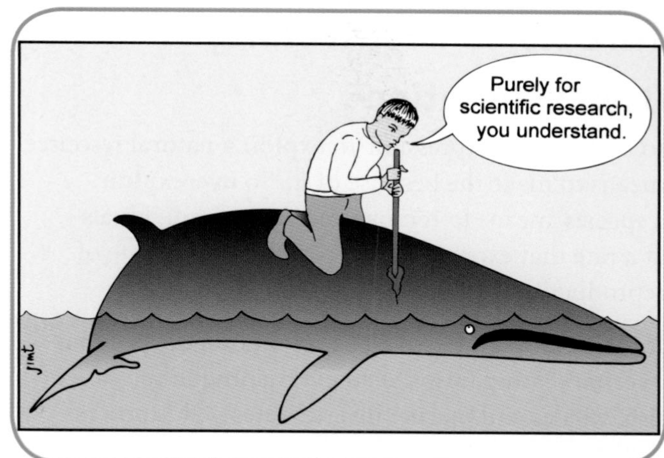

Figure 20.10 Overexploitation of whales?

Bottleneck effect

A significant percentage of a population may be wiped out by a disaster such as a fire, flood or earthquake acting in an unselective way. If the surviving population is very small, it may have lost much of the genetic variation needed to adapt to future environmental change. Such a population is said to have suffered the **bottleneck effect** as a result of the disaster, which is called the **bottleneck event**.

In Figure 20.11 the bottle containing coloured beads represents **genetic variation** among the members of the original population. The container representing the surviving population is found to possess:

- less genetic variation
- allele frequencies that are different from the original population.

The loss of genetic diversity produces a population whose members are so similar that reproduction among them is genetically equivalent to inbreeding. Since inbreeding results in poorer rates of reproduction, the population may become extinct or it may slowly recover (see Figure 20.12) and survive. Some species, such as the northern elephant seal, possess a naturally low level of genetic diversity among their members, yet still manage to remain viable.

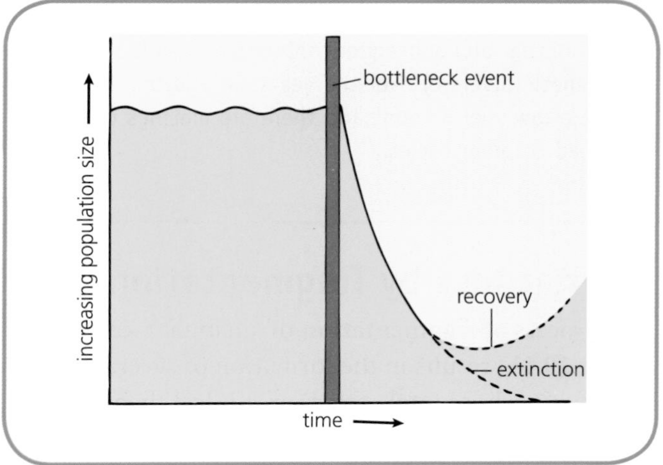

Figure 20.12 Aftermath of bottleneck event

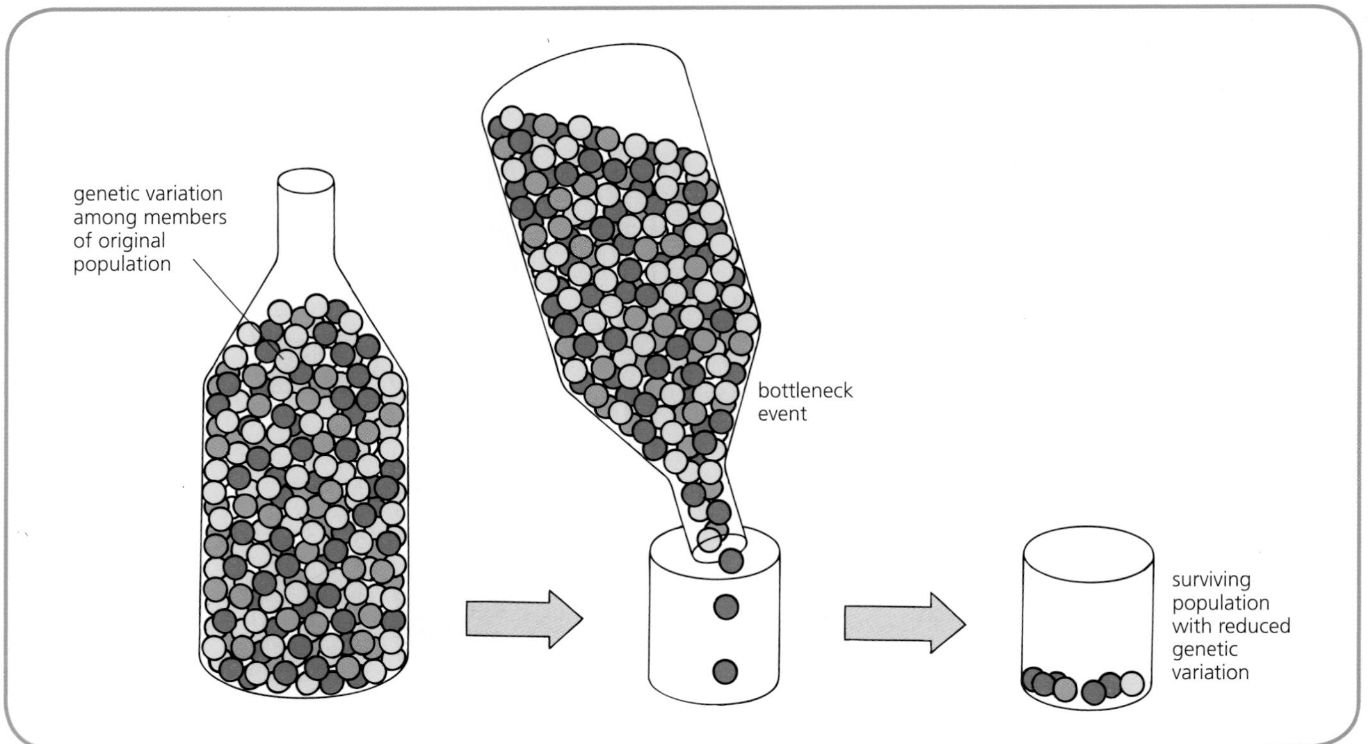

Figure 20.11 Bottleneck effect

Research Topic Cheetah

There are only a few thousand cheetahs in the world and they are so closely related genetically that a skin graft from one cheetah to another is not rejected. It is thought that extreme climatic change acting as a bottleneck event about 10 000 years ago wiped out all but a handful of cheetahs. Close relatives had no option but to mate with one another and subsequent inbreeding has led to a loss of genetic diversity over the years. In addition, males have a low sperm count and therefore matings often fail to lead to pregnancies.

However, it has been discovered recently that, unlike other big cats, female cheetahs are surprisingly promiscuous and often mate with more than one partner per litter of cubs. The benefit of this behaviour is that the female undergoes **induced ovulation** (release of an egg) each time she mates. She could therefore produce a litter of cubs each with a different father. This process helps to ensure that any traces of genetic diversity survive in the population. It may also serve to protect the cubs from being killed by an adult male cheetah, several of whom are the parents of the litter.

Habitat loss by fragmentation

The process of **fragmentation** of a habitat (see Figure 20.13) results in the formation of several **habitat fragments** whose total surface area is less than that of the original habitat.

Degradation of the edges of the fragment leads to a further decrease in size of the fragment and loss of habitat. Normally the fragments possess limited resources and are only able to support a lower species richness than that of the original habitat. In addition, small fragments can only support small populations and these are more vulnerable to extinction than large populations possessing greater genetic diversity.

Edge to interior ratio

Fragmentation and subsequent degradation of the edges of the fragments result in an **increase in the ratio** of the total length of a fragment's edge to the total surface area of its interior (see Figure 20.14). As a result, species adapted to the original habitat's edge may prosper at the edge of a fragment. However, as their numbers increase, they may be driven to invade the interior of the fragment and compete with the species adapted to the interior. Therefore small fragments do not tend to favour species that need an interior habitat and this results in a loss of biodiversity.

Human causes

Habitat fragmentation often occurs when natural ecosystems such as forests are cleared for agriculture, housing or hydroelectric dams. Often the remaining fragments are tiny 'islands' of natural forest isolated from one another by farmland, housing estates or motorways.

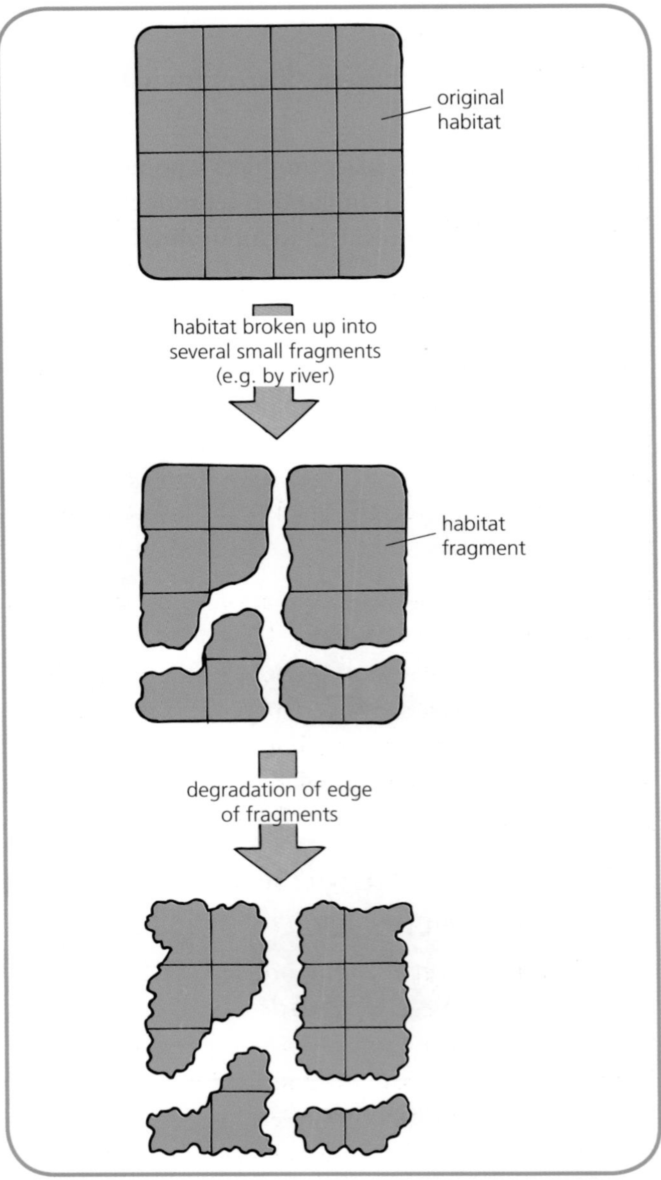

Figure 20.13 Habitat fragmentation

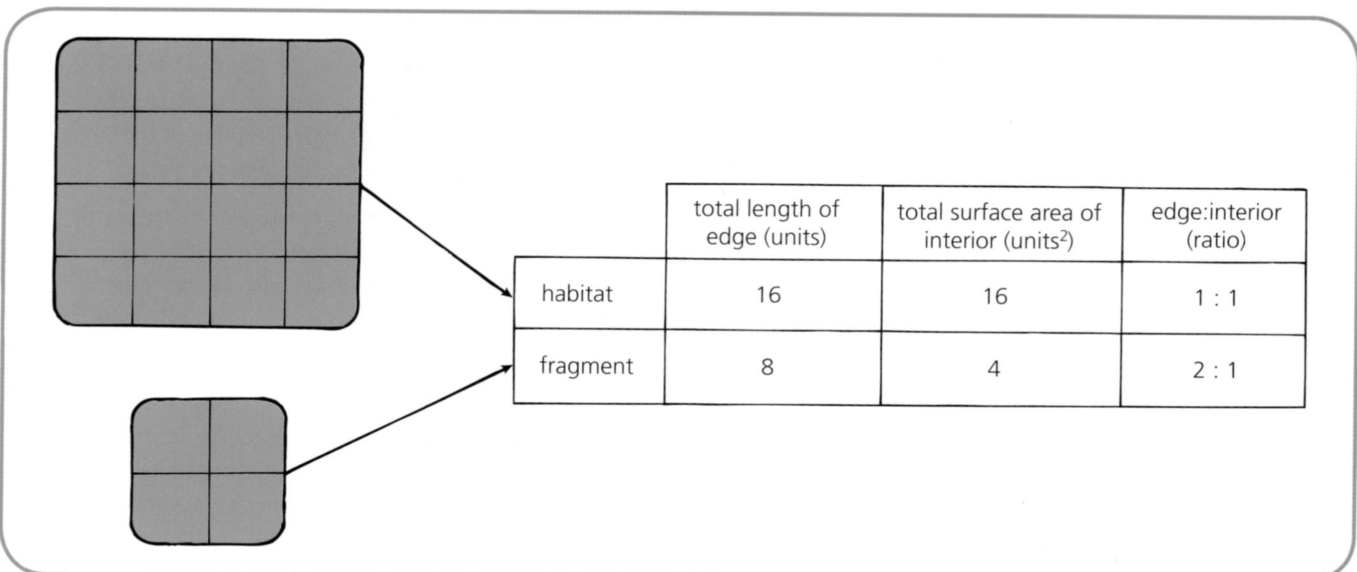

	total length of edge (units)	total surface area of interior (units^2)	edge:interior (ratio)
habitat	16	16	1 : 1
fragment	8	4	2 : 1

Figure 20.14 Ratio of edge to interior

Habitat corridors

One possible solution to the problem is to link isolated fragments with **habitat corridors**. A habitat corridor is composed of a narrow strip or a 'stepping-stone' series of clumps of quality habitat by which species can move between disconnected fragments of habitat. For example, it may take the form of a streamside habitat, a wooded strip between two forest fragments or an underpass beneath a motorway. This would enable the members of a species to feed, mate and recolonise habitats after local extinctions.

It is not yet clear whether the creation of habitat corridors increases biodiversity. They do overcome the problem of isolation but not of loss of interior habitat. Some scientists have suggested that they could even be harmful because they could allow the spread of disease among small vulnerable populations possessing limited genetic diversity.

Introduced, naturalised and invasive species

The relationship between **introduced**, **naturalised** and **invasive** species is summarised in Figure 20.15. Over the years many foreign species have been introduced to the UK. Only a minority of these non-invasive species have become invasive and have had a negative impact on the local native communities and the economy.

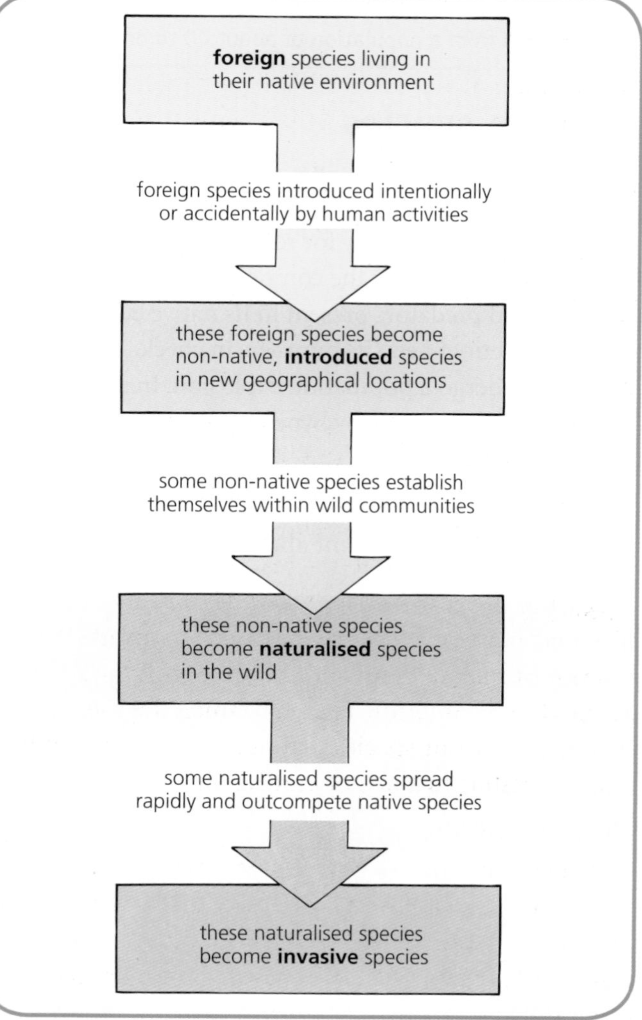

Figure 20.15 Emergence of invasive species

Related Topic

Habitat corridors for tigers

During the twentieth century, the world's population of tigers was reduced by 95% as a result of hunting and poaching. Three subspecies became extinct and a fourth, the South China tiger, has not been seen for many years. It is estimated that there are about 3000 tigers left in the wild.

To thrive, tigers need large territories containing abundant prey. However, many of these habitats have been cleared by humans for timber, industry, urbanisation and the building of new roads. The remaining tigers in the wild are largely restricted to habitat fragments and nature reserves. Efforts have been made to connect these areas with **habitat corridors** to allow the tigers to find new mates in other regions and **increase genetic variation** among future generations. The presence of a corridor in the far east of Russia, for example, has enabled the Siberian tiger to increase from a population of about 40 to one of around 500. However, these tigers are so closely related that they are equivalent to just 14 genetically different individuals. Therefore their future remains, at best, uncertain.

Tiger habitat corridors are also present in India but they are not protected legally and exist alongside a dense and fast-growing human population. Some of the corridors have been cut off by dams, highways or urban developments and funds are unavailable for the building of underpasses.

On the positive side, the wild cat conservation group Panthera has set up the Tiger Corridor Initiative and has assessed the potential for a multinational Eastern Himalayan corridor. This would involve several countries including India, Bhutan, Nepal and Thailand. However, the success of any tiger corridor that runs through a region heavily populated by humans depends upon the understanding and acceptance by the local people combined with excellent maintenance and management of the corridor.

Invasive species

An invasive species normally succeeds because it is able to prey upon or to outcompete the other native members of its adopted community for resources. This is made possible by the fact that the competitors, pathogens, parasites and predators present in its native ecosystem are not present to keep its numbers in check. Therefore it is free to undergo a population explosion. In some cases invasive species succeed by hybridising with native species.

Economy

Japanese knotweed (see Figure 20.16) was introduced to the UK as a garden plant about 200 years ago. It is now widespread throughout Britain. It is capable of growing through hard structures such as foundations of buildings and car parks. Its activities affect the economy because vast sums of money need to be spent each year to eradicate it.

Health

Giant hogweed (see Figure 20.17) was also introduced as a garden plant but it has become invasive. Its sap is

Figure 20.16 Japanese knotweed

Figure 20.17 Giant hogweed

poisonous and causes severe burning and blistering of the skin (see Figure 20.18).

Loss of biodiversity

Many invasive species pose a huge threat to the biodiversity of the region that they invade. In the UK, for example, red squirrels are rapidly facing extinction because they are being outcompeted by their rival, the grey squirrel introduced from North America. Similarly, water voles are being wiped out by American minks.

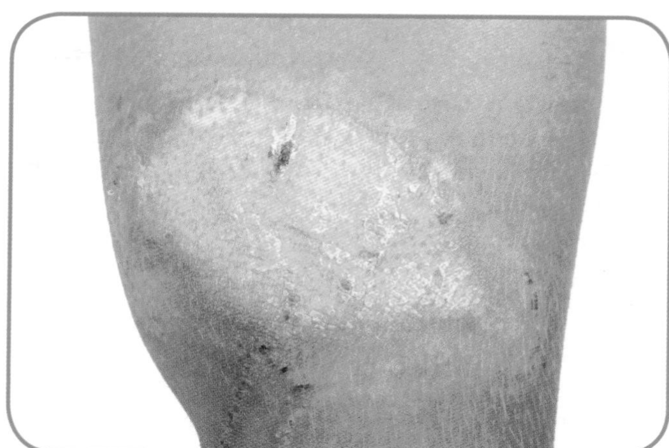

Figure 20.18 Effect of giant hogweed

Testing Your Knowledge 2

1 Explain the difference between *exploitation* and *overexploitation* with reference to a named natural resource. (2)

2 a) In what way could a *bottleneck effect* alter the quantity of genetic variation present in a small population of a species? (1)

b) Why would this decrease the population's chance of survival? (1)

3 a) Copy and complete Table 20.5. (4)

b) By what means can small populations isolated in habitat fragments be reunited? (1)

4 By what means does an introduced species become:

a) *naturalised*? (1)

b) *invasive*? (1)

	Type of habitat	
	Island fragment	**Unfragmented land mass**
relative size of habitat (large/small)		
species richness supported (high/low)		
edge : interior (large/small)		
chance of edge species invading interior (high/low)		

Table 20.5

Applying Your Knowledge and Skills

1 The data in Table 20.6 refer to the catch of plaice in the North Sea by a fleet of trawlers over a 5-year period.

a) Which age of plaice is the most common in each year's catch? (1)

b) i) What trend is shown by the data when read vertically downwards from 5-year-old fish?
ii) Explain why. (2)

c) Suggest why the '12+ years' entry is always greater than the '11 years' entry. (1)

d) Make a generalisation about the way in which the catch data for years 4 and 5 differ from those for years 1–3. (2)

e) The fishermen were pleased with their catch in years 4 and 5. Suggest why they should be concerned about the future (if the trend shown by the data continues). (2)

f) What is the percentage decline in number of 10-year-old fish in the 5-year study? (1)

2 Several equal-sized areas of river bed from a non-polluted region and a polluted region were examined. The abundance of species for each area was recorded as shown in Table 20.7.

a) i) Do the two types of area differ in species richness?
ii) Explain your answer. (2)

Species	Mean number of individuals	
	Non-polluted	Polluted
rat-tailed maggot	2	21
sludge worm	3	29
mayfly nymph	15	0
caddisfly larva	11	0
stonefly nymph	12	0
freshwater mussel	7	0

Table 20.7

b) Calculate the biodiversity index for each area using the formula: (2)

$$D = \frac{N\ (N-1)}{\Sigma^{n}\ (n-1)}$$

where D = biodiversity index
N = total number of individuals of all species
n = number of individuals per species
Σ = sum of

3 Every species of invertebrate living on each of the five small islands shown in Figure 20.19 was identified and recorded. No vertebrates were found on the islands. Next a tent was erected over each island and it was fumigated with pesticide to kill every animal. Recolonisation of the islands was then monitored over a period of a year. The results are shown in Table 20.8 where a tick indicates presence of a species.

Age of fish (years)	Total number of plaice caught				
	Year 1	Year 2	Year 3	Year 4	Year 5
2	89	78	83	421	573
3	547	625	602	2127	2268
4	2485	2331	2491	2683	2594
5	2137	2027	2076	1527	1332
6	891	766	750	363	257
7	417	389	428	215	209
8	235	201	217	92	76
9	67	94	86	41	38
10	50	81	63	20	15
11	43	57	38	17	12
12+	77	94	69	32	28

Table 20.6

→

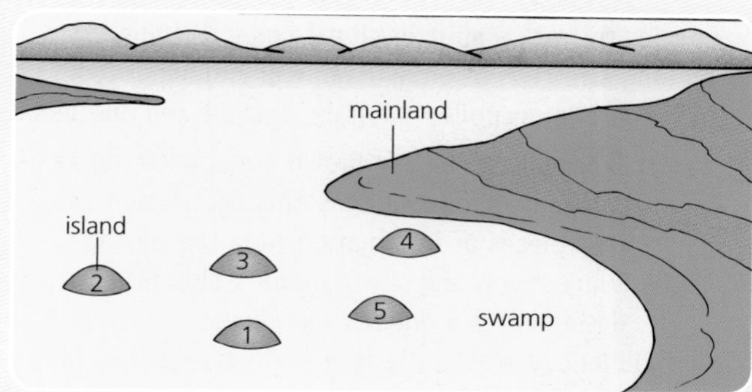

Figure 20.19

Species	Island									
	1		2		3		4		5	
	b	a	b	a	b	a	b	a	b	a
A	✓			✓	✓	✓	✓	✓	✓	
B	✓	✓	✓	✓	✓		✓			✓
C						✓			✓	✓
D		✓				✓	✓	✓		
E							✓		✓	✓
F					✓	✓		✓	✓	
G	✓	✓	✓				✓	✓		
H		✓			✓	✓		✓	✓	✓
I			✓	✓			✓	✓		✓
J	✓				✓	✓	✓	✓	✓	
K	✓	✓			✓		✓	✓		✓

(b = before fumigation, a = after fumigation)

Table 20.8

a) Do the data refer to the richness of species or relative abundance of species present on these islands? (1)

b) Which island possesses
   i) the largest number of different species?
   ii) the smallest number of different species?
   iii) Suggest why this is the case. (3)

c) What effect did recolonisation have on the equilibrium species number of an island? (1)

d) Following fumigation, which species managed to colonise
   i) the most islands?
   ii) the fewest islands? (2)

4 The data in Table 20.9 refer to estimated numbers of six species of whale. Calculate the data missing from boxes a)–e). (5)

5 Read the passage and answer the questions that follow it.

The **zebra mussel** is a freshwater invertebrate with a yellow and brown striped shell that varies in length from 6 to 50 mm. It is native to Eastern European waters but was introduced to North America in the ballast water of freighters.

The female can produce several hundred thousand eggs per year. These become larvae that attach themselves to any firm surface and grow into adults. One such surface is the shell of the local species of mussel which dies and becomes extinct in the area affected.

Zebra mussels are now widespread in the USA and are responsible for blocking pipes and water intakes to hydroelectric schemes. Their shells are sharp enough to shear fishing lines and injure swimmers. They feed on plankton by filtering lake water, which increases its clarity and enables algae to grow at deeper depths.

Zebra mussels are of high nutritional value to their natural predators, crayfish and fish such as roach, common in Eastern European lakes. Zebra mussels

Species of whale	Estimated number		Estimated percentage remaining
	Before commercial whaling	Present day	
Blue	200 000	5 000	2.50
Fin	450 000	33 000	**a)**
Humpback	**b)**	28 000	20.00
Right	50 000	**c)**	16.80
Sei	250 000	40 000	**d)**
Sperm	**e)**	360 000	24.00

Table 20.9

→

are credited with increasing the population numbers of fish such as smallmouth bass and yellow perch in some American lakes.

a) i) Which of the following terms best describes the zebra mussel – *introduced, naturalised* or *invasive*?
ii) Justify your choice. (2)

b) Give an example of a way in which the spread of the zebra mussel in the USA has impacted negatively on:
i) recreational activities
ii) a utility. (2)

c) The filtering action of zebra mussels helps to remove pollutants from lake water.
i) Which member of the lake ecosystem benefits from this effect?
ii) Explain why. (2)

d) Name:
i) a vertebrate
ii) an invertebrate
that is a natural predator of the zebra mussel in its native ecosystem. (2)

e) Suggest the means by which an attempt could be made to exert biological control of zebra mussels in a small lake in the USA. (1)

f) Does shell length in zebra mussels show discrete or continuous variation? (1)

6 Read the passage and answer the questions that follow it.

Butterflies are very sensitive to environmental change. The slightly warmer conditions that have occurred in the UK in recent years have favoured some species but not others. The fortunate ones are able to change their distribution in response to climate change. The **brown argus**, for example, has rapidly expanded its geographical range and the variety of flowering species upon which it feeds. However, the range occupied by the **silver-studded blue** has decreased significantly and the **chequered skipper** is confined to a few parts of Scotland, having vanished completely from England.

Some butterfly species are found to respond to climate change by moving their range. The **scotch argus**, for example, is found to be heading northwards while the **mountain ringlet** is moving its range to a higher altitude.

The rate at which a species of butterfly can change its range is affected by availability of suitable new habitats. Where a habitat has become severely fragmented by human activities, the ability of a species to track a shift in climate may be impeded by long stretches of unfavourable environment lying between one unspoiled habitat fragment and another.

a) i) Experts describe the brown argus butterfly as a *mobile generalist*. Justify this description using TWO pieces of information from the passage.
ii) Suggest why the silver-studded blue is described as a *habitat specialist*.
iii) Identify a butterfly from the passage that is probably extinct from England. (4)

b) Even when an area becomes warmer and suitable for occupancy by a particular species of butterfly, often that species is unable to extend its range into the new area. Suggest why. (2)

c) i) Explain why 'moving its range to a higher altitude' solves the mountain ringlet butterfly's problem.
ii) Why might this response to climatic change be disastrous in the long run? (2)

d) Suggest ONE human activity that could be responsible for '*... a habitat ... severely fragmented by human activities ...*' (1)

7 Figure 20.20 shows box plots of shell length for three populations of zebra mussel. (See Appendix 2 for help.)

a) What percentage of data is contained in a box in a box plot? (1)

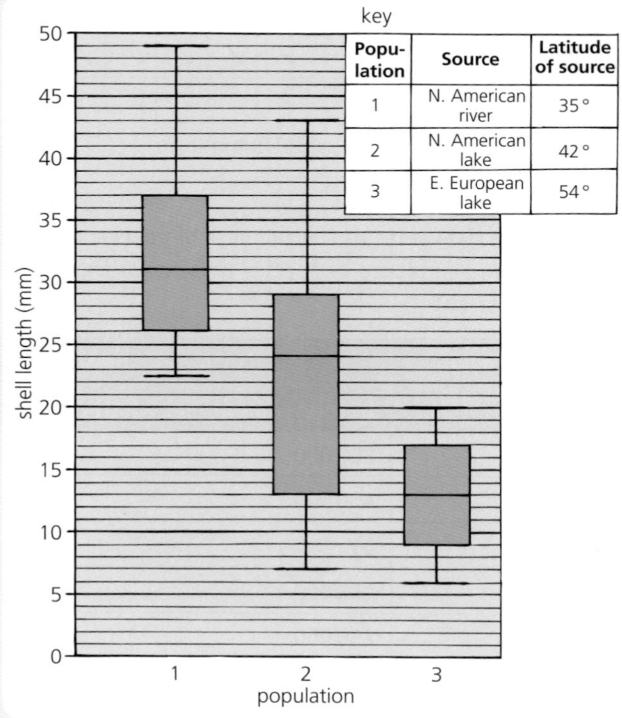

Figure 20.20

b) i) By what means is the median value of the data indicated in a box plot?
ii) State the median value for each of the populations. (2)

c) i) Which box set shows the widest distribution of values between its median and its upper quartile?
ii) Which box set shows the widest overall distribution of values? (2)

d) Does a whisker represent a 95% level of confidence or an actual value? (1)

e) What was the lowest value recorded? (1)

f) If the data were plotted as a graph of number of individuals against shell length, which population would give a symmetrical bell-shaped curve? (1)

g) Construct an hypothesis to account for the effect of latitude on shell size. (1)

What You Should Know

Chapter 20

(See Table 20.10 for word bank)

abundance	dominated	invasive
alleles	edge	naturalised
bottleneck	evolutionary	outcompete
change	fragmentation	overexploitation
corridors	frequency	richness
diversity	genetic	species

Table 20.10 Word bank for chapter 20

1 Components of biodiversity that can be measured are ______ diversity, species ______ and ecosystem diversity.

2 The genetic diversity of a population consists of the variation shown by the number and ______ of all the ______ of all the genes possessed by its members.

3 ______ diversity of an ecosystem depends on the number of different species present, called the species ______ and the proportion of each species present, called the relative ______. If a community is ______ by one species, it has a lower species diversity than one that is equally rich but not dominated by one species.

4 Overharvesting is a form of ______ of a population. If the process is brought to a halt, the population may recover.

5 As a result of the ______ effect, a population may be reduced to a small group that lacks the genetic diversity needed to make ______ responses to environmental ______.

6 ______ of a habitat results in the formation of fragments that support a lower level of species richness than the original large area. Species adapted to the ______ of a fragment may invade the interior and ______ the species living there. Isolated habitat fragments can be linked by ______.

7 When a non-native species is introduced to a new geographical location and establishes itself in the wild, it is said to become ______. When it then spreads rapidly and impacts negatively on native species, it is described as an ______ species.

Appendix 1

The genetic code

Table Ap1.1 shows DNA's bases grouped into 64 (4^3) triplets of bases.

first letter of triplet		second letter of triplet				third letter of triplet
		A	G	T	C	
A		AAA	AGA	ATA	ACA	A
		AAG	AGG	ATG	ACG	G
		AAT	AGT	ATT	ACT	T
		AAC	AGC	ATC	ACC	C
G		GAA	GGA	GTA	GCA	A
		GAG	GGG	GTG	GCG	G
		GAT	GGT	GTT	GCT	T
		GAC	GGC	GTC	GCC	C
T		TAA	TGA	TTA	TCA	A
		TAG	TGG	TTG	TCG	G
		TAT	TGT	TTT	TCT	T
		TAC	TGC	TTC	TCC	C
C		CAA	CGA	CTA	CCA	A
		CAG	CGG	CTG	CCG	G
		CAT	CGT	CTT	CCT	T
		CAC	CGC	CTC	CCC	C

Table Ap1.1 DNA's bases as 64 triplets
(A = adenine, G = guanine, T = thymine, C = cytosine)

Abbreviation	Amino acid
ala	alanine
arg	arginine
asp	aspartic acid
asn	asparagine
cys	cysteine
glu	glutamic acid
gln	glutamine
gly	glycine
his	histidine
ile	isoleucine
leu	leucine
lys	lysine
met	methionine
phe	phenylalanine
pro	proline
ser	serine
thr	threonine
trp	tryptophan
tyr	tyrosine
val	valine

Table Ap 1.2 Key to amino acids

Appendix 2

Box plots

The data in Table Ap2.1 refer to three groups of golden delicious apples randomly sampled from orchards A, B and C. It is difficult to compare the variability between the three groups from the data table alone.

Apple number	Mass of apple (g)		
	Group A	Group B	Group C
15	142	131	119
14	130	130	116
13	127	127	115
12	126	125	114
11	122	124	112
10	119	121	110
9	116	120	109
8	110	118	108
7	109	117	106
6	105	115	102
5	100	115	99
4	99	112	96
3	97	110	91
2	96	108	83
1	82	106	76

Table Ap2.1

A **box plot** is a way of presenting information which allows differences between groups, sets, populations etc. to be compared easily. Each box plot shows the **median** which is the **central value** in the series of values when they are arranged in order. A box plot also displays the **upper quartile** (in this case the value 25% above the median) and the **lower quartile** (the value 25% below the median). The maximum and minimum values are called **upper and lower whiskers**. Figure Ap2.1 shows how the data for group A are converted into a box plot.

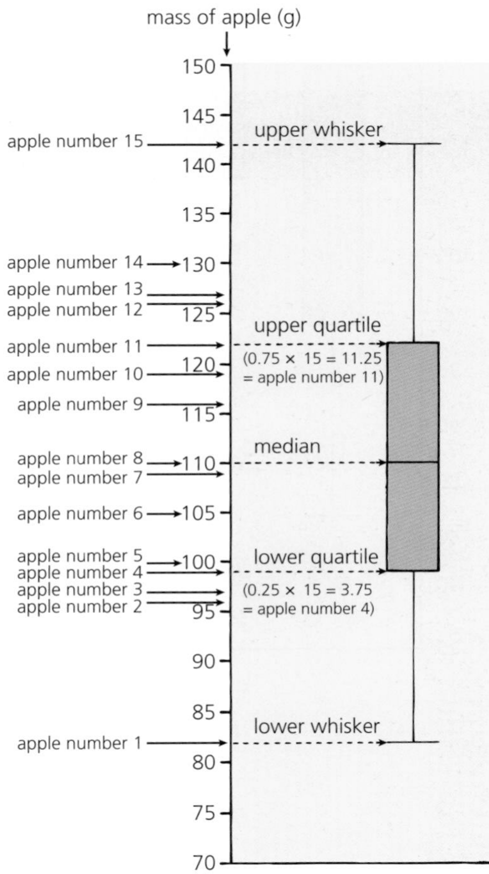

Figure Ap2.1

Figure Ap2.2 shows group A's box plot drawn alongside those for groups B and C. The box plots give a clear visual representation that allows the variability between the three groups to be compared more easily than by studying the table of data.

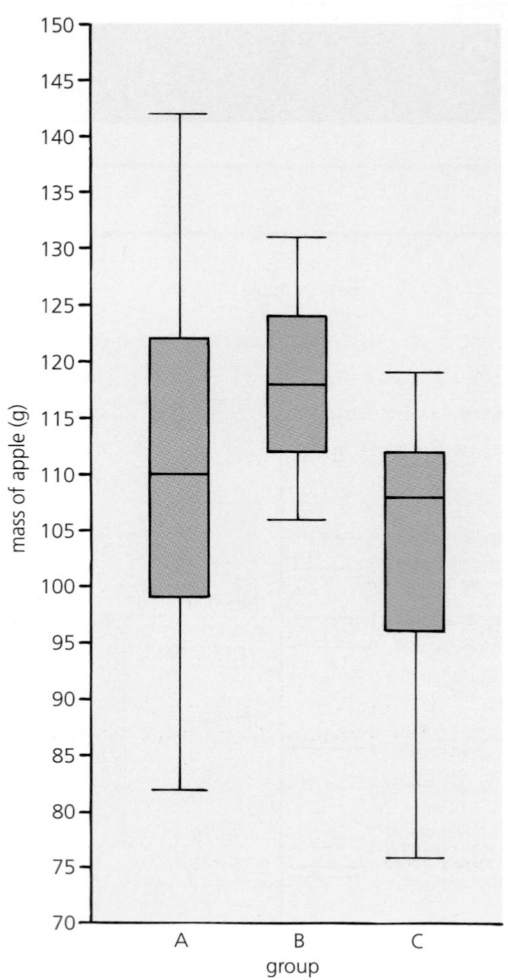

Figure Ap2.2

From these box plots, group B appears to be the best group with respect to mass and uniformity of size. It has a high median value and 50% of the apples are clustered around this value. Although it does not have the highest upper whisker, it contains no apples below a mass of 106 g. Group C appears to be the poorest group with the lowest median and many of its apples featuring in the lower range of mass. It also has the lowest whisker.

Appendix 3

Statistical concepts

A scientist needs to organise the data collected as results from an investigation into a manageable form from which conclusions may be drawn.

Mean

The **mean** is often referred to as the average. It is the most widely used measure of the **central tendency** of a set of data. It is found by adding up all the values obtained and dividing them by the total number of values. For example, for the two populations of seedlings shown in the scatter graphs in Figure Ap3.1, the mean for population A = 2100/70 = 30 mm and the mean for population B = 4900/70 = 70 mm.

Range

The **range** is the difference between the two most extreme values in a set of data. For example, for population A the range = 42 − 14 = 28 mm and for population B the range = 92 − 44 = 48 mm.

Standard deviation

Standard deviation is a measure of the spread of individual data values around their mean and shows how much variation from the mean exists. A normal distribution of results can be divided into intervals of standard deviation as shown in Figure Ap3.2. 68% of the values fall within plus or minus one standard

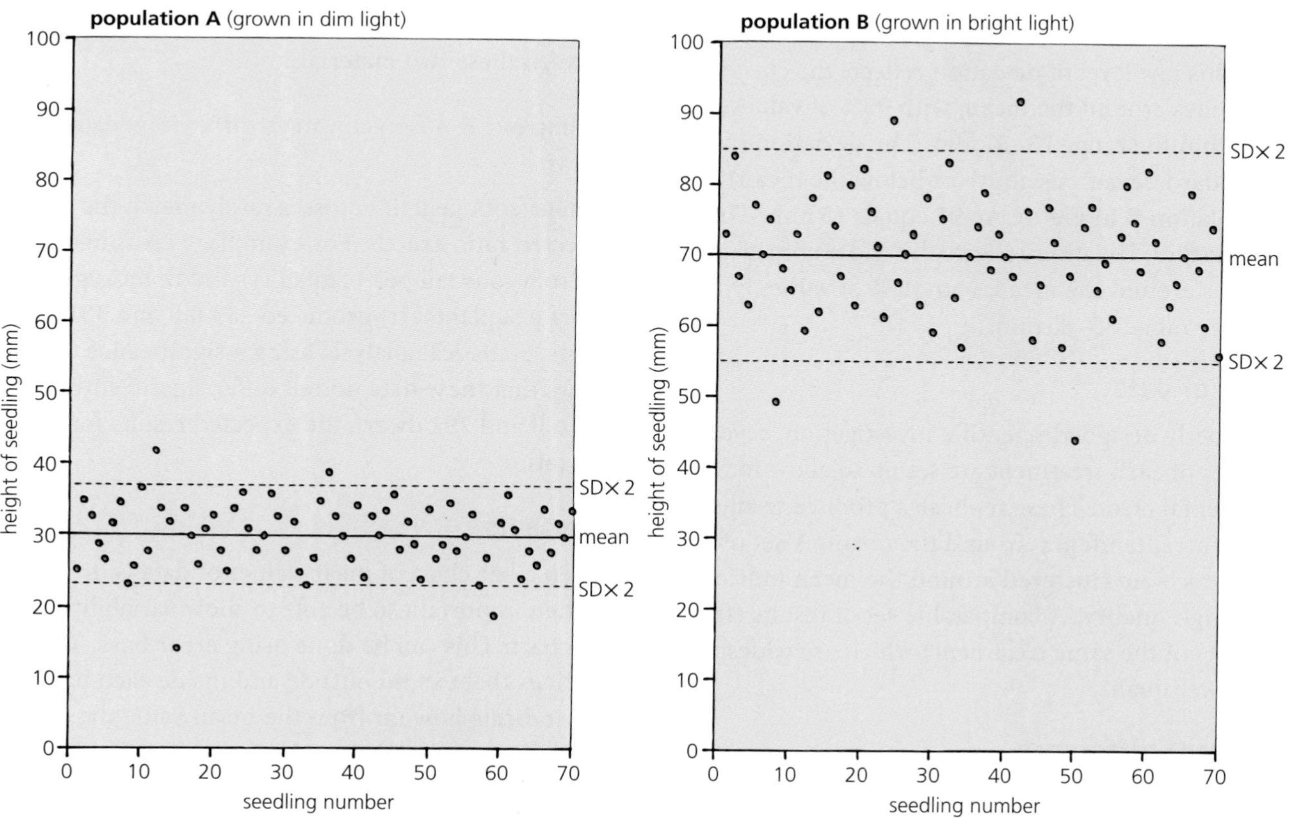

Figure Ap3.1 (SD × 2 = 2 standard deviations)

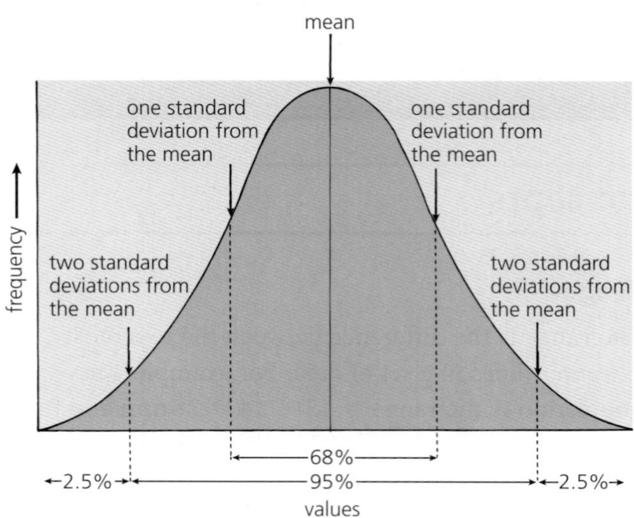

Figure Ap3.2

deviation of the mean; 95% of the values fall within plus or minus two standard deviations of the mean.

The standard deviation of a set of data is calculated using a mathematical formula (often with the aid of an appropriate calculator or computer software). The deviation (as two standard deviations above or below the mean) for population A in Figure Ap3.1 equals 7 mm. This low level of deviation reflects the clustering of the values around the mean, with 95% of values lying within the range 23–37 mm. The deviation (as two standard deviations above or below the mean) for population B in Figure Ap3.1 equals 15 mm. This higher level of deviation reflects the wider spread of the values around the mean, with 95% of values lying within the range 55–85 mm.

Quality of data

In a properly designed scientific investigation, several **replicates** of each treatment are set up to allow for experimental error. These replicates produce results with a **central tendency** around the **mean**. A set of results which are **clustered** around the mean indicates data of **high quality**. A comparable set of results (from a replicate of the same treatment) which are widespread are of lower quality.

Significant difference

In biology, an experiment is carried out to test an hypothesis. Once results have been obtained, the scientist needs to know whether these data (which rarely conform 'exactly to the expected outcome') support the hypothesis or not.

Testing the difference between two means

A **significance test** (a type of statistical analysis) can be used to find out whether the observed differences between two sets of data are statistically significant or simply the result of chance.

The data in Figure 17.9 on page 260, for example, refer to the results of an investigation to find out which flooring material was preferred by pigs in their sleeping area. A **plus sign** after a result indicates that the significance test shows the value to be **significantly higher** than would be expected by chance alone; a **minus sign** after a result indicates a value **significantly lower** than would be expected by chance alone.

In pen type 1, where the pigs spent 99 out of 168 hours on material P and only 69 out of 168 hours on material S, a significant difference is found to exist between the mean times spent on the two materials. Therefore the pigs can be said to prefer material P. In pen type 2, no such significant difference is found to exist between the mean times spent by pigs on materials P and M. Therefore the pigs can be said to show no preference between these two materials.

Finding out if observed values differ from expected values

Results from genetics crosses rarely match the expected ratio exactly. For example, a cross between heterozygous tall pea plants (Tt) and homozygous dwarf pea plants (tt) produced 283 tall and 301 dwarf plants. Statistical analysis using a significance test shows that these data do not differ significantly from 292 tall and 292 dwarf, the expected results for an exact 1 : 1 ratio.

Error bars

When a bar chart of mean values of data is drawn, it is often important to be able to show variability on the chart. This can be done using **error bars**. These are lines that extend outside and inside each bar and indicate how far from the mean value the true error-free value is likely to be. Error bars can be based on aspects of variability such as 95% level of confidence and standard deviation.

Figure Ap3.3 shows a bar chart of the results from a survey carried out on 10 000 young people in a country to estimate the incidence of asthma. Each bar represents a mean value with a 95% level of confidence whose range is indicated by error bars. Based on the information in the bar chart, health care experts could be 95% confident that the percentage number of asthma cases for the whole population would be 13–21% for 2–4-year-old males, 7–14% for 9–15-year-old females etc.

Significant difference

Error bars also allow a comparison to be made between two means to determine whether they are **significantly different** from one another. If their error bars (based on 95% level of confidence) do not overlap, the difference between the two means is regarded as being significant.

In the example shown in Figure Ap 3.4, it can be said with a 95% level of confidence that in blue light, the rate of photosynthesis did not drop below 10.6 oxygen bubbles min^{-1} and that in orange light, it did not rise above 9.2 bubbles min^{-1}. Therefore the two means are significantly different.

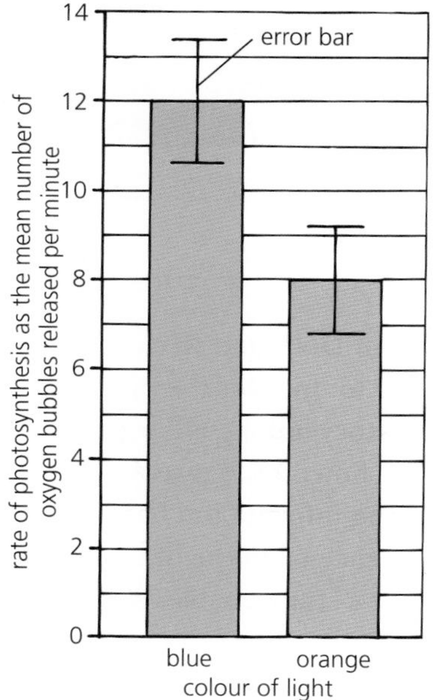

Figure Ap3.4

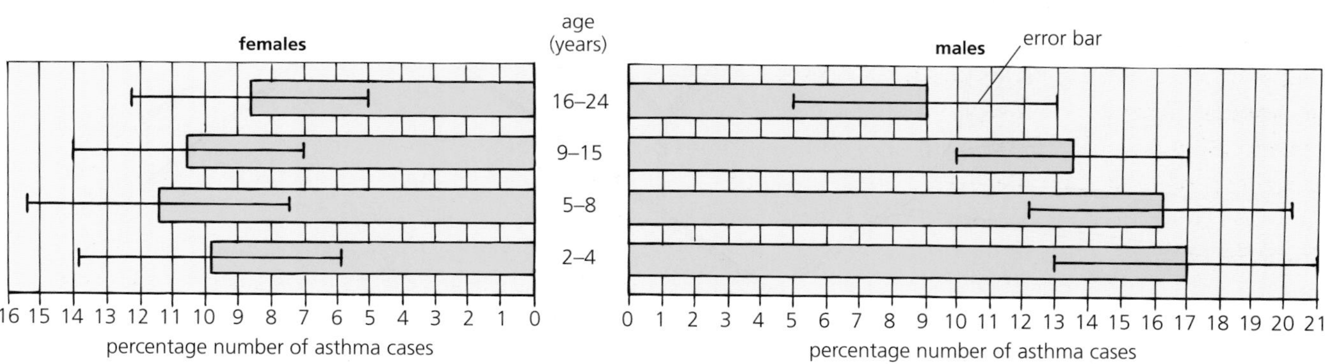

Figure Ap3.3

Appendix 4

Tree of life

Information from DNA and RNA sequences has been used to try to uncover the primary lines of evolutionary history among all organisms on Earth. The **tree of life** shown in Figure Ap4.1 has been constructed using information based on a study of the nucleotide sequence of a type of rRNA possessed by all organisms. Such a study enables scientists to assess the relationships between a wide range of organisms.

Trees of life constructed using data obtained from other nucleotide sequences are found to be very similar though the exact relationships between the three domains and the 'root' of the tree are still under debate.

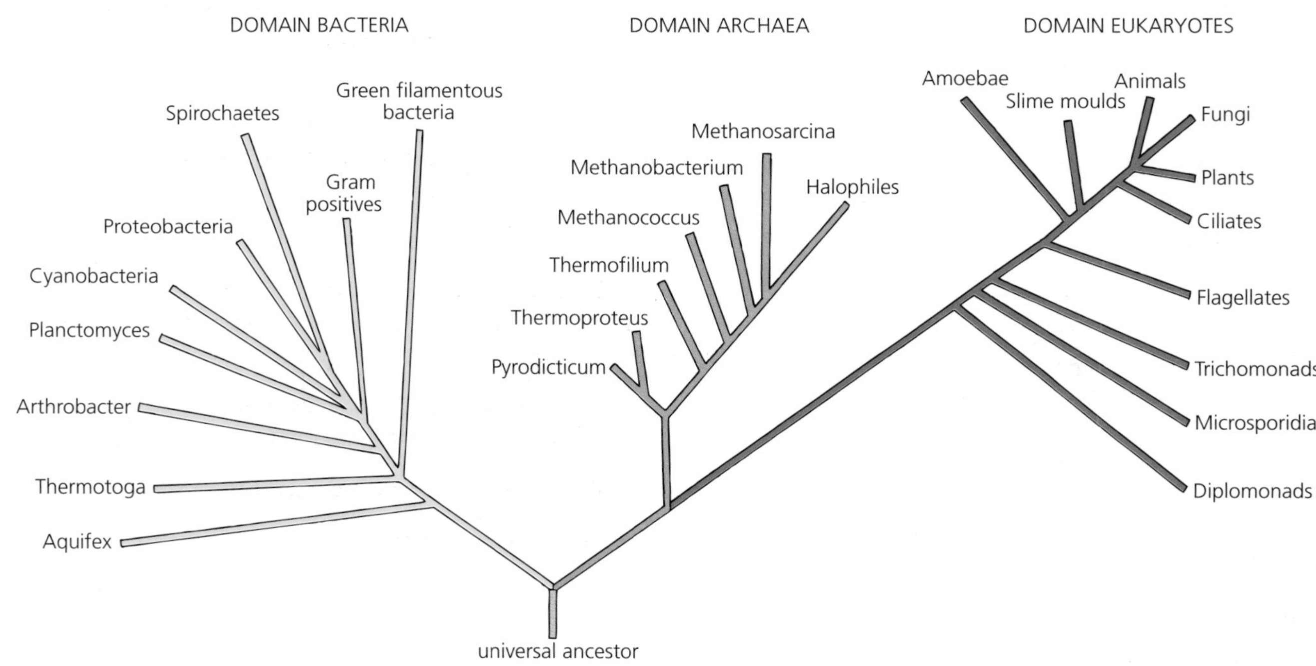

Figure Ap4.1 Tree of life

Testing Your Knowledge Answers

1 Structure of DNA

Testing Your Knowledge

1 a) (i) 4 (ii) adenine (A), thymine (T), guanine (G), cytosine (C) (3)

b) Hydrogen (1)

c) Each base can only join up with one other type of base: A with T and G with C. (1)

2 a) (i) and (ii) See Figure An1.1. (2)

b) Antiparallel (1)

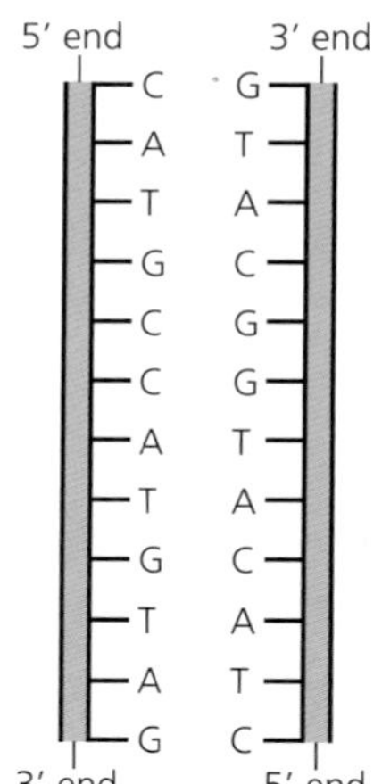

Figure An1.1

3 a) Double helix (1)

b) (i) Base pairs (ii) Sugar-phosphate backbones (2)

4 a) (i) Bacterium (ii) Yeast (iii) A tiny ring of DNA. (3)

b) (i) In a chloroplast (or mitochondrion) of a green plant cell. (ii) In the nucleus of the same green plant cell. (2)

2 Replication of DNA

Testing Your Knowledge

1 a) T (1)

b) F Cytosine (1)

c) F Hydrogen (1)

d) T (1)

e) T (1)

2 a) 5 (1)

b) 1 (1)

c) 4 (1)

d) 6 (1)

e) 3 (1)

f) 2 (1)

3 a) DNA, the four types of nucleotide, appropriate enzymes and an energy supply. (4)

b) DNA replication ensures that an exact copy of a species' genetic information is passed on from cell to cell during growth and from generation to generation during reproduction. (2)

4 a) Many copies of a DNA sample (1)

b) A piece of single-stranded DNA complementary to a target sequence at the end of the DNA strand to be replicated (2)

c) To break hydrogen bonds between base pairs and separate the DNA strands. (1)

d) To allow the primer to bind to its target sequence. (1)

e) It is heat tolerant. (1)

3 Gene expression

Testing Your Knowledge 1

1 A molecule of DNA is double-stranded and contains deoxyribose sugar and the base thymine whereas a molecule of RNA is single-stranded and contains ribose sugar and the base uracil. (3)

2 a) They differ by the sequence of the bases in their DNA. (1)

b) 3 (1)

c) Triplets (1)

3 a) See Figure An3.1. (2)

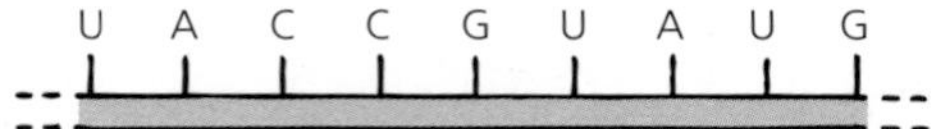

Figure An3.1

b) RNA polymerase (1)

4 (a) An exon is a coding region of DNA; an intron is a non-coding region.

(b) Introns (c) Splicing (3)

Testing Your Knowledge 2

1 a) 1 (1)

b) An amino acid molecule. (1)

2 a) One of many tiny, roughly spherical structures in a cell's cytoplasm (where translation of genetic information into protein occurs). (1)

b) (i) 3 (ii) To a codon on mRNA. (2)

c) Peptide (1)

d) It is discharged from the ribosome and reused. (2)

3 a) See Table An3.1. (2)

Stage of synthesis	Site in cell
formation of primary transcript of mRNA	nucleus
modification of primary transcript of mRNA	nucleus
collection of amino acid by tRNA	cytoplasm
formation of codon-anticodon links	ribosomes

Table An3.1

b) Ribosomal RNA (rRNA) (1)

4 a) Codons (1)

b) Transcription (1)

c) Intron (1)

d) Anticodons (1)

e) Ribosome (1)

5 a) 20 (1)

b) Polypeptide (1)

c) The sequence of bases in DNA. (1)

d) Arranged in long parallel strands or folded and coiled into a spherical shape. (2)

4 Cellular differentiation

Testing Your Knowledge

1 a) Differentiation is the process by which an unspecialised cell becomes altered and adapted to perform a special function as part of a permanent tissue. (1)

b) A meristem is a group of unspecialised plant cells capable of dividing repeatedly throughout the life of a plant. (1)

2 a) It has cilia which sweep dirt in mucus up and away from the lungs making the cell well suited to its function. (1)

b) During differentiation most genes, including those that code for insulin, were switched off so the goblet cell only expresses the genes left switched on that control the characteristics of that type of cell, such as the secretion of mucus. (2)

3 a) They can reproduce themselves while remaining undifferentiated. They can differentiate into specialised cells when required to do so. (2)

b) (i) Embryonic and tissue (adult) stem cells.
(ii) Embryonic in a blastocyst; tissue in bone marrow. (4)

c) Embryonic (1)

4 a) (i) Leukemia/cancer of the blood (ii) Bone marrow (2)

b) Parkinson's disease (1)

c) Because bone marrow cells can only produce more bone marrow cells (and not nerve cells) since many of their genes are switched off. (2)

5 Because some people believe that a human embryo, even in its very early stages of development, is already a person and that it is morally wrong that it dies when stem cells are extracted from it. (2)

5 Genome and mutations

Testing Your Knowledge 1

1 A mutation is a change in structure or composition of an organism's genome. A mutant is an individual or allele affected by a mutation. (2)

2 a) They occur spontaneously (and at random) but very rarely. (2)

b) (i) No **(ii)** On very rare occasions a mutant allele may confer an advantage on an organism that receives it. (2)

3 a) Substitution, insertion and deletion. (3)

b) (i) Substitution **(ii)** It only affects one amino acid in the protein expressed and does not cause the frameshift effect. (2)

4 a) F Genome (1)

b) T (1)

c) F Eukaryote (1)

d) T (1)

e) F Non-coding (1)

f) F Post (1)

Testing Your Knowledge 2

1 a) (i) Deletion **(ii)** Harmful **(iii)** Essential genes will be lost. (3)

b) (i) Inversion **(ii)** They become reversed. (2)

2 a) (i) Duplication **(ii)** The extra copies may mutate and produce new useful DNA sequences. (2)

b) (i) Translocation **(ii)** It normally increases the number on one and decreases the number on the other. (2)

6 Evolution

Testing Your Knowledge 1

1 (b) (1)

2 a) It compensated for high rate of gene loss and it led to rapid spread of new genetic sequences. (2)

b) The genetic sequence gained might be harmful. (1)

3 a) Lack of food and overcrowding. (2)

b) (i) Natural selection

(ii) Natural selection is the process that results in the increase in *frequency* among a *population* of those *genetic sequences* that confer an *advantage* on members of the population. (3)

4 Stabilising selection favours intermediate versions of the trait and acts against the extreme variants. Disruptive selection, on the other hand, favours extreme versions of the trait at the expense of the intermediates. (2)

Testing Your Knowledge 2

1 a) No (1)

b) The hybrid is sterile. (1)

2 a) Speciation is the formation of new biological species by evolutionary change. (1)

b) D, A, C, E, B (1)

3 a) (i) Allopatric **(ii)** Mountain range and river. (2)

b) Sympatric (1)

7 Genomic sequencing

Testing Your Knowledge 1

1 a) The sequence of nucleotide bases in a genome. (1)

b) (i) It is a fusion of molecular biology, statistical analysis and computer technology. **(ii)** They use it to analyse DNA sequences and compare them. (2)

2 a) Yeast and fruit fly. (2)

b) Their genome contains genes equivalent to genes in the human genome that are responsible for diseases. Therefore they may provide understanding of how these genes work. (2)

3 To discover which genetic sequences are only present in the pathogenic strain since some of these will be the ones responsible for disease. (2)

4 a) It means that the same or very similar DNA sequences are present in the genome of a wide range of organisms. (1)

b) Because it codes for an important protein that is needed by almost all plant species. (1)

Testing Your Knowledge 2

1 a) It is the study of evolutionary relatedness among different groups of organisms. (1)

b) Molecular information shows how similar or different their genomes are. Structural features, on the other hand, may show no differences although underlying genetic differences are present. (2)

2 Two related groups of organisms known from fossil records to have <u>diverged</u> at a certain point in <u>geological</u> time are chosen. Many genetic <u>sequences</u> for the two groups are compared and the number of nucleotide <u>substitutions</u> by which they differ is determined. The quantity of molecular change in their DNA that has occurred is a measure of how long ago the groups diverged from a common ancestor. The DNA can therefore be used as a molecular clock. (3)

3 a) **(i)** Bacteria, archaea and eukaryotes
(ii) True nucleus bounded by a double membrane and membrane-enclosed organelles. (3)

b)

photosynthetic land plants
↑
multicellular green plants
↑
photosynthetic eukaryotes
↑
photosynthetic prokaryotes
↑
last universal ancestor (2)

4 a) The sequencing and analysis of an individual's genome using bioinformatics. (2)

b) Drug choice and dosage may be customised to suit the individual. The risk of a genetic disease or disorder may be predicted in time to take preventive measures. (2)

8 Metabolic pathways

Testing Your Knowledge 1

1 a) Metabolism is the collective term for the thousands of enzyme-controlled chemical reactions that occur in a living cell. (2)

b) One breaks down complex molecules to simpler ones and normally releases energy; the other builds up simpler molecules into complex ones and consumes energy. (2)

2 a) A channel-forming protein allows certain molecules to pass through its pore by diffusion which is a passive process requiring no energy. (2)

b) A carrier protein, acting as a pump, actively transports ions across the cell membrane against a concentration gradient and requires energy to do so. (2)

Testing Your Knowledge 2

1 They lower the activation energy needed for the chemical reaction to proceed. They speed up the reaction. They remain unchanged at the end of the reaction. (3)

2 a) The chemical structure of the protein of which the enzyme is made and the bonding between its component amino acids. (1)

b) The chemical attraction between them. (1)

c) Induced fit (1)

d) The shape of the active site ensures that the reactants are correctly orientated so that the reaction can take place. This is made possible by the fact that the enzyme decreases the activation energy needed by the reactants to reach the transition state. (3)

3 a) Quantity of chemical change that occurs per unit time. (1)

b) **(i)** Initially it causes an increase in rate but at higher concentrations no further increase in rate occurs.
(ii) At low concentrations of substrate there are not enough molecules of substrate to occupy all the active sites on the enzymes. At higher concentrations of substrate, all the active sites on the enzyme molecules are occupied. (4)

Testing Your Knowledge 3

1 a) Its molecular shape is similar to that of the substrate. (1)

b) **(i)** It brings about an increase in the rate of the reaction. **(ii)** Substrate molecules eventually outnumber those of the competitive inhibitor causing more and more sites on the enzyme molecules to become occupied with substrate rather than inhibitor. (3)

2 Non-competitive (1)

3 a) **(i)** P **(ii)** Q (2)

b) **(i)** Q **(ii)** R (2)

c) Left to right (1)

d) **(i)** If a high concentration of R built up some of it would bind to some molecules of enzyme X and slow down the conversion of P to Q. **(ii)** It keeps the pathway under finely-tuned control. (3)

9 Cellular respiration

Testing Your Knowledge

1 a) Adenosine triphosphate (1)

b) ATP has three phosphate groups whereas ADP has two. (1)

c) ADP + P_i + energy → ATP (2)

2 a) Two molecules of ATP are used during the energy investment phase so the net gain is only two ATP. (1)

b) As soon as oxaloacetate is formed, it combines with acetyl CoA to form citrate. Therefore there is never very much present at any given moment. (1)

3 a) G (1)

b) E (1)

c) C (1)

d) E (1)

e) G (1)

f) G and C (1)

g) C (1)

h) G, C and E (1)

4 a) **(i)** glucose → ethanol + carbon dioxide

(ii) glucose → pyruvic acid → lactic acid (2)

b) Equation **(ii)** (1)

5 ATP is used to transfer energy to cellular processes such as protein synthesis and transmission of nerve impulses that require energy. (4)

10 Metabolic rate

Testing Your Knowledge 1

1 a) Metabolic rate is the quantity of energy consumed by an organism per unit time. (1)

b) Oxygen consumption per unit time or carbon dioxide production per unit time. (2)

2 The heart of a fish contains two chambers. Blood is pumped at high pressure to the gills and then on to the body's capillary beds at low pressure. The heart of a mammal contains four chambers. Blood is pumped to the mammal's lungs at high pressure and to the body's capillary beds at high pressure. (3)

3 a) T (1)

b) F Once (1)

c) F Amphibian/reptile (1)

Testing Your Knowledge 2

1 See Table An10.1. (3)

Feature	Conformer	Regulator
ability to control internal environment by physiological means	not able to do so	able to do so
relative metabolic costs of lifestyle	low	high
extent of range of ecological niches that can be exploited	narrow	wide

Table An10.1

2 a) Physiological homeostasis is the maintenance of the body's internal environment within certain tolerable limits despite changes in the body's external environment. (2)

b) **(i)** When the body's internal environment deviates from its normal level, this change is detected by receptors which communicate with effectors. These trigger responses which return the system to normal. This corrective mechanism is called negative feedback control. **(ii)** It is of advantage to the organism because it provides the stable conditions needed by its body to function efficiently despite wide fluctuations in the external environment. (5)

3 a) Endotherm (1)

b) Human beings are able to maintain constant internal body temperature despite fluctuations in external temperature. (1)

4 a) The hypothalamus has central thermoreceptors which receive nerve impulses from skin thermoreceptors and are sensitive to changes in temperature of blood. (2)

b) Skin and skeletal muscles (2)

11 Metabolism and adverse conditions

Testing Your Knowledge 1

1 Extreme cold and lack of food. (1)

2 a) **(i)** Decrease of metabolic rate to a minimum.
(ii) Winter buds present but not growing or absence of leaves. (2)

b) Predictive dormancy means the organism becomes dormant before the adverse conditions arrive whereas consequential dormancy means the organism becomes dormant after the adverse conditions arrive. (2)

c) **(i)** Predictive **(ii)** They shed their leaves and become dormant in response to decreasing day length before winter arrives. (2)

3 a) **(i)** They both involve a decrease in metabolic rate. **(ii)** Hibernation is used to survive a period of extreme cold whereas aestivation is used to survive a period of intense heat or drought. (2)

b) Daily torpor as shown by the hummingbird (for example) means that the animal's rate of metabolism becomes greatly reduced for part of each 24-hour cycle during the night. This helps the animal to conserve energy at times when searching for food would be unsuccessful. (2)

Testing Your Knowledge 2

1 a) Migration is the regular movement by the members of a species (e.g. Arctic skua) from one place to another over a relatively long distance and then back again months later. (2)

b) It relocates the bird to a more favourable environment for part of the year enabling it to avoid metabolic adversity. (1)

2 a) Ringing and tagging (2)

b) transmitter implanted under skin

↓

transmitter gives out signals

↓

signals picked up by receivers on satellites

↓

signals sent to ground station

↓

information relayed to scientists (3)

3 Innate behaviour is inherited and inflexible whereas learned behaviour is gained from experience after birth and is flexible. (2)

12 Environmental control of metabolism

Testing Your Knowledge 1

1 They are easy to cultivate and they grow quickly. (2)

2 a) Nutrient agar is a solid whereas nutrient broth is a liquid. (1)

b) Nutrient broth (1)

3 a) **(i)** For protein synthesis **(ii)** For synthesis of ATP **(iii)** For energy (3)

b) To keep the conditions sterile in order to eliminate contaminants. (1)

4 a) Sensors monitoring temperature, pH and oxygen concentration. (3)

b) 1 = C (acid in), 2 = E (cold water in), 3 = D (oxygen in), 4 = A (products out), 5 = F (cold water out), 6 = B (waste gases out) (6)

c) **(i)** Motor **(ii)** To mix the contents of the fermenter. **(iii)** Stainless steel (3)

Testing Your Knowledge 2

1 Growth is an irreversible gain in dry biomass. Mean generation time is the time needed for a population of unicellular organisms to double in number. (2)

2 a) During the lag phase, intense metabolic activity occurs in preparation for cell division. During the stationary phase, on the other hand, metabolism slows down and secondary metabolites are produced. (1)

b) The cells are multiplying at the maximum rate. (1)

c) Lack of nutrients and accumulation of toxic metabolites. (2)

3 a) Antibiotic (1)

b) It may inhibit the growth of bacteria that would compete with the fungus for resources in the soil ecosystem. (1)

13 Genetic control of metabolism

Testing Your Knowledge 1

1 Genetic stability of Y, ability of Y to grow on low cost nutrients and ability of Y to vastly overproduce the useful product y. (3)

2 a) Mutagenesis is the creation of mutants. (1)

b) Ultra-violet light and mutagenic chemicals. (2)

3 a) In the absence of outside influences, mutations arise very rarely. (1)

b) An agent that increases the rate of mutagenesis is called a mutagen. (1)

c) Mutagenesis can be used to create a mutant strain that lacks a particular undesirable characteristic. (1)

Testing Your Knowledge 2

1 a) Gene for insulin cut out of the human genome and transferred to and inserted into the genetic material of a bacterium. (2)

b) The insulin produced is the same as the human type. (1)

c) Plasmid (1)

2 a) A restriction endonuclease is an enzyme that cuts open DNA leaving sticky ends whereas a restriction site is a location on a plasmid that gets cut open by an endonuclease. (2)

b) Ligase is an enzyme which seals complementary sticky ends of DNA together enabling a DNA fragment to be sealed into a bacterial plasmid. (2)

c) It enables scientists to tell whether or not a host cell has taken up the plasmid vector. (1)

3 a) The polypeptide molecule could be inactive because it is incorrectly folded. (1)

b) Recombinant yeast (1)

14 Food supply, plant growth and productivity

Testing Your Knowledge 1

1 Ability of a human population to access food of adequate quantity and quality. (2)

2 Identification of a limiting factor (such as shortage of minerals in the soil) and then increasing the supply of that factor. Replacement of an existing crop strain with a higher-yielding cultivar. (2)

3 Energy loss is reduced therefore more is available to feed humans. (1)

Testing Your Knowledge 2

1 a) By passing a beam of white light through a prism. (1)

b) (i) Violet/blue (ii) Red (2)

2 a) An absorption spectrum shows the degree of absorption that occurs at each wavelength of visible light by pigment(s). An action spectrum, on the other hand, indicates the effectiveness of each wavelength of visible light at bringing about photosynthesis. (2)

b) Blue and red (2)

c) Because it absorbs the other colours and gives green back out by transmission and reflection. (2)

3 It extends the range of wavelengths of light that can be used by the plant for photosynthesis. (1)

Testing Your Knowledge 3

1 a) It is transferred to electrons which become excited and raised to a higher energy level. (1)

b) $\text{ADP} + \text{P}_i \xrightarrow{\text{ATP synthase}} \text{ATP}$ (2)

c) The oxygen is released. The hydrogen combines with NADP to form NADPH. (2)

2 a) (i) Rubisco (ii) ATP and NADPH (3)

b) (i) RuBP (ii) Sugar (2)

3 a) (i) Cellulose (ii) Starch (2)

b) Proteins and DNA (2)

15 Plant and animal breeding

Testing Your Knowledge 1

1 Increase in yield, increase in nutritional value and resistance to pests. (3)

2 To compare the performance of two different cultivars. To investigate the effect of environmental factors on a new crop cultivar. (2)

3 a) A plot is one of several equal-sized portions of a field whereas a treatment is the way in which one plot is treated compared to other plots with respect to the variable factor being investigated. (2)

b) See Table An15.1. (2)

Design feature	Reason
randomisation of treatments	to prevent bias existing in the system
selection of treatments involving one variable factor	to ensure that a fair comparison can be made
inclusion of several replicates	to take experimental error (uncontrolled variability) into account

Table An15.1

Testing Your Knowledge 2

1 **a)** During inbreeding, closely related members of a species are selected and bred over many generations to produce a uniform strain.
During crossbreeding, members of two different strains of a species are crossed to try to produce an improved strain. (2)

b) Homozygous describes a genotype that contains two identical alleles of a particular gene whereas heterozygous refers to a genotype that contains two different alleles of a particular gene. (2)

2 **a)** T (1)

b) F Heterozygosity (1)

c) T (1)

d) F Genotype (1)

3 By crossbreeding it with a different variety. (1)

4 **a)** Because they inherit desirable characteristics from both parents. (1)

b) **(i)** Heterozygous **(ii)** Uniform **(iii)** Varied (3)

c) They are maintained for the purpose of crossing them with one another to produce crossbreed animals that express hybrid vigour. (1)

5 **a)** **(i)** The assembly of overlapping DNA fragments from an organism's genome into a sequence of nucleotide bases. **(ii)** It can be used to identify organisms that possess a particular allele for a desirable characteristic and then use these organisms in a breeding programme. (3)

b) They have become resistant to insect pests following the insertion of the Bt toxin gene into their genome. (1)

16 Crop protection

Testing Your Knowledge 1

1 **a)** Natural ecosystem (1)

b) **i)** Agricultural ecosystem **(ii)** The pest can feed and reproduce repeatedly without running out of food. (2)

2 Reduction in crop productivity and contamination of grain crop with their seeds. (2)

3 **a)** Short life cycle and production of many seeds. (2)

b) They are already established and have storage organs that supply food until environmental conditions are favourable for photosynthesis. (2)

4 **a)** They suck sugar from the plant which is therefore denied some of its energy supply. This affects the crop plant's vigour and yield adversely. (2)

b) Nematode and mollusc. (2)

c) By being airborne or being carried by vectors. (2)

Testing Your Knowledge 2

1 **a)** It refers to a traditional non-chemical method that has evolved over time by trial and error. (1)

b) A series of different crop plants are grown in turn on the same piece of ground. If a pest can only attack a certain type of host plant, it is controlled because it cannot survive for years until its host returns. (3)

c) Ploughing and early removal of weeds. (2)

2 **a)** It kills the broad-leaved weeds but not the narrow-leaved cereal crops. (1)

b) The systemic type is absorbed and transported to all plant parts giving overall protection whereas the contact type acts as a protective layer to those plant parts above ground only and may be washed off by rain. (2)

3 **a)** See Figure An16.1. (1)

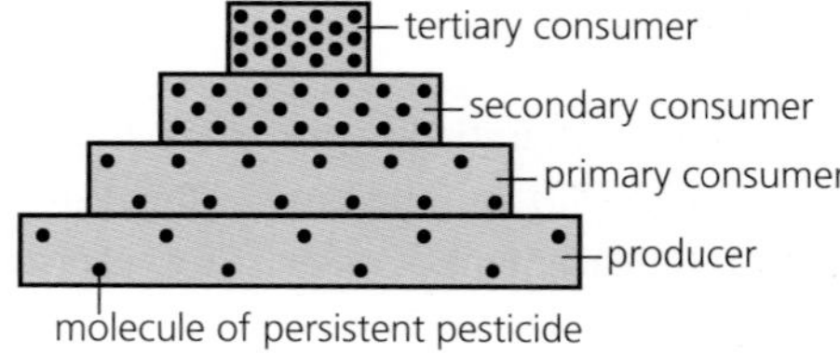

Figure An16.1

b) **(i)** See Figure An16.1. **(ii)** They are densest at the top because the molecules of persistent chemical accumulate in the food chain and become most concentrated in the tertiary consumer. (3)

4 **a)** Continued use of pesticide exerts a <u>selection</u> pressure on a population producing a <u>resistant</u> population of pests. (2)

b) The reduction of a <u>pest</u> population by the deliberate introduction of one of its natural enemies is called <u>biological</u> control. (2)

c) The use of a combination of techniques such as chemical and biological control and host plant resistance is called integrated pest management. (2)

d) The form of management referred to in **c)** aims to make minimum use of chemicals on the farm and to control pests. (2)

17 Animal welfare

Testing Your Knowledge

1 a) Contented animals breed more successfully and generate products of higher quality. (1)

b) In their opinion it is wrong to eat meat and eggs from hens that have had their beaks trimmed and lived in overcrowded conditions. (1)

2 a) **(i)** B **(ii)** D **(iii)** A **(iv)** C (4)

b) Improve the animals' welfare by enlarging their habitat and including some features present in their natural habitat. (2)

18 Symbiosis

Testing Your Knowledge

1 a) Symbiosis is an ecological relationship between organisms of two different species that live in direct contact with one another. (2)

b) See Table An18.1. (2)

Type of symbiosis	Species 1	Species 2
mutualism	+	+
parasitism	–	+

Table An18.1

2 By direct contact. For example, body lice being passed from one person to another.
Released as resistant stages able to survive adverse conditions. For example, resistant larvae of cat flea.
Carried by a vector. For example, *Plasmodium* carried by mosquitoes. (6)

3 a) It lacks a means of locomotion (and a digestive system). (1)

b) Light needed by its symbiotic partner zooxanthella for photosynthesis does not reach water at depths of 100 metres or more from the surface. (1)

c) It can use the secondary/intermediate host as a site of asexual reproduction. (1)

19 Social behaviour

Testing Your Knowledge 1

1 a) Head raised, eyes staring, teeth bared, hackles raised. (2)

b) Head lowered, eyes averted, teeth covered, hackles lowered. (2)

c) **(i)** Social hierarchy **(ii)** Real fighting is kept to a minimum. Experienced leadership is guaranteed. (3)

2 All dogs including the subordinate ones obtain food. Large prey can be tackled that would be too big and strong for one dog to overpower. (2)

3 They form a protective group with the cows and calves at its centre and use a combined charge called mobbing to drive off the wolves. (2)

4 a) Altruism is 'unselfish' *behaviour* where a *donor* animal behaves in a way that may be *harmful* to itself but *helpful* to another animal, the *recipient.* (3)

b) Reciprocal altruism occurs when one individual at some cost to itself helps another in the knowledge that the favour will be returned. (1)

c) Although the individual does not survive, many of its genes will be present in the genomes of its offspring and close relatives. Therefore the genes will survive. (2)

Testing Your Knowledge 2

1 A division of labour exists among social insects. Food is gathered by numerous sterile members of the group but reproduction is carried out by a few fertile members only. (2)

2 a) Recognising danger and foraging for food. (2)

b) It reduces conflict to a minimum enabling the weaker members to live close to the stronger ones. (1)

3 a) F Large (1)

b) F Friendship/co-operation/alliances (1)

c) T (1)

d) F Appeasement/submission (1)

e) T (1)

f) F Alliances (1)

20 Components of and threats to biodiversity

Testing Your Knowledge 1

1 a) Genetic diversity and species diversity. (2)

b) Because all the alleles that it contains are also found in other populations. (1)

2 a) The one not dominated by one species. (1)

b) Woodland, freshwater pond and river (3)

3 a) T (1)

b) F Richness (1)

c) F Abundance (1)

d) T (1)

Testing Your Knowledge 2

1 Exploitation means making the best use of a resource such as fish. Overexploitation means removing individuals such as fish at a rate greater than their maximum rate of reproduction. (2)

2 a) It could reduce it. (1)

b) It would be less able to adapt to environmental change. (1)

3 a) See Table An20.1. (4)

	Type of habitat	
	island fragment	unfragmented land mass
relative size of habitat (large/small)	small	large
species richness supported (high/low)	low	high
edge : interior (large/small)	large	small
chance of edge species invading interior (high/low)	high	low

Table An20.1

b) By habitat corridors. (1)

4 a) When it survives in the wild. (1)

b) When it outcompetes native species in the wild. (1)

Applying Your Knowledge and Skills Answers

1 Structure of DNA

1 a) X = 22.0; Y = 0.98 (2)

b) i) The number of adenine bases in DNA equals the number of thymine bases and the number of guanine bases equals the number of cytosine bases. (A:T = 1:1 and G:C = 1:1).

ii) Yes

iii) Because the percentage of A is always very nearly equal to the percentage of T but not close to that of G or C. Similarly the percentage of G is always close to the percentage of C but not close to that of A or T. (3)

c) C (1)

2 a) 30% (1) **b)** 3200 (1)

3 a) i) 1= chromosome, 2= DNA, 3= base

ii) 1

iii) 3 (5)

b) i) 10000:1

ii) Because this is a constant reliable measurement whereas length measured in μm varies according to degree of coiling. (2)

2 Replication of DNA

1 See Figure An KS 2.1. (4)

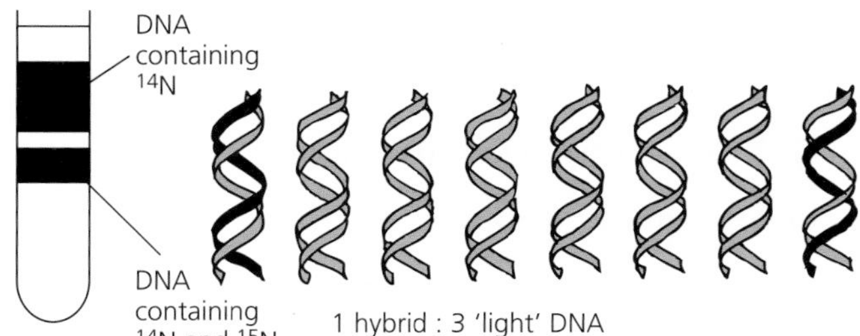

Figure An KS 2.1

2 a) See Figure An KS 2.2. (6)

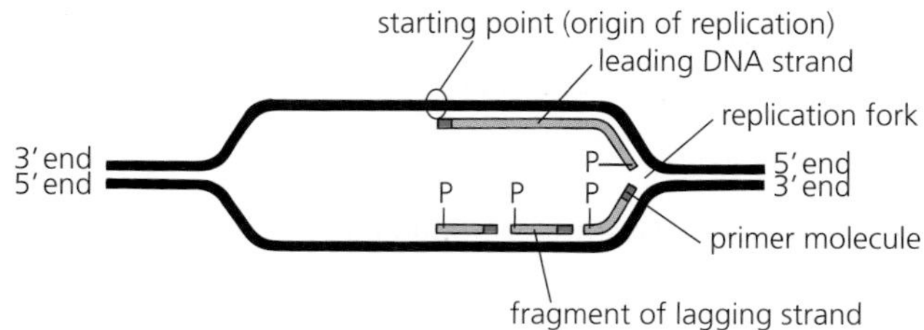

Figure An KS 2.2

b) i) 20000 minutes **ii)** During replication, many replication forks operate simultaneously which ensures speedy copying of the DNA. (2)

3 a) i) Semi-logarithmic **ii)** To accommodate the very high numbers involved. (2)

b) 23 (1) **c)** 10 (1)

d) i) 1000000000 **ii)** One billion (2)

4 See core text pages 11, 15–18. (9)

3 Gene expression

1 a) 1 = C, 2 = T, 3 = T, 4 = A, 5 = U, 6 = A, 7 = G, 8 = C, 9 = G (2)

b) P = transcription and release of mRNA; Q = translation of mRNA into protein (1)

c) See Table An KS 3.1. (2)

Amino acid	Codon	Anticodon
alanine	**GCG**	CGC
arginine	CGC	**GCG**
cysteine	**UGU**	ACA
glutamic acid	GAA	**CUU**
glutamine	**CAA**	GUU
glycine	GGC	**CCG**
isoleucine	**AUA**	UAU
leucine	CUU	**GAA**
proline	**CCG**	GGC
threonine	ACA	**UGU**
tyrosine	**UAU**	AUA
valine	GUU	**CAA**

Table An KS 3.1

d) CAA (1)

e) U = proline, V = glutamine, W = glutamic acid, X = cysteine, Y = arginine, Z = isoleucine (2)

f) i) ACACUUGCGGGC **ii)** TGTGAACGCCCG (2)

2 a) The increase in relative numbers of ribosomes in cells during days 1 to 5 indicates that this is the period of most rapid protein synthesis and growth of the new leaf. After day 5, growth slows down and eventually comes to a halt at day 11 when the leaf has reached its full size. (2)

b) A basic number of ribosomes will always be needed by a cell of a fully grown leaf to make proteins such as enzymes essential for biochemical pathways (e.g. photosynthesis). (1)

3 See core text pages 30–33. (9)

4 a) X = phenylalanine, Y = threonine, Z = lysine (3)

b) i) X **ii)** Z **iii)** The most soluble one is carried to the highest position by the solvent. (2)

5 See Figure An KS 3.1. (1)

Figure An KS 3.1

6 a) i) Lower molecular weight. **ii)** This is suggested by the fact that they have travelled the greatest distance from the negative electrode. (2)

b) i) Alpha-2-globulins **ii)** Iron-deficiency anaemia (2)

c) i) D **ii)** Because gamma-globulins also increase in concentration for other reasons such as response to viral invasion. (2)

7 a) Casein contains them all. Group 2 rats gained weight throughout the experiment. Zein lacks two essential amino acids. Group 1 rats lost weight throughout the experiment. (4)

b) i) Zein **ii)** Their diet could have been changed to casein or to zein supplemented with the two essential amino acids that it lacks. (3)

c) 35 g (1) **d)** 20% (1)

4 Cellular differentiation

1 a) i) = T **ii)** = P **iii)** = S **iv)** = Q

v) = U **vi)** = R (6)

b) B (1)

c) i) Cow **ii)** Because it is considered inappropriate to use cells that contain even a tiny amount of cow material to repair human tissue in this way. (2)

2 a) Nuclear transfer technique (1)

b) Because she is a genetic copy of another sheep. (1)

c) i) White **ii)** Because her genetic material came from a white-faced sheep. (2)

d) i) A **ii)** The DNA that she received came from a sheep not a ram. Therefore she could not have received a Y chromosome necessary to become a male. (2)

3 a) Jill (1) **b) i)** Unconvincing **ii)** Because some embryos are lost naturally, does not justify using embryos for stem cell research. (2)

4 See core text pages 48–49. (9)

5 Genome and mutations

1 a) A = mRNA, B = rRNA, C = tRNA (3)

b) W (1)

c) i) = 2 **ii)** = 1 **iii)** = 3 (2)

2 a) Deletion **b)** Substitution **c)** Insertion (3)

3 a) i) Substitution **ii)** One of the codons is now GAC instead of AAC. (2)

b) i) glutamic acid–serine–leucine–threonine

ii) As answer to **i)** (2)

c) The altered codon refers to the same amino acid as the original codon and does not lead to a change in the amino acid sequence. (1)

d) i) Missensical **ii)** It still makes sense but not the original sense because the amino acid has changed to valine and this results in the production of a different type of haemoglobin. (2)

4 a) i) Mean root length **ii)** Radiation dosage (1)

b) See Figure An KS 5.1. (4)

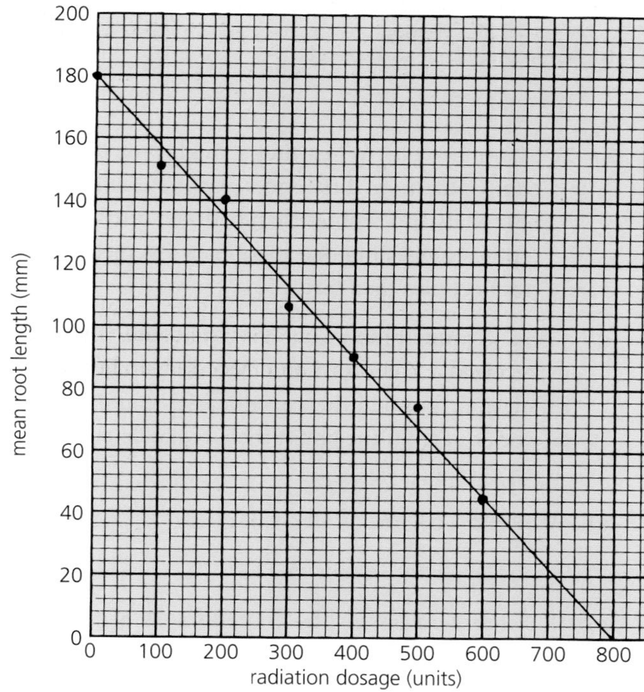

Figure An KS 5.1

c) i) An inverse relationship ii) Radiation damages the cells and their genetic material. The greater the dose of radiation, the worse the damage inflicted. (2)

d) 800 units (1)

e) To increase the reliability of the results. (1)

f) 500 (mutations at gene site per million gametes) (1)

5 a) i) See Figure An KS 5.2.

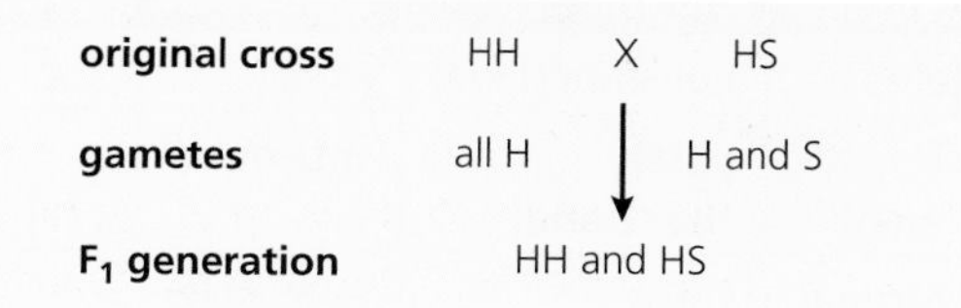

Figure An KS 5.2

ii) See Figure An KS 5.3. (4)

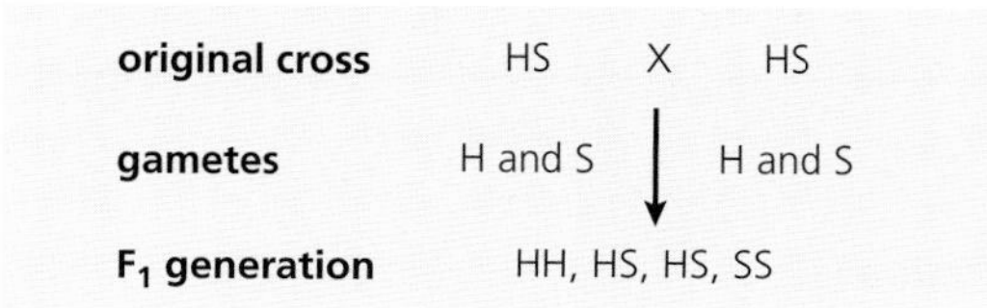

Figure An KS 5.3

b) A higher percentage of local people in malarial areas have sickle cell trait as they are resistant to malaria and are favoured by natural selection. This selective advantage is lost in non-malarial areas therefore the percentage of those with sickle cell trait is very low. (2)

6 a) Cell 1 = deletion; cell 2 = duplication (2)

b) i) Cell 1 ii) Essential genes would probably have been lost. (2)

7 a) i) Y ii) X (1)

b) i) Y ii) TUV (2)

6 Evolution

1 a) Vertical (1) b) See Figure An KS 6.1. (3)

Figure An KS 6.1

2 a) The black one (1) b) i) See Figure An KS 6.2.

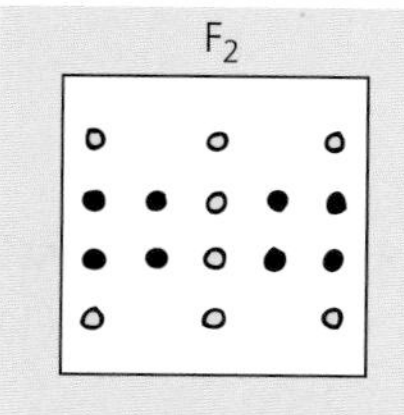

Figure An KS 6.2

ii) 1 : 1 (2)

c) See Figure An KS 6.3.

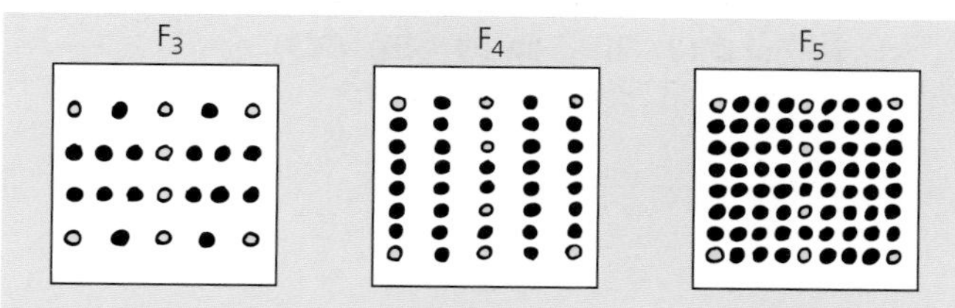

Figure An KS 6.3

i) F_3 ii) F_5 (2)

3 a) Consumption of warfarin interferes with the blood-clotting mechanism and the rat bleeds to death if it becomes cut or injured. (1)

b) It thins their blood and prevents the formation of unwanted internal clots. (1)

c) Mutation (1)

d) It has been favoured by natural selection since members of the sensitive strain normally die after eating the warfarin. (1)

e) Extinction (1)

f) Very few W^rW^r rats will be fortunate enough to find a continuous supply of food containing the large amount of vitamin K that they need. (2)

4 a) 3 (1)

b) Isabella and Espanola (2)

c) Santa Cruz (1)

d) X = large seed-eating from Pinta; Y = large cactus-eating from Pinta; Z = medium seed-eating from Culpepper (2)

e) i) Type X (large seed-eating)

ii) One type might become extinct (or two species might survive as small populations living in fierce competition or …?). (2)

5 See core text pages 78–9. (9)

7 Genomic sequencing

1 AACCGATCAGCGCAGCGCTTGATCAGATCGCGCTAG (1)

2 a) TACTGGTACT (1) **b)** ATGACCATGA (1)

3 a) i) It is a variation in DNA sequence that affects a single base pair in a DNA chain. **ii)** Site 4 (2)

b) 5 (1)

c) i) 7 and 12 **ii)** 5 and 9 **iii)** 8 (3)

d) 17 (1)

e) i) CTTATG **ii)** 45% (2)

f) 10 (1)

g) 4 and 11 (1)

h) Increase the number of people sampled and include more sites in the study. (2)

4 a) Fish and reptiles (1)

b) i) Vertebrates and insects

ii) 600 million years ago (2)

c) i) Mammals **ii)** When their cytochrome c is compared, more differences exist between mammals and reptiles than between birds and reptiles suggesting that mammals diverged earlier from reptiles than birds. (2)

d) Fish and amphibian specimens could have their cytochrome c compared and the number of differences found in their amino acids could be used to work out from the graph their probable point of divergence. (2)

5 a) i) 39 996 **ii)** 8322 **iii)** 1.23 : 1 (3)

b) *Bradyrhizobium japonicum* (1)

c) i) *Neurospora crassa* **ii)** Its genome is 'gene sparse' compared with the other two so the spaces between the protein-coding genes probably contain a large amount of repetitive DNA. (2)

d) 1 = eukaryotes because it has a nucleus with a double membrane

2 = bacteria because it has no introns

3 = archaea because it has few introns but no nuclear membrane (3)

6 a) See Table An KS 7.1. (4)

Alleles of gene present in genome	State of enzyme	Person's metabolic profile
two null alleles	non-functional	**poor**
one null allele and one inferior allele	**partly functional**	**intermediate**
one or two normal alleles	**fully functional**	extensive
more than two copies of normal allele	highly functional	**ultra-rapid**

Table An KS 7.1

b) Duplication (1)

c) i) Poor metabolisers **ii)** Their bodies will be so slow to clear the drug that it may do them harm. (2)

d) i) Ultra-rapid metabolisers **ii)** Their bodies would remove the drug so quickly that it would not have time to bring about the desired effect. (2)

e) If the personal genome sequencing becomes routine then knowledge of a person's DNA profile may enable doctors to customise medical treatments to suit an individual's exact requirements. (2)

8 Metabolic pathways

1 a) i) 5 **ii)** 3 (2)

b) Some of I would be converted to G by enzyme 5 and then G would be converted to H by enzyme 4. (2)

c) i) H could become L and M by the action of enzyme 8 and then L and M could become J and K by the action of enzyme 7.

ii) $H + I \xrightarrow{\text{enzyme 6}} J + K \xrightarrow{\text{enzyme 7}} L + M$

iii) $G \xrightarrow{\text{enzyme 4}} H \xrightarrow{\text{enzyme 8}} L + M \xrightarrow{\text{enzyme 7}}$

$J + K \xrightarrow{\text{enzyme 6}} H + I$ (3)

d) It allows finely-tuned control and prevents build-ups and bottlenecks. (1)

2 a) i) Outside **ii)** Inside **iii)** Inside **iv)** Outside

v) The sodium/potassium pump actively transports sodium to the outside and potassium to the inside. (3)

b) 55 times (1)

c) i) Brings about an increase. **ii)** Sugar, needed for energy, has become a limiting factor.

iii) Inverse relationship. Energy is needed for ion uptake, therefore as ions are taken up, the number of units of sugar present decreases. (4)

3 a) P and S (1)

b) i) Q, S, P, R **ii)** R, P, S, Q (2)

4 a) i) Concentration of substrate **ii)** Independent

iii) It caused an increase in reaction rate. (3)

b) Concentration of enzyme (1)

c) i) A **ii)** C **iii)** B (3)

d) More enzyme could be added (1)

5 a) i), ii) and **iii)** See Figure An KS 8.1. (4)

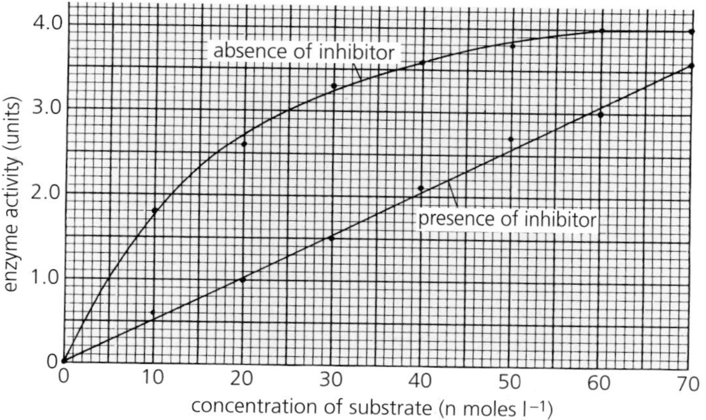

Figure An KS 8.1

b) A (1)

c) 3 times (1)

d) i) 61.54% **ii)** 10% (2)

e) There would always be a few enzyme sites blocked by inhibitor. (1)

6 a) 1, 2 and 3 (1)

b) Substrate concentration (1)

c) 4, 5 and 6 (1)

d) Experiment = 4, 5 and 6; controls = 1, 2 and 3 (2)

e) i) Iodine solution **ii)** Non-competitively

iii) If it had been a competitive inhibitor, the inhibitory effect would have decreased as substrate concentration increased and this would have resulted in some yellow colour appearing in tube 6 and maybe a faint yellow colour in tube 5. However, iodine completely inhibited the enzyme at all concentrations of ONPG showing that it acted non-competitively. (4)

7 a) i) Carbamyl phosphate and aspartate **ii)** Carbamyl aspartate and phosphate **iii)** Cytidylic acid (3)

b) i) The concentration of carbamyl phosphate will increase. **ii)** Fewer molecules of P will be free to act on carbamyl phosphate. (2)

c) i) Decreased **ii)** There will be so little cytidylic acid present that very few molecules of enzyme P will be affected by the negative feedback process. (2)

d) All three (1)

9 Cellular respiration

1 a) See Figure An KS 9.1. (4)

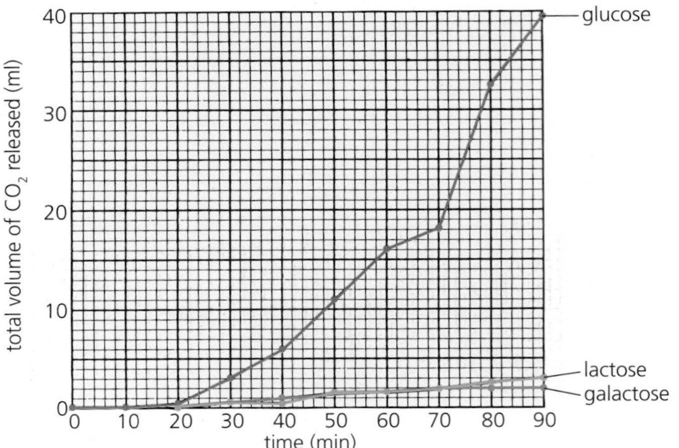

Figure An KS 9.1

b) i) The result at 70 min **ii)** It is much lower than would be expected from the general trend. (2)

c) 6400% (1)

d) It is able to make very good use of glucose but not able to make good use of galactose or lactose. (1)

e) i) It is hardly able to break lactose down.

ii) If it had been able to do so, glucose would have been released from lactose and rapidly used as a respiratory substrate. (2)

f) Repeat the experiment (1)

2 a) See Table An KS 9.1. (5)

Stage of respiratory pathway	Principal reaction or process that occurs	Products
glycolysis	splitting of glucose into [pyruvate]	[ATP], NADH and pyruvate
[citric] acid cycle	removal of [hydrogen] ions from molecules of respiratory [substrate]	[CO_2], [NADH] and ATP
[electron] transport chain	release of [energy] to form ATP	ATP and [water]

Table An KS 9.1

b) Glycolysis (1)

c) Electron transport chain (1)

d) Citric acid cycle and electron transport chain (2)

3 a) Repeat the experiment with several concentrations of the substrate (succinic acid) as the independent variable factor. (1)

b) If low concentrations give little or no colour change in the presence of malonic acid but higher concentrations do bring about decolourisation of DCPIP, then this shows that malonic acid is less effective at higher concentrations of substrate and is acting as a competitive inhibitor. (2)

4 a) i) Intermembrane space **ii)** A flow of high-energy electrons from NADH and $FADH_2$ pumps H^+ ions across the membrane against a concentration gradient. (3)

b) i) ATP synthase **ii)** The return flow of H^+ ions to the region of lower H^+ ion concentration via molecule X makes part of it rotate and catalyse the synthesis of ATP. (3).

c) By stopping electron flow, cyanide brings the movement of H^+ ions to a halt therefore no ATP is synthesised and the organism lacks access to energy and dies. (2)

5 a) and **b)** See Figure An KS 9.2. (4)

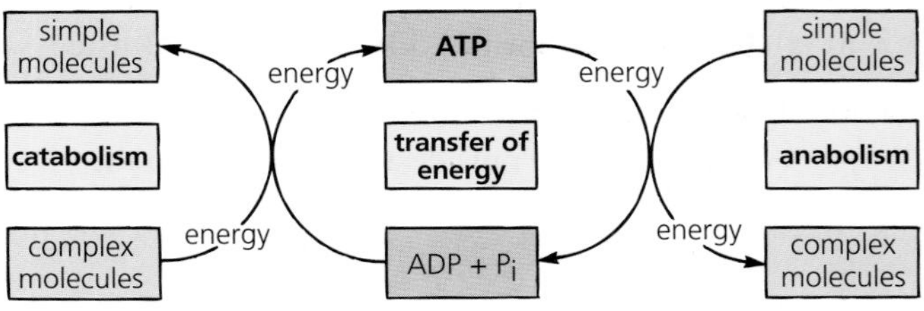

Figure An KS 9.2

6 a) C

b) A

c) C

d) C

e) A (5)

7 a) 56% (1)

b) 1267.2 kJ (1)

c) Synthesis of protein from amino acids; contraction of muscles (2)

8 a) Production of ATP (1)

b) ADP (1)

c) i) It increased **ii)** The process of respiration (which releases CO_2) was no longer limited by shortage of inorganic phosphate to make ATP. (2)

d) Ethanol (formed during fermentation) began to poison some of the yeast cells. (1)

9 See core text pages 126–7 and 128–9. (9)

10 Metabolic rate

1 a) BMR is lower in females. BMR decreases with age. (2)

b) 1.5 times (1)

c) 27.78% (1)

2 a) i) TSA = $2400\,cm^2$, TV = $8000\,cm^3$

ii) TSA = $4800\,cm^2$, TV = $8000\,cm^3$ (2)

b) i) 0.3 : 1 ii) 0.6 : 1 (2)

c) Small cube (1)

d) i) A ii) F (1)

e) i) F ii) A (1)

f) i) The smaller the body size, the higher the metabolic rate and vice versa.

ii) The smallest animal needs the highest metabolic rate because it has the largest surface area relative to its body size from which heat can be lost. (2)

3 a) See Table An KS 10.1. (5)

b) i) Incomplete double ii) Frog iii) A = high, B = low, C = low. (5)

4 a) i) 30 °C ii) 32 °C (1)

b) i) 23 °C ii) 30 °C (1)

c) i) Y ii) X (1)

d) i) X ii) Behavioural (2)

e) If members of population Y sought one of the rare sunny spots in the forest they would run the risk of:

i) expending more energy in movement than they would gain from the sun and

ii) increasing their chance of being caught by a predator. (2)

5 a) Hypothalamus (1)

b) It sends nerve impulses to them. (1)

c) i) It would become constricted.

ii) Less blood would flow to the skin surface so less heat would be lost by radiation. (2)

d) i) It would increase its rate of sweat production.

ii) When the liquid sweat coated the outside of the skin, excess body heat would be used to convert it to water vapour thereby cooling the body. (2)

6 a) and b) See Figure An KS 10.1. (7)

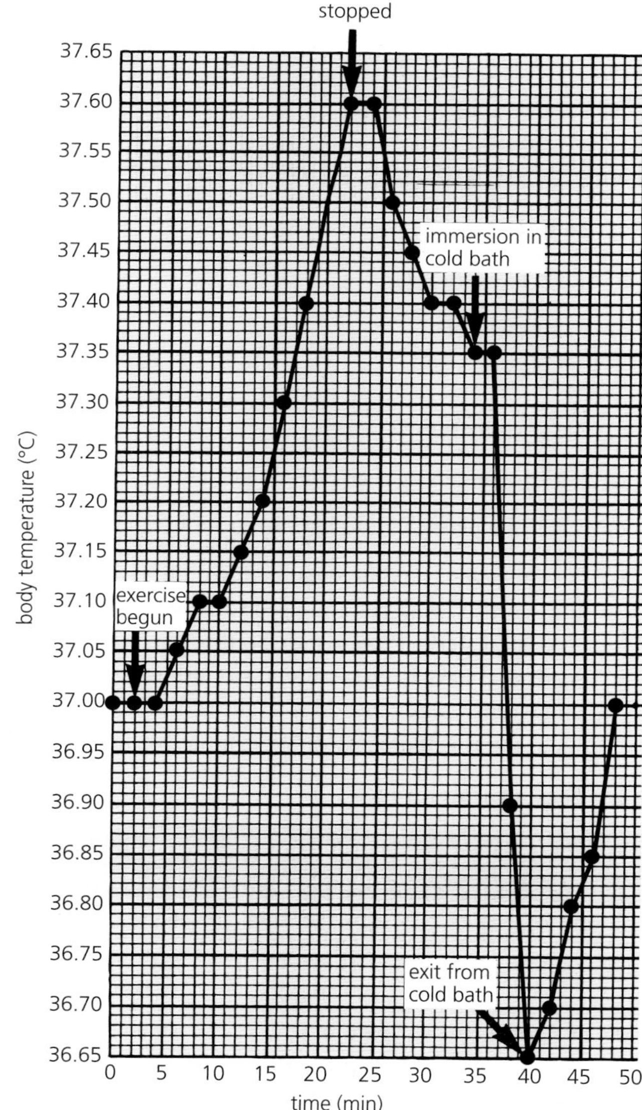

Figure An KS 10.1

Vertebrate group	Type of circulation	Number of chambers in heart	Pressure of blood arriving at skeletal muscles	Evolutionary level of circulatory system
fish	**single**	**2**	**low**	**primitive**
amphibian/reptile	incomplete double	**3**	**high**	intermediate
mammal	**complete double**	**4**	**high**	**advanced**

Table An KS 10.1

c) **i)** Increase **ii)** Decrease **iii)** Time is required for heat gain or loss by tissues to affect blood temperature. (3)

d) **i)** Vasodilation has occurred. **ii)** Overheating of the body is corrected by excess heat being lost by radiation from extra blood at the skin surface. (2)

e) **i)** D **ii)** Overcooling of the body is corrected by the heat generated by muscular contraction. (2)

7 See core text pages 144–9. (9)

11 Metabolism and adverse conditions

1 **a)** 16 times (1)

b) 1.2 $kJ\,m^{-2}\,min^{1}$ (1)

c) 130 times (1)

d) 91·67% (1)

2 **a)** No germination (1)

b) A dish containing seeds free of gel. (1)

c) Replicate plates should have been set up. (1)

d) The juicy gel could be separated from the seeds of the Scottish tomatoes and used to soak blotting paper in several sterile Petri dishes. South American tomato seeds that have been separated from their own gel and washed could be added to these dishes and incubated to find out if they germinate or not. (3)

3 **a)** See Table An KS 11.1. (5)

Feature	European hedgehog	Desert hedgehog
habitat	wooded area	edge of desert
food source	earthworms and insects	scorpions and snakes
time of dormancy	winter	summer
type of dormancy	hibernation	aestivation
extent of drop in metabolic rate	large	small
relative number of times animal wakes up during dormancy	few	many

Table An KS 11.1

b) Relatively more heat is lost by an animal with a smaller body size and large thin extremities which makes the desert hedgehog perfectly adapted to survival in the desert. (2)

c) **i)** European **ii)** It needs this fat to act as insulation and to generate energy, enabling it to survive lengthy periods of extreme cold when food is scarce. (3)

4 **a)** It increases (1)

b) **i)** A minimum of 14 hours of light in each 24-hour cycle. **ii)** Mid-April (2)

c) They had not been exposed to the length of photoperiod that triggers this behaviour until mid-May. (1)

d) Because birds received 12 hours of light in mid-March but there was no response. (1)

5 **a)** Q (1)

b) **i)** Inherited **ii)** P and S (2)

c) R (1)

d) They were young birds that had not flown the route before and were not able to recognise familiar geographical features to help them alter their course towards the correct destination. (2)

6 **a)** A and B (and D) (1)

b) **i)** Only one starling was used.

ii) Repeat the experiment using many starlings. (2)

c) Because no record is left by the bird of its movements. (1)

d) A and B (1)

e) D and F (1)

f) C and E (1)

g) Transparent windows, autumn, overcast sky. Mirrored windows, spring, clear sky. (2)

12 Environmental control of metabolism

1 **a)** D (1)

b) **A** The air-in tube lacks a filter/the end of the air-out tube is immersed in solution.

B The end of the tube from the syringe for taking samples is not immersed in the solution.

C The air-in and air-out tubes are the wrong way round and one tube lacks a filter. (3)

2 **a)** **i)** A **ii)** It follows the normal growth pattern and eventually the cells die at 15 hours once a high concentration of alcohol builds up. (2)

b) i) C **ii)** It continues to decrease as it is used up by the yeast cells. (2)

c) i) B **ii)** It does not appear for a few hours until the yeast has passed its lag phase. It levels off once the yeast cells go into decline. (2)

3 a) See Table An KS 12.1. (4)

Time (in 20-minute intervals)	Cell number ($\times 10^3$)	Cell number (correct to two decimal places)
0	3	3.00×10^3
1	6	**6.00×10^3**
2	12	1.20×10^4
3	24	2.40×10^4
4	48	**4.80×10^4**
5	**96**	**9.60×10^4**
6	**192**	1.92×10^5
7	**384**	**3.84×10^5**
8	768	7.68×10^5
9	**1536**	**1.54×10^6**
10	**3072**	**3.07×10^6**
11	6144	6.14×10^6
12	**12 288**	**1.23×10^7**

Table An KS 12.1

b) 3 h 20 min (1)

4 a) i) $q = p \times 2^n$

$= 3 \times 10^3 \times 2^n$

$= 3000 \times 2^3$

$= 3000 \times 8$

$= 24\,000$ bacteria

ii) $g = \dfrac{t}{n} = \dfrac{60}{3} = 20$ min (2)

b) $q = p \times 2^n$

$\therefore 16 \times 10^3 = 10^3 \times 2^n$

$\therefore 16 = 2^n$

$\therefore 2^4 = 2^n$

$\therefore n = 4$

$\therefore g = \dfrac{t}{n} = \dfrac{120}{4} = 30$ min (2)

5 a) To eliminate experimental error and improve the reliability of the results. (1)

b) This arrangement gives a larger surface area of liquid medium (exposed to air) upon which fungal growth can occur. (1)

c) A (1)

d) (i) As glucose concentration increases so also does dry weight of mycelium. **(ii)** The more carbohydrate that is available, the more protoplasm the fungus can build. **(iii)** Sporulation increases to an optimum at 1.0% glucose and then drops at the higher concentration of 10%. (3)

e) (i) They mass produce the vegetative mycelium when they want to obtain the antibiotic and they promote maximum sporulation when they need spores for the next inoculum. **(ii)** Vegetative 5 10%, sporulation 5 1% **(iii)** Repeat the experiment using many more concentrations of glucose, e.g. 2%, 3%, 4% etc. (3)

6 a) $P = 8 \times 10^6$, $Q = 1 \times 10^6$ (2)

b) i) 16×10^6

ii) Sixteen million (2)

c) i) 28×10^6

ii) 32 times

iii) 45 min

iv) $g = \dfrac{t}{n} = \dfrac{225}{5} = 45$ min (4)

7 a) and **b)** See Figure An KS 12.1. (6)

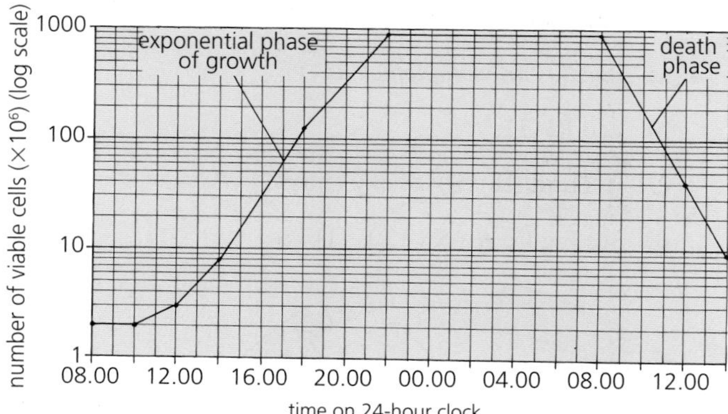

Figure An KS 12.1

c) 30 (1)

d) i) 22.00–08.00

ii) The number of cells may have varied during this 10-hour period when no readings were taken.

iii) Take more readings e.g. every 2 hours. (3)

8 See core text pages 175–7. (9)

13 Genetic control of metabolism

1 a) i) Mutations (%) **ii)** Dosage of X-rays (2)

b) As the dosage of X-rays increases so does the percentage of mutations. (1)

c) i) C **ii)** B (2)

d) There is a good chance that the site-specific mutation will bring about the exact improvement required. The random nature of exposing the culture to a mutagen makes the chance of creating a strain with the desired property very remote. (2)

e) 0.1 per million cells (1)

2 a) B, D, F, E, A, C (1)

b) i) B and D

ii) So that the cut ends would be compatible and stick together. (3)

c) F (1)

3 a) i) Y **ii)** X (1)

b) i) 5 hours **ii)** 2 hours after injection (1)

c) 20 hours (1)

d) The original biosynthetic insulin is slow to start acting and stops after 20 hours whereas X takes effect immediately and Y lasts a full 24 hours. (2)

4 a) *Bacillus amyloliquefaciens* (1)

b) Eco RI (1)

c) i) Hind III **ii)** Sticky ends (2)

d) See Figure An KS 13.1. (1)

Figure An KS 13.1

e) See Figure An KS 13.2. (1)

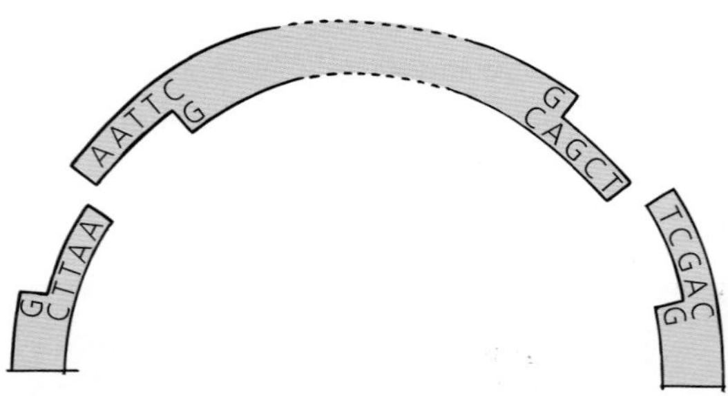

Figure An KS 13.2

5 a) i) High **ii)** High **iii)** Z (3)

b) i) Moderate **ii)** Low **iii)** X (3)

c) i) 2 **ii)** 3 (2)

d) 5 (1)

e) Containment of a risk 4 (but not risk 3) microbe must include controlled negative air pressure and presence of air locks and a compulsory shower for staff. (2)

6 See core text pages 185–7. (9)

14 Food supply, plant growth and productivity

1 a) 3 (1)

b) 6 (1)

c) 4 (1)

d) 6 (1)

2 a) W (1)

b) 4% (1)

c) Heat lost from the body. Energy used for movement. (2)

d) Faeces and urine contain chemical substances that provide energy for decomposers. (1)

e) 9.6 kJ (1)

3 a) 0.5 (1)

b) See Figure An KS 14.1. (1)

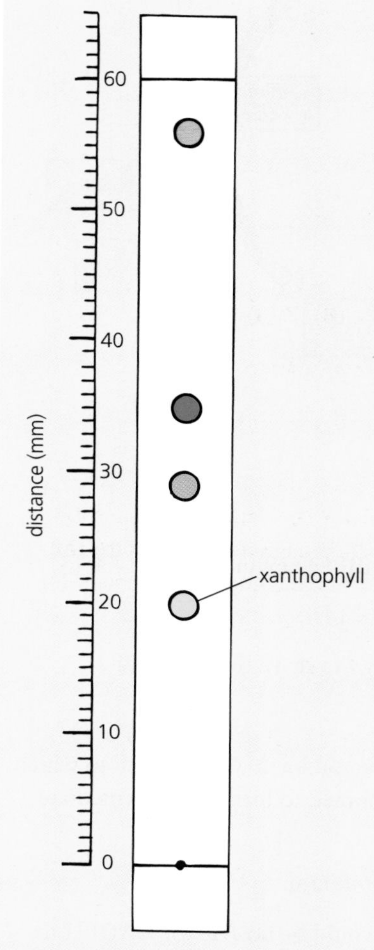

Figure An KS 14.1

c) Carotene; 0.95 (2)

d) To produce a concentrated spot of pigments. (1)

e) To prevent pigments from dissolving in the main bulk of the solvent at the bottom of the tube. (1)

f) Propanone (1)

4 a) Colour/wavelength of light (1)

b) They are absorbed by the filter. (1)

c) To make the results more reliable. (1)

d) To allow the plant to return to equilibrium. (1)

e) See Figure An KS 14.2. (2)

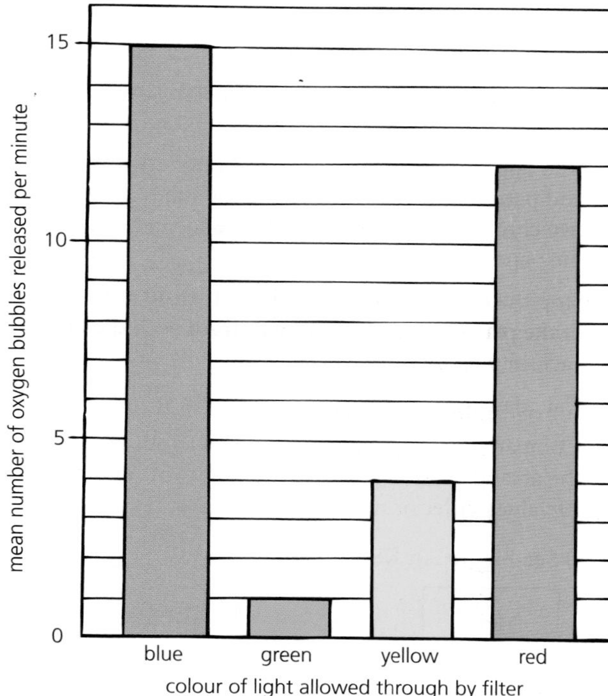

Figure An KS 14.2

f) Photosynthetic rate was greatest in the blue light. (1)

g) See Table An KS 14.1. (3)

Unit of length	Abbreviation	Fraction of one metre
metre	m	1
millimetre	mm	10^{-3}
micrometre	μm	10^{-6}
nanometre	nm	10^{-9}

Table An KS 14.1

5 a) Chlorophyll a and b (2)

b) i) Carotene and xanthophyll **ii)** Yes

iii) In this case it accounts for the difference between the action spectrum (rate of photosynthesis at different light wavelengths) and absorption spectrum (percentage of light absorbed by chlorophyll at different light wavelengths). (4)

6 a) Blue and red (2)

b) Most photosynthesis occurs at these regions on the strand of alga. These sites therefore release most oxygen which in turn attracts many aerobic bacteria. (2)

7 • Grind up equal masses of both leaf types in sand and propanone.
- Filter to give two pigment extracts.
- Repeatedly spot and dry each extract onto its own thin layer strip or length of chromatography paper.
- Add appropriate solvent (e.g. propanone, cyclohexane and petroleum ether) to two boiling tubes and run the two chromatograms by allowing the solvent to ascend the paper or strips.
- Stop the process when the solvent is close to the top of the papers or strips and mark the solvent front on each in pencil.
- Calculate Rf values for all the pigment spots.
- Compare the chromatograms and find out whether the arrangement and number of pigments and their Rf values differ or not. (9)

8 a)–g) See Figure An KS 14.3. (6)

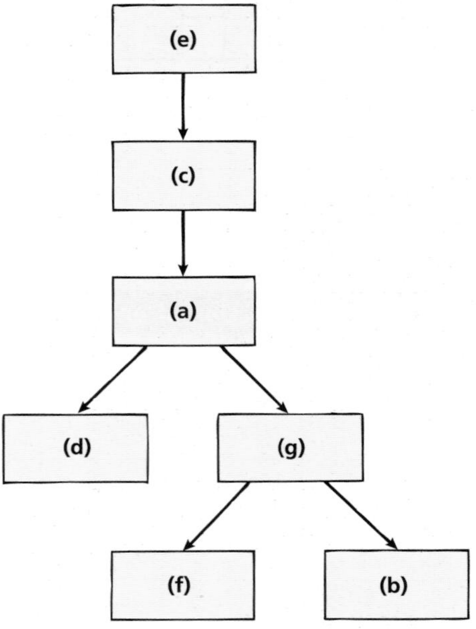

Figure An KS 14.3

9 a) See Figure An KS 14.4. (2)

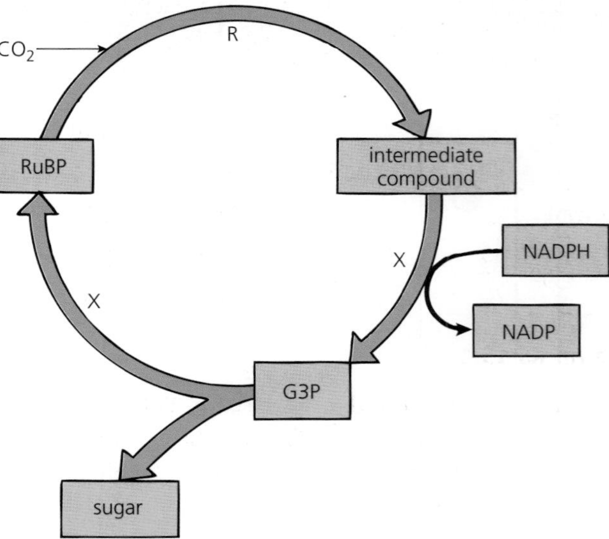

Figure An KS 14.4

b) i) RuBP **ii)** and **iii)** See Figure An KS 14.4. (3)

c) i) See Figure An KS 14.4.

ii) To provide energy to drive the cycle. (2)

d) i) RuBP

ii) Because there would be no CO_2 present to which it could be combined to form the intermediate compound. (2)

e) i) Intermediate compound

ii) Because there would be no ATP or NADPH to convert it to G3P. (2)

15 Plant and animal breeding

1 a) $0 + 0 + 8 + 0 + 0 + 0 + 10 + 0 = 18$ kg (1)

b) $6 + 6 + 0 + 0 + 3 + 0 + 10 + 10 = 35$ kg (1)

c) $W^1w^1W^2w^2W^3w^3W^4w^4$ (1)

2 a) 2 (1)

b) i) 3 **ii)** 2 (2)

c) The plots were of equal size. (1)

d) Three replicates were set up. (1)

e) The treatments were randomised. (1)

f) i) See Table An KS 15.1.

Fungicide or no fungicide	Fertiliser treatment (kg/acre)	Replicate 1	Replicate 2	Replicate 3
no fungicide	5	201	**211**	**207**
	35	304	**312**	**305**
	70	316	**327**	**329**
fungicide	5	252	**258**	**249**
	35	371	**366**	**364**
	70	379	**383**	**386**

Table An KS 15.1

ii) It is effective and helps the crop. This is demonstrated by the fact that the dry mass of plant material at harvest is greater for all plots with fungicide.

iii) As they stand, the results are not sufficiently different from one another to justify a conclusion being drawn in the absence of statistical analysis. (6)

g) The inclusion of plots without fertiliser. (1)

3 a) 50% (1)

b) 25% (1)

c) A (1)

d) F_8 (1)

4 a) F_1 (1)

b) F_1 is taller, has longer mean ear length and has higher mean yield than either parent. (3)

c) P_1 (1)

d) Inbreeding depression means the decline of a certain characteristic as a result of continuous selfing as shown by plant height in this example. (1)

e) i) The yield has increased dramatically and less land is needed.

ii) By doing so they are guaranteed a bumper F_1 crop whereas grains kept back by themselves would give poorer and poorer plants. (3)

5 a) They were developed (by breeding and selection) as a crop suited to the warm conditions of Southern Europe. (2)

b) By a complicated breeding programme involving crosses between European and North American strains. (2)

c) It matures early therefore it is able to make the most of the short growing season. (1)

d) i) Cattle and sheep

ii) Because it produces good solid cobs with high starch content that makes good animal feed. (3)

e) i) Detach the male flower from A and shake the pollen onto the female flower of A, then isolate A in a greenhouse or inside a large transparent plastic bag. Repeat for B.

ii) Detach the male flower from A and shake the pollen onto the female flower of B. Detach the male flower from B and shake the pollen onto the female flower of A. Isolate A and B as before. (4)

6 a) See Figure An KS 15.1. (2)

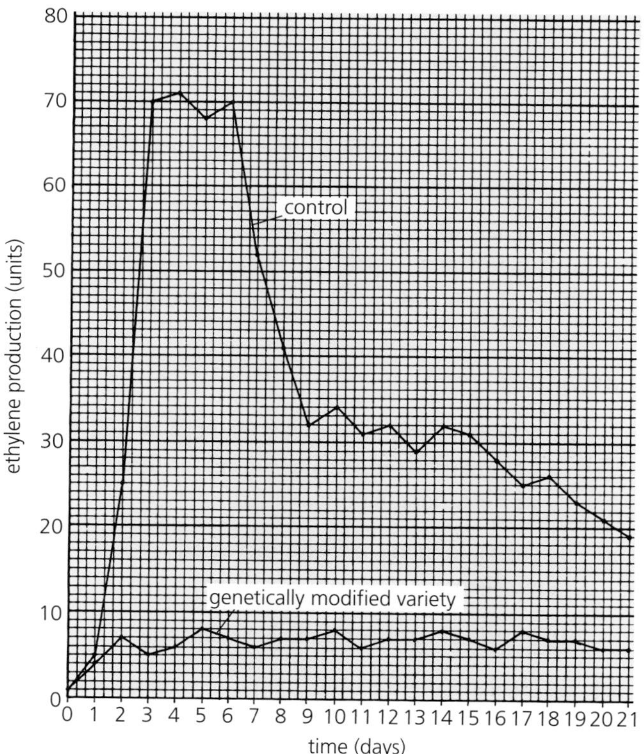

Figure An KS 15.1

b) X = genetically modified; Y = control (1)

c) i) 7 **ii)** 70 (1)

d) 90% (1)

e) The tomato would fail to ripen. (1)

f) Genetically modified tomatoes have a longer shelf life. (1)

16 Crop protection

1 See Table An KS 16.1. (7)

Weed species	Related non-weed species
can grow well on poor soil	needs fertile soil to grow well
able to flower in any day length	requires short days to flower
quick to flower and produce many tiny seeds	slow to flower and produces a few large seeds
self-pollinated	cross-pollinated
tolerant of drought	intolerant of drought
tolerant of water-logged soil	intolerant of water-logged soil
short life cycle	long life cycle

Table An KS 16.1

2 a) i) = Q **ii)** = S **iii)** = P **iv)** = R (3)

b) The wet areas could be drained and sprinklers could be used in the dry areas. (2)

c) Farmers could spray the chemical into the ground well in advance of planting the potatoes. The cysts would hatch and the larvae would emerge but die of starvation. (2)

d) i) 7 years

ii) It would be more difficult for PCN to survive for 7 years than for 4 years waiting for the return of its host, the potato plant. (2)

3 a) i) The yield increased with increased level of nitrogen.

ii) Nitrogen has been acting as a limiting factor. (2)

b) i) The later the weeding, the poorer the grain yield.

ii) Early season

iii) If the weeds are removed early, the cereal has time to establish itself in the ground and then compete successfully with any weeds that return, therefore yield is affected less. (4)

c) 31.99% (1)

d) Other soil nutrients and water (2)

e) 340 g m^{-2} (1)

4 a) 12 (1)

b) 550 (1)

c) 19 times (1)

d) 166.67% (1)

e) i) R

ii) Fewer plants were infected. (2)

5 a) *Erysiphe graminis* and *Puccinia graminis* (1)

b) To improve the reliability of the results. (1)

c) No significant effect by either A or B compared with the control. (1)

d) Fungicide A had no effect at 5 ppm but worked to some extent at 50 ppm. Fungicide B had no effect at either concentration. (1)

e) i) A **ii)** A **iii)** 50 ppm of A (2)

f) i) 99% **ii)** 38% (2)

g) They are effective on fungi W and X (that cause mildews) but have almost no effect on the other two fungi. (1)

h) A is more effective than B at killing spores of fungi Wand X and it is less toxic to wildlife. (2)

6 a) seed → seed-eating bird → peregrine falcon (1)

b) Dieldrin is non-biodegradable and therefore it builds up in concentration along the food chain as each organism eats a large number of the previous organism in the chain. (2)

c) C (1)

d) 65% (1)

e) Many thin-shelled eggs break during incubation so fewer young survive to become breeding adults. (1)

f) The number of breeding pairs is rising because use of dieldrin has been stopped and its effects are gradually disappearing. (1)

7 a) 1:20 (1)

b) 1600 (1)

c) i) 21.25 **ii)** 4.25

iii) No because there is a large overlap between the ranges covered by their error bars.

ii) Yes because there is no overlap between the ranges covered by their error bars. (6)

d) Because the predators would escape and not eat the prey. (1)

8 a) See core text page 235. (3)

b) See core text pages 235, 238–40. (4)

c) See core text pages 243. (2)

17 Animal welfare

1 a) i) 1, 4, 5 and 7 ii) 2, 3, 6 and 8 (2)

b) i) Freedoms 3 and 4

ii) They are not able to express their normal behaviour. They may injure themselves. (2)

c) Allow the cows out to graze in pastures for short periods over a few weeks in warm weather. (1)

2 a) i) They are forms of behaviour that cannot be performed easily in an overcrowded cage but can be performed freely out of doors.

ii) 4 and 7 (2)

b) 10 (risk of external parasites) (1)

c) i) Uncaged

ii) It includes an outside area where birds may peck at materials infected with parasites from wild birds. (2)

d) i) Uncaged

ii) They can move about freely therefore their bones develop strength. (2)

e) i) The risk of it happening increases.

ii) The caged birds express their stress by pecking one another's feathers and are able to do so freely if their beaks are left untrimmed. (2)

f) i) Conventional = 11, furnished = 11, uncaged = 8

ii) Install well-designed perches

iii) It would make the furnished and uncaged hen houses carry less risk, leaving the conventional cages carrying the most overall risk and the uncaged carrying the least overall risk. (4)

3 a) Three (freedom from pain, injury and disease) (1)

b) i) To kill bacteria that might cause disease.

ii) These would survive and might cause the disease which would then not be treatable by the antibiotic. (3)

c) i) They must wear masks and protective clothing.

ii) In the absence of a protective layer, the chemical could enter their bodies and cause serious harm. (2)

d) i) $0.25\,g\,l^{-1}$ ii) $25 \times 10^4\,\mu g\,l^{-1}$ (2)

e) i) Draining the pasture and spraying it with molluscicide is preventative. Using drugs to kill the parasite inside the sheep is curative.

ii) Prevention is better because it 'nips the problem in the bud', avoids stress and misery to the animal and is usually cheaper than tackling a disease epidemic. (3)

4 a) Because they are housed in an environment that is overcrowded and lacking in basic facilities. (1)

b) See Figure An KS 17.1. (5)

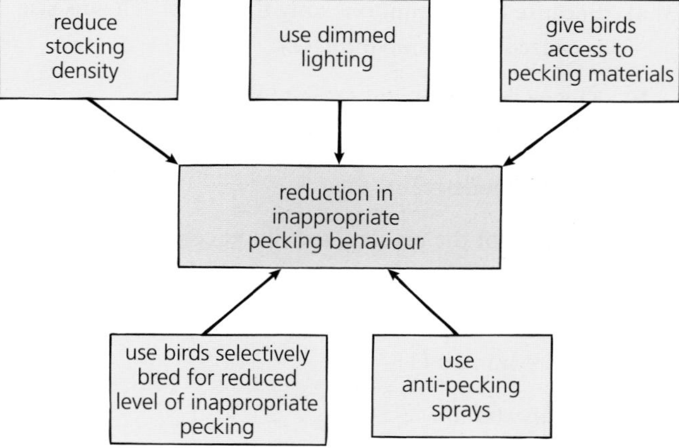

Figure An KS 17.1

c) i) Anti-pecking spray

ii) Chain reaction of injurious pecking and cannibalism (2)

5 a) To reduce the effect of experimental error. (1)

b) To increase the reliability of the results. (1)

c) i) 'left for one week to become habituated to their pen'

ii) At the start they might spend more time on a novel substrate out of curiosity but not make it their preferred substrate in the end. (2)

d) Size of pen; mass/volume of substrate used (2)

e) P = M > S > W (1)

6 See core text pages 254–6. (9)

18 Symbiosis

1 Mutation = **a)**, **d)**, **f)** and **g)**; parasitism = **b)**, **c)**, **e)** and **h)** (4)

2 **a)** **i)** It brings about an increase in growth rate.

ii) Two weeks after the start of the experimental diet.

iii) It brought about a decrease in the growth rate.

iv) Because the parasite is using some of the protein for its own growth. (4)

b) The adults took from week 2 to week 6 to breed and produce larvae. (1)

c) 6 times (1)

d) 166.67% (1)

e) Week 19 (1)

3 **a)** Because the mosquito is only the vector. The disease is caused by *Plasmodium*. (1)

b) The proboscis is a two-way tube. (1)

c) **i)** Because the toxic substances are also released into the bloodstream.

ii) Many of the person's red blood cells have been destroyed leaving them anaemic. (2)

d) Primary host = mosquito; secondary/intermediate host = human (1)

e) **i)** *Plasmodium*

ii) Mosquito

iii) Human (3)

4 **a)** 27 °C (1)

b) **i)** 1 and 2 = 100% healthy

ii) 1 = 50% healthy, 25% bleached, 25% dead; 2 = 75% bleached, 25% dead (2)

c) **i)** $0.85 \times 10^6\,cm^{-2}$ **ii)** 0 (2)

d) **i)** $1.4–1.9 \times 10^6\,cm^{-2}$ **ii)** No

iii) Their ranges overlap. (3)

e) **i)** $1.3–1.8 \times 10^6\,cm^{-2}$ **ii)** Yes

iii) Their ranges do not overlap. (3)

f) Initially there is no change then there is a sudden decrease at 32 °C. (1)

g) There is a continuous decrease with increasing temperature. (1)

h) It is able to acquire some tolerance if it contains type A zooxanthellae but not if it contains type B. (1)

19 Social behaviour

1 **a)** See Table An KS 19.1. (1)

Winner	Net number of contest won
T	14
R	14
P	16
R	20
S	4
R	6
P	8
T	10
R	18
P	12

Table An KS 19.1

b) **i)** Q

ii) Q was never an overall contest winner against any other bird. (2)

c) R, P, T, S, Q (1)

2 **a)** 56 (1)

b) 15 (1)

c) 15 hours (1)

d) 6 hours (1)

e) The recipient's gain (1)

f) Reciprocal altruism (1)

3 **a)** **i)** 2, 9 and 14 **ii)** 3, 8, 10, 16 and 17

iii) 4, 5, 11, 12 and 18

iv) 1, 6, 7, 13, 15, 19 and 20 (4)

b) **i)** 4, 5, 9, 11, 12 and 18 **ii)** 3, 5, 10 and 12 (2)

c) 60 (1)

d) **i)** See Figure An KS 19.1.

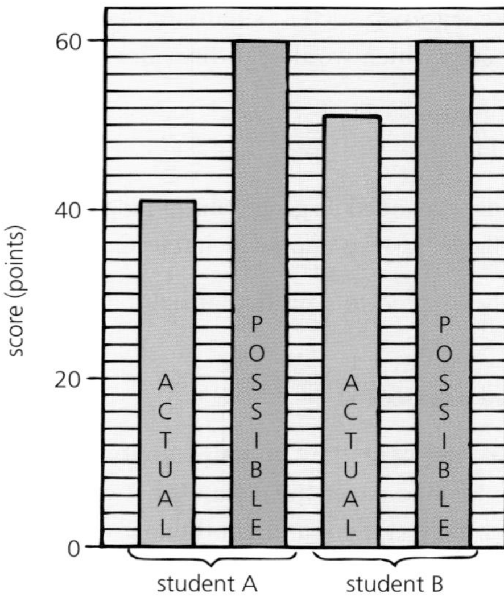

Figure An KS 19.1

ii) B beat A at the game. Both students would have achieved a higher score if they had co-operated throughout. (5)

e) Even knowing that *co-operation* results in the best outcome for the group as a whole, many players of the prisoner's *dilemma* find it difficult to reciprocate acts of *altruism* by choosing to *reduce* the sentence of their accomplice at the cost of staying a little *longer* in jail themselves. In the hope of achieving their personal optimal outcome, they tend to make *self-interested* decisions and continue to *betray* their accomplice which results in both players ending up *worse* off than they would have been by acting co-operatively. (4)

4 a) C, E, A, D, B (1)

b) Day 4 = C and E; day 17 = A and D (2)

c) 24 hours (1)

d) 20% (1)

e) i) Day 23 = much foraging (70% of time) and little resting (20% of time). Day 24 = little foraging (20% of time) and much resting (70% of time).

ii) If it does a lot of foraging in a day then it only has time for a short rest that day therefore the next day it needs a longer rest and only has time for a little foraging. (4)

5 a) i) The first female to become part of the group has the highest rank and so on to the last female to join the troop, who has the lowest status.

ii) The female with the lowest status (and any offspring that she has).

iii) He would be unable to protect three females and their offspring all at once, so it is better if he concentrates on guarding one female and her young. (3)

b) i) Young gorillas, on reaching puberty, emigrate from their troop of birth therefore they do not breed with their close relatives.

ii) When a son inherits a father's group. (2)

6 a) i) As age increases so does the ability to read emotions.

ii) Both genders of adult ape, but not the juveniles, chose the banana box more often than would be expected by chance alone.

iii) As apes grow older, they become more experienced at interpreting emotional expressions of the face. (3)

b) i) There is no difference.

ii) The results for males and females do not appear to be significantly different for either age group (though further statistical analysis would be needed to verify this). (2)

c) So that an ape has to depend on his/her ability to read emotions to choose the banana box and does not learn by association which box contains the banana. (2)

7 See core text pages 278–80. (9)

20 Components of and threats to biodiversity

1 a) 4 years (1)

b) i) Decrease in the total number of plaice caught.

ii) The stocks are being overfished. (2)

c) The '11 years' entry refers to one year only; the '12+ years' entry refers to the sum of several years. (1)

d) A greater number of younger fish were caught in years 4 and 5. (2)

e) If the catch continues to get younger, eventually there will be no adult fish left to produce future stocks. (2)

f) 70% (1)

2 a) i) Yes **ii)** Non-polluted has a higher number of different species. (2)

b) Non-polluted = 4.88; polluted = 1.99 (2)

3 a) Richness (1)

b) i) 4

ii) 2

iii) 4 is nearest to the mainland whereas 2 is furthest away from the mainland. (3)

c) No change (1)

d) i) H **ii)** E (2)

4 a) 7.33 (1)

b) 140 000 (1)

c) 8400 (1)

d) 16.00 (1)

e) 1 500 000 (1)

5 a) i) Invasive

ii) It has brought about the extinction of native fresh water mussels. (2)

b) i) It cuts fishing lines/injures swimmers.

ii) It blocks water intakes to hydroelectric schemes. (2)

c) i) Algae

ii) They grow better because more light for photosynthesis reaches them through the clearer water. (2)

d) i) Roach **ii)** Crayfish (2)

e) Stock the lake with fish such as smallmouth bass and yellow perch which would eat the zebra mussels. (1)

f) Continuous (1)

6 a) i) It has rapidly expanded its geographical range and the variety of flower species upon which it feeds.

ii) It is closely adapted to a particular habitat and is unable to change.

iii) Chequered skipper (4)

b) The area is so fragmented by human activity that the butterfly is unable to cover the distance between one unspoiled habitat fragment and another. (2)

c) i) As the climate becomes warmer, the butterfly seeks cooler habitats up the hills.

ii) If global warming continues, it will eventually run out of high altitude habitats and face extinction. (2)

d) Agriculture/urbanisation (1)

7 a) 50% (1)

b) i) By a horizontal line inside the box

ii) 1 = 31 mm, 2 = 24 mm, 3 = 13 mm (2)

c) i) 1

ii) 2 (2)

d) An actual value (1)

e) 6 mm (1)

f) 3 (1)

g) As latitude decreases, temperature increases promoting growth. (1)